The Geochemistry of Reservoirs

Geological Society Special Publications
Series Editor A. J. FLEET

GEOLOGICAL SOCIETY SPECIAL PUBLICATION NO. 86

The Geochemistry of Reservoirs

EDITED BY

J. M. CUBITT

Geochem Group, Chester, UK

W. A. ENGLAND

BP-Statoil Alliance, Norway

1995

Published by

The Geological Society

London

THE GEOLOGICAL SOCIETY

The Society was founded in 1807 as the Geological Society of London and is the oldest geological society in the world. It received its Royal Charter in 1825 for the purpose of 'investigating the mineral structure of the Earth'. The Society is Britain's national society for geology with a membership of 7500 (1995). It has countrywide coverage and approximately 1000 members reside overseas. The Society is responsible for all aspects of the geological sciences including professional matters. The Society has its own publishing house, which produces the Society's international journals, books and maps, and which acts as the European distributor for publications of the American Association of Petroleum Geologists, SEPM and the Geological Society of America.

Fellowship is open to those holding a recognized honours degree in geology or cognate subject and who have at least two years' relevant postgraduate experience, or who have not less than six years' experience in geology or a cognate subject. A Fellow who has not less than five years' relevant postgraduate experience in the practice of geology may apply for validation and, subject to approval, may be able to use the designatory letters C Geol (Chartered Geologist).

Further information about the Society is available from the Membership Manager, The Geological Society, Burlington House, Piccadilly, London W1V 0JU, UK. The Society is a Registered Charity No. 210161

Published by the Geological Society from:
The Geological Society Publishing House
Unit 7
Brassmill Enterprise Centre
Brassmill Lane
Bath BA1 3JN
UK
(*Orders*: Tel 01225 445046
Fax 01225 442836

First published 1995

British Library Cataloguing in Publication Data

A catalogue record for this book is available from the British Library ISBN 1-897799-26-8

Typeset by EJS Chemical Composition,
Midsomer Norton, Bath, UK

Printed by Alden Press, Oxford, UK

Distributors

USA
AAPG Bookstore
PO Box 979
Tulsa
OK 74101-0979
USA
(*Orders*: Tel (918) 584-2555
Fax (918) 548-0469)

Australia
Australian Mineral Foundation
63 Conyngham Street
Glenside
South Australia 5065
Australia
(*Orders*: Tel (08) 379-0444
Fax (08) 379-4634)

India
Affiliated East-West Press pvt Ltd
G-1/16 Ansari Road
New Delhi 110 002
India
(*Orders*.: Tel (11) 327-9113
Fax (11) 326-0538

Japan
Kanda Book Trading Co.
Tanikawa Building
3–2 Kanda Surugadai
Chiyoda-Ku
Tokyo 101
Japan
(*Orders*: Tel (03) 3255-3497
Fax (03) 3255-3495)

Contents

Geochemistry of reservoirs, an introduction

WILLIAM A. ENGLAND[1,3] & JOHN M. CUBITT[2]
[1] *BP Exploration Operating Company Ltd., BPX Technology Provision, Chertsey Road, Sunbury-on-Thames, Middlesex, TW16 7LN, UK*
[2] *Geochem Group Ltd., Chester Street, Saltney, Chester CH4 8RD, UK*
[3] *Present Address: BP–Statoil Alliance, N-7004, Trondheim, Norway*

We believe this is the first collection of papers to be devoted entirely to reservoir geochemistry, which is an area of growing scientific and economic importance. The main aim of reservoir geochemistry is to understand the distributions and origin(s) of the petroleums, waters and minerals in the reservoir and account for their possible spatial and compositional variation. This is ideally related to basin history and location of source-rock kitchens and migration pathways.

As well as being of interest in its own right, reservoir geochemistry has many important practical applications during petroleum exploration, appraisal and development. The most important uses are related to proving or disproving connectivity between different regions of a particular well or horizon. During exploration, reservoir geochemistry can indicate the direction from which a field filled, pointing the way for future wells. During production, studies of variations in composition with time may also be made, although this is a little studied area at present.

Published accounts of the origins and applications of reservoir fluid differences appear to be relatively recent. England *et al.* (1987) noted the phenomenon and were probably the first to relate it to migration and subsequent time-scales of in-reservoir mixing. Workers, particularly in the USA, have developed the use of 'gas chromatography (GC) fingerprinting' to distinguish different oil families, often in reservoirs with numerous thin hydrocarbon-bearing intervals (Kaufman *et al.* 1990). Effects due to the Earth's gravitational field were first considered by Sage & Lacey (1939). An extensive bibliography is presented by Larter & Aplin (this volume).

Although many techniques can be used to study the fluids in a reservoir, the most useful for reservoir geochemistry appear to be:

(1) *For hydrocarbons*
- conventional and high resolution gas chromatography–mass spectrometry
- high resolution chromatography ('GC fingerprinting')
- gas analysis and isotope determination
- fluid inclusion studies
- type analysis on extracts of (to locate/study) tar mats
- thermovaporization (to eliminate conventional core extraction steps)

(2) *For waters*
- conventional multi-element analysis
- oxygen/hydrogen isotope determinations
- strontium isotope ratios (especially on core)

Note, however, that other available non-geochemical data should also be included in any interpretations.

In order to classify the articles in a volume of this kind, we have attempted a broad split between (a) more general reviews and discussions of techniques, and (b) papers with a greater case study flavour.

General reviews and new techniques

Larter & Aplin give a comprehensive and readable review of the current 'state of the art', discussing the origins and natures of water and petroleum heterogeneities in reservoirs and their rates of dissipation over geological time. They also provide a useful account of the growing importance of statistics, which is vital for finding real (as opposed to imaginary) differences in the large datasets often collected. Finally, they discuss the importance of interactions between organic compounds and formation waters during migration.

Bjørlykke *et al.* review the geochemistry of North Sea and Gulf Coast formation waters. They emphasize the thermodynamic and kinetic controls on the possible reactions involving, for example, dissolved potassium and potassium-containing minerals. These factors are vital when considering the complex interactions involved during the emplacement of petroleum in reservoirs which may be simultaneously undergoing diagenesis.

Anissimov discusses geochemical criteria of value in reservoir characterization and in particular the inorganic compounds in formation waters. This short paper is illustrated by the use of benzene and

From CUBITT, J. M. & ENGLAND, W. A. (eds), 1995, *The Geochemistry of Reservoirs*, Geological Society Special Publication No. 86, pp. 1–3.

dissolved hydrocarbon gas in formation waters as an aid to exploration in the Volgograd Region.

Smalley *et al.* show how measurements of $^{87}Sr/^{86}Sr$ isotope ratios (on residual salts redissolved from cores) can be used to detect potential reservoir flow barriers. The possible heterogeneity in formation water resistivity/salinity is also discussed; this can potentially have a large influence on reserve estimates made from wireline logs.

Philp *et al.* move the focus of attention to high molecular weight organic compounds (with more than 40 carbon atoms). These are often neglected because they are difficult to analyse. However, as they often precipitate as solids in the reservoir pore space, they can have a serious effect on recovery. Pipes and valves may also be blocked in the well or at the surface. Wilhelms & Larter continue the high molecular weight theme, with a discussion of the tar mats found in marine-sourced oils in the North Sea and North America. Various formation mechanisms are discussed, with the authors favouring those in which tar mats are formed from stable asphaltene precursors. The precursors can be precipitated as tar mats by thermal degradation of a migrating oil (e.g. Ula Field, Norwegian North Sea) or an increase in dissolved gas content (Oseberg Field, Norwegian North Sea). Li *et al.* describe a study of pyrollic nitrogen compounds in the measurement of approximate migration distances in reservoired oils. The method exploits the differences in relative oil/water fractionation of a family of these compounds. If successfully developed, 'migration' parameters could become a valuable new addition to the geochemist's standard 'source' and 'maturity' parameters.

Case studies

Case studies are important because they let us test our theories against natural systems and provide details of the approach used by oil companies during field development and appraisal.

Two papers make good use of fluid inclusions found in authigenic cements. Inclusions can reveal the fluids present during key stages of reservoir development. By their use, McNeil *et al.* are able to constrain the temperature and salinity of the fluids responsible for cementation in the Rotliegend Sandstones of the Southern North Sea. The presence of methane in the inclusions shows that cementation occurred predominantly in the presence of gaseous hydrocarbons. Oxtoby *et al.* present a detailed study of the relative abundances of petroleum-bearing inclusions in cores from the Ula Field of the Norwegian North Sea. They deduce that filling was synchronous with cementation, and they relate the local maxima in petroleum-containing inclusions to the way in which the field filled with oil. The relationship of their work to the variable oil–water contact in Ula is also covered.

Horstad *et al.* discuss the variations in geochemical properties in a number of fields in the Tampen Spur region of the Norwegian North Sea. They identify five main charging systems, and suggest the presence of an additional carrier bed. This new concept has been tested by two exploration wells, one of which has successfully tested hydrocarbons.

England *et al.* show how a reservoir simulator can be used to interpret variations in fluid density, which are often observed in oil accumulations. If one can estimate the approximate time available since the reservoir filled, then the strength of internal flow barriers can be estimated. The method is illustrated by a case history from the Forties Field (UK North Sea).

Karlsen *et al.* present an extensive study of 33 petroleums from the Haltenbanken area of the Norwegian North Sea. They use organic geochemistry to group the oils according to their source rocks, taking maturity into account. Stoddard *et al.* give a good example of sharp variations in geochemical properties observed in several wells from the Eldfisk Field in the Norwegian North Sea. They relate these measurements on core extracts to barriers located by RFT pressure logging. This gave the authors a chance to test their geochemical discontinuities against the likely position of production barriers. Mason *et al.* stress the role of biodegradation in their study of Tertiary reservoired North Sea oils. The relationship to meteoric water influx and oxygen budget is discussed. Finally Aplin & Coleman show how H_2S production can change during the life of a producing field. Geochemical data, including isotopic measurements, were used to shed light on the possible mechanisms involved in this economically important process.

Discussion

This book gives a snapshot of academic and industrial reservoir geochemistry. We think that we can detect a growing maturity in this new area of geochemistry. In the future, we expect to see the following trends develop:

(1) further comparisons of production data with predictions (e.g. of barriers) made by reservoir geochemistry, which will allow the reliability of geochemical predictions to be tested;
(2) the collection of larger datasets and the application of sophisticated statistical analysis in order to identify real groupings and to measure success rates objectively;

(3) increased cooperation between reservoir geochemistry and engineering/modelling disciplines, including the increased application of reservoir flow simulators over geological timescales;
(4) the development of cheaper, faster and novel analytical methods, particularly on core samples (e.g. thermal extraction GC and GCMS, Compound Specific Isotope Analysis);
(5) an increased interest in water/oil/mineral interactions and the relation between organic and inorganic reservoir geochemistry;
(6) further attempts to relate reservoir geochemistry to rock 'wettability' (an important factor in oil recovery).

We acknowledge the efforts of the referees and authors who have helped to assemble this publication.

References

ENGLAND, W. A., MACKENZIE, A. S., MANN, D. M. & QUIGLEY, T. M. 1987. The movement and entrapment of petroleum in the subsurface. *Journal of the Geological Society, London*, **144**, 327–347.

KAUFMAN, R. L., AHMED, A. S. & ELSINGER, R. L. 1990. Gas chromatography as developed as a development and production tool for fingerprinting oils from individual reservoirs: applications in the Gulf of Mexico. *In*: *9th Annual Research Conference, Proceedings*. Gulf Coast Section of the Society of Economic Paleontologists and Mineralogists Foundation, 263–282.

SAGE, B. H. & LACEY, W. N. 1939. Gravitational concentration gradients in static columns of hydrocarbon fluids. *Transactions of the American Institute of Mining and Metallurgical Engineers*, **132**, 120.

Reservoir geochemistry: methods, applications and opportunities

S. R. LARTER & A. C. APLIN

Fossil Fuels & Environmental Geochemistry (Postgraduate Institute): NRG, Drummond Building, University of Newcastle, Newcastle upon Tyne, NE1 7RU, UK

Abstract: Petroleum reservoirs often contain compositionally heterogeneous petroleum, water and sometimes gas columns which can be interpreted in geologically useful ways. Kilometre-scale lateral compositional heterogeneities in fluid columns suggest petroleum emplacement directions, directions of water flow in biodegraded oilfields and the presence of major barriers to fluid flow within reservoirs. Vertical fluid column heterogeneities on a 10 m scale suggest that reservoirs are vertically compartmentalized. These data can be used as a tool in reservoir description and are thus relevant to many aspects of reservoir appraisal and development. The data supplement those generated by reservoir geologists and engineers, and can be obtained immediately after drilling, at reasonable cost.

Much recent work has focused on polar nitrogen (N) and oxygen (O) containing compounds such as quinolines, carbazoles and phenols. These compounds interact relatively strongly with water, rock minerals and each other, and may significantly influence the viscosity and pressure–volume–temperature (PVT) properties of petroleum, and also reservoir wettability. There are obvious links here between reservoir geochemistry and the more traditional disciplines of petroleum and reservoir engineering which we expect to be exploited in the coming years.

In the absence of phase changes during migration, the composition of reservoired oil reflects the expelled oil plus changes induced by adsorption, especially of polar compounds, onto solid phases or by partitioning into porewaters along secondary migration pathways. The extent of change, as measured by the relative abundances of selected compounds in reservoired oil and associated water, is potentially a direct measure of migration volume and thus of the relative amounts of water, oil and rock on a migration pathway. We present a simplified, equilibrium model which may allow one to assess migration volumes, and use *n*-hexane/benzene ratios of some North Sea oils to suggest that these oils have contacted fewer than one hundred volumes of water on their migration pathways. Future work should aim to quantify the interactions of migrating and reservoired petroleum with rock and water, and should yield a more quantitative assessment of both migration volumes and the controls of reservoir wettability.

Historically, the role of organic geochemistry in the petroleum industry has been overwhelmingly geared towards exploration. Identifying (or predicting) the hydrocarbon source rock, regional maturity and volumes of petroleum generated were the major aims of the petroleum geochemist. Since 1985, however, the focus of petroleum geochemists has moved away from exploration and increasingly towards production and reservoir-related problems.

The focus of inorganic geochemistry in the petroleum industry has always been towards reservoir quality, with an important role in describing and quantifying rock properties. There is a clear link here with sedimentology, and most inorganic geochemical studies have dealt exclusively with the rock matrix, ignoring the geological and geochemical information contained within reservoir fluids. It is this niche which 'reservoir geochemists', both organic and inorganic, seek to exploit.

One of the key observations that underpins reservoir geochemistry is that fluids (water, gas, oil) are often compositionally heterogeneous, in both a vertical and lateral sense, in petroleum reservoirs. A central aim of reservoir geochemistry is to understand the origin of these heterogeneities and to use them to enhance exploration and, more particularly, production and development strategies (Fig. 1). A second application of reservoir geochemistry is to utilize our knowledge of the chemistry of petroleum to understand the way in which petroleum, rock and water interact and thus influence petroleum wettability, viscosity and the formation of features in reservoirs such as tar mats. Although reservoir geochemistry can stand alone as a sub-discipline within petroleum science, it is also

From Cubitt, J. M. & England, W. A. (eds), 1995, *The Geochemistry of Reservoirs*, Geological Society Special Publication No. 86, pp. 5–32.

Reservoir Geochemistry

Applications to:

Exploration

- Determination of Source Rock Type and Maturity from well test oils
- Definition of field fill points/migration routes to aid in new-field satellite location and regional migration routes
- Assessment of basin palaeohydrology/oil alteration controls
- Assessment of seal efficiency

Appraisal

- Determination of fluid contacts
- Determination of reservoir continuity
- Assessment of oil column quality and history for equity and unitization decisions
- Sw calculation
- Location of tar mats
- Identification of potential production problems associated with deasphaltation

Production

- Barrier location and input to production model
- Production monitoring to assess production plan
- Assessment of tubing leakage - mixed production problems
- Assessment of injection water breakthrough
- Reservoir souring mechanism

Fig. 1. Some applications of reservoir geochemistry.

useful to think of it as a link between reservoir geology and engineering. Indeed, the results generated from geochemical studies must be integrated with engineering and geological data if the benefits of the studies are to be fully exploited.

The dual aims of this paper are to review the current status of reservoir geochemistry and to discuss new developments and applications. In particular, we wish to point out applications which may become important to the petroleum industry but which cannot be utilized without improvements in basic understanding and/or analytical techniques. In this respect, reservoir geochemistry can and should link science and engineering, academia and industry.

Fundamentals

Although this volume testifies to the current interest in the applications of geochemistry to reservoir-related problems, few publications existed prior to 1985. A computer search of all Society of Petroleum Engineers publications for the period 1951–1991 (SPE Masterdisk 1951–1991) reveals only a couple of publications concerning reservoir geochemistry. A recent overview of 'the future of reservoir characterisation for petroleum engineering purposes' by Haldorsen & Damsleth (1993) fails to mention fluid geochemistry at all. The driving force behind this shift of emphasis has been the movement of many companies towards a more production-oriented enterprise.

Development of field filling and mixing models

Although the occurrence of compositionally heterogeneous fluids in reservoirs has been known for decades (Sage & Lacey 1939; Espach & Fry 1951; Schulte 1980; Slentz 1981; Larter *et al.* 1991), their current geological interpretation is largely based on the work of England and co-workers (England *et al.* 1987; England & Mackenzie 1989; England 1990). Much of the summary of field filling and mixing outlined subsequently is based on their work.

Petroleum initially enters a potential reservoir through pores, with the lowest pore-entry pressures most often associated with the coarser lithologies. Continued migration leads to increasingly buoyant pressures, the filling of smaller pores with petroleum and the trapping of residual formation water. Freshly generated petroleum, if it is arriving from one side of the trap, advances into the trap as a series of 'fronts', displacing previously generated petroleum both laterally and vertically (England & Mackenzie 1989; England 1990) and preventing widespread mixing of the petroleum column. Because the physical and chemical nature of petroleum changes with increasing maturity (or if a second source becomes mature during the filling process), this ensures that lateral and vertical compositional variations exist in petroleum columns as reservoir filling is complete. Compositional variation can also be introduced by local biodegradation or water washing of a petroleum column.

Once the reservoir is full, density-driven mixing and molecular diffusion seek to eliminate inherited compositional variations in an attempt to establish mechanical and chemical equilibrium in the petroleum column. Thermal convection appears not to be a significant mechanism for mixing liquid-petroleum compositional variations (Horstad *et al.* 1990). England & Mackenzie (1989) established order-of-magnitude estimates of the time-scales of these mixing mechanisms. Their basic conclusions were that:

(1) diffusive mixing is geologically rapid vertically within individual oil columns, leading to gravitationally segregated concentration gradients being established on a scale of *c.* 100 m in a 1 Ma time-frame;

(2) diffusive mixing of petroleum columns is geologically slow laterally across large fields, compositional heterogeneities persisting for tens of Ma;

(3) density-driven mixing of petroleum columns is geologically rapid (time-scales of 10^4–10^6 years) in fields with high fieldwide permeability (good reservoir quality, absence of extensive shale/carbonate bands);

(4) density-driven mixing is geologically slow in fields with low fieldwide permeability.

An important implication of England's work is that, in many cases, lateral compositional gradients inherited from the process of reservoir filling should not have been eliminated. In some cases, lateral density gradients may have been eliminated without homogenizing the chemical composition of the petroleum, because many chemical compositions could produce fluids with identical densities. Furthermore, mechanical instability within single petroleum columns is unlikely; its occurrence would suggest the occurrence of barriers to vertical mixing.

Several case histories (England & Mackenzie 1989 – several unidentified fields; England 1990 – Forties and Magnus Fields, North Sea; Larter *et al.* 1991 – Ula Field, North Sea; Karlsen & Larter 1989 – Draugen Field, North Sea; Horstad *et al.* 1990 – Gullfaks Field, North Sea) have broadly substantiated this model for the hydrocarbon fraction of petroleums in clastic reservoirs of high permeability (>100 mD). It appears that in high quality, connected reservoirs in giant fields, the petroleum columns are vertically homogeneous, laterally heterogeneous and in mechanical equilibrium unless either barriers are present or recent field tilting, filling or leaking has occurred (Stoddart *et al.* 1995; Horstad *et al.* 1995).

Figure 2 shows a summary of the principal scales at which fluid heterogeneities occur in petroleum reservoirs. At the kilometre to tens of kilometres scale, lateral gradients in composition suggest regional petroleum emplacement directions (gas/oil ratio (GOR), maturity markers), regional water flow directions in biodegraded oilfields (*n*-alkane concentrations), and the presence of large-scale barriers to fluid flow (compositional steps). At the 10 m scale, vertical compartmentalization of the reservoir can be confirmed by heterogeneities in, for example, aqueous species (salinity or $^{87}Sr/^{86}Sr$ ratios) or molecular marker (biomarker) parameters. At the tens of micrometres to molecular scale, reservoir geochemistry is concerned with the surface chemistry of reservoirs and the interpretation of palaeofluids in fluid inclusions.

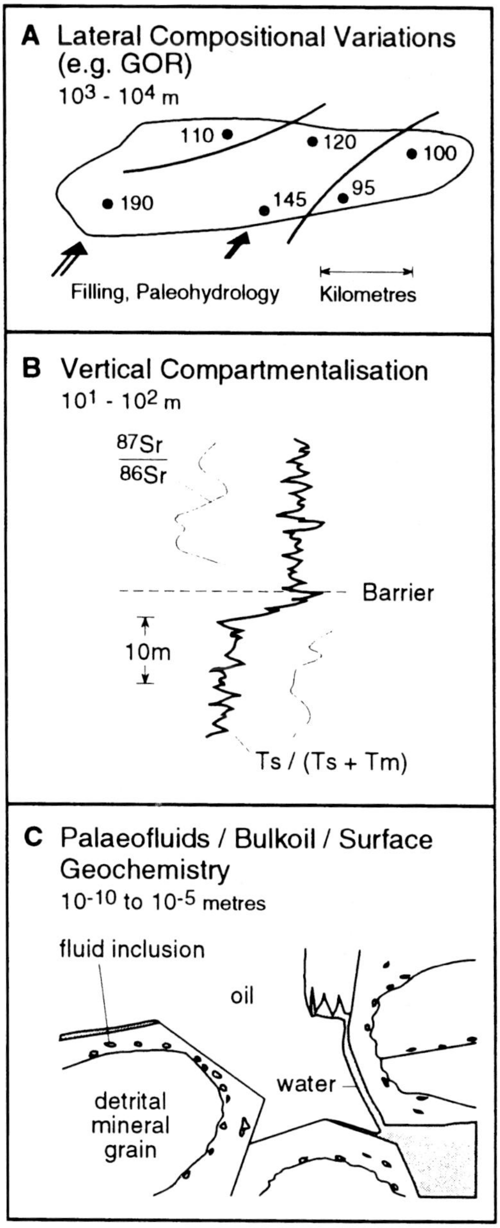

Fig. 2. Scales at which compositional heterogeneities occur in petroleum reservoirs. **(A)** At kilometre scales laterally in a reservoir, gradients in GOR or maturity parameters may indicate filling. **(B)** Vertically on scales of a few to a hundred metres, compositional steps may indicate barriers. **(C)** At the pore scale to molecular scale on reservoir surfaces, or within fluid inclusions, compositional variations may inform us about reservoir state or palaeofluid composition.

Key analytical developments

Another major factor in the development of reservoir geochemistry has been the development of new procedures enabling the gross, isotopic and detailed molecular composition of reservoir fluids and core extracts to be obtained rapidly from small samples (a few grams). Many of these advances stem from the modification of the liquid chromatography (LC), gas chromatography (GC) and GC–mass spectrometry (GC–MS) techniques developed by petroleum and organic geochemists in the 1970s and 1980s. There are three key analytical methods.

(1) Thin layer chromatography–flame ionization detection (TLC–FID), commercially available as the Iatroscan instrument. This has permitted high resolution screening of reservoir core solvent extracts (Karlsen & Larter 1991) and has complemented RockEval screening and evaluation (Wilhelms 1992; Ghenima *et al.* 1991).

(2) Thermovaporization–gas chromatography–FID or –MS. This has evolved from the early, simple systems enabling the analysis of single source rock samples (Larter & Horsfield 1993) to sophisticated, fully automated devices capable of obtaining high quality GC and GC–MS data on free petroleum in reservoir cores (Bjorøy *et al.* 1985, 1991*b*; Stoddart *et al.* 1995). These devices are now available commercially as the GHM and GHM-MS instruments and enable routine acquisition of GC–MS data at a rate of up to a sample an hour directly from unprocessed, milligram-sized core/cuttings samples.

(3) Determination of the $^{87}Sr/^{86}Sr$ isotopic composition of salts precipitated from formation water or residual water evaporated during core storage (Smalley *et al.* 1992; Smalley & England 1992). This technique, known as residual salt analysis (RSA), permits one to obtain some information about the composition of waters originally present in the reservoir, at high spatial resolution.

Samples and methods

The basic strategy outlined here is based on work performed at the Universities of Oslo and Newcastle from 1987 to present (Larter *et al.* 1991; Horstad *et al.* 1990; Karlsen *et al.* 1993; Stoddart *et al.* 1995; Willhelms & Larter 1994*a*). The procedures employed are based on geochemical methods developed in the 1970s and 1980s, modified to cope with increased sample throughput. Much of reservoir geochemistry is simply traditional petroleum geochemistry applied to some new goals!

Sampling programmes should encompass all available sample types including well-test fluids (DST/RFT), produced gases, oils and waters, plus cores and cuttings. A typical sample set consists of 10 to 20 DST/RFT oils, cores from 5 to 11 wells with 50 to 100 samples per well. Cores are usually sampled in conjunction with available core and well logs, samples being taken near porosity/permeability core plug holes to facilitate integration of geochemical and petrophysical data. At least a few wells should be sampled every 1 to 3 m to identify small features such as minimats (small tar mats).

The analytical strategy depends on the question to be addressed, but generally starts with rapid screening of large numbers of samples followed by a more detailed analysis of selected samples (Fig. 3). Table 1 shows the types of samples and main analytical techniques used in reservoir geochemical studies and the types of application that they have been used for. As in any study, care should be taken to match analytical technique (and cost) to desired information.

A typical organic geochemical analytical programme might screen 450 to 500 core extracts by Iatroscan and thermal desorption–GC followed by 150 to 200 thermal desorption GC–MS analyses on selected core samples. All gases and produced oils would be analysed by appropriate LC, GC and GC–MS methods. Specialized analyses, such as GC–IR–MS or nitrogen compound and phenol analysis, would be performed on DST oils and perhaps a few core samples (Bjorøy *et al.* 1990, 1991*a*; Stoddart *et al.* 1995). Profiles such as those shown in Fig. 4 require a sampling density of approximately 20 to 30 samples/100 m reservoir section.

Fluid samples are extremely valuable though often inadequately sampled. This is especially true for formation waters. It is particularly important to obtain and then archive the fluids initially present in the reservoir. Changes in the composition of produced fluids during time (Dahl & Speers 1985, 1986) can have serious economic consequences and it is often difficult to determine the cause of the change without reference to the initial state of reservoir fluids. Fluid samples allow a detailed analysis of fluid that is mobile under well-test or production conditions. In contrast, reservoir core extracts, obtained by extraction with organic solvents for petroleums or with deionized water for precipitated salts, provide information on the residual fluids in the reservoir. These may include immobile phases such as tar mats and adsorbed reservoir surface coatings, extracts of indigenous organic matter in the reservoir plus the remains of the fluids that were in the reservoir porosity during core recovery. Although one cannot completely reconstruct fluid composition from a core extract, it is possible to build a detailed spatial picture of

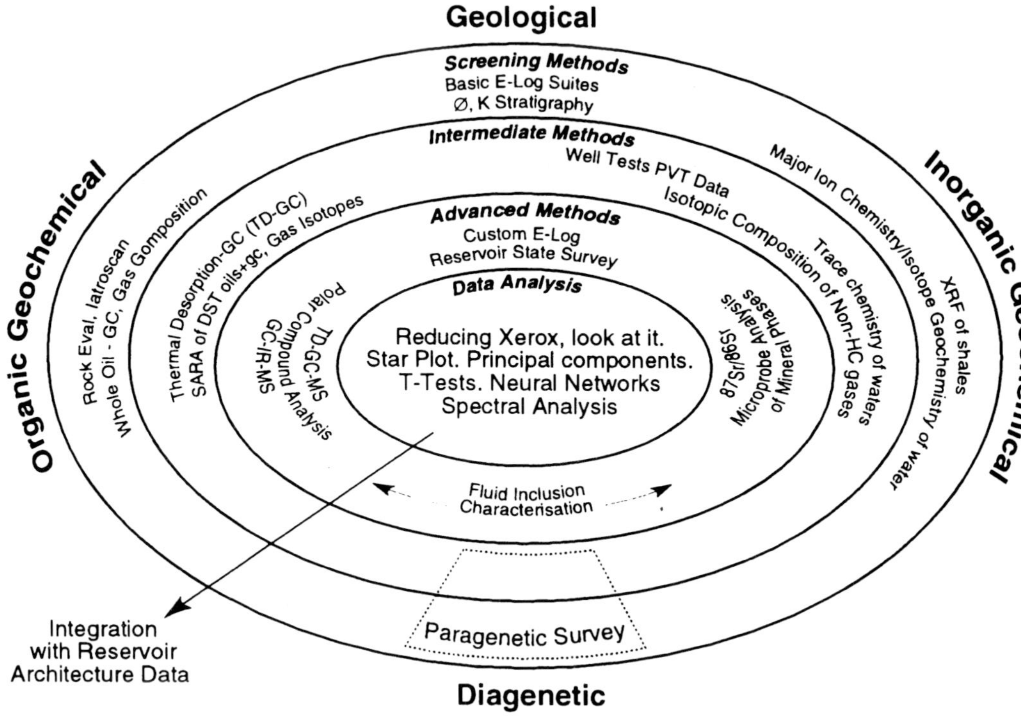

Fig. 3. Anatomy of a reservoir geochemical study. The concentric fields represent progressively higher level, more sophisticated and costly analyses chosen on the basis of lower-level screening. The number of samples analysed decreases at each level.

compositional variations. In contrast, well test or production fluids may be produced from a significant depth range within the reservoir, possibly covering several productive reservoir zones. They allow complete compositional characterization, but give more limited information about spatial variability in petroleum composition.

Differences between the composition of DST fluids and reservoir core extracts have been frequently reported (Horstad *et al.* 1990; Leythaeuser & Ruckheim 1989; Hillebrand & Leythaeuser 1992; Stoddart *et al.* 1995). Core extracts using solvents such as dichloromethane/methanol (93 : 7) are invariably enriched in non-hydrocarbons compared to corresponding well-test fluids (Fig. 5). In general, however, it seems that both sample types have similar hydrocarbon fingerprints.

Aqueous and petroleum palaeofluids can be sampled in cores and cuttings samples by analysing the contents of fluid inclusions (Horsfield & McLimans 1984; Nedkvitne *et al.* 1993; Karlsen *et al.* 1993). Recent work (Macleod *et al.* 1993*a*) suggests that, like the core extracts, petroleum trapped within fluid inclusions may also be enriched in non-hydrocarbons compared to the bulk fluid from which they were formed.

Data analysis

Conventional approaches to visualization and analysis of geochemical data (reducing xerox + simple observation of GC, GC–MS distributions, etc.) can certainly be valuable and are recommended for routine use. Plotting petroleum-fraction concentrations and molecular parameters on both vertical logs and maps should typically be the first step, core-extract data and produced fluid data being plotted separately. Cross-plots of petroleum concentration obtained from core-extract analysis (Iatroscan or RockEval) or conductivity logs versus depth and porosity data may allow the detection of oil/water contacts (OWC) and tar mats, or the assessment of reservoir filling–draining scenarios (Fig. 6; cf. Karlsen & Larter 1991). Visual analysis is often as effective as other approaches, but should be confirmed with statistical analyses.

Table 1. *Methods and applications of reservoir geochemistry*

Application	Sample	Method	References
Screening methods			
• Determination of petroleum concentration and bulk composition; detection of tar mats, OWCs and OBM contamination	Cores/cuttings	RockEval Iatroscan (TLC–FID)	Karlsen & Larter (1989, 1991) Ghenima *et al.* (1991) Wilhelms & Larter (1995)
• Assessment of filling versus draining	Cores	Yield/porosity	Augustson (1992)
• Determination of reservoir continuity, production tubing leakage, production well siting	Oils	High resolution GC fingerprints	Kaufman *et al.* (1987, 1990) Slenz (1981) Ross & Ames (1988)
• Cap rock integrity analysis of cap rock	Cuttings	Head space gas	Leith *et al.* (1993)
Intermediate methods			
• Determination of reservoir continuity and field filling directions; tar mat studies	Oils/hydrocarbon and non-hydrocarbon gases	Conventional LC, GC, GC–MS	England (1990); Larter *et al.* (1991); James (1990)
	Core extracts	Isotopes	Leythaeuser & Ruckheim (1989 Hillebrand & Leythaeuser (1992) Horstad *et al.* (1990, 1992) Wilhelms & Larter (1995)
	Cores/cuttings	Thermal desorption	Bjorøy *et al.* (1991*b*)
		GC, GC–MS	Stoddart *et al.* (1995)
• Reservoir biodegradation	Oils/cores	LC, GC, GC–MS	Horstad *et al.* (1990, 1992); Horsfield *et al.* (1991)
• Wax problems	Cores/oils	High temperature GC	Del Rio *et al.* (1992)
• Reservoir continuity/correlation	Cores	$^{87}Sr/^{86}Sr$ residual salt analysis	Smalley *et al.* (1992)
• Injection water breakthrough	Produced water	$^{87}Sr/^{86}Sr$; δO; δD	Smalley & England (1992)
• Reservoir souring	Waters Gases	$\delta^{34}S$ SO_4, in water $\delta^{34}S$ H_2S in gas	Aplin & Coleman (1995)
• S_w Calculation	Waters	Major ion chemistry	Coleman (1992)
• Reserves estimates	Waters		
• Scale prediction	Waters	Ca, Ba, ΣSO in waters	
Advanced methods			
• Palaeofluid distributions	Cores	Fluid inclusion geochemistry	Horsfield & McLimans (1984) Karlsen *et al.* (1993); Nedkvitne *et al.* (1993) Burruss (1989)
Developing methods			
• Petroleum population definition	Oils/rock extracts	GC–IR–MS	Bjorøy *et al.* (1991*b*); Stoddart *et al.* (1995)
• Migration range parameters	Oils/waters	Nitrogen compound analysis Phenol analysis	Li *et al.* (1992, 1995) Yamamoto (1992) Macleod *et al.* (1993*b*); Taylor *et al.* (1993) Ioppolo *et al.* (1992)
• Engineering data	PVT data/well logs	Various applicarions	England (1990); Thompson (1988); Stoddart *et al.* (1995); Larter *et al.* (1991); Larter & Mills (1992)

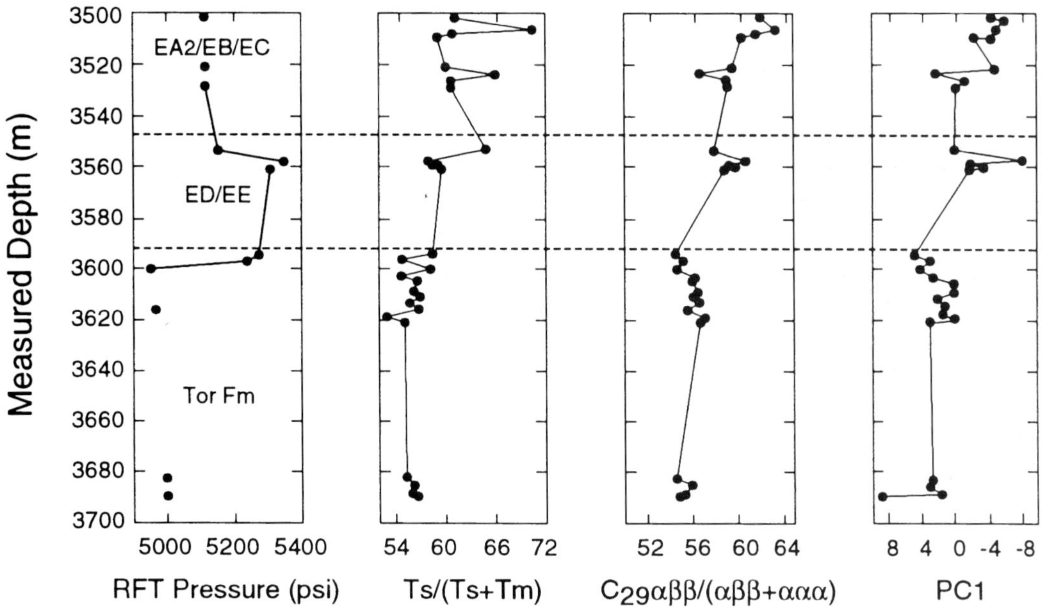

Fig. 4. Geochemical and engineering data profiles through a well in the Eldfisk Field. Shown are Repeat Formation Test (RFT) pressures, biomarker parameters and a summary parameter (PCI) derived by principal components analysis of a multivariate biomarker dataset. Breaks or steps are visible in all logs between the ED/EE and Tor Formation boundary after Stoddart *et al.* (1995). The two petroleum populations are statistically different.

In many cases, the sheer size of datasets and the subtle differences within them necessitate the use of statistical tests to validate conclusions or to reduce datasets to interpretable sizes (Horstad 1989; England 1990; Stoddart *et al.* 1995). Table 2 shows the types of problems typically encountered and some plausible approaches. Three scenarios are described and no attempt is made to give a complete coverage of methods. Many of the techniques discussed are designed to detect changes or steps in compositional data profiles. Good summaries of the applications of statistics and multivariate data analysis to Earth science data in general, and geochemistry in particular, are given in Davis (1973), Marsal (1987), Kvalheim (1987*a*,*b*) and Kvalheim *et al.* (1985, 1987).

Kvalheim (1987*a*) considers that the aims of geochemical multivariate data analysis can be broadly divided into problems associated with interpretation/identification, classification/correlation and calibration/prediction. Problem 1 in Table 2 broadly involves recognition and classification of samples into different petroleum populations, whilst Problem 2 involves confirmation of the significance level of the division. Kaufman *et al.* (1987) have shown how Star Plots can successfully be used to display and compare data from high resolution GC fingerprints of crude oils. The results can be applied to a wide variety of production tubing leakage and reservoir continuity problems.

There are many more sophisticated multivariate data analysis techniques that can be used to cluster samples from compositional data. There are many hierarchical clustering methods based on various nearest- or furthest-neighbour distance algorithms that are applicable to geochemical data (Milligan 1980) and which are suitable for the definition of different petroleum populations. Although not usually considered as a 'clustering' procedure, principal components analysis (PCA) has the advantage that it allows the geochemist to look at large multivariate data sets (e.g. > 50 samples and > 40 variables), revealing compositional similarities between the samples (scores plot) and relationships between the variables (loadings plot). Our experience with natural and synthetic datasets is that recognition and separation of different petroleum populations is quick and easy on a scores plot, the compositional differences between the populations being chemically definable on the loadings plot (Stoddart *et al.* 1995). False separation of synthetic datasets has been observed under unusual circumstances. A further advantage of

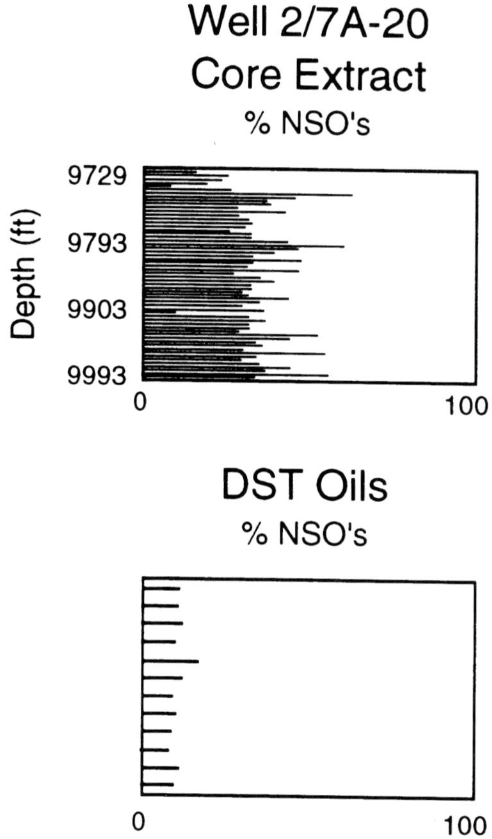

Fig. 5. Comparison of the polar compound (NSO) concentrations in a typical reservoir core extract and DST oils from the same reservoir zone. Yields are expressed as percentages of total C_{14+} material analysed. After Stoddart *et al.* (1995).

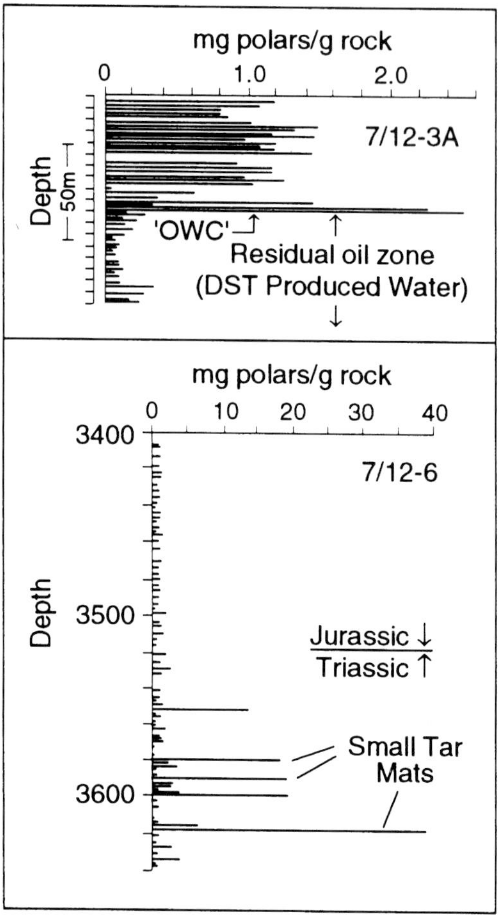

Fig. 6. Examples of geochemical logs. The top diagram shows a log of the concentration of polar compounds extracted from a core in the Ula Field, showing a distinct oil–water contact (OWC). The lower diagram shows a log of the polar compound yields in the Ula and Skaggerak Formations of Well 7/12-6 of the Ula Field (after Karlsen & Larter 1991; Wilhelms & Larter 1994*a*). Several zones in the Triassic oil column have high absolute yields of polar compounds (= resins + asphaltenes). These zones represent small tar mats ('minimats') within the oil column that have accumulated on sedimentologically defined features.

PCA is that reduced or latent variables (principal components) are produced which describe the variance of the data in terms of linear combinations of existing variables. These principal component scores can be plotted as a condensed variable downhole or across a reservoir, summarizing all variance in the dataset.

Figure 4 shows molecular parameters ($T_s/(T_s + T_m)$); sterane-C29 ($\alpha\beta\beta/(\alpha\beta\beta + \alpha\alpha\alpha)$) obtained from thermal desorption–GC–MS (Bjorøy *et al.* 1985, 1991*b*) analysis of core samples plotted against depth within a production well from the Eldfisk Field (Stoddart *et al.* 1995). Also plotted are the scores for each sample on principal component 1 obtained by PCA of a biomarker dataset. The vertical distribution of the molecular data suggests that the Tor Formation has a different petroleum population compared with the overlying Ekofisk Formation, and this is confirmed by PCA scores. Also plotted on Fig. 4 are RFT pressure data for the well, which show that the ED/EE layer is acting as a pressure seal (Stoddart *et al.* 1995). Both visual and PCA analysis suggest that the two petroleum populations are separated by the Ekofisk/Tor boundary. Such equivalence of geochemical and engineering data lends credence to the use of

Table 2. *Statistical techniques in reservoir geochemistry*

Problem 1:	Are there any compositional differences among the fluid data I have?
Techniques:	Star plots (Slentz 1981; Kaufman *et al.* 1987); cluster analysis; principal components analysis (PCA; Stoddart *et al.* 1995).
Problem 2:	Compositional differences in fluid compositions are suspected across a reservoir or down hole within a well. Are they real?
Techniques	CA (Stoddart *et al.* 1995); significance tests such as Student *t*-test (Horstad *et al.* 1990; England 1990; Stoddart *et al.* 1995); automated roving tests.
Problem 3:	We have identified several petroleum populations in a well. How do these populations distribute themselves across a reservoir or field?
Techniques:	Discriminant function analysis; neural network applications.

(relatively) cheap molecular data for the early identification of flow barriers in reservoirs.

The Student *t*-test can be used to determine whether two sample sets are part of the same population. In the previous case, the two sample sets (above and below the barrier) are significantly different at a 99% confidence level. One technique we have found useful is to use the power of modern computerized spreadsheets to automate the *t*-test and search for significantly different populations (roving *t*-test). Figure 7 shows a spreadsheet output for another $T_s/(T_s + T_m)$ data set. The two plots show the actual data and the *t*-statistic computed for the whole dataset split into two at each sample, the group above being compared with the rest of the dataset below. The critical *t*-statistic for such a dataset is 2.05 (95% confidence) or 2.76 (99% confidence). The maximum *t* value in this case is

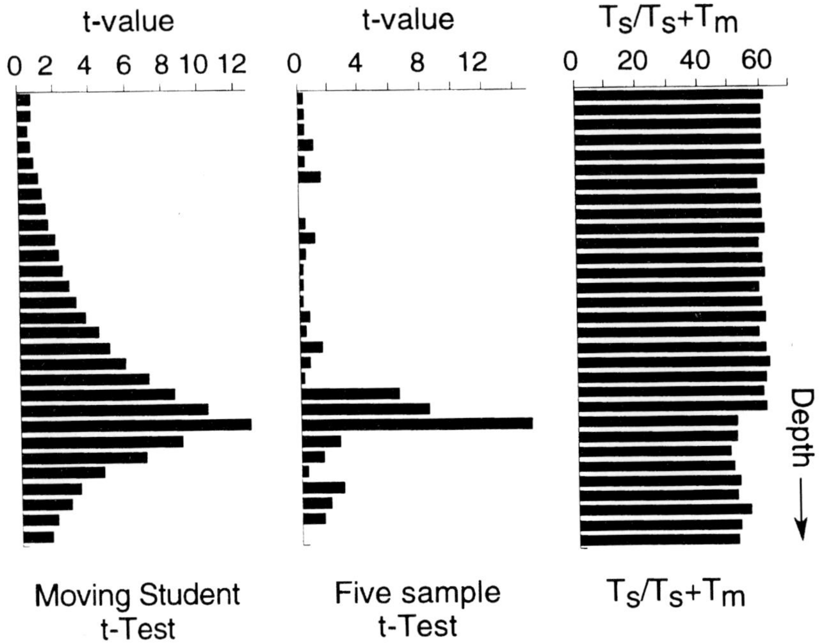

Fig. 7. Automated *t*-tests using spreadsheets. The right trace shows a depth sequence of $T_s/(T_s + T_m)$ data from core-extract analysis by GHM–MS. The reservoir is approximately 100 m thick. The middle trace shows a five-sample roving *t*-test in which the data are compared at each depth by dividing the two populations into the five samples above and below the test horizon, a *t*-statistic being computed at each point. The left trace shows the same procedure in which the whole well is divided into two populations, one above and one below the specified depth. The steps are obvious and statistically significant.

13, indicating that the two petroleum populations are significantly different at the Tor/Ekofisk boundary. Other roving *t*-tests can compare datasets a fixed distance (say five samples) above and below a moving horizon. In large datasets (many wells) these techniques can aid in spotting potential population change points which may indicate several barriers in single wells (Fig. 7).

Once petroleum populations have been defined locally, it can be useful to search for the same populations elsewhere in the reservoir or in other reservoirs. This is a developing area and several relatively new techniques are available. Of these, artificial neural networks (Hertz *et al.* 1990) may have some application as predictive pattern recognition techniques. Neural networks have been successfully trained to recognize sediment facies and depositional environment signatures from electric-log suites (Baldwin *et al.* 1989; Baldwin 1991; Cardon *et al.* 1991) and we have successfully trained networks to recognize features such as tar mats from reservoir core-extract data.

Results and interpretation

Reservoir geochemical studies are likely to be most effective as a reservoir description tool when used in conjunction with geological, well-test and engineering data (England 1990; Saigal *et al.* 1992; Stoddart *et al.* 1995; Fig. 3). However, one of the benefits of reservoir geochemistry is that data collection and interpretation can commence directly after wildcat drilling or early in field appraisal, as soon as cores and/or fluids (and in some cases cuttings) are obtained. As discussed below, each sample type contains slightly different types of information.

Fluids

There are now many published examples of lateral and vertical compositional variations in reservoired petroleum, indicating variations on both a bulk and molecular level (see Table 1 for references). These variations have been used to (1) infer the maturity and source facies of petroleum populations within reservoirs; (2) identify maturity gradients across fields and thus infer fill points of reservoir segments and locate satellite fields; (3) infer laterally extensive, in-reservoir barriers; (4) detect leaking production strings; and (5) help in the siting of production wells.

It is less well established that water columns can also be compositionally heterogeneous. As for petroleum, vertical compositional variability in the water leg can indicate the presence of barriers to fluid flow. This will be important for the correct location of perforations within water injection wells. Variation in brine chemistry or CO_2 content across a field (James 1990) could impact the solution of scaling problems during production, while unrecognized salinity variations can result in reserve estimate errors. Compositional variations in residual water within the oil leg cannot presently be interpreted in the same way because we are unclear about the connectivity of water within the oil leg.

One example of a compositionally variable water column has been provided by Coleman (1992), for an unnamed North Sea field. Coleman suggested that two compositionally distinct waters occur in the reservoir: the present-day formation water and a palaeoformation water, trapped within the oil leg when the reservoir filled. The trapped palaeowater contains about 25 000 mg $l^{-1}Cl^-$ whilst the present-day formation water contains about 60 000 mg l^{-1} Cl^-. This example shows that we should beware of the assumption used in S_w calculations that the salinity of residual water in the oil leg is identical to that in the water leg. If reservoir filling occurred 10 to 50 Ma years ago, as is common in the North Sea, there is no clear-cut reason to assume that the formation water present in the reservoir at that time is the same as that now in the aquifer. In the preceding example, assuming that residual water contained 60 000 mg l^{-1} Cl^- rather than 25 000 mg l^{-1} would lead to a 10% overestimate of oil in place.

Core extracts

The first stage in a study of core extracts is to determine the extent of contamination from drilling muds. Clearly, oil-based muds (OBM) can be problematical for petroleum analysis, whilst water-based muds (WBM) are potentially worse for residual salt analysis (RSA) (Smalley *et al.* 1992). Contamination by diesel-rich OBM can be recognized by reduced concentrations of petroleum-derived polyaromatic hydrocarbons or by GC fingerprints (Larter *et al.* 1991). In general, mud contamination of reservoir cores does not seem to be a major problem for organic geochemistry, and we have found that reliable biomarker data can be obtained from heavily contaminated cores. For RSA, it is important to assess the extent of penetration of drilling mud into cores. This can be achieved by determining the $^{87}Sr/^{86}Sr$ ratio of salts leached from samples taken from transects across cores with a range of permeabilities (Smalley *et al.* 1992). Establishing the extent of penetration gives confidence to a sampling programme. Samples for analysis are generally taken from the heart of cores from lower permeability horizons. RockEval can be used to screen reservoir cores for bulk petroleum concentration (S1 yield) and to provide some information on the location of heavy-oil zones or tar mats (Karlsen & Larter 1989; Ghenima *et al.* 1991; Wilhelms & Larter 1994*a*; Wilhelms *et al.* 1995). However, direct group type analysis of core-extract petroleums (C_{14+} fraction) provides more information than bulk-flow pyrolysis techniques, albeit more slowly. RockEval petroleum yields are systematically lower than core-extract C_{14+} petroleum yields due to the involatility of higher boiling fractions, resulting in their pyrolysis and partial charring during S2 analysis.

Karlsen & Larter (1989, 1991) adapted a commercial TLC–FID system and proposed a calibration and analytical protocol to enable the rapid, quantitative characterization of petroleum group type composition. The method provides data on saturated, mono- and diaromatic hydrocarbon, polyaromatic hydrocarbon, resin and asphaltene concentrations from up to 70 reservoir core samples daily. This technique greatly expands the capabilities of RockEval systems for the characterization of the fluids in reservoir cores, providing direct chemical data on the petroleum and enabling the precise location of oil–water contacts and small tar mats (Fig. 6). Small tar mats ('minimats') seem to be common, if previously unrecognized features of many light oil reservoirs and may inhibit vertical fluid flow in the reservoir (Wilhelms & Larter 1994*b*).

The rapidity with which data can be collected by the Iatroscan allows high resolution geochemical logs to be made. Figure 8 shows a comparison between fluid distributions from a section of the Eldfisk chalk reservoir estimated from a geochemical log (Iatroscan) and MSFL (induction log) (Stoddart *et al.* 1995). In general, the logs are similar but the geochemical logs give greater resolution. Applications of geochemical logs might therefore include locating perforations in low quality reservoirs, or assessing fluid composition prior to testing. It is emphasized, however, that integration of geochemical and geophysical/engineering data inevitably provides the best approach to core logging.

Thermovaporization or thermal-desorption gas chromatography (TD–GC) is an important technique which allows the direct acquisition of gas chromatograms of petroleum retained on reservoir core (Horsfield 1984; Bjorøy *et al.* 1985, 1991*b*). Chromatograms can be obtained from material volatile at 300°C and from the pyrolysates of material obtained by heating the sample from 300 to 600°C. The latter is useful in distinguishing free petroleum from indigenous organic matter or tar mats. Coupled directly with a mass spectrometer (Bjorøy *et al.* 1991*b*), the system is capable of producing high quality GC–MS data directly from cores at a sampling density sufficiently high to allow statistically interpretable datasets (20 to 30 analyses/100 m reservoir core). The data in Fig. 4 were obtained in less than two days with an automated system, and can be used to detect barriers and thus to locate perforation intervals in wells where RFT pressure measurements are absent.

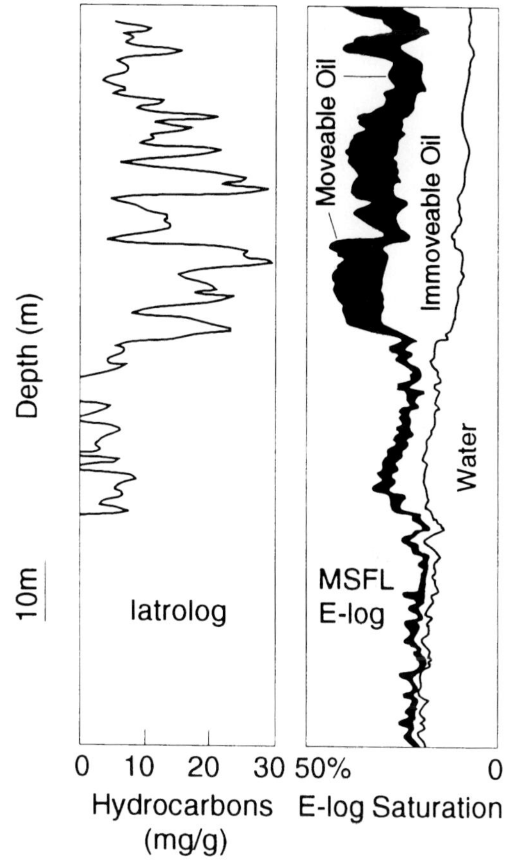

Fig. 8. Determination of petroleum saturations by geochemical and electric logs. The logs are broadly correlative but the direct measurement of the geochemical log permits greater near-well resolution. While in this case it is not significant, in marginal reservoirs use of geochemical logs to locate perforations may reduce water cut during production (Stoddart *et al.* 1995).

Whilst it is theoretically possible to reconstruct the composition of formation water by redissolving residual salts, trials have been unsuccessful. However, Smalley *et al.* (1992) have shown that it is possible to determine the Sr isotopic composition of the water. As with the petroleum studies, the benefit of RSA is that compositional maps may be generated from analyses made on readily available dried core, both from the oil and water legs (Smalley & England 1992).

Smalley *et al.* (1992) showed an example from Ekofisk with a similar stratigraphy to the nearby Eldfisk reservoir. Oil is reservoired in the Ekofisk and Tor Formations, separated by a chalk hardground (Ekofisk ED layer). The $^{87}Sr/^{86}Sr$ isotope ratios of residual salts in the Ekofisk Formation are essentially constant, implying that residual waters are well mixed now and probably were at the time of oil emplacement. In contrast, $^{87}Sr/^{86}Sr$ isotope ratios within the Tor Formation change systematically towards the hardground separating the two formations. It is interesting to compare these data with the biomarker data obtained from the geologically equivalent Eldfisk reservoir (Fig. 4). Both techniques pick up the barrier between the Ekofisk and Tor Formations. In other examples, baffles or barriers may be inferred from the Sr data but not the petroleum data, emphasizing the need for complementary analysis of waters and oils. England *et al.*'s (1987) oil mixing model can be used to broadly estimate barrier lengths from oil column heterogeneities (Stoddart *et al.* 1995), but the lack of a calibrated model for solute transport in residual water in oil legs prevents a quantitative interpretation of RSA data. Nonetheless, the technique is powerful and has proven utility.

If reservoirs contain isotopically stratified waters, Sr isotopes may be used, under favourable circumstances, to identify the pathways taken by injection waters (Smalley *et al.* 1988). If injected water moves evenly through all parts of the reservoir, then produced waters will have the $^{87}Sr/^{86}Sr$ ratio of the average formation water. Breakthrough of injected waters can be detected by a change in the $^{87}Sr/^{86}Sr$ ratio of produced fluids towards the value of the injection water. If, in contrast, flow is channelled through a high permeability 'thief' zone with a specific $^{87}Sr/^{86}Sr$ ratio, then breakthrough will be characterized by a different ratio.

Geochemical alteration of a reservoir fluid may take place over human as well as geological timescales. Reservoir souring, defined as 'an increase in the concentration of H_2S in produced gas during the lifetime of a reservoir', is a common problem in the North Sea. It is usually ascribed to the reduction of injected seawater sulphate by sulphate-reducing bacteria. This assertion is difficult to prove and may not always be true. H_2S may be produced chemically, or may be naturally present in the reservoir's water leg, in which case it may not be detected until water cuts increase (Whittingham & Jones 1987; Aplin & Coleman 1995). Increasing water cuts will lead to increasing masses of H_2S in lower volumes of produced gas (Fig. 9). Clearly, cost-effective remedial strategies depend on determining the correct souring mechanism; there is no point biociding the reservoir if sulphate-reducing bacteria did not produce the H_2S!

Aplin & Coleman (1995) describe how the isotopic composition of sulphur in the H_2S can be used to identify its source. Unfortunately, the isotopic composition of sulphur in H_2S does not uniquely constrain its origin, except in fortuitous cases. Successful application of sulphur isotopic analysis depends on obtaining samples of H_2S, where relevant, from wells which have not suffered

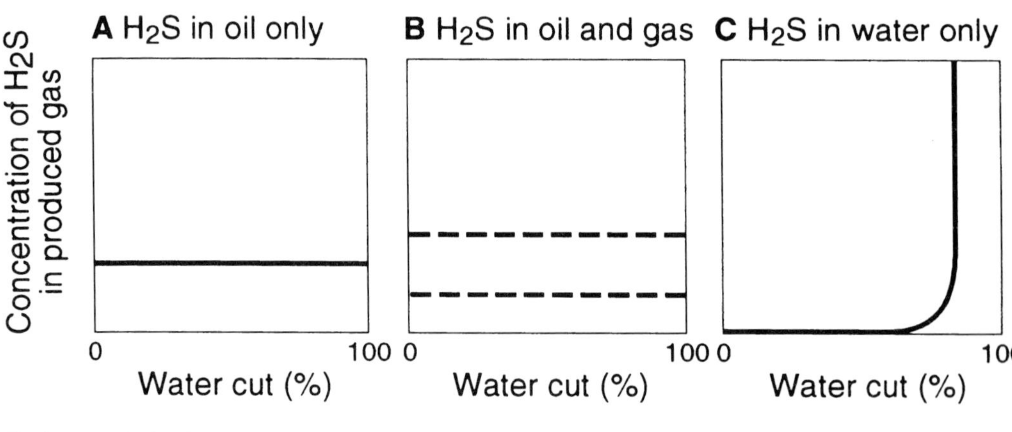

Fig. 9. Control of H_2S concentration in produced gases as a function of the partition of H_2S in reservoir fluids. If the H_2S is solely in the oil phase (**A**) then the H_2S content will depend only on the GOR of the produced fluids and might be expected to be relatively constant with time. If the H_2S occurs in both reservoired oil and gas (**B**), the H_2S content of produced gas will only vary between a restricted range of values (dashed lines). If the H_2S is dominantly in the water phase (**C**), 'souring' may occur independently of man-induced effects as the water cut increases with time. After Aplin & Coleman (1995).

water or injected water breakthrough, and obtaining samples of dissolved sulphate for isotopic analysis. Successful studies require continuous monitoring of fluid composition, a strategy which operators are often loath to follow for short-term financial reasons.

Fluid inclusions

Fluid inclusions are tiny samples of fluids previously present in the reservoir which have been trapped in diagenetic cements. Aqueous inclusions give information on the salinity of the water present when the host mineral precipitated. The timing of mineral precipitation can be estimated from homogenization temperatures of the fluid inclusions combined with a time–temperature curve for the formation (McLimans 1987). Comparison with the salinity of the water presently in the reservoir may indicate whether water has been replaced in the reservoir since the fluid inclusion formed.

With care and luck, petroleum fluid inclusions can be analysed in the same manner as core extracts and produced oils (Horsfield & McLimans 1984; McLimans 1987; Burruss *et al.* 1985). Information on source, maturity and composition can be derived if sufficient sample can be extracted by crushing or thermally decrepitating extensively cleaned samples. Inclusions in carbonates can be analysed by thermal decrepitation–GC–MS (Horsfield & McLimans 1984), but those within silicates require careful cleaning and crushing of separated cement phases in order to produce interpretable data (Karlsen *et al.* 1993).

Fluorescence spectra of included petroleums have been used to assess maturity, but must be calibrated within a specific petroleum province using geochemical data (McLimans 1987). It is more satisfactory, however, to assess source and maturity by direct geochemical measurement. An example of the type of data which can be produced has been given by Nedkvitne *et al.* (1993), who used biomarker data to show that a carbonate or marl-sourced oil was present in the Ula Field (North Sea) when the reservoir was at about 50°C (present temperature = 145°C). The present oil was sourced from a clastic marine source (Mandal Formation), suggesting the presence of a second minor source rock in the area which was mature long before the marine source, thus revealing the complexity of field filling. These authors also showed that a fluid-inclusion study could substantially enhance a conventional reservoir geochemical study by revealing the distribution of petroleum (as inclusions) in the developing or proto-Ula Field. Such information on how the field actually filled provides input into decisions relating to field connectivity and drainage, because most fields do not fill in a simple tank-like manner.

Homogenization temperatures, combined with time–temperature–pressure histories, can be used to estimate the time at which the inclusion formed. This is important if one is to assess, where relevant, when fluids were displaced from a reservoir block or when the composition of fluids changed. However, determination of the true trapping temperature or pressure of the inclusion from the homogenization temperature depends on the pressure–volume–temperature (PVT) properties of the petroleum, which are much more difficult to estimate than those of water. Usually, it is assumed that the composition of the petroleum is identical to that of locally reservoired oil. Recent work implying that petroleum in inclusions may be enriched in NSO compounds suggests that this approach may not always be realistic (Macleod *et al.* 1993*a*).

NSOs: new directions in reservoir geochemistry

NSOs are defined here as petroleum compounds containing the heteroatoms N, S and O (resins + asphaltenes). In other papers, these compounds have been variously called polar compounds, polars, or heterocompounds. We have chosen NSOs here to avoid the complications of the varying polarities of species within and beyond this broad compound class. Despite their relatively low abundance (5 to 25% NSOs in North Sea petroleums), the often polar nature of NSOs means that they exert disproportionately large effects on both the viscosity and potentially the phase behaviour of petroleum. They are also the species involved in the formation of emulsions with waters during production (Acevedo *et al.* 1992). Through adsorption onto mineral surfaces, NSOs may affect reservoir wettability (Cuiec 1977, 1984; Crocker & Marchin 1988; Dubey & Waxman 1989; Clementz 1976). Finally, some NSOs, such as phenols (or even the hydrocarbon benzene), are significantly water-soluble (Table 3) and may be depleted from petroleum during secondary migration. This raises the possibility that the relative abundances of water-soluble and water-insoluble compounds in petroleum may be used to assess migration volumes and, in a different context, the quantitative extent of water-washing of petroleum (Lafargue & Barker 1988; Macleod *et al.* 1993*b*; Li *et al.* 1995). These developing applications link exploration and production geology with reservoir engineering and petroleum engineering and are discussed in this section. Of particular current interest are those O- and N-bearing compounds which, through the formation of hydrogen and ionic bonds, interact

Table 3. *Distribution properties of crude oil components between hydrocarbon, water and minerals*

Water solubility (mg l^{-1})
Carbazole 1.8; hexane 9.5; benzene 1750; quinoline 6100; *p*-cresol 21 520; acetic acid unlimited.
References: McAuliffe (1978); Smith *et al.* (1978).

Hexane solubility
1,4-Dimethyl, 2,4,8-trimethyl, 1,3,5,8-tetramethylcarbazole: 3.5–4.5 mg ml^{-1}.
2-Methyl, 3-methyl, 2,4-dimethyl, 2,6-dimethyl, 2,4,6-trimethyl, 2,4,6,7-tetramethylcarbazole: 0.2–0.3 mg ml^{-1}.
Reference: Frolov *et al.* (1989).

Oil–water partition coefficients

Compound	Coefficient	Reference
Short chain carboxylic acids	<<<1	Barth & Bjorlykke (1992)
C_0–C_3 Alkylphenols	0.1–200	Taylor (1994)
3,4-Dimethylphenol	28	
2,5-Dimethylphenol	46	
C_0–C_2 Alkylcarbazoles	500–30 000	
3,4-Dimethylcarbazole	7250	
1,8-Dimethylcarbazole	26 733	
C_0–C_1 Benzenes and naphthalenes (heptane/water)		Southworth *et al.* (1983)
Benzene	180	
Toluene	570	
Naphthalene	2800	
1-Methylnaphthalene	8000	

Approximate oil/'mineral' distribution coefficients (ml g^{-1}), *estimated from adsorption isotherms*
Reference: Taylor (1994).

System	Distribution coefficient
Carbazole on montmorillonite	2.1
Carbazole on quartz	0.34
8-Methylquinoline on montmorillonite	80
8-Methylquinoline on quartz	15
3,5-Dimethylphenol on quartz	7.5
2,6-Dimethylphenol on quartz	7.3
3,5-Dimethylphenol on coal	53
2,6-Dimethylphenol on coal	16

strongly both with each other and with rock minerals and water.

Nitrogen and oxygen compounds in petroleum

The nitrogen content of crude oils varies between 0.1 and 2.0 wt%. Nitrogen in both crude oils and asphaltenes is present mainly as aromatic heterocycles, with a predominance of neutral pyrrolic structures over basic pyridinic forms (Richter *et al.* 1952; Dorban *et al.* 1984; Wilhelms *et al.* 1992).

Some key physicochemical properties of selected petroleum N compounds are given in Table 3. Although there are important differences between individual compounds, they are all strongly adsorbed onto mineral surfaces and solid-phase organic matter (Bakel & Philp 1990; Charlesworth 1986; Brothers *et al.* 1991; Yamamoto 1992). Experimental work using synthetic model compounds and natural petroleum fractions suggests that basic (pyridinic) species are more strongly adsorbed than neutral (pyrrolic) species (Charlesworth 1986; Table 3). The experiments also suggest that for a typical fraction such as the carbazoles, adsorption onto rock minerals can be described in terms of normal phase chromatographic fractionation with variable, irreversible adsorption of different isomers (i.e. adsorption from a mobile, low polarity organic phase onto a more polar solid phase; Li *et al.* 1994). Fractionation is chiefly controlled by steric effects related to the alkylation position relative to the active pyrrole functionality (Fig. 10), nitrogen-

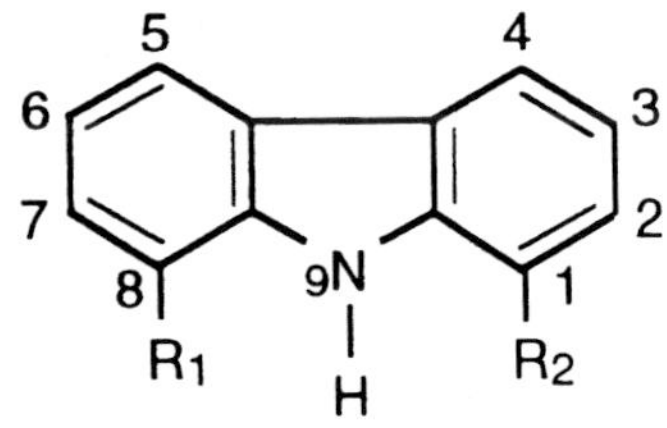

G1 Shielded pyrrolic hydrogen:
$R_1, R_2=C_nH_{2n+1}$, where $n \geq 1$

G2 Partially shielded pyrrolic hydrogen:
$R_1=H, R_2=C_nH_{2n+1}$ or,
$R_2=H, R_1=C_nH_{2n+1}$ where $n \geq 1$

G3 Exposed pyrrolic hydrogen:
$R_1=R_2=H$

Fig. 10. Alkylcarbazole nomenclature showing the alkylation positions described in the text. In this paper, all discussion concerns methylcarbazoles where R_1 and R_2 are methyl groups (when not hydrogens). In the text, the Group 1 isomers such as 1,8 dimethylcarbazole, where both 1 and 8 positions are methyl-substituted, are contrasted with Group 3 isomers, such as 3,5 dimethylcarbazole, where no alkylation is present adjacent to the active pyrrole >N–H group, shielding it from interaction with other species. After Li *et al.* (1995).

shielded isomers (e.g. 1,8-dimethylcarbazole) adsorbing less strongly than nitrogen-exposed isomers (e.g. 3,5-dimethylcarbazole; Table 3).

The oxygen content of crude oils is generally much less than 2.0%. Oxygen is present mainly as either carboxyl or phenolic hydroxyl species (Tissot & Welte 1984). Carboxylated and hydroxylated species are those most likely to be involved in strong interactions with non-petroleum phases or with other petroleum components. Table 3 also presents some physicochemical properties of selected O compounds. As for nitrogen species, the aqueous solubility, oil–water partition and solid-phase adsorption properties are controlled by the extent and position of alkylation (Taylor *et al.* 1993; Taylor 1994).

Tar mats

While hydrocarbon compositional variations are quickly equilibrated vertically within connected petroleum columns, the same is not true for polar species such as resins and asphaltenes (Stoddart *et al.* 1995). Particularly high relative concentrations of NSOs are found in petroleum-lean zones in reservoirs and in water legs, whilst high absolute concentrations are found in tar mats (Table 4). Tar mats are sharply defined zones of asphaltene-rich petroleum, up to a few tens of metres thick, usually occurring in light paraffinic oil reservoirs (Jones & Speers 1976; Dahl & Speers 1986; Wilhelms & Larter 1994*a,b*, 1995). Tar mats typically occur close to geological discontinuities such as OWCs or above permeability contrasts. The ocurrence of tar mats is generally detrimental to reservoir quality because they contain immovable reserves, present low permeability intra-reservoir flow barriers, and suggest that deasphaltation problems are likely to occur during production (Wilhelms & Larter 1994*b*).

Iatroscan analysis of core extracts has shown that small (10 cm to 1 m thick) tar mats ('minimats') are common features in light-oil reservoirs. Recent work has shown that tar mats result from the precipitation of asphaltenes from oil charges due to subtle chemical and physical changes in the solubility factors of crude oils induced by gas solution, thermal alteration and perhaps pressure and temperature changes during petroleum migration (Wilhelms & Larter 1995). Future work must produce accurate predictive models for tar-mat formation by developing an understanding of the subtle phase behaviour of asphaltenes in crude oils (Hirschberg 1984; Leontaritis & Mansoori 1987).

Table 4. *C_{14}^+ Polar compound (resins + asphaltenes) concentrations in petroleum reservoirs*

Situation	Relative concentration (% extractable OM)	Absolute concentration (mg g^{-1} rock)	Reference
Tar mats	30–60	5–40	Wilhelms & Larter (1995)
Lean zones	≤ 90 +	< 1	Horstad *et al.* (1990); Wilhelms & Larter (1995)
Water legs	≤ 90 +	> 0.1	Karlsen & Larter (1991)
Oil leg reservoir Extracts (N. Sea)	≤ 40	2–20	
DST oils (N. Sea)	≤ 25		

Reservoir wettability and NSOs

Many authors have reported that reservoir core extracts are substantially enriched in NSOs compared to oils produced from the same reservoir zone (Leythaeuser & Ruckheim 1989; Karlsen & Larter 1989; Larter *et al.* 1991; Horstad *et al.* 1990; Hillebrand & Leythaeuser 1992; Wilhelms 1992; Stoddard *et al.* 1995; Fig. 5). Although it is impossible to rule out the possibility that this enrichment occurred during core storage, circumstantial evidence suggests that this is not the whole story. Horstad *et al.* (1990) considered the possibility that enrichment in NSOs was caused by evaporative loss of light material from the cores during storage, but could generate no experimental evidence to support this. They concluded that the polar species must be naturally retained during oil production. On the other hand, Horstad (1989) did note the occurrence of a polar-rich rim on some reservoir cores, suggesting that core recovery may sometimes produce enrichments on the edge of the core. Stoddart *et al.* (1995) have shown that in the Eldfisk Field the molecular composition of the neutral nitrogen compounds is heterogeneous on a metre scale and is clearly out of equilibrium with compositionally homogeneous produced (mobile) oils. It would appear that in this case some nitrogen species extracted from cores are largely immobile and thus effectively isolated from, and out of equilibrium with, the bulk oil charge.

The state, nature and distribution of these apparently non-producible NSOs is not known, but they may be widespread. Simple volume calculations indicate that the NSOs in a petroleum containing 2 mg g^{-1} NSOs could form a 10^{-9}m (1 nm) thick layer over the majority of a reservoir with a surface area of 2 m^2 g^{-1}. For comparison, sand-grade quartz grains have specific surface areas of around 2 m^2 g^{-1} and typical reservoir clays reach values of 50 m^2 g^{-1}. These calculations suggest that NSOs could potentially coat a substantial proportion of the surface of clean reservoir sands, but less than half the surface of sands containing 10% or more clay minerals.

These calculations are supported by Mitchell *et al.*'s (1990) direct observations of organic coatings on grains in a variety of reservoirs. The atomic N/C ratio of this material was around 0.037, was fairly consistent for a wide variety of reservoirs and was essentially independent of the AMOTT wetting index. However, the quantity of organic material on reservoir surfaces correlated broadly with the AMOTT wetting index, more surface organic material giving more oil-wet reservoirs. The nitrogen content of the surface organic matter was much higher than that of asphaltenes from marine-sourced oils (N/C = 0.003–0.0147; Wilhelms *et al.* 1992) and also of produced oils (N/C = 0.001–0.02).

The nature of the adsorbed species is unclear. Mitchell *et al.* (1990) used electron-impact MS to show that the material on reservoir surfaces was probably acidic and alkylated with chains of a length of > C_{14} (note that this technique may not have detected nitrogen species). Figure 11 shows ^{13}C nuclear magnetic resonance (NMR) spectra of various fractions of a crude oil along with estimates of *n*-alkyl chain length made using the method of O'Donnell *et al.* (1980). Both resin and asphaltene fractions are extensively *n*-alkylated, with average chain lengths in the range indicated for reservoir surface organic matter by Mitchell *et al.* (1990). The asphaltenes appear to have slightly longer average *n*-alkyl chain lengths than resins or aromatic hydrocarbons (Wilhelms & Larter 1995). Adsorption of resins/asphaltenes on mineral or reservoir surfaces has been demonstrated from model systems by Cuiec (1977, 1984), Dubey & Waxman (1989), Trabelsi (1987) and Crocker & Marchin (1988), but it would appear that the surface films in reservoirs are actually enhanced in organic nitrogen compared even to asphaltenes. This suggests that subsets of these fractions are adsorbed. Cuiec (1977, 1984) has demonstrated that both mineralogy and oil composition control adsorption of NSOs on reservoir surfaces.

It is clear that, despite some success with detailed molecular analysis of a few low molecular weight homologues of a few classes of NSOs, detailed analysis of non-hydrocarbon species by GC/GC-MS is unlikely to provide solutions to the details of such adsorption processes. It is sobering to consider that many of the key advances in the understanding of the adsorption of NSOs on reservoir surfaces have been made using bulk analyses. It is thus clear that reservoir geochemists may have limited scope for the application of GC-based technology to the higher molecular weight long chain alkylated polar fractions most likely to be altering reservoir surfaces. New methods and approaches to the characterization of the whole polar fraction are needed.

How might the observation of NSO adsorption on reservoir surfaces influence our thinking on reservoir wettability and the controls thereof? Although there is strong debate over the physical meaning of wettability determinations, there is abundant literature evidence to indicate that many reservoirs are of intermediate or, more probably, mixed wettability (Treiber *et al.* 1972), with a significant oil-wet fraction. The observations and calculations described previously support these ideas, and suggest that strongly adsorbing N and other compounds may influence wettability (Mitchell *et al.* 1990). It is not clear if there are

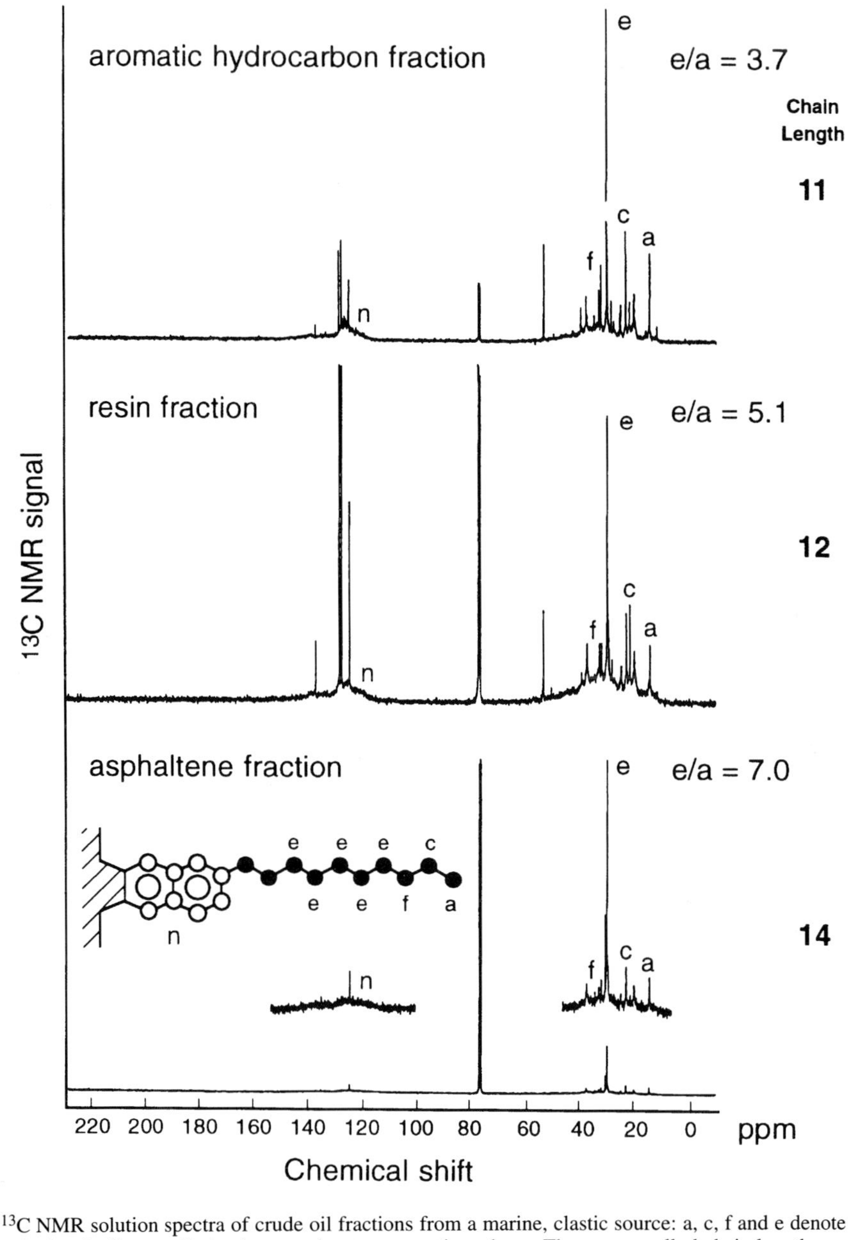

Fig. 11. ^{13}C NMR solution spectra of crude oil fractions from a marine, clastic source: a, c, f and e denote resonances due to terminal and adjacent alkylcarbons; n denotes aromatic carbons. The average alkyl chain lengths are calculated according to O'Donnell *et al.* (1980). Average alkyl chain lengths for the various fractions are indicated (Wilhelms 1992).

sufficient NSOs to completely coat reservoir surfaces, and indeed, recent data suggest that at least the molecular composition of neutral N compounds may be heterogeneous on a metre scale (Stoddart *et al.* 1995). Do these data imply variable reservoir wettability on a metre scale or even a pore scale? Such an idea is not outrageous when one considers the mechanical and chemical complexity of a reservoir rock.

PVT

The application of NSO geochemistry to petroleum engineering is an area of potential development in the next five years. We have already shown that the NSO content of core extracts may be much greater than that of DST oils (Table 4). Increasing the NSO content of an oil by 15% could reduce its API gravity from 35° to 25° and could change its bubble point by several hundred psi (Glasø 1980). This would mean that the PVT properties of reservoired petroleum (core extracts) could be substantially different to those of well-test (DST, RFT) oils. Macleod *et al.* (1993*a*) have suggested that petroleum trapped within fluid inclusions may be substantially enriched in polar species over their mother oil. Enrichment in NSOs may affect the assessment of the true trapping pressures and temperatures of fluid inclusions, which are currently made under the assumption that petroleum within inclusions is compositionally identical to that of produced oil.

When reservoir engineers attempt to match the actual production history of a field with their simulations, they often find discrepancies. These errors may be attributable to incomplete knowledge of the rock properties, but fluid compositional variation may be equally important. This may be especially so if it is true that well-test oils are compositionally distinct from the total petroleum in the reservoirs. The discrepancy between the nitrogen species distribution and concentration in core extracts and produced oils of one reservoir (Stoddart *et al.* 1995) reassures us that we are probably describing different phases in the reservoir (i.e. core extracts and DST oils do not represent the same materials) and that conventional PVT procedures are valid. In this case, an enrichment of NSOs on the surfaces of minerals is suggested, with implications for reservoir wettability. It appears that, however we look at the problem, reservoir geochemistry may have major implications for reservoir engineering procedures.

Assessment of migration volumes: oil–water–rock interaction

During secondary migration, petroleum typically moves updip or upsection over distances ranging between 10 and 10^6 m. Significant volumes of petroleum are lost in the carrier system and strongly affect the prediction of potential reserves (Mackenzie & Quigley 1988; England *et al.* 1987; Larter & Horstad 1992). Mass balance calculations suggest that the oil-saturated portions of the carrier

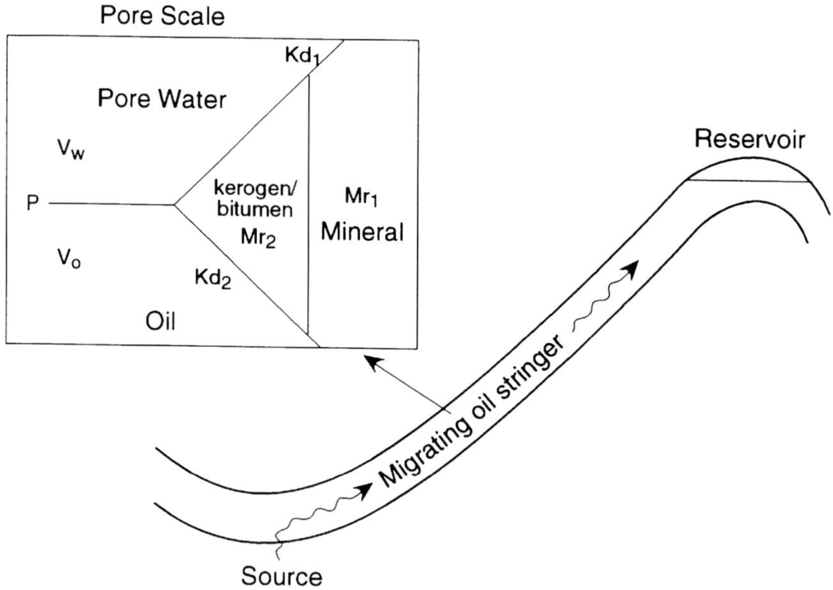

Fig. 12. Pore-scale interactions between oil, water and solid phases on a secondary migration pathway. Equilibrium between the phases is controlled by an oil–water partition coefficient (P), two distribution coefficients describing equilibrium between oil-wet organic matter and oil (K_{d1}) and between water and water-wet mineral phases (K_{d2}), and the relative masses or volumes of the four phases present: oil (V_o), water (V_w), oil-wet organic matter (M_{r2}), water-wet minerals (M_{r1}).

bed are on the order of 1 to 10% of carrier porous volume (England *et al.* 1987). Because a percentage of the carrier system appears to retain oil, well constrained estimates of total carrier volume are essential for the accurate assessment of migration losses.

The key question is to determine how much water and rock a petroleum charge 'sees', or interacts with, during its passage from source to trap. Migration pathways are rarely drilled and drainage volumes are generally inferred from seismic data and from calculations of petroleum mass balance (Macgregor & Mackenzie 1986). In theory, the chemistry of petroleum and associated water can be used to determine migration volumes directly. While bold attempts have been made previously to use petroleum composition to assess migration distances (Seifert & Moldowan 1978), the resulting geochemical indicators have been based on hydrocarbon species, are qualitative and have found little application. A better approach is to use a range of compound classes exhibiting the range of chemical properties appropriate for the task. We require a range of compounds which partition differently between oil, water and rock during the migration process and are fortunate that petroleum contains them.

Some compounds, for example longer chain *n*-alkanes, are strongly retained in oil. Others, for example phenolic compounds, are partitioned between oil, water and rock whilst some compounds, for example alkylquinolines and alkylcarbazoles, are strongly adsorbed onto the surfaces of minerals and solid organic matter. A few compounds, for example short-chain carboxylic acids, are strongly partitioned into the aqueous phase (Barth & Riis 1992; Barth & Bjorlykke 1992). The composition of reservoired petroleum is thus a function of the composition of generated petroleum and of changes induced by a variety of possible interactions with water and rock along the primary and secondary migration pathways, or in the reservoir itself (Lafargue & Barker 1988; Fig. 12). The extent of change, as indicated by the relative abundances of selected compounds in reservoired petroleum, is a potential measure of the migration volume. Equally, the absolute and relative concentrations of petroleum-derived compounds in formation waters is potentially an indication of water/oil ratios (V_w/V_o; Macleod *et al.* 1993*b*, Fig. 12), and it is likely that the most fruitful approach to the assessment of migration volumes will involve the study of both waters and oils. Some progress has also been made in this area using non-petroleum species. Ballentine *et al.* (1991) have elegantly and successfully used noble gas data to assess the fluxes of water involved in the accumulation of gas in the Pannonian Basin.

Evidence for oil–water–rock interaction

The species in petroleum most likely to interact strongly with minerals and water are those with functional groups capable of strong acid/base or hydrogen bond interactions, i.e. carboxylic acids, phenols, carbazoles and quinolines. Most aromatic and non-hydrocarbon polar petroleum components will also partition significantly into solid organic matter (Sandvik *et al.* 1992).

There is strong experimental and empirical evidence both for the adsorption of N species onto minerals and for the fractionation of individual N compounds during petroleum migration (Charlesworth 1986; Bonilla & Engel 1988; Brothers *et al.* 1991; Yamamoto 1992; Li *et al.* 1992, 1995; see Krooss *et al.* 1992, for a review of geochromatography in general). Further, enhanced concentrations of some NSO species are found in migration pathways (Miles 1990). Experimental work suggests that pyridinic N compounds adsorb more strongly than pyrrolic compounds (Li & Larter 1993), and that the extent to which individual pyrrolic compounds are adsorbed from a moving organic phase onto a polar surface is determined by steric effects related to the alkylation position relative to the active pyrrole functionality. Compounds in which the pyrrole functionality is shielded by adjacent methyl groups (e.g. 1,8-dimethylcarbazole) are adsorbed less strongly than compounds in which the pyrrole functionality is exposed (e.g. 3,4-dimethylcarbazole; Li *et al.* 1995; Fig. 10).

The experimental work detailed previously underpins similar observations within petroleum systems. Compared to source-rock extracts, petroleums appear to be depleted in C_0–C_3 carbazoles by factors of five to ten in concentration and to have significantly higher alkylcarbazole/(alkylcarbazole + alkylbenzocarbazole) ratios, lower percentages of nitrogen-exposed isomers and higher C_3/C_2 substituent ratios for substituted alkylcarbazoles (Fig. 13; Li *et al.* 1995; Stoddart *et al.* 1995; Chen *et al.* pers. comm.). However, the differences in the amounts and distribution of N compounds between source-rock extracts and petroleum could result from several processes, not all of which have been fully evaluated.

Figure 12 shows a schematic system for the equilibration of NSOs between oil, porewater, oil-wet organic matter and initially water-wet mineral phases in a reservoir or carrier system. No separate gas phase is present into which petroleum components may partition (Thompson 1988; Larter & Mills 1992). Studies of the restoration of cores for laboratory core analysis (Cuiec 1977) and of the rate of partition of petroleum compounds from oil spills into water (Southworth *et al.* 1983) indicate

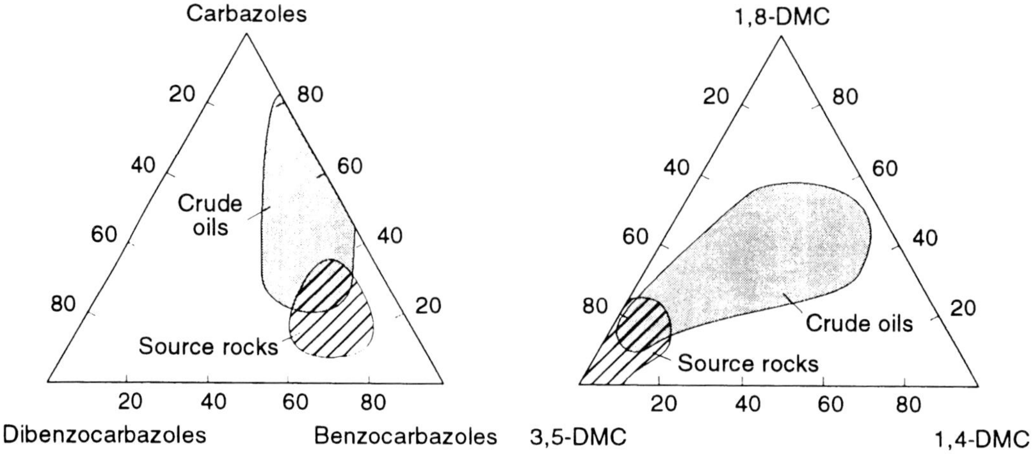

Fig. 13. Composition of carbazoles in associated source-rock extracts and reservoired petroleums. Source rocks are enriched in benzocarbazoles and Group 3 (exposed) carbazoles (3,5 dimethylcarbazole) compared to crude oils (DMC = dimethylcarbazoles). See also Fig. 10. After Li *et al.* (1995).

that this complex system should attain local (centimetre–metre) equilibrium within a maximum of a few years within a static oil column. In contrast, equilibration of petroleum species with water and rock masses surrounding the migrating oil stringer will depend on the relative rates of migration of the petroleum stringer and of diffusion from the stringer into the surrounding water–rock system. Assuming local equilibrium in the oil-saturated section, factors affecting the distribution will be (1) the aqueous solubility of the component; (2) an oil–water partition coefficient (P); (3) two distribution coefficients describing equilibrium between oil-wet organic matter and oil (K_{d1}) and between water and water-wet mineral phases (K_{d2}); (4) the relative masses or volumes of the four phases present (oil, V_o, water, V_w, oil-wet organic matter, M_{r2} and water-wet minerals, M_{r1}).

Table 3 shows partition and distribution coefficients for some selected petroleum components. A consistent theme is that properties such as aqueous and hydrocarbon solubilities, oil–water partition coefficients and distribution coefficients are related to the extent of alkylation and the position of the alkylation relative to the active functional group. With increasing alkyl substitution near the active pyrrole or phenol functional group, partition coefficients increase, distribution coefficients decrease, hydrocarbon solubilities increase and aqueous solubilities decrease (Lyman *et al.* 1982).

Source and maturity effects may influence the absolute abundances of nitrogen compounds in petroleum but are less likely to affect their relative abundances. Partition between petroleum and water can also be ruled out as a major factor affecting the ratios of carbazole isomers (Li *et al.* 1995). Using a simple equilibrium partition model it is possible to determine the volume of water which a volume of oil must interact with in order to change the relative amounts of chosen compounds in the oil between specified values. For example, the Chinese source rocks studied by Li *et al.* (1995) have 1,8-dimethylcarbazole/3,4-dimethylcarbazole ratios of less than 0.2, compared to ratios of more than 2 in equivalent migrated oils. Figure 14 shows that the oils must equilibrate with vast and unrealistic volumes of water if the 1,8/3,4 ratios are to be increased to values found in oils solely by interaction with water. It is concluded that retention on solid phases is required to account for the observed distributions of N compound isomers in oils.

The extent to which nitrogen compounds are retained within the source rock or secondary migration pathway, and the extent to which retention is by minerals or solid-phase organic matter, remain to be determined. In aquifers, adsorption of hydrophobic species is mainly onto organic matter where TOC exceeds 0.1% (Lee *et al.* 1988; Perlinger & Eisenreich 1991). While much of the environmental literature concerns adsorption onto humic substances, kerogens also have a high sorption capacity for petroleum species (Pepper 1992; Sandvik *et al.* 1992). Thus, solid-phase organic matter in deeply buried sediments may be important for adsorbing hydrophobic compounds (Sandvik *et al.* 1992) whilst minerals will affect both hydrophobic and hydrophilic species.

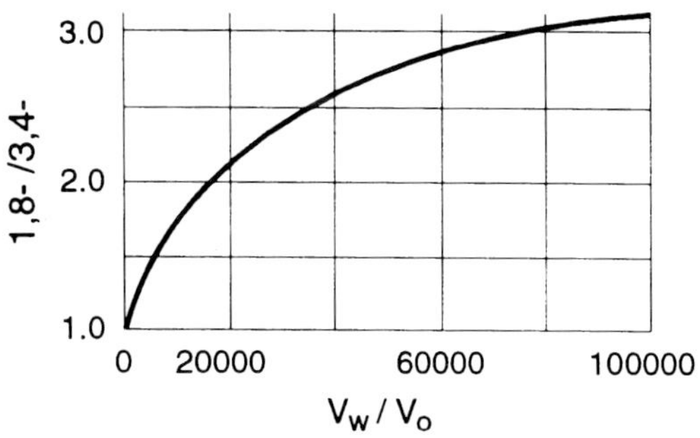

Fig. 14. The relative concentrations of a Group 1 carbazole (1,8 dimethylcarbazole) versus a Group 3 carbazole (3,4 dimethylcarbazole) plotted as a function of water-washing with different relative volumes of water and oil (V_w/V_o). Because the oil–water partition coefficients are different (Table 3) the ratio, initially set as one, increases but the volumes of water required to reach the values found in crude oils (greater than two) are absurdly large (many tens of thousands of water volumes equilibrating with each volume of oil). Compound fractionation along the migration pathway is inferred to occur by adsorption onto solid phases.

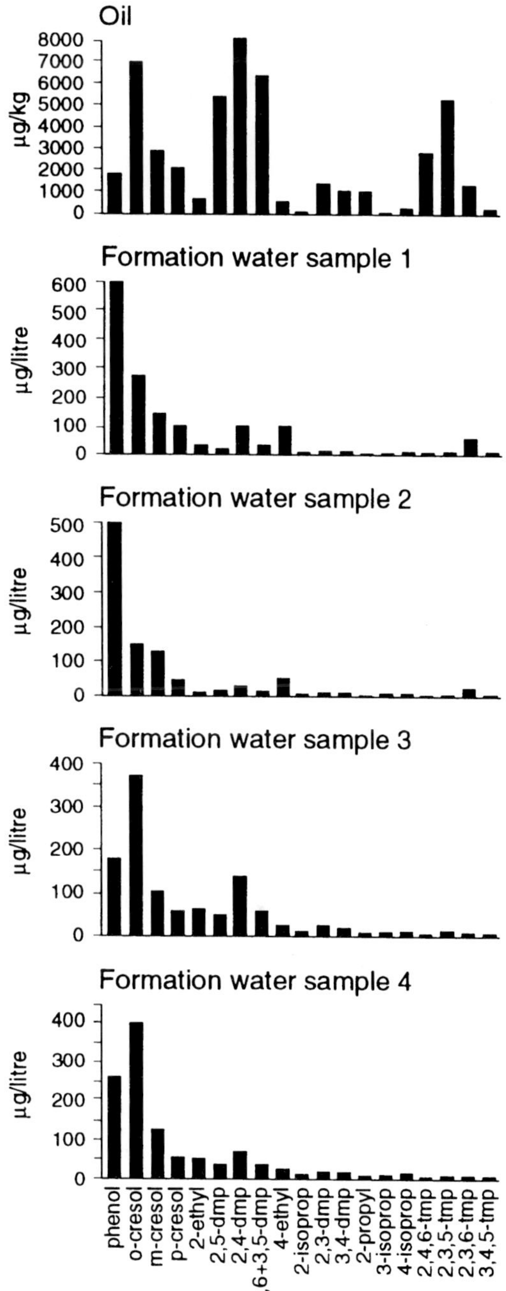

Fig. 15. C_0–C_3 alkylphenols in North Sea formation waters and a crude oil. Compared to the waters, the oil is enriched in the more alkylated phenols (C_2, C_3), relative to phenol and cresols. Ortho-substituted isomers are enriched in the oil relative to meta (*m*) and para (*p*) isomers. This distribution suggests that phenols in the water–oil system may be close to equilibrium (dmp = dimethyl phenols and tmp = trimethyl phenols). After Macleod *et al.* (1993*b*).

Certainly, it has been proposed that the partitioning of petroleum components into solid organic matter plays an important role in primary migration (Pepper 1992; Sandvik *et al.* 1992) and may also be a major factor in controlling the distribution of NSOs within reservoirs, as many reservoirs contain abundant indigenous organic matter.

Alkylphenol compounds have much lower oil–water partition coefficients than do carbazoles, and thus the distribution of these compounds may be affected by partition between oil and water (Table 3). Figure 15, from Macleod *et al.* (1993*b*), shows the distribution of alkylphenols in formation waters and an oil from the North Sea. Two of the waters are dominated by phenol, and two by *o*-cresol. The oil contains around 5000 $\mu g\ kg^{-1}$ individual phenolic compounds, with much higher concentrations of the more highly alkylated phenols. This is consistent with the increase in oil–water partition coefficient as the extent of alkylation increases. Even for simple species, such as alkylphenols, the organic–aqueous phase partition process is complex (Bizek *et al.* 1992), but the relatively greater affinity of phenol and the cresols for water compared to higher alkylphenol homologues is accurately reflected in the relative compositions of the oils and formation waters. While the actual water–oil partition coefficients will change with temperature, pressure, salinity and solution-gas content (Knaepen *et al.* 1990), these initial results lend confidence to the idea that the phenolic compound distributions in coexisting oils and water may be controlled by partition processes.

A simple model for determining oil–water ratios

The distribution of solutes between a moving fluid (migrating oil), other mobile fluids (water) and rock surfaces (minerals/kerogens and bitumens) can be described by complex sets of differential equations, potentially involving non-equilibrium descriptions of diffusion and partitioning processes in addition to advection (Luchner & Schestakow 1991; Bear & Buchlin 1991). Recognizing this complexity, a simplified equilibrium partition model is presented below which illustrates the main factors involved for an equilibrium three-phase partition process.

Assuming equilibrium between oil and water and water and rock (i.e. a simplified water-wet porous medium), the mass balance and partition of a component within the system are:

$$C_i V_o = C_r M_r + C_o V_o + C_w V_w \qquad (1)$$

$$P = \frac{C_o}{C_w} \qquad (2)$$

$$K_d = \frac{C_r}{C_w} \qquad (3)$$

where C_i = initial component concentration in oil (kg m^{-3}); V_o = volume oil (m^3); C_r = concentration adsorbed on rock (kg kg^{-1}); M_r = mass rock (kg); C_o = final concentration in oil (kg m^{-3}); C_w = concentration in water (kg m^{-3}); V_w = volume of water (m^3); P = partition coefficient between oil and water; K_d = distribution coefficient between rock and water.

Solving for C_o, C_w and C_r we obtain:

$$C_o = \frac{C_i P V_o}{K_d M_r + P V_o + V_w} \qquad (4)$$

$$C_w = \frac{C_i V_o}{K_d M_r + P V_o + V_w} \qquad (5)$$

$$C_r = \frac{C_i K_d V_o}{K_d M_r + P V_o + V_w} \qquad (6)$$

Where adsorptive interactions are negligible ($K_d \rightarrow 0$) the equation for C_o becomes:

$$C_o = (C_i P V_o)/(P V_o + V_w) \qquad (7)$$

or if single volumes of water are contacted and equilibrated one at a time:

$$C_o = C_i ((P V_o)/(P V_o + 1))^{V_w} \qquad (8)$$

More complex expressions are obtained for the more realistic situation in which oil, water, oil-wet rock and water-wet rock are simultaneously in equilibrium. These will be discussed elsewhere.

Figure 16 shows a cross-plot of partition coefficient versus the concentration of a solute in crude oil, normalized to an initial concentration equal to one. The contours on the plot are values of V_w/V_o, generated using equation (7). Useful migration tracers will have final concentrations in reservoired petroleum between 0.1 and 0.5, and a range of tracers may be needed, depending on the migration volume. Petroleum contains the required range of compounds (Table 3). This plot, of course, describes idealized tracers that interact only with water.

The model can be illustrated by *n*-hexane and benzene. Neither are strongly adsorbed onto mineral surfaces, but benzene partitions much more significantly into water (P = 180; Perlinger & Eisenreich 1991; Southworth *et al.* 1983; Lafargue & Barker 1988). Changes in *n*-hexane/benzene ratios along the migration pathway will thus be

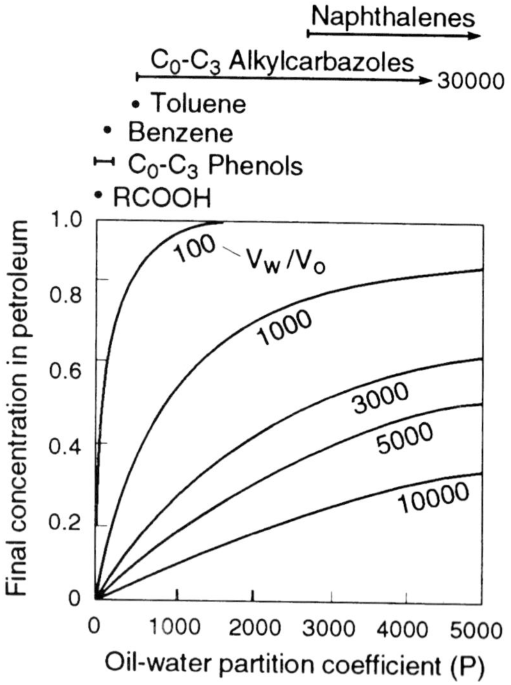

Fig. 16. Reduction in the normalized concentration of a petroleum component due to partition with water, as a function of the oil–water partition coefficient (P) and the relative volumes of water and oil (V_w/V_o) (equation 7).

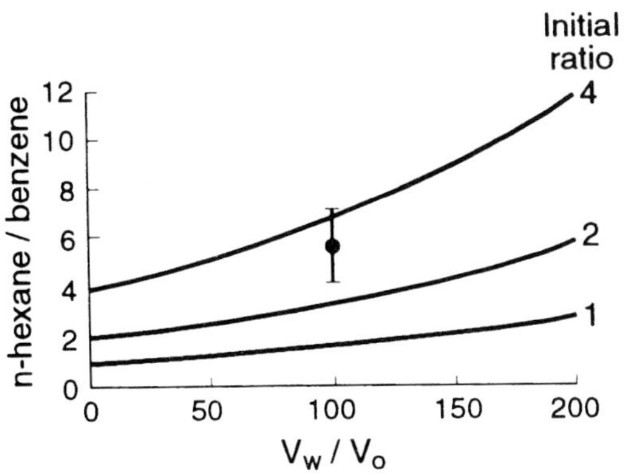

Fig. 17. Evolution of the *n*-hexane/benzene ratio of oil as it interacts with increasing volumes of water (V_w/V_o). Each curve represents the path taken by oils with a different initial *n*-hexane/benzene ratio. The data point represent actual values (± 3 sigma) from a North Sea oil field.

significantly controlled by partitioning between oil and water, assuming oil–gas phase equilibrium to be unimportant. Figure 17 shows how *n*-hexane/benzene ratios would change with increasing water/oil ratios, for starting ratios between one and four. Also plotted are summarized data from 15 North Sea oils. In this case, a starting hexane/benzene ratio of four to seven is not unreasonable (Horsfield pers. comm.), suggesting a water/oil volume ratio of less than 100 for this migration path. Mass balance estimates of residual oil saturation levels in carriers would suggest maximum V_w/V_o values of around 10 to 100 (England *et al.* 1987), with maximum values in the North Sea's Brent province of around 30 (Larter & Horstad 1992). Using noble gas data, Ballentine (pers. comm.) derived even lower values for the Magnus oilfield, and our value is almost certainly a ceiling value.

Such small values of V_w/V_o and the occurrence in reservoired oils of water-soluble and/or highly adsorbtive components such as quinolines suggest that there is relatively little interaction between migrating oil and carrier rock phases. This implies a few efficient oil rivers connecting source to trap, with limited access to large water/rock volumes. It is probable that each migration pathway is used many times, consistent with the idea that oil reservoirs fill through identifiable field fill points ('keyholes') and the view that local carrier-bed geometry and sedimentology are as crucial to controlling oil emplacement as is regional structure. This places substantial constraints on the ability to model local secondary migration directions prior to drilling.

Summary and conclusions

The focus of petroleum-related geochemistry has shifted in recent years away from exploration and increasingly towards reservoir appraisal, development and production. The resulting sub-discipline of reservoir geochemistry involves the study of reservoir fluids and their interaction with reservoir rocks, and can be viewed as a link between reservoir geology and engineering.

The increased use of rapid screening analyses of both fluids and core extracts has shown, using a variety of statistical techniques, that hydrocarbons and waters within petroleum reservoirs are often compositionally heterogeneous on scales ranging from metres to kilometres. Heterogeneities are inherited during field filling and usually result in reservoir fluids that are out of chemical equilibrium. Diffusion and density-driven mixing seek to mechanically and chemically equilibrate the fluids at rates which depend on the distances over which the instabilities exist and the permeability structure of the reservoir. In general, chemical equilibrium will be rapidly achieved within individual fluid columns unless there are barriers to vertical mixing. In contrast, large-scale lateral compositional gradients inherited during filling can persist for geologically significant periods of time. Mechanical equilibrium in fluid columns is rapidly achieved in well connected reservoirs. Compositional heterogeneities of reservoired hydrocarbons and waters thus contain regional and reservoir-scale geological information and can be used as a tool in reservoir description, appraisal and development, and also in regional exploration. They can also be used as a tool for production, for example to monitor the zonal contributions to total produced fluids, the occurrence of tubing leakage and the flowpaths of injected water. These data support rather than supercede more common geological and well-test data but have the important benefit of being collectable immediately a core has been taken, and at relatively low cost.

N- and O-bearing compounds occur in relatively low abundance in petroleum but, because of their often polar nature, interact strongly with minerals, water and each other. The viscosity, density and phase behaviours of petroleum are all influenced by the abundance of polar species. This is exemplified by the occurrence and physical properties of polar-rich tar mats, which can be readily identified by Iatroscan analysis and which can act as barriers to vertical fluid flow. Future work should be aimed at producing accurate predictions of tar-mat formation through an understanding of the chemistry and phase behaviour of asphaltenes.

There is good field and laboratory evidence that N compounds adsorb strongly onto the surface of minerals and organic matter. Reservoir core extracts are universally but variably enriched in nitrogen compared to produced oil, and in some cases appear to be out of equilibrium with the oil. Pyridines adsorb more strongly than pyrroles and the extent of adsorption within a compound class appears to be controlled by the alkylation position relative to the nitrogen functional group. Some implications of these data include the likelihood that N compounds are both adsorbed and fractionated along migration pathways and the possibility that reservoir wettability may be locally variable and significantly influenced by the occurrence of adsorbed N species. Furthermore, should reservoired oil be enriched in polar species compared to produced oil, this would alter its PVT properties and the use of these properties in reservoir simulators. There are obvious links here between geochemistry and reservoir/petroleum engineering which are likely to be exploited in the next few years.

Petroleum contains a wide variety of compounds which partition differently between petroleum, water and the mineral matrix. In the absence of phase changes, the composition of reservoired petroleum is a function of the composition of expelled petroleum plus changes induced by partition into water and solid phases along the migration pathway. The extent of change is a measure of the migration volume which can thus be assessed by determining the relative abundance of carefully selected compounds in reservoired petroleum and formation water. Initial work in this area is encouraging and has shown that (1) carbazoles are strongly retained on solid phases; (2) N compound fractionation occurs between source and reservoir; and (3) phenols in some North Sea oils and formation waters are probably close to equilibrium. In order to create quantitative indicators of migration volumes, one needs to describe the distribution of solutes between a migrating oil, water and mineral surface. A rigorous description of this system requires a complex set of differential equations. However, some progress has been made by using a simple equilibrium model describing the partition of solutes between oil and water. Using *n*-hexane/benzene ratios in the model, we make preliminary estimates of maximum water/oil ratios of 100 for one suite of North Sea petroleums. These small water/oil ratios, plus the occurrence in petroleum of water-soluble and/or adsorptive compounds, support the idea that petroleum sources and traps are connected by relatively few, but efficient rivers of oil.

Note added in proof: Recent work (Taylor 1994; Larter *et al.* 1994) has suggested that adsorbtion of hydrophobic components on solid organic matter in carrier beds is a very important process. Thus estimates of migration volumes (V_w/V_o) made using oil/water partition alone will be conservatively large. Three-phase partition processes (oil–water–rock) seem necessary to explain observed distributions of alkylphenols in migrated crude oils.

This paper is based on ideas generated through discussions with many scientists and the following are acknowledge gratefully: K. Bjørlykke, A. Wilhelms, I. Horstad, D. Karlsen, T. Nedkvitne and D. Leythaeuser at Oslo; Maowen Li, G. Macleod, P. Taylor, D. Stoddart, Y. Frolov, M. Jones, Mei Chen, S. Petch and A. Douglas at Newcastle. C. Smalley, P. Hall, M. Bjorøy, W. England, B. Carpentier, J. Fox, J. Senftle and B. Horsfield are also thanked for useful discussion. We thank the sponsors of our reservoir geochemical studies carried out since 1986 (BP Norway, Norwegian Oil Directorate, Statoil, Phillips Petroleum, Unocal, Saga Petroleum, PSTI). This review was compiled as part of a technology demonstration carried out under the EC Thermie Programme. The manuscript was prepared by Y. Hall and the figures by C. Jeans and B. Brown. We thank C. Cornford and W. England for their helpful and constructive reviews.

References

ACEVEDO, S., ESCOBAR, G., GUTIERREZ, L. & RIVAS, H. 1992. Isolation and characterization of natural surfactants from extra heavy crude oils, asphaltenes and maltenes. Interpretation of their interfacial tension–pH behaviour in terms of ion pair formation. *Fuel*, **71**, 619–623.

APLIN, A. C. & COLEMAN, M. L. 1995. Sour gas and water chemistry of the Bridport Sands reservoir, Wytch Farm, UK. *This volume*.

AUGUSTSON, J. H. 1992. A method of classification of oil traps based on heavy oil contents in cores with relevance to filling and draining of Barents Sea oil bearing structures. *In*: *Arctic Geology and Petroleum Potential*. Proceedings of NPF Conferences, Tromsø, Norway, 15–19 August 1990.

BAKEL, A. J. & PHILP, R. P. 1990. The distribution and quantitation of organonitrogen compounds in crude oils and rock pyrolysates. *Organic Geochemistry*, **16**, 353–367.

BALDWIN, J. L. 1991. *Using a simulated bidirectional associative neural network memory with incomplete prototype memories to identify facies from intermittent logging data acquired in a siliciclastic depositional sequence: a case study*. Society of Petroleum Engineers, Paper No. **22843**, 273–286.

——, OTTE, D. N. & WHEATLEY, C. L. 1989. *Computer emulation of human mental processes: application of neural network simulators to problems in well log interpretation*. Society of Petroleum Engineers, Paper No. **19619**, 481–493.

BALLENTINE, C. J., O'NIONS, R. K., OXBURGH, E. R., HORVATH, F. & DEAK, J. 1991. Rare gas constraints on hydrocarbon accumulation, crustal degassing and groundwater flow in the Pannonian Basin. *Earth Planetary Science Letters*, **105**, 229–246.

BARTH, T. & BJØRLYKKE, K. 1992. Organic acids from source rock maturation. Generation potentials, transport mechanisms and relevance for mineral diagenesis. *Applied Geochemistry*, **8**, 325–337.

—— & RIIS, M. 1992. Interactions between organic acid anions in formation waters and reservoir mineral phases. *Organic Geochemistry*, **19**, 455–482.

BEAR, J. & BUCHLIN, J.-M. (eds) 1991. *Modelling and applications of transport phenomena in porous media.* Kluwer Academic Publishers, 380 p.

BIZEK, V., HORACEK, J. & PROCHAZKA, J. 1992. Extraction of phenols: I. Distribution of individual phenols between organic solvents and water or aqueous solvents. *Canadian Journal of Chemical Engineering*, **70**, 341–349.

BJORØY, M., HALL, K., GILLYON, P. & JUMEAU, J. 1991*a*. Carbon isotope variations in n-alkanes and isoprenoids of whole oils. *Chemical Geology*, **93**, 13–20.

——, ——, HALL, P. B., LEPLAT, P. & LØBERG, R. 1991*b*. Biomarker analysis of oils and source rocks using a thermal extraction-GC-MS. *Chemical Geology*, **93**, 1–11.

——, —— & JUMEAU, J. 1990. Stable carbon ratio analysis on single components in crude oils by direct GC- isotope analysis. *Trends Analytical Chemistry*, **9**, 331–337.

——, SOLLI, H., HALL, K. & LEPLAT, P. 1985. Analysis of source rocks, reservoir rocks and cap rocks by combined thermal extraction and pyrolysis-gas chromatography. *In*: THOMAS, B. M. *et al.* (eds) *Petroleum Geochemistry in Exploration of the Norwegian Shelf.* Norwegian Petroleum Society, Graham & Trotman, London, 327–337.

BONILLA, J. V. & ENGEL, M. H. 1988. Chemical alteration of crude oils during simulated migration through quartz and clay minerals. *Organic Geochemistry*, **13**, 503–512.

BROTHERS, L., ENGEL, M. H. & KROOS, B. M. 1991. The effects of fluid flow through porous media on the distribution of organic compounds in synthetic crude oils. *Organic Geochemistry*, **17**, 11–24.

BURRUSS, R. C. 1989. Palaeotemperatures from fluid inclusions: advances in theory and technique. *In*: NAESER, N. D. & MCCULLOCH, T. H. (eds) *Thermal History of Sedimentary Basins – Methods and Case Histories*. Springer, New York, 119–131.

——, CERCONE, K. R. & HARRIS, P. M. 1985. Timing of hydrocarbon migration: Evidence from fluid inclusions in calcite cements, tectonic and burial history. *In*: SCHNEIDERMANN, N. & HARRIS, P. M. (eds) *Carbonate Cements*. Society of Economic Paleontologists and Mineralogists, Special Publication, **26**, 277–289.

CARDON, H., VAN HOOGSTRATEN, R. & DAVIES, P. 1991. A neural network application in geology: identification of genetic facies. *In*: KOHONEN, T. ET AL. (eds). *Artificial Neural Networks*. Elsevier Science Publishers B.V., North-Holland, 809–813.

CHARLESWORTH, J. M. 1986. Interaction of clay minerals with organic nitrogen compounds released by kerogen pyrolysis. *Geochimica et Cosmochimica Acta*, **50**, 1431–1435.

CLEMENTZ, D. M. 1976. Interaction of petroleum heavy ends with montmorillonite. *Clays and Clay Minerals*, **24**, 312–319.

COLEMAN, M. L. 1992. Water composition variations within one formation. *In*: KHARAKA, Y. K. & MAEST, A. S. (eds) *Water Rock Interaction*. Balkema, Rotterdam, 1109–1112.

CROCKER, M. E. & MARCHIN, L. M. 1988. Wettability and adsorption characteristics of crude-oil asphaltene and polar fractions. *Journal of Petroleum Technology*, **16**, 470–474.

CUIEC, L. 1977. Study of the problems related to the restoration of the natural state of core samples. *Journal of Canadian Petroleum Technology*, Oct-Dec, 69–80.

—— 1984. *Rock/crude-oil interactions and wettability: an attempt to understand their interrelation.* Society of Petroleum Engineers, Paper No. **13211**.

DAHL, B. & SPEERS, G. C. 1985. Organic geochemistry of the Oseberg Field (I). *In*: THOMAS, B. M. ET AL. (eds) *Petroleum Geochemistry in Exploration of the Norwegian Shelf.* Norwegian Petroleum Society, Graham & Trotman, London, 185–195.

—— & —— 1986. Geochemical characterization of a tar mat in the Oseberg field Norwegian sector, North Sea. *Organic Geochemistry*, **10**, 547–558.

DAVIS, J. C. 1973. *Statistics and Data Analysis in Geology*. Wiley, New York.

DEL RIO, C. J., PHILP, R. P. & ALLEN, J. 1992. Nature and geochemistry of high molecular weight hydrocarbons (above C_{40}) in oils and solid bitumens. *Organic Geochemistry*, **18**, 541–554.

DORBAN, M., SCHMITTER, J. M., GARRIGUES, P. IGNATIADIS, I., EDWARD, M., ARPINO, P. & GUIOCHON, G. 1984. Distribution of carbazole derivatives in petroleum. *Organic Geochemistry*, **7**, 111–120.

DUBEY, S. T. & WAXMAN, M. H. 1989. *Asphaltene adsorption and desorption from mineral surfaces.* Society of Petroleum Engineers, Paper No. **18462**, 51–62.

ENGLAND, W. A. 1990. The organic geochemistry of petroleum reservoirs. *Organic Geochemistry*, **16**, 415–425.

—— & MACKENZIE, A. S. 1989. Geochemistry of petroleum reservoirs. *Geologische Rundschau*, **78**, 291–303.

——, ——, MANN, D. M. & QUIGLEY, T. M. 1987. The movement and entrapment of petroleum fluids in the subsurface. *Journal of the Geological Society, London*, **144**, 327–347.

ESPACH, R. H. & FRY, J. 1951. Variable characteristics of the oil in the Tensleep sandstone reservoir, Elk Basin Field, Wyoming and Montana. *Transactions of the American Institute of Mining and Metallurgical Engineers*, **192**, 75–83.

FROLOV, Y. B., SMIRNOV, M. B., VANYUKOVA, N. A. & SANIN, P. I. 1989. Carbazoles of crude oils. *Petroleum Chemistry USSR*, **29**, 87–102, in English.

GHENIMA, R., ESPITALIE, J., BARSONY, I., TRABELSI, K., GRIBAA, R., BERTERO, L. & VOLPI, B. 1991. Organic geochemistry study of tar mat deposits in a Triassic sandy reservoir. *In*: *3rd European Association of*

Petroleum Geologists Conference, Abstracts, European Association of Petroleum Geologists, Amsterdam, Netherlands.

GLASØ, O. 1980. Generalized pressure–volume–temperature correlations. *Journal of Petroleum Technology*, **32**, 789–795.

HALDORSEN, H. H. & DAMSLETH, E. 1993. Challenges in reservoir characterisation. *American Association of Petroleum Geologists, Bulletin*, **77**(4), 541–551.

HERTZ, J., KROGH, A. & PALMER, R. G. 1990. *Introduction to the Theory of Neural Computation.* Addison-Wesley, New York.

HILLEBRAND, T. & LEYTHAEUSER, D. 1992. Reservoir geochemistry of Stockstadt oilfield: compositional heterogeneities reflecting accumulation history and multiple source input. *Organic Geochemistry*, **19**, 119–133.

HIRSCHBERG, A. 1984. *The role of asphaltenes in compositional grading of a reservoir's fluid column.* Society of Petroleum Engineers, Paper No. **13171**.

HORSFIELD, B. 1984. Pyrolysis studies and petroleum exploration. *In*: BROOKS, J. & WELTE, D. H. (eds) *Advances in Petroleum Geochemistry. Vol.* **1**. Academic Press, London, 247–292.

—— & MCLIMANS, R. K. 1984. Geothermometry and geochemistry of aqueous and oil-bearing fluid inclusions from Fateh Field, Dubai. *Organic Geochemistry*, **6**, 733–740.

——, HECKERS, J., LEYTHAEUSER, D., LITTKE, R. & MANN, U. 1991. A study of the Holzener Asphaltkalk, Northern Germany: Observations regading the distribution, composition and origin of organic matter in an exhumed petroleum reservoir. *Marine and Petroleum Geology*, **8**, 198–211.

HORSTAD, I. 1989. *Petroleum composition and heterogeneity within the Middle Jurassic reservoirs in the Gullfaks field area, Norwegian North Sea.* Candidat Scient. Thesis, Dept. of Geology, Univ. of Oslo.

——, LARTER, S. R., DYPVIK, H., AAGAARD, P., BJØRNVIK, A. M., JOHANSEN, P. E. & ERIKSEN, S. 1990. Degradation and maturity controls on oil field petroleum column heterogeneity in the Gullfaks field, Norwegian North Sea. *Organic Geochemistry*, **16**(1–3), 497–510.

——, —— & MILLS, N. 1992. A quantitative model of biological petroleum degradation within the Brent Group reservoir in the Gullfaks Field, Norwegian North Sea. *Organic Geochemistry*, **19**, 107–117.

——, —— & —— 1995. Migration of hydrocarbons in the Tampen Spur area, Norwegian North Sea: A reservoir geochemical evaluation. *This volume.*

IOPPOLO, M., ALEXANDER, R. & KAGI, R. I. 1992. Identification of C_0-C_3 phenols in some Australian crude oils. *Organic Geochemistry*, **18**(5), 603–609.

JAMES, A. 1990. Correlation of reservoired gases using the carbon isotopic compositions of wet gas components. *American Association of Petroleum Geologists, Bulletin*, **74**, 1441–1458.

JONES, H. P. & SPEERS, R. G. 1976. *Permo-Triassic reservoir of Prudhoe Bay field, North Slope, Alaska.* American Association of Petroleum Geologists, Memoir, **24**, 23–50.

KARLSEN, D. & LARTER, S. R. 1989. A rapid correlation method for petroleum population mapping within individual petroleum reservoirs – application to petroleum reservoir description. *In*: COLLINS, J. (ed.) *Correlation in Hydrocarbon Exploration.* Norwegian Petroleum Society, Graham & Trotman, London, 77–85.

—— & —— 1991. Analysis of petroleum fractions by TLC-FID: applications to petroleum reservoir description. *Organic Geochemistry*, **17**, 603–617.

——, NEDKVITNE, T., LARTER, S. R. & BJØRLYKKE, K. 1993. Hydrocarbon composition of authigenic inclusions: Application to elucidation of petroleum reservoir filling history. *Geochimica et Cosmochimica Acta*, **57**, 3641–3659.

KAUFMAN, R. L., AHMED, A. S. & ELSINGER, R. L. 1990. Gas chromatography as a development and production tool for fingerprinting oils from individual reservoirs: applications in the Gulf of Mexico. *In*: *Gulf Coast Section of the Society of Economic Paleontologists and Mineralogists Foundation 9th Annual Research Conference Proceedings*, 263–282.

——, —— & HEMPKINS, W. B. 1987. A new technique for the analysis of comingled oils and its application to production allocation calculations. *In*: *Proceedings of the 16th Annual Convention of the Indonesian Petroleum Association*, Indonesian Petroleum Association, Jakarta, 247–268.

KNAEPEN, W. A. I., TIJSSEN, R. & VAN DEN BERGEN, E. A. 1990. Experimental aspects of partitioning tracer tests for residual oil saturation determination with FIA-based laboratory equipment. *Society of Petroleum Engineers, Reservoir Engineering*, **5**, 239–244.

KROOSS, B. M., BROTHERS, L. & ENGEL, M. H. 1992. Geochromatography in petroleum migration: A review. *In*: ENGLAND, W. A. & FLEET, A. J. (eds) *Petroleum Migration.* Geological Society, London, Special Publication, **59**, 149–166.

KVALHEIM, O. M. 1987*a*. Oil-source correlation by the combined use of principal component modelling, analysis of variance and a coefficient of congruence. *Chemometrics and Intelligent Laboratory Systems*, **2**, 127–136.

—— 1987*b*. *Methods for the interpretation of multivariate data.* PhD Thesis, University of Bergen, 437 p.

——, AKSNES, D. W., BREKKE, T., EIDE, M. O., SLETTEN, E. & TELNAES, N. 1985. Crude oil characterisation and correlation using principal component analysis of ^{13}C NMR spectra. *Analytical Chemistry*, **57**, 2858–2864.

——, CHRISTY, A. A., TELNAES, N. & BJORSETH, A. 1987. Maturity determination of organic matter in coals using the methyl-phenanthrene distribution. *Geochimica et Cosmochimica Acta*, **51**, 1883–1888.

LAFARGUE, E. & BARKER, C. 1988. Effect of water washing on crude oil compositions. *American Association of Petroleum Geologists, Bulletin*, **72**, 263–276.

LARTER, S. R. & HORSFIELD, B. 1993. Determination of structural components of kerogens using analytical pyrolysis methods. *In*: ENGEL, M. & MACKO, S. (eds) *Organic Geochemistry.* Plenum Press, New York, 271–287.

—— & HORSTAD, I. 1992. Migration of hydrocarbons into

Brent Group Reservoirs – some observations from the Gullfaks Field, Tampen Spur Area North Sea. *In*: MORTON, A. C. ET AL. (eds) *Geology of the Brent Group*. Geological Society, London, Special Publication, **61**, 441–452.

—— & MILLS, N. 1992. Phase-controlled molecular fractionations in migrating petroleum charges. *In*: ENGLAND, W. A. & FLEET, A. J. (eds) *Petroleum Migration*. Geological Society, London, Special Publication, **59**, 137–147.

—, APLIN, A. C., CORBETT, P., EMENTON, N., TAYLOR, P. & MEI, C. 1994. *Reservoir geochemistry. A link between reservoir geology and engineering?* Society of Petroleum Engineers, Paper No. **28849**.

——, BJØRLYKKE, K. O., KARLSEN, D. A. ET AL. 1991. Determination of petroleum accumulation histories: examples from the Ula Field, Central Graben, Norwegian North Sea. *In*: BULLER, A. T., BERG, E., HJELMELAND, O., KLEPPE, J., TORSAETER, O. & AASEN, J. O. (eds) *North Sea Oil & Gas Reservoirs II*. Graham & Trotman, London, 319–330.

LEE, L. S., RAO, P. S., BRUSSEAU, M. L. & OGWADA, R. A. 1988. Nonequilibrium sorption of organic contaminants during flow through columns of aquifer materials. *Environmental and Toxicological Chemistry*, **7**, 779–793.

LEITH, T. L., KAARSTAD, I., CONNAN, J., PIERRON, J. & CAILLET, G. 1993. Recognition of caprock leakage in the Snorre Field, Norwegian North Sea. *Marine & Petroleum Geology*, **10**, 29–41.

LEONTARITIS, K. J. & MANSOORI, G. A. 1987. *Asphaltene flocculation during production and processing: A thermodynamic colloidal model*. Society of Petroleum Engineers, Paper No. **16258**.

LEYTHAEUSER, D. & RUCKHEIM, J. 1989. Heterogeneity of oil composition within a reservoir as a reflectance of accumulation history. *Geochimica et Cosmochimica Acta*, **53**, 2119–2123.

LI, M. & LARTER, S. R. 1993. Interactions of organic nitrogen species with mineral/water and organic networks: Implications to petroleum geochemistry. *205th American Chemical Society Meeting, Abstracts*, American Chemical Society.

——, —— & FROLOV, Y. B. 1994. Adsorptive interactions between petroleum nitrogen compounds and organic/mineral phases in subsurface rocks as models for compositional fractionation of pyrrolic nitrogen compounds in petroleum during petroleum migration. *Journal of High Resolution Chromatography*, **17**, 230–236.

——, ——, STODDART, D. & BJORØY, M. 1992. Practical liquid chromatographic separation schemes for pyrrolic and pyridinic nitrogen aromatic heterocycle fractions from crude oils suitable for rapid characterisation of geochemical samples. *Analytical Chemistry*, **64**, 1337–1344.

——, ——, —— & —— 1995. Fractionation of pyrrolic nitrogen compounds in petroleum during migration: derivation of migration-related geochemical parameters. *This volume*.

LUCHNER, L. & SCHESTAKOW, W. M. 1991. *Migration processes in the soil and groundwater zone*. Lewis Publishers Inc., Chelsea, Michigan, USA, 485 p.

LYMAN, W. J., REEHL, W. F. & ROSENBLATT, D. H. 1982. *Handbook of chemical property estimation methods. Environmental behaviour of organic compounds*. McGraw-Hill, New York.

MCAULIFFE, C. D. 1978. Oil and gas migration – chemical and physical constraints. *American Association of Petroleum Geologists, Bulletin*, **63**, 761–781.

MACGREGOR, D. S. & MACKENZIE, A. S. 1986. Quantification of oil generation and migration in the Malacca Strait Region. *In*: *Proceedings of the 15th Annual Convention of the Indonesian Petroleum Association*, Indonesian Petroleum Association, Jakarta.

MACKENZIE, A. S. & QUIGLEY, T. 1988. Principles of geochemical prospect appraisal. *American Association of Petroleum Geologists, Bulletin*, **72**, 399–415.

MACLEOD, G., PETCH, G. S., LARTER, S. R. & APLIN, A. C. 1993a. Investigations on the composition of hydrocarbon fluid inclusions. *205th American Chemical Society Meeting, Abstracts*, American Chemical Society.

——, TAYLOR, P. N., LARTER, S. R. & APLIN, A. C. 1993b. Dissolved organics in formation waters: insights into water–oil–rock ratios in petroleum systems. *In*: PARNELL, J., RUFFELL, A. & MOLES, N. (eds) *Geofluids 93: International conference on fluid evolution, migration and interaction in rocks*. Geological Society, London, 18–20.

MCLIMANS, R. K. 1987. The application of fluid inclusions to migration of oil and diagenesis in petroleum reservoirs. *Applied Geochemistry*, **2**, 585–603.

MARSAL, D. 1987. *Statistics for Geoscientists*. Pergamon, Oxford, 176 p.

MILES, J. A. 1990. Secondary migration routes in the Brent sandstones of the Viking Graben and East Shetland basin: Evidence from oil residues and subsurface pressure data. *American Association of Petroleum Geologists, Bulletin*, **74**, 1718–1735.

MILLIGAN, G. W. 1980. An examination of the effect of six types of error perturbation on fifteen clustering algorithms. *Psychometrica*, **45**, 325–342.

MITCHELL, A. G., HAZELL, L. B. & WEBB, K. J. 1990. *Wettability determination: Pore surface analysis*. Society of Petroleum Engineers, Paper No. **2050**, 351–360.

NEDKVITNE, T., KARLSEN, D. A., BJØRLYKKE, K. & LARTER, S. R. 1993. Relationship between reservoir diagenetic evolution and petroleum emplacement in the Ula Field, North, Sea. *Marine and Petroleum Geology*, **10**, 255–270.

O'DONNELL, D. J., SIGLE, S. O., BERLIN, K. D., STURM, G. P. & VOGH, J. W. 1980. Characterisation of high boiling petroleum distillate fractions by proton and ^{13}C NMR spectroscopy. *Fuel*, **59**, 166–174.

PEPPER, A. S. 1992. Estimating the petroleum expulsion behaviour of source rocks – a novel quantitative approach. *In*: ENGLAND, W. A. & FLEET, A. J. (Eds) *Petroleum Migration*. Geological Society, London, Special Publication, **59**, 9–32.

PERLINGER, J. A. & EISENREICH, S. J. 1991. Sorption of alkylbenzenes to mineral oxides. *In*: BAKER, R. A. (ed.) *Organic Substances and Sediments in Water*.

Lewis Publishers Inc., Chelsea, Michigan, USA, 40–79.

RICHTER, P. P., CEASSER, P. D., MEISEL, S. L. & OFFENHAUSER, R. D. 1952. Distribution of nitrogen according to basicity. *Industrial Engineering Chemistry*, **44**, 2601–2605.

ROSS, L. M. & AMES, R. L. 1988. Stratification of oils in Columbus Basin off Trinidad. *Oil & Gas Journal*, September 26, 72–76.

SAGE, B. H. & LACEY, W. N. 1939. Gravitational concentration gradients in static columns of hydrocarbon fluids. *Transactions of the American Institute of Mining and Metallurgical Engineers*, **132**, 120–131.

SAIGAL, G. C., BJØRLYKKE, K. & LARTER, S. R. 1992. The effects of oil emplacement on diagenetic processes: Examples from the Fulmar Reservoir Sandstones, Central North Sea. *American Association of Petroleum Geologists, Bulletin*, **76**, 1024–1033.

SANDVIK, E. I., YOUNG, W. A. & CURRY, D. J. 1992. Expulsion from hydrocarbon sources: the role of organic adsorbtion. *Organic Geochemistry*, **19**(1), 77–88.

SCHULTE, A. M. 1980. *Compositional variations within a hydrocarbon column due to gravity*. Society of Petroleum Engineers, Paper No. **9235**.

SEIFERT, W. K. & MOLDOWAN, J. M. 1978. Applications of steranes, terpanes, and monoaromatics to the maturation, migration, and source of crude oils. *Geochimica et Cosmochimica Acta*, **42**, 77–95.

SLENTZ, L. W. 1981. *Geochemistry of reservoir fluids as unique approach to optimum reservoir management*. Society of Petroleum Engineers, Paper No. **9582**.

SMALLEY, P. C. & ENGLAND, W. A. 1992. *Assessing reservoir compartmentalisation during field appraisal: how geochemistry can help*. Society of Petroleum Engineers, Paper No. **25005**.

——, LØNØY, A. & RÅHEIM, A. 1992. Spatial $^{87}Sr/^{86}Sr$ variations in formation water and calcite from the Ekofisk chalk oil field: Implications for reservoir connectivity and fluid composition. *Applied Geochemistry*, **7**, 341–350.

——, RÅHEIM, A., DICKSON, J. A. D. & EMERY, D. 1988. Strontium isotope variations in waters from the Lincolnshire limestone aquifer, England, and the potential of natural strontium as a tracer for a secondary recovery seawater injection processes in oilfields.*Applied Geochemistry*, **3**, 591–600.

SMITH, J. H., MABEY, W. R., BOHNUS, N. ET AL. 1978. *Environmental pathways of selected chemicals in freshwater systems, Part II Laboratory Studies*. US EPA Contract No. **600/7-78-074**, Athens, Georgia, USA.

SOUTHWORTH, G. R., HERBES, S. E. & ALLEN, C. P. 1983. Evaluating a mass transfer model for the dissolution of organics from oil films into water. *Water Research*, **17**, 1647–1651.

STODDART, D. P., HALL, P. B., LARTER, S. R., BRASHER, J., LI, M. & BJORØY, M. 1995. The reservoir geochemistry of the Eldfisk Field, Norwegian North Sea. *This volume*.

TAYLOR, P. N. 1994. *The petroleum geochemistry of phenols*. PhD thesis, University of Newcastle upon Tyne.

——, JONES, D. M., LARTER, S. R. & BROMLEY, B. W. 1993. Phenol distributions in heavy oil reservoirs: Assessment of geochemical and geological controls. *In*: *205th American Chemical Society Meeting, Abstracts*, American Chemical Society.

THOMPSON, K. F. M. 1988. Gas-condensate migration and oil fractionation in deltaic systems. *Marine and Petroleum Geology*, **5**, 237–246.

TISSOT, B. P. & WELTE, D. H. 1984. *Petroleum Formation and Occurrence*. 2nd Edition, Springer Verlag, Berlin, 699 p.

TRABELSI, K. 1987. Study of rock-fluid interactions. *In*: DOLIGEZ, B. (ed.) *Migration of Hydrocarbons in Sedimentary Basins*. Editions Technip, Paris, 313–328.

TREIBER, L. E., ARCHER, D. L. & OWENS, W. W. 1972. Laboratory evaluation of the wettability of 55 oil producing reservoirs. *Society of Petroleum Engineers Journal*, **12**(6), 531–540.

WHITTINGHAM, K. P. & JONES, T. J. 1987. Reservoir souring. *Royal Society of Chemistry*, **67**, 228–243.

WILHELMS, A. 1992. *An investigation into the factors influencing tar mat formation in petroleum reservoirs*. PhD Thesis, University of Oslo, Norway.

—— & LARTER, S. R. 1994*a*. On the origin of tar mats in petroleum reservoirs: Part I. Introduction and case studies. *Marine and Petroleum Geology*, **11**, 418–441.

—— & —— 1994*b*. On the origin of tar mats in petroleum reservoirs: Part II. Formation mechanisms for tar mats. *Marine and Petroleum Geology*, **11**, 442–456.

—— & —— 1995. The geochemistry of tar mats. *This volume*.

——, CARPENTIER, B. & HUC, A. Y. in press. New methods to detect tar mats in petroleum reservoirs. *Journal of Petroleum Science and Engineering*.

——, PATIENCE, R. L., LARTER, S. R. & JORGENSEN, S. 1992. Nitrogen functionality distributions in asphaltenes isolated from several oils from different source rock types. *Geochimica et Cosmochimica Acta*, **56**, 3745–3750.

YAMAMOTO, M. 1992. Fractionation of azarenes during oil migration. *Organic Geochemistry*, **19**, 389–402.

Geochemical constraints from formation water analyses from the North Sea and the Gulf Coast Basins on quartz, feldspar and illite precipitation in reservoir rocks

KNUT BJØRLYKKE[1], PER AAGAARD[1], PER K. EGEBERG[2] & SCOTT P. SIMMONS[1]

[1] *Department of Geology, Box 1047, University of Oslo, 0316 Oslo, Norway*

[2] *Department of Chemistry, ADH, Postbox 607, 4601 Kristiansand, Norway*

Abstract: Formation water analyses from North Sea and Gulf Coast Basin reservoirs provide important constraints on burial diagenetic processes. Thermodynamic calculations show that aluminium solubilities are low (< 1 ppm at 130°C) even in the presence of organic acids. Illite therefore forms mainly at the expense of dissolving aluminous minerals such as smectite and kaolinite. The potassium values of the formation waters are always lower than K-feldspar saturation during burial diagenesis (> 1.5 km burial depth), suggesting that K-feldspar is continually being dissolved. The potassium concentrations indicate, in most cases, supersaturation with respect to illite. This is evidence that the failure of illite to form is probably because aluminium is not available. In the North Sea Basin, potassium concentrations approach equilibrium with kaolinite and illite at higher temperatures (140°C), indicating that illitization at that point keeps up with K-feldspar dissolution. In the subsiding parts of the Gulf Coast Basin, where kaolinite and smectite are being heated, the potassium values are clearly lower than in the uplifted (cooling) parts of the basin where illite is not forming. This suggests that the potassium concentrations are to a large extent controlled by the rate of illite precipitation, which again depends on the availability of aluminium from dissolving Al-minerals. There is considerable evidence that both illitization and quartz cementation may continue in the water-saturated parts of the pore after oil emplacement. This is also consistent with an almost closed-system burial diagenesis. The primary mineralogical composition of the sediments and early diagenetic processes strongly influence deeper burial diagenesis and in particular the precipitation of illite.

For the last 15 years, research in clastic diagenesis has attempted to establish models which can predict changes in reservoir quality as a function of sedimentation and burial history. Early progress in this field gave high expectations that this research would soon be of practical use in exploration and production geology. It is therefore now pertinent to assess to what extent we have succeeded in establishing generally accepted theories that are useful in practical predictions of porosity and permeability.

The Jurassic Brent Group from the North Sea forms important reservoirs and is probably one of the most intensively studied sandstones in the literature. The Geology of the Brent Group (Morton *et al.* 1992) contains several important studies of the diagenesis of the Brent sandstone, and the amount of petrographic, mineralogical and geochemical data on the sandstone is now impressive. There is, however, a striking absence of agreement between many of the authors regarding fundamental diagenetic processes. With the variety of theories and conclusions in the different papers, the predictions made regarding reservoir quality depend very much on which theories were followed. Here, we shall concentrate on the most fundamental processes affecting reservoir quality:

(1) formation of secondary porosity from feldspar dissolution and precipitation of diagenetic kaolinite;
(2) quartz cementation and the source of quartz cement;
(3) precipitation of illite and its dependence on precursor minerals;
(4) effects of hydrocarbon saturation on diagenesis.

Kaolinite has been interpreted as forming early in diagenesis by meteoric water-flushing (Glasmann 1992; Bjørlykke 1984; Bjørlykke *et al.* 1992). Other workers have concluded that diagenetic kaolinite forms mainly as a result of leaching by organic acids at greater depths (Surdam *et al.* 1984, 1989; Burley *et al.* 1985) or that there are two 'episodes' of kaolinite precipitation: one early, during meteoric water-flushing, and another

From CUBITT, J. M. & ENGLAND, W. A. (eds), 1995, *The Geochemistry of Reservoirs*, Geological Society Special Publication No. 86, pp. 33–50.

at greater depth, i.e. due to the effects of organic acids (Haszeldine *et al.* 1992; Kantorowicz *et al.* 1992).

Quartz cementation is the main cause of porosity reduction in well sorted sandstones. If we are to be able to predict the rate of quartz cementation as a function of burial depth, the source of silica must first be identified and the mechanism for its dissolution and precipitation must be understood. Quartz cement has commonly been interpreted to have formed relatively early at shallow depth (Gluyas 1985; Dutton & Diggs 1990). Several authors have suggested that quartz cement is introduced through the episodic flow of hot fluids. If this is the case, predictions must be made based on possible patterns of such flow (Hogg *et al.* 1992; Gluyas & Coleman 1992; Robinson & Gluyas 1992). Other authors consider the sources of quartz cement to be rather local and suggest that the solubility of silica is too low for significant long-distance transport of silica (Bjørlykke *et al.* 1992; Bjørlykke & Egeberg 1993; Giles 1987; Giles & De Boer 1990). The temperature and pressure history of the basin can be modelled, but often a single formation of well sorted sand can have large variations in porosity and permeability at any depth intervals. Prediction of the rate of quartz cementation must therefore also be based on textural relationships and the presence of factors that inhibit the dissolution of quartz (clay coating (Ehrenberg 1993)) or enhance dissolution (e.g. mica).

Diagenetic growth of illite is one of the most important factors controlling permeability in reservoir rocks. The growth of illite has been interpreted to be related to hot potassium-rich fluids (Jourdan *et al.* 1987), and age dating suggests that it formed relatively early (40–50 Ma), at a time when the Brent Group was buried to only about 2 km or less (Scotchman *et al.* 1989). The interpretation of the K–Ar ages from North Sea illites is difficult. At depths of less than 3.3 km, clastic clay minerals are often found to be of Jurassic age (Burley & MacQuaker 1992). At greater depths, ages between 20 to 100 Ma are measured (Burley & MacQuaker 1992; Hamilton *et al.* 1992; Liewig *et al.* 1987), suggesting a different timing of illitization or that mixtures of diagenetic illite and clastic illite or feldspar have been analysed (Lieweg *et al.* 1987). If illite is formed at relatively shallow depth by the introduction of hot fluids along faults, as suggested by Burley & MacQuaker (1992), the prediction of the distribution of diagenetic illite must be based on the flow patterns of such fluids and the proximity to the faults, in order to use this in practical prediction of reservoir properties, the timing of flow and which faults would be permeable. Age dating of illite formed by this mechanism will date the episode of hot fluid flow (due to tectonic rifting or other mechanisms), but will probably not be related to the time of oil emplacement. Oil emplacement does not seem to retard illitization significantly, but the distribution of illite and thereby the permeability may be different because illitization is restricted to water-saturated pores (Ehrenberg & Boassen 1993).

A significant increase in the illite content in the fine fraction is found at 3.7 to 4.0 km present-day burial depth both in the North Sea (Bjørlykke *et al.* 1992; Burley & MacQuaker 1992) and at Haltenbanken (Bjørlykke *et al.* 1986; Ehrenberg & Nadeau 1989; Ehrenberg 1990). This corresponds to a temperature of 120 to 140°C. Other studies of the Brent Group have found increases in fine-grained illite at depths of about 3.2 km (Giles *et al.* 1992). The fact that the increase in illite content seems to be related to present-day burial depths in areas of different burial history suggests that much of the illite observed in the North Sea and Haltenbanken reservoirs formed during the Pliocene/Pleistocene subsidence, which may have exceeded 1 km.

Illite may form both from smectite and from kaolinite and potassium feldspar. In Jurassic reservoirs of the North Sea, where kaolinite is most abundant, the local availability of potassium in the form of potassium feldspar is critical for the formation of illite (Bjørlykke *et al.* 1992). Loss of potassium feldspar during burial, and particularly when approaching 4 km, is consistent with this model (Harris 1989; Bjørlykke *et al.* 1992).

The lack of consensus about important diagenetic processes, as discussed previously, may be due to the fact that the interpretations are based on different types of observations and data, and that a theoretical framework for diagenesis is poorly developed. Observation of the textural relationships between mineral phases has been an important method for establishing the relative timing of precipitation of authigenic minerals. This method cannot, however, give any indication about the absolute timing or temperature of precipitation, and the interpretation of a 'cement stratigraphy' based on textural evidence is not always straightforward. Individual studies of reservoir rocks, even when based on cores from a large number of wells, rarely provide a basis for drawing 100% unique interpretations of the processes involved. In this situation, it may be useful to use a more theoretical approach and establish models based on theoretical considerations, and working hypotheses based on thermodynamics, kinetics, porewater compositions and fluid flow models.

Pore-filling minerals such as kaolinite are often interpreted as having precipitated later than cements precipitated on grains (i.e. quartz overgrowths), but this does not agree with interpretation based on isotopic data (Glasmann 1992). The abundance of diagenetic kaolinite and near-absence of diagenetic quartz in shallow wells also indicate that kaolinite predates quartz overgrowth (Bjørlykke *et al.* 1992; Giles *et al.* 1992).

Thermodynamic theory combined with formation water data provide another method of constraining diagenetic reactions. In this paper we use previously published data from formation water in the North Sea and Gulf Coast Basins, combined with thermodynamically calculated mineral stabilities, to make predictive models for the growth of illite in reservoirs during burial diagenesis. The quality of the formation water analyses has been discussed in the publications quoted.

Mineral buffering constraints

The calculations of equilibrium between minerals and their surrounding porewater were based on the thermodynamic database of Helgeson and coworkers with its recent updates reported by Johnson *et al.* (1992). All calculations were made using an average North Sea temperature gradient taken from Smith & Ehrenberg (1989).

In order to compare mineral stabilities directly with observed element concentrations in formation water, we have constructed a series of concentration–temperature diagrams where the position of mineral stability field boundaries are implicitly corrected for aqueous activity coefficients. This approach will always have some uncertainties in the location of phase boundaries, but has the advantage of depicting the true scatter in the amount of dissolved elements on a linear scale. Some of the phase boundaries are also based on *in situ* pH estimates. We have assumd the average North Sea trend reported by Aagaard *et al.* (1992) which again was based on speciations of North Sea formation water using SOLMINEQ 88 (Kharaka *et al.* 1988). Activity coefficient calculations were made according to the scheme of Truesdell & Jones (1974) for the less concentrated waters, and by the Pitzer formulation in SOLMINEQ 88 (Kharaka *et al.* 1988) for the most concentrated range. Estimations of aluminium concentration in formation waters were based on SOLMINEQ 88 calculations for a typical North Sea formation water saturated with kaolinite and quartz (Fig. 1). To include acetate complexation in the calculations, a constant value of 800 mg l^{-1} was used for the whole pressure–temperature range.

The amount of dissolved aluminium in formation water is a critical factor influencing aluminium mobility. Gluyas & Coleman (1992) have advocated dissolved transport from external sources to

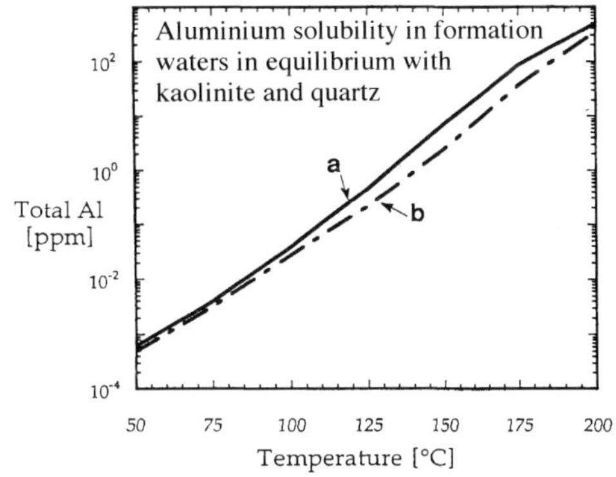

Fig. 1. Calculated solubilities of aluminium in equilibrium with kaolinite and quartz as a function of burial temperature (SOLMINEQ 88 (Kharaka *et al.* 1988)). The calculations are based on pCO_2 values from Smith & Ehrenberg (1989). Line **a** corresponds to formation water without organic acids. Line **b** is calculated assuming a constant acetate value of 800 mg l^{-1}.

explain formation of authigenic illite in sandstone reservoirs. However, the critical capacity of porewater to transport aluminium will be a direct function of its solubility, so we have tried to evaluate aluminium concentration in formation water based on kaolinite saturation. In order to fix the silica activity, we have assumed quartz saturation, a condition which seems to be appropriate in sedimentary basins at temperatures above 60 to 70°C. Because aluminium concentrations in equilibrium with kaolinite will always be higher than those in equilibrium with illite, these calculations will give upper estimates if illite controls the aluminium solubility. Complexing with organic acid anions may also be of importance in increasing aluminium solubility. To simplify the calculation, we have considered only acetate, and used a value of 800 mg l^{-1}. This amount represents a typical high value for North Sea formation waters (Barth 1991). The results of these calculations by SOLMINEQ 88 are summarized in Fig. 1, and indicate very low aluminium solubility (< 1 ppm) below 120 to 130°C. Calculated pH at calcite saturation varies from about 6.0 at 50°C to 5.2 at 100°C and 4.2 at 200°C. Temperatures above 170°C are required to bring dissolved aluminium up to 20 to 30 ppm. These calculated solubility values for the temperature range 100 to 150°C are in general agreement with analytical Al data from the Stevens Formation, California, reported by Fisher & Boles (1990). They found Al concentrations increasing from 0.2 to 5 ppm with burial depth. This low solubility considerably restricts the mobility of aluminium, and precludes any major import/export in solution.

Table 1. *Formation water analyses, offshore Norway*

Well	*D* (m)	Lith.	pH	Alk.	δ O	δ D	$^{87}Sr/^{86}Sr$	*T* (°C)	Ca (ppm)	Na	K	Cl	Mg	Sr	Ba	Br	SO_4
15.9.7	3672	ss	6.36	6.7				135	5226	45 681	2080	87 810	1495	315	385		0
15.9.11	2406	ss	6.05	6.7				95	2084	21 013	211	36 159	272	350	69		19
25.4.5	3987	ss	6.07	10.3				130	5162	32 577		61 329	576	675	288		0
29.6.1	4294	ss	6.5	8.6	2.20	–46.00		152	3030	24 392	235	43 745	175	333	220		0
29.6.1	4258	ss	6.27	14.2	2.60	–32.00		151	2705	24 323	262	42 753	175	263	41		0
30.3.4	2267	ss	6.47	17.1	–1.00	–31.30	0.7143	93	152	6920	145	10 990	19	26	41	55	0
30.6.6	2926	ss	6.05	11.5				113	1259	14 989	477	24 425	173	149	261		0
30.6.7	2735	ss	6.14	12.2				110	814	14 599	395	24 425	112	175	137		0
30.6.15	3352	ss	5.88	11.2				99	1747	16 047	477	29 175	175	228	137		0
30.9.1	2752	ss	6.04	13.6				107	894	14 162	516	23 929	112	166	110		0
30.9.2	2640	ss	5.98	14.5	–2.10	–28.50		101	894	13 725	274	22 794	112	175	82		0
30.9.3A	3572	ss	6.4	4.4				125	3727	22 047	1040	43 249	423	464	494		0
30.9.4	4207	ss	5.92	7.1				100	3976	23 289	680	44 773	224	631	508		0
31.2.1	1595	ss	6.81	10.1				160	1908	13 265		24 390	297	175			538
31.2.8	1845	ss	5.64	13.1				55	2441	16 162		29 884	321	272	14		67
31.2.11	1683	ss	6.42	2.6				77	1856	18 461	438	33 288	408	324	206		0
31.2.15	1576	ss	6.2	8.5				65	854	18 599	516	31 940	445	228	179		10
31.4.3	2160	ss	5.93	9.1				71	1567	17 518		30 593	284	315	96		0
31.4.3	2160	ss	5.69	13.2				71	2056	16 323	438	29 991	321	280	151		58
31.5.2	1579	ss	6.2	6.1	–2.50	–25.80	0.711	55	910	10 185	278	17 796	165	175	0	79	0
31.5.3	1608	ss	6.04	8.2	–1.50	–23.70	0.711	68	1339	16 898	461	30 062	513	254	179	160	0
33.9.8	2747	ss	6.62	18.5				99	120	8943	203	13 436	51	0	398		0
33.9.9	2850	ss	6.18	5.9				91	10500	8161	121	14 286	68	0	41		29
33.9.A7	3600	ss	6.3	4.8	–0.20	–28.20	0.713	107	890	7541	113	13 187	53	96	41	66	0
33.9.A14	3727	ss	6.76	5				101	305	9081	192	14 676	53	9	55		0
33.9.A23	3750	ss	6.76	5.2	–0.20	–24.20	0.712	103	886	7863	121	13 861	58	44	69	77	0
33.12.B10	2795	ss	6.19	19.2	–0.30	–24.80	0.712	91	72	9449	164	14 428	51	35	41	70	0
33.12.B26	4117	ss	6.63	20.3				149	76	9564	152	15 988	56	35	55		10

34.4.1	2611	ss	5.93	12.6				98	1259	13 150	155	22 156	246	0			874
34.4.4	2622	ss	5.84	7.5	–1..10	–31.80	0.711	96	3659	12 805	184	27 155	173	342	0	124	163
34.4.7	2588	ss	7.54	4.6	–1.30	–30.00	0.709	100	2076	11 840	356	23 468	204	280	0	97	86
34.7.5	2508	ss	6.3	13.6	–5.00	–38.40	0.711	90	52	9886	8	15 527	90	35	0	77	96
34.7.6	2683	ss	7.14	6.4	–4.10	–29.20	0.711	93	934	12 277	121	21 235	165	158	41	88	0
34.7.12	2730	ss	7.05	15.5	–0.50	–25.00	0.712	82	377	11 288	86	17 725	102	44	0	83	0
34.10.3	1992	ss	7.11	6.9				75	934	14 162	348	25 311	294	0	14		0
34.10.17	2885	ss	5.96	18.2				98	689	12 047	274	19 108	97	53	69		o
6406.3.1	3785	ss	6.43	4.9				111	1267	20 576	837	35 344	139	158	577		0
6507.7.1	4365	ss	6.78	6.5	1.30	–18.80		160	3419	30 117	282	54 522	474	280	632		29
6507.7.2	2525	ss	6.17	15.3				83	489	20 185	555	32 791	197	114	151		0
7120.6.1	2462	ss	5.91	5.7				89	4637	59 705	782	102 202	562	237	55		586
7120.12.2	1988	ss	5.77	5.7				66	5916	42 187	610	79 018	885	526	288		19
2/4-A2	3468	ls	6.41	8.5	6.70	–30.60	0.708	136	2858	20 530	145	38 605	357	394	41	174	10
2/4-A3	4018	ls	6.24	5.5	6.90	–32.40	0.708	157	5138	26 232	188	52 112	613	675	247	351	0
2/4-A7	3269	ls	6.7	7.1	3.60	–31.60	0.708	128	1455	17 909	117	31 373	194	228	14	225	10
2/4-B9	4132	ls	6.15	3.7	9.30	–21.90	0.708	161	4689	28 209	180	54 026	525	789	96	252	10
2/4-B12	3632	ls	5.99	3.1	7.60	–26.50	0.71	142	14 894	39 221	266	95 750	1293	1060	700	562	0
2/4-B17	3820	ls	6.34	6.5	6.30	–30.30	0.707	149	1511	21 427	106	36 797	175	315	55	166	10
2/4-B18	3221	ls	6.3	4.7	7.10	–29.30	0.707	126	605	23 473	192	44 880	343	648	27	201	29
2/4-F9	4150	ls	6.46	8.6	6.70	–31.90	0.708	162	4613	23 979	215	47 751	469	762	14	227	38
2//44-E7	4403	ls	6.48	2.5	8.10	–34.40	0.708	171	16 457	35 060	430	92 489	1337	1270	27	582	19
2/7-C5	3481	ls	6.61	6.7	6.30	–30.00	0.707	136	565	12 553	66	20 561	44	105	0	80	106
2/7-A11	3431	ls	6.27	6.2	4.40	–29.30	0.708	134	449	25 358	117	45 128	224	289	14	274	19

Data from Egeberg & Aagaard (1989).

Experimental studies by Stoessel & Pittman (1990) give higher values, but they are not directly comparable to our data because they are carried out at much lower and higher pH conditions. They also concluded, however, that Al in formation water had a limited mobility due to low solubility.

Porewater geochemistry in the North Sea Basin

The chemical composition of some North Sea reservoirs published by Egeberg & Aagaard (1989) is presented in Table 1. The potassium distribution in North Sea formation waters shows that the concentration of potassium varies greatly with burial depth (Fig. 2). Comparing these data with the calculated stability fields of kaolinite, illite and K-feldspar, it is evident that the porewater is saturated/supersaturated with respect to illite and undersaturated with respect to K-feldspar. A few samples plot close to the kaolinite–illite boundary but most of the porewaters fall outside the stability field of kaolinite. Illite appears to be the stable phase in formation waters from the North Sea Basin.

K-feldspar overgrowths, which are commonly observed in North Sea reservoir rocks, are often interpreted, on a textural basis, to have formed during early diagenesis, often predating carbonate cement (Bjørlykke *et al.* 1992). The formation water geochemistry, however, suggests that the porewater in the reservoir sandstones is clearly undersaturated with respect to K-feldspar. The

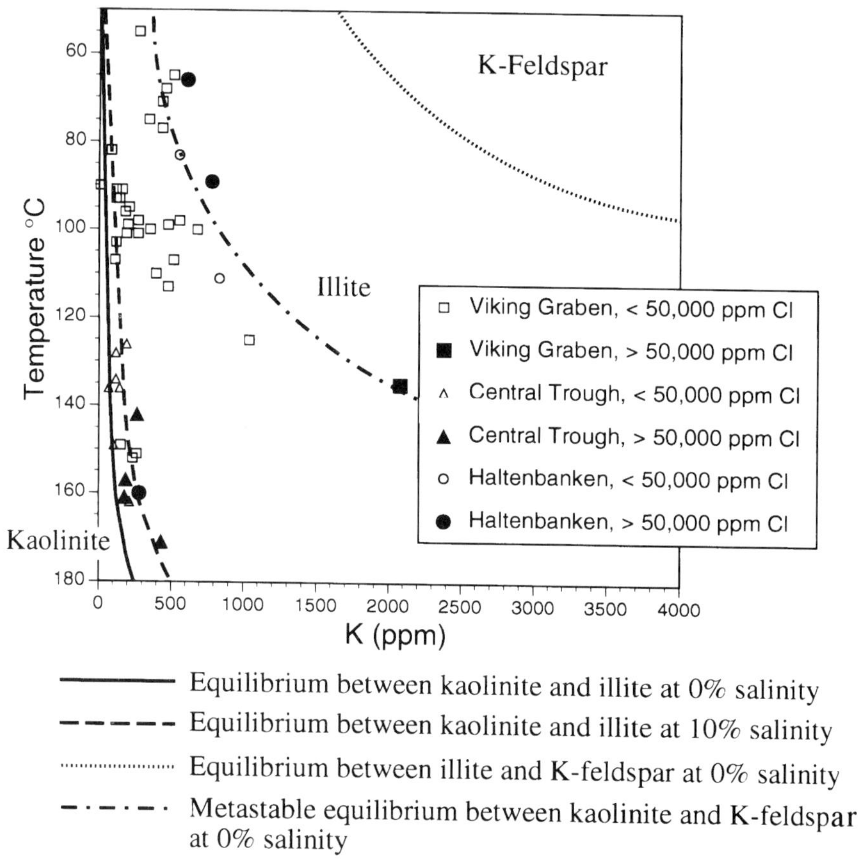

Fig. 2. Potassium concentration in porewaters from offshore Norway as a function of temperature. Open symbols indicate samples with chlorine concentrations below 50 000 ppm, whereas solid symbols represent chlorine concentrations in excess of 50 000 ppm. Equilibria between kaolinite and illite and illite and K-feldspar are indicated for 0% and 10% salinity. Mineral equilibrium calculations assume quartz saturation, and a pH trend from Aagaard *et al.* (1992) (i.e. pH = 6.2 at 50°C, 5.6 at 100°C, 4.8 at 150°C and 4.3 at 180°C). Data from Egeberg & Aagaard (1989).

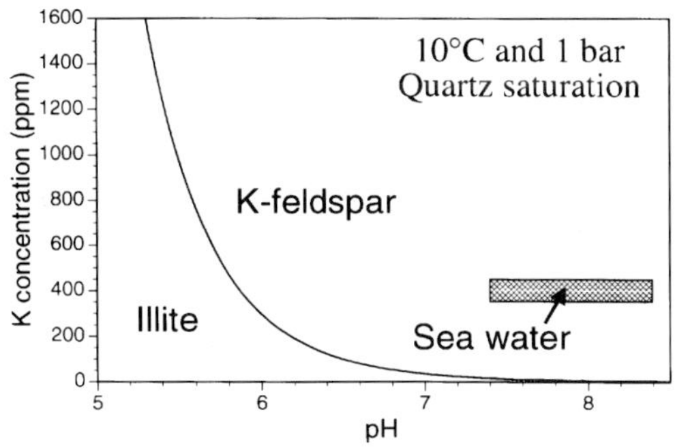

Fig. 3. Stability field of K-feldspar and illite as a function of pH and potassium, assuming potassium activity coefficient as in normal seawater. This illustrates that seawater has sufficient potassium to form K-feldspar if silica and aluminium are available. K-feldspar may therefore form close to the sea floor where the pH is high.

conditions are different in marine sedimentary environments close to the sea floor. Figure 3 depicts the same illite–K-feldspar phase boundary, but at 10°C and expressed as a function of potassium concentration and pH. The boundary is drawn at quartz saturation; an increase in aqueous silica would expand K-feldspar's stability field. Seawater, with its average potassium concentration close to 400 ppm (Chester 1990), is clearly supersaturated with respect to K-feldspar. Porewaters evolving during early diagenesis are known to increase silica and decrease pH, but are still supersaturated with respect to K-feldspar. It is likely that K-feldspar overgrowths could form in the uppermost few centimetres from the sediment–water interface, where potassium can be supplied by diffusion from seawater. Small amounts of amorphous aluminium hydroxides and biogenic silica in the sediment may provide aluminium and silica for the crystallization of K-feldspar close to the sea floor. At shallow depth (< 1.5 km), the pH in the porewater is normally relatively high in the presence of carbonate due to the low pCO_2 pressure. The silica concentration will be high if biogenic silica is present, and the porewater is then likely to fall in the stability field of K-feldspar. The degree of supersaturation of silica with respect to quartz strongly influences reactions involving kaolinite and illite (Bjørkum & Gjeldsvik 1988). This would be consistent with frequent petrographic observations of K-feldspar predating carbonate cementation (Bjørlykke *et al.* 1986). Glasmann (1992) described very early feldspar overgrowth in the Brent Formation and suggested that it may have formed almost syndepositionally. K-feldspar overgrowths from the Ula Field, however, contain fluid inclusions of oil which are not biodegraded, suggesting formation of the overgrowths at least some metres below the sea floor (Nedkvitne *et al.* 1993).

At higher temperatures, albite is the stable feldspar phase and K-feldspar is often replaced by albite (Saigal *et al.* 1988; Aagaard *et al.* 1990). the main cause of albitization may be that, as the rate of illite formation increases approaching a temperature of 100°C, dissolution of smectite and kaolinite and precipitation of illite will remove potassium from the porewater so that the potassium concentration falls below the concentration that represents equilibrium with K-feldspar. This may serve as a driving force for albitization (Aagaard *et al.* 1990). Phase relations involving albite are shown in Fig. 4, where two mineral reactions delimit the stability of albite + illite and the halite saturation line marks the upper sodium solubility. The lower curve represents the metastable equilibrium between the reactant minerals, K-feldspar and kaolinite, and the

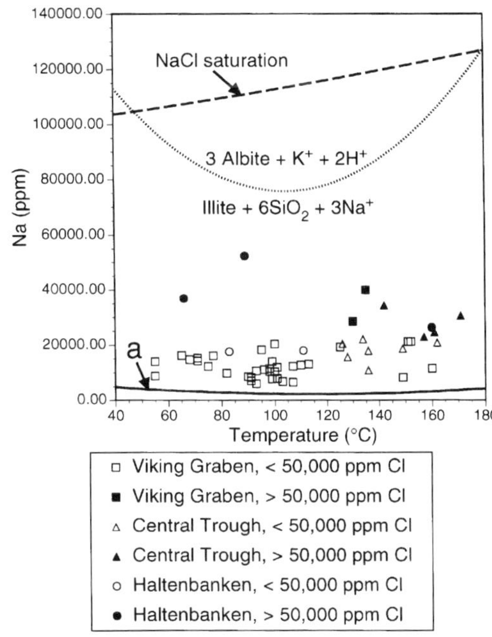

Fig. 4. Sodium concentration in porewaters from offshore Norway as a function of temperature. Open symbols indicate samples with chlorine concentrations below 50 000 ppm, whereas solid symbols represent chlorine concentrations in excess of 50 000 ppm. Line **a** corresponds to the metastable phase assemblage, K-feldspar–kaolinite–albite–illite, while the dotted line denotes the illite out reaction boundary. Halite saturation is also shown for comparison. The boundaries are based on the same pH values as in Fig. 2 and on the aK^+ trends reported by Aagaard *et al.* (1992) for North Sea formation water. Data from Egeberg & Aagaard (1989).

authigenic albite and illite. All the formation water samples from the North Sea plot in the albite + illite stability field (Fig. 4).

Utilizing the principles of aluminium conservation, the illitization of kaolinite may take place through two different overall reactions:

$$\underset{\text{Kaolinite}}{Al_2Si_2O_5(OH)_4} + \underset{\text{K-feldspar}}{KAlSi_3O_8} = \underset{\text{Illite}}{KAl_3Si_3O_{10}(OH)_2} + 2SiO_2 + H_2O \quad (1)$$

$$\underset{\text{Kaolinite}}{3Al_2Si_2O_5(OH)_4} + \underset{\text{K-feldspar}}{2KAlSi_3O_8} + 2Na^+ = \underset{\text{Illite}}{2Kal_3Si_3O_{10}(OH)_2} + \underset{\text{Albite}}{2NaAlSi_3O_8} + 2H^+ + 3H_2O \quad (2)$$

In the first case, aluminium from both dissolving feldspar and kaolinite is conserved and precipitated in illite, often replacing kaolinite and leaving secondary pores after K-feldspar. Excess silica is likely to precipitate as quartz. However, if the porewater is also supersaturated with respect to albite, reaction (2) may take place. The K-feldspar is often pseudomorphically replaced by albite and little secondary porosity is developed (Saigal *et al.* 1988). In this case, no excess silica is released. Reaction (1) implies that aluminium is transported (by diffusion) from the dissolving K-feldspar to the site of illite growth. If illite grows primarily on dissolving kaolinite, the distance from the K-feldspar grains to the kaolinite minerals may be important in determining whether K-feldspar is dissolved or albitized.

Potassium concentrations far above the kaolinite–illite boundary indicate that the kinetics of illite precipitation are too slow to remove potassium at a significant rate. As pointed out by Bjørkum & Gjelsvik (1988), the coexistence of kaolinite and K-feldspar should indicate a narrow range of thermodynamic conditions. Because of the slow kinetic reaction rates at low temperatures, the porewater is often out of equilibrium with these minerals. Precipitation of authigenic K-feldspar probably occurs at low temperatures when the pH is high (Fig. 3), but not at the same time as the meteoric water leaching and precipitation of kaolinite.

At temperatures higher than 130°C, the potassium concentrations are lower and closer to the stability field of kaolinite. This may be explained by the contemporaneous dissolution of kaolinite and precipitation of illite when the temperature is high enough for illitization to occur rapidly. The removal of potassium from the porewater makes K-feldspar more unstable at corresponding depths (3.5 to 4.0 km). In the Brent Group of the North Sea, the concentration of K-feldspar is found to decrease steeply and it is often nearly absent at depths greater than 4 km. The concentration of plagioclase shows no similar trend (Bjørlykke *et al.* 1992). In the Gulf Coast Basin, similar losses of K-feldspar with depth are observed (Sharp *et al.* 1988; Hower *et al.* 1976).

The small amounts of illitization of kaolinite observed in certain hot reservoirs where K-feldspar is practically absent, such as the Hild Field of the North Sea (Lønøy *et al.* 1986), suggest that potassium concentration is below illite saturation and that there is slow advective or diffusive supply of potassium from other adjacent formations where K-feldspar is present. The fact that formation water in the North Sea Basin nearly always has potassium concentrations saturated/supersaturated with respect to illite is indirect evidence that illite fails to precipitate due to low concentrations of aluminium or slow kinetic reaction rates. The formation water data (Fig. 2) suggests that compositions fall within the stability field of illite and represent conditions where kaolinite cannot form. We have no samples from the shallow (< 1.5 km) part of the basin where there may be recent meteoric water-flushing. Studies of onshore sandstones suggest that recent meteoric water precipitates kaolinite (Bath *et al.* 1987). During meteoric water-flow, potassium and silica released during feldspar dissolution are removed by the water.

At greater depths, however, the potassium released by dissolution of K-feldspar is not removed. Acids, if present, can only leach small amounts of K-feldspar before the porewater composition is back in the illite field, bcause the pH is buffered by carbonate and silicate minerals (Hutcheon *et al.* 1993). Focused flow driven by compaction are likely to flow from deeper and hotter parts of the basin, where K-feldspar is more soluble and the potassium concentration is higher. If kaolinite is to form from K-feldspar, the released potassium must be removed, and it is difficult to find a sink for potassium other than illite. Illite has a much lower K/Al ratio than K-feldspar, and excess potassium will increase the potassium concentration in the porewater until it is in equilibrium with K-feldspar. Formation of kaolinite at depths below the influence of meteoric water-flow should therefore not be expected to occur in the presence of K-feldspar, but should theoretically be able to form from plagioclase in the absence of K-feldspar. This is because the activity of potassium may then be low enough so that the porewater composition falls in the stability field of kaolinite.

The Gulf Coast Basin

Published analyses of formation waters from the Gulf Coast Basin show a wide scatter in the potassium values, but all of them are below the level of K-feldspar saturation (Fig. 5). It is clear, however, that most of the high values (> 1000 ppm) come from limestones, whereas most of the analyses that show lower potassium concentrations are from sandstones. It can also be seen that pore-waters from the younger, actively subsiding parts of the basin have potassium concentrations that fall close to the boundary of the stability field of kaolinite and illite. This is the pattern that should be expected if illite is being formed at a relatively high rate as smectite and kaolinite are being heated and dissolved during burial. In the part of the basin where there has been recent tectonic uplift (middle and lower Texas coast), or at least no recent subsidence, the potassium concentrations are higher than in the actively subsiding parts (Louisiana and easternmost Texas). When plotted as potassium/chlorine ratios, this difference becomes even more clear (Fig. 6). It is clear that the potassium concentration is poorly correlated with salinity when all analyses are taken together. Limestones and the two types of sandstones, however, show relatively good correlation separately.

This pattern of potassium distribution suggests that the potassium concentration is to a large extent controlled by the rate of illite precipitation, which again depends on the availability of illite precursor minerals such as kaolinite and smectite. In the actively subsiding parts of a basin, sedimentary rocks are continually being heated so that kaolinite

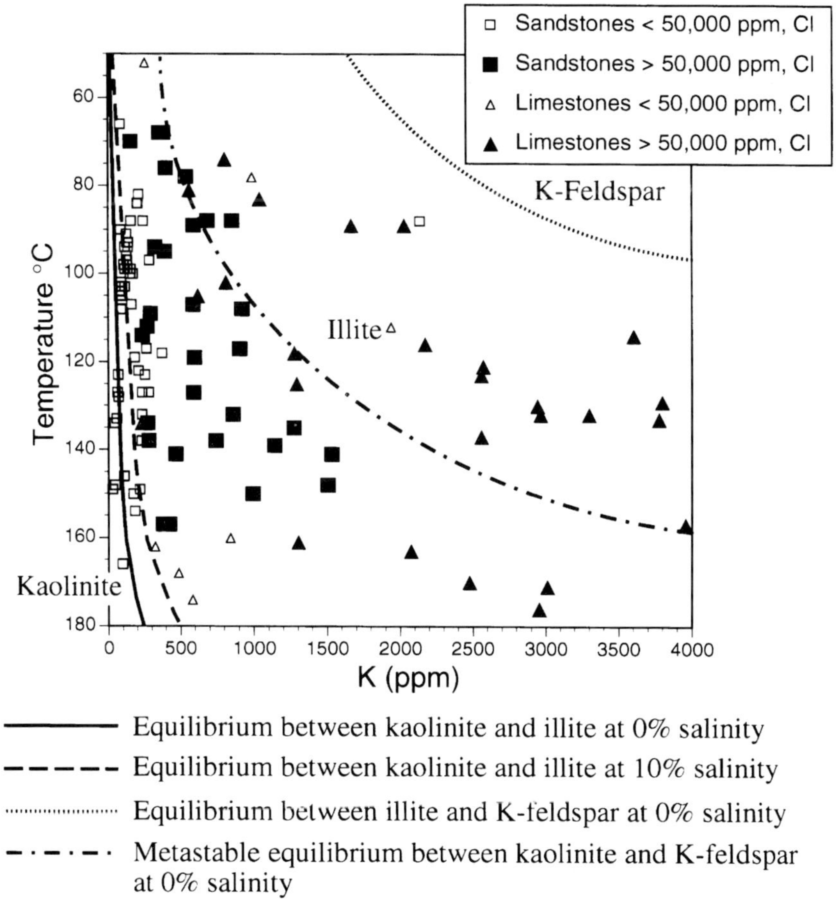

Fig. 5. Potassium concentration in porewaters from the Gulf of Mexico Basin as a function of temperature. Equilibria between kaolinite and illite and illite and K-feldspar are indicated for 0% and 10% salinity. Mineral equilibrium calculations assume quartz saturation and the same pH conditions as in Fig. 2. Data from Dickey *et al.* (1972); Gautier *et al.* (1985); Jessen & Rolhausen (1944); Kharaka *et al.* (1977); Kraemer & Kharaka (1986); Land & Prezbindowski (1981); Macpherson (1992); Schmidt (1973).

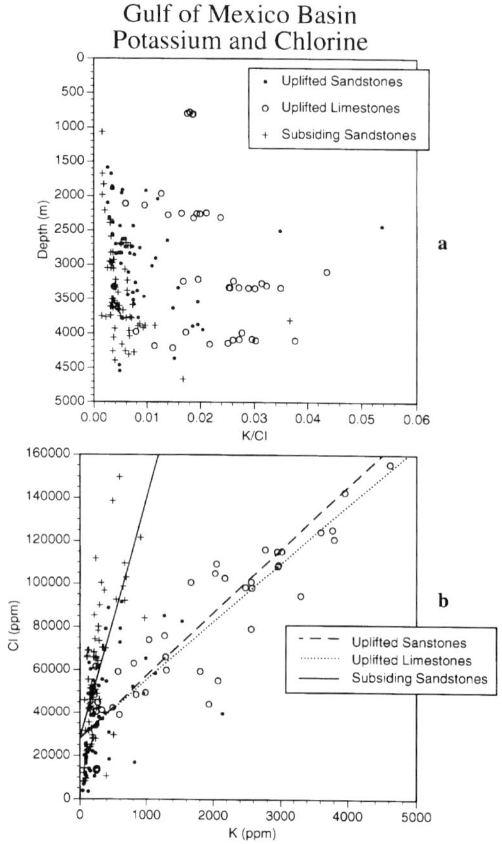

Fig. 6. **(a)** Potassium/chlorine ratio vs. depth for porewaters from the Gulf of Mexico Basin. Circles represent samples from areas of the basin with no active subsidence, crosses indicate samples from the actively subsiding portion of the basin. **(b)** Potassium concentration as a function of chlorinity, Gulf of Mexico. Symbols are the same as for **(a)**. R^2 values for the regression lines are: uplifted sandstones 0.246; uplifted limestones 0.826; subsiding sandstones 0.457. Data from Dickey *et al.* (1972); Gautier *et al.* (1985); Jessen & Rolhausen (1944); Kharaka *et al.* (1977); Kraemer & Kharaka (1986); Land & Prezbindowski (1981); Macpherson (1992); Schmidt (1973).

and smectite are dissolved and the released aluminium precipitated as illite. The concentration of potassium in the porewater is therefore low, because potassium from dissolving K-feldspar and from evaporites is being consumed by illitization as the sediments sink through the porewater (Bjørlykke 1993).

Petrographic and mineralogical analyses show that K-feldspars have undergone dissolution in all Cenozoic units in the Gulf Coast and that this process has often gone to completion at temperatures exceeding 100°C (Sharp *et al.* 1988). The consumption of potassium is also reflected in the process of albitization which begins at these temperatures (100–120°C, Sharp *et al.* 1988). In sandstones which have been subject to tectonic uplift, illite may have precipitated at the time of maximum burial, but during cooling or constant temperatures, remaining kaolinite and smectite will not continue to dissolve at any significant rate, thus releasing little aluminium to form illite. Dissolution of K-feldspar may, however, continue as long as the porewater is undersaturated with respect to K-feldspar. Without a sink, potassium would be expected to build up to higher concentrations in the porewater, thus explaining the observed data. The porewater composition falls in the stability field at coexisting albite and illite (Fig. 7).

In limestones, the concentration of minerals that could serve as sinks for potassium, such as kaolinite and smectite, is normally low, and this may explain the high potassium concentrations in these rocks. This requires, however, that some

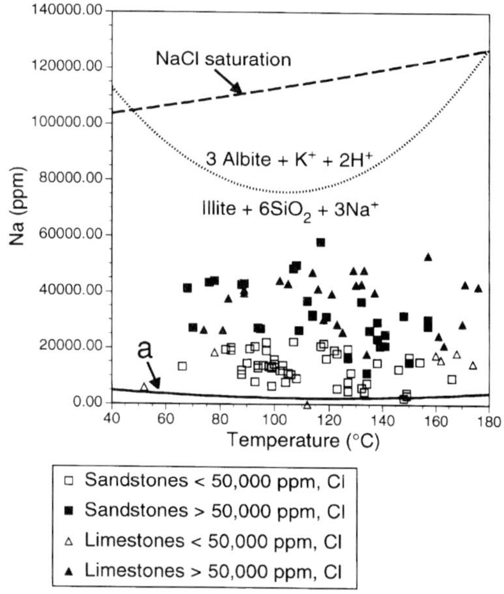

Fig. 7. Sodium concentration in porewaters from the Gulf of Mexico Basin as a function of temperature. Line **a** corresponds to the metastable phase assemblage, K-feldspar–kaolinite–albite–illite, while the dotted line denotes the illite out reaction boundary. Halite saturation is also shown for comparison. pH and potassium assumptions as in Fig. 4. Data from Dickey *et al.* (1972); Gautier (1985); Jessen & Rolhausen (1944); Kharaka *et al.* (1977); Kraemer & Kharaka (1986); Land & Prezbindowski (1981); Macpherson (1992); Schmidt (1973).

K-feldspar is present. Porewaters from limestones show variable potassium concentrations which may reach high values (1000–5000 ppm) at depths exceeding 4 km (Fig. 5). The porewater is then supersaturated with respect to illite. However, illite does not precipitate, probably because of the low concentration of aluminium and precursor minerals, such as smectite and kaolinite, in limestones. The presence of porewater with relatively high potassium concentrations is indirect evidence that the Al concentration in these porewaters is low.

Mississippi Salt Basin

Porewater analyses from the Mississippi Salt Basin fall within the stability field of illite, and there is a trend towards increasing potassium values with temperature (Fig. 8). These porewaters are found in rocks which have been uplifted and subjected to recent cooling and now have excess potassium relative to their illite saturation. This is probably because there is no available aluminium and because the moderate temperatures cause slow kinetic precipitation rates for illite. Sodium concentrations are between 40 000 and 60 000 ppm, corresponding relatively closely to the upper stability of albite + illite at *c.* 100°C (Fig. 9).

Illite precipitation in reservoir rocks

Because illite has a major effect on reservoir properties, it is important to understand the factors controlling its growth in the intergranular pore space. In the basins discussed previously, potassium concentrations are in many cases in

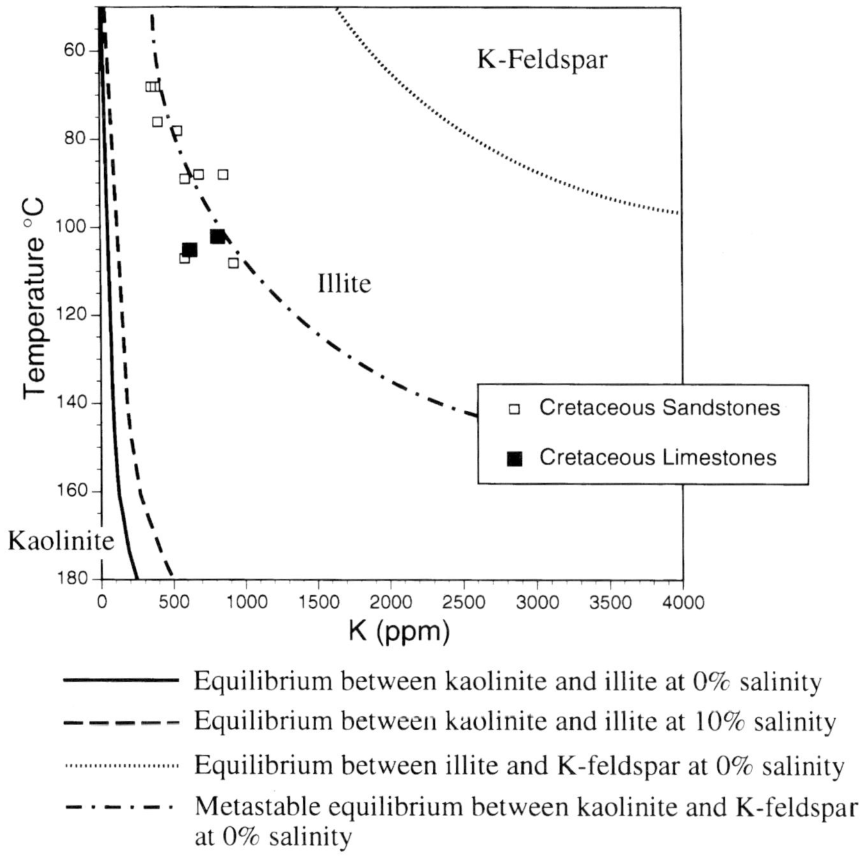

Fig. 8. Potassium concentration in porewaters from the Mississippi Salt Basin as a function of temperature. Equilibria between kaolinite and illite and illite and K-feldspar are indicated for 0% and 10% salinity. Mineral equilibrium calculations assume quartz saturation and pH assumptions as in Fig. 2. Data from Kharaka *et al.* (1987).

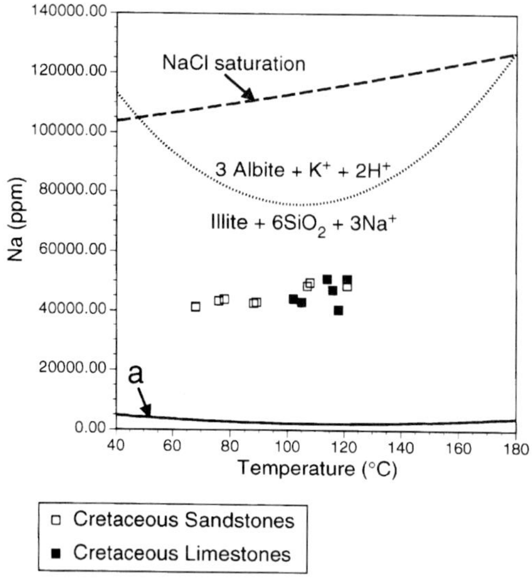

Fig. 9. Sodium concentration in porewaters from the Mississippi Salt Basin as a function of temperature. Line **a** corresponds to the metastable phase assemblage, K-feldspar–kaolinite–albite–illite, while the dotted line denotes the illite out reaction boundary. Halite saturation is also shown for comparison. pH and potassium assumptions as in Fig. 4. Data from Kharaka *et al.* (1987).

excess of the level of illite saturation. This illustrates that the supply of potassium is not the limiting factor controlling illite growth and that illite fails to form due to lack of available aluminium or due to a slow kinetic reaction rate. The reaction rate of illite precipitation is then closely linked to the rate of dissolution of precursor minerals such as kaolinite and smectite. Only when these minerals start to become unstable at 100–140°C does the supply of potassium become critical, as aluminium is now readily available. Analyses from both the North Sea Basin and the subsiding parts of the Gulf Coast Basin suggest that when sandstones are subsiding (heating), potassium is being removed from the porewater so that it approaches equilibrium with both kaolinite and illite. In this situation, the supply of potassium is critical and if K-feldspar is not present, kaolinite may remain stable to much higher temperatures (Bjørlykke *et al.* 1992). There are usually no other significant sources of potassium in sandstones and shales when zeolites are not present. Illitization in the Brent sandstone of the North Sea Basin has been interpreted to have been caused by the introduction of hot fluids in early Tertiary time (Jourdan *et al.* 1987; Liewig *et al.* 1987). It is true that anomalously high temperatures at moderate burial depth would cause illitization if both kaolinite and K-feldspar were present. Introduction of hot fluids could explain the Tertiary illite dates which are found in most Jurassic reservoirs, but a high geothermal gradient on a regional scale must be assumed for that time. This would have caused thermal uplift and inversion of the basin in Palaeocene or Eocene times, yet this is not observed (Glennie 1990). In presently subsiding basins, such as the North Sea and parts of the Gulf Coast Basin, thermal anomalies are presently small. The geothermal gradients in the North Sea Basin vary from 30 to 40°C km^{-1} (Hermanrud *et al.* 1991). If hot fluid were introduced along faults and fractures in lower Tertiary time, the flow rate should have been expected to be reduced by the overlying clay-rich Cretaceous and lower Tertiary sediments. Based on this mechanism one would expect the amount of illite to decrease away from the faults (Fig. 10, Bjørlykke 1993). Such relationships have, however, not been demonstrated. The Permian Zechstein salt of the North Sea has mostly low potassium values (Table 2). Only locally are potassium-rich evaporites found which could serve as a source of potassium. Close to this source the K^+/Al^{3+} ratio in the porewater may be high, so that K-feldspar will precipitate rather than illite. K-feldspar seems, therefore, the most important source of potassium for illite growth.

In the Lousiana Gulf Coast, the isotherms dip at low angles and there is little evidence of hot fluids driven by compaction (Harrison & Summa 1991). It is highly questionable whether hot fluids in sedimentary basins should be rich in potassium because the porewater would be in equilibrium with illite and feldspar. The potassium concentration in the porewater seems to be controlled by the rate of subsidence and heating of sediments containing precursor minerals for illite, such as smectite and kaolinite (Fig. 11). When there is little or no subsidence and heating of the sediments, illite is not being formed continuously but K-feldspar may still dissolve as long as the porewater remains undersaturated with respect to this mineral. The examples discussed previously illustrate the fact that during burial diagenesis, the composition of porewater is, to a large extent, controlled by the mineralogy and by the temperature and kinetics of mineral reactions. It has little meaning to say that a mineral precipitated from the porewater without indicating the source of the ions in therms of local dissolution of minerals or advective transport from a more distant source. When authigenic minerals precipitate, we should also look for a source of the critical ions, in particular Al^{3+}, Si^{4+} and K^+.

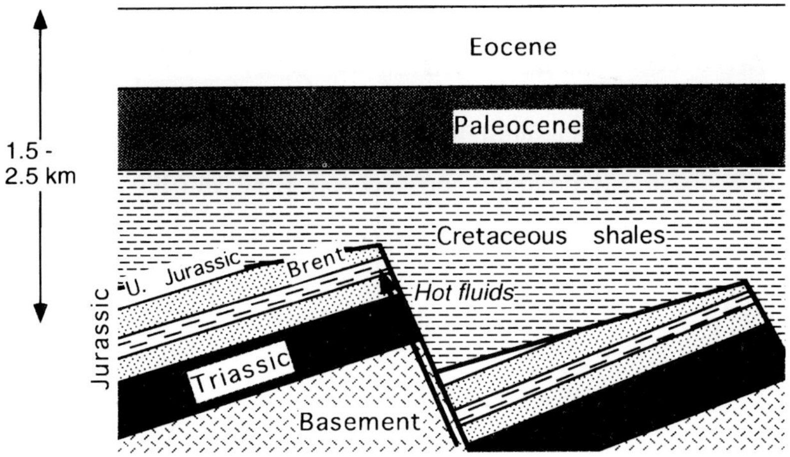

Fig. 10. Illitization in the Brent Group by hot fluid along fault in lower Tertiary time. This model would predict that the distribution of illite was controlled by faults and fractures and less by burial depth. The present authors consider this model problematic because : (1) overlying soft Cretaceous and Tertiary shales would reduce the flow even if underlying faults should be permeable; (2) evidence of active faulting in lower Tertiary strata is relatively rare; (3) hot porewater is not likely to be potassium-rich unless it is derived from potassium-rich salts; (4) a regional heating event should be expected to be associated with thermal uplift and this is not observed; (5) illite is not only formed close to major fault and fractures.

Effects of oil and gas emplacement in reservoirs on diagenesis

The assumption that the emplacement of hydrocarbons would halt diagenetic reactions may seem reasonable, but an increasing amount of recently published evidence suggests that this is not the case (Giles *et al.* 1992; Saigal *et al.* 1992; Walderhaug 1990). In the Brent Group, there is no systematic difference in the porosity of reservoirs in the hydrocarbon zones as compared to the water-bearing zones (Giles *et al.* 1992, p. 298). Fluid inclusions in quartz cement often record temperatures higher than those estimated for the time of reservoir filling (Walderhaug 1990; Saigal *et al.* 1992; Nedkvitne *et al.* 1993). Although data from the Fulmar Field suggest that the rate of quartz cementation is lower in the oil zone than the water zone, more studies are necessary to determine to what degree quartz cementation is inhibited by high hydrocarbon saturation. This may vary with the lithology, texture and mineralogy of different sandstones. Even if there is an effect, it may be difficult to detect above and below the oil–water contact if the hydrocarbon filling continued up to the present time. If precipitation of quartz and illite continue in the remaining water-saturated zones after oil emplacement, this would be a good example of closed-system diagenesis. The fact that illite can be observed in the water-filled pores above the oil-water contact is

Table 2. *Average Zechstein salt compositions*

Well	Unit	Depth (m)	No. of samples	Na (ppm × 1000)	Cl (ppm × 1000)	Br (ppm)	K (ppm)
17.4.1	Z3	2710-2715	2	338	535	136	75
17.4.1	Z2	2739-3445	84	369	579	220	181
17.4.1	Z1	3505-3824	39	380	594	218	137
17.11.1	Z3	2560-2591	4	380	591	108	143
17.11.1	Z2	2637-3152	41	371	580	202	389
17.11.1	Z1	3155-3267	18	373	585	217	508
8.3.1		2309-2950	36	375	588	215	223

Data from Storli (1980).

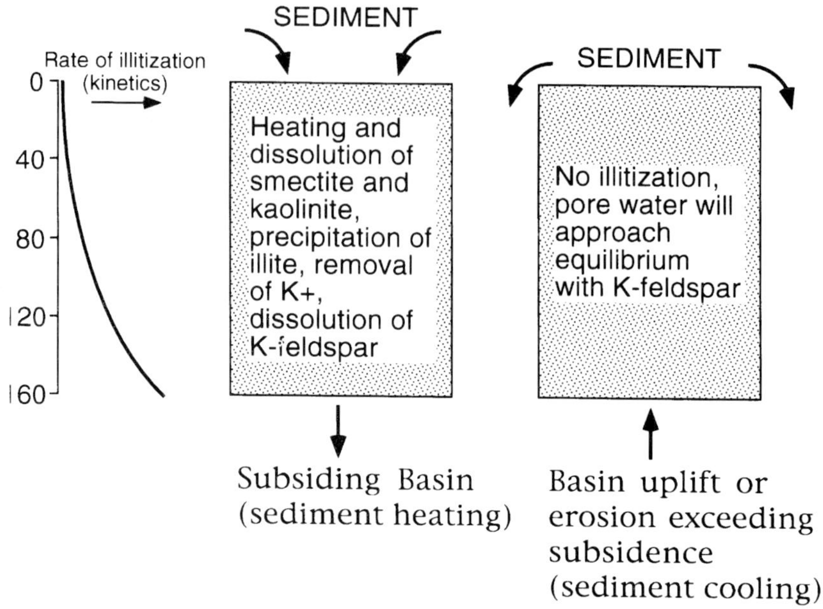

Fig. 11. Schematic illustration of the continuous removal of potassium from porewater by illitization during subsidence and heating of subsiding sediments. When there is no subsidence or uplift the process of illitization stops. Potassium released from K-feldspar or introduced from saline water is not removed. The potassium concentration in porewater is controlled by the rate of K^+ removal by illitization and by the rate of K^+ supply from K-feldspar dissolution.

good evidence of isochemical formation of illite from K-feldspar and kaolinite (Ehrenberg & Boassen 1993).

Oil migrating from source rock to reservoir may also transport organic acids dissolved in the oil phase, but many of these acids are highly water-soluble and may diffuse into the water phase during migration (Barth & Bjørlykke 1993).

During hydrocarbon filling there will normally be no water-flow into the reservoir because the oil leg at the top of the reservoir is not permeable to water. Water must be displaced downwards in order to accommodate the oil. Along the migration pathways, locally high oil saturation will reduce the relative permeability of the water phase to close to zero.

Mobility of silica and aluminium

The low solubility of the main components of silicate minerals puts important constraints on mass transport during diagenetic reactions. Large fluxes (10^8 cm^3 cm^{-2}) are required to introduce significant volumes of silica (Bjørlykke & Egeberg 1993). In the case of aluminium, the concentrations may be on the order of 1 ppm at 130°C and the solubility gradient is much lower than in the case of quartz (Fig. 1). Even if the aluminium concentration should be significantly increased by complexing with organic acids, as suggested by Surdam *et al.* (1989), large porewater fluxes would be required in order to precipitate significant amounts of Al-minerals such as illite in the absence of a dissolving aluminium-bearing precursor mineral. The distribution of illite should then be predictable primarily on the basis of the pattern of such high rates of porewater flow and on the processes causing the breakdown of the complexing of aluminium by organic acids.

One of the highest measured concentrations of aluminium in formation water is from the San Joaquin Basin, where concentrations of up to 6 ppm have been reported (Fisher & Boles 1990). Petrographic studies in this basin suggest that some aluminium may have been exported from sandstones during meteoric water-flushing due to low temperatures and low kinetic reaction rates. During burial diagenesis at greater depths, aluminium appears to be conserved (Hayes & Boles 1992). Calculations by Giles & De Boer (1990) also suggest low solubility and mobility of aluminium so that it is redistributed locally when feldspar and mica are dissolved and secondary porosity formed (Giles & De Boer 1990).

Modelling of diagenetic reactions

The net effects of diagenesis as observed in reservoir rocks are the result of a complex set of processes. The main variables are:

(1) initial textural, mineralogical and geochemical composition;
(2) burial history (temperature and pressure);
(3) 'import' or 'export' of components through fluid flow or diffusion

The present authors presume that diagenetic reactions during burial, below the reach of active meteoric water-flow, must be close to isochemical. As a consequence of this, the initial mineral composition at the time of deposition and the degree of meteoric water-induced reactions at shallow depth are the most important factors controlling mineral reactions at greater burial. Modelling diagenetic reactions as part of basin modelling programmes is limited by the difficulties in predicting the primary composition of source rocks and depositional environments. Vertical sections through reservoir rocks often show extremely variable porosities, and diagenetic processes such as quartz and illite cementation are often different in adjacent sandstone beds. These beds may have experienced the same temperature–pressure history and the differences in porosity cannot be modelled without a detailed knowledge of the primary composition and textures of the sandstones. As demonstrated previously, the primary K-feldspar content of the sandstones, which is related to provenance, may be the most important factor controlling illite growth.

Conclusions

A review of recent literature on reservoir sandstones, such as the Brent Group of the North Sea, shows that there is a lack of consensus between authors on very fundamental diagenetic processes.

Available analyses of formation waters from reservoirs may help to constrain diagenetic models and provide a better basis for practical prediction of porosity and permeability.

The porewater composition in the North Sea Basin and Haltenbanken (mid-Norway) can be explained in terms of equilibrium reactions with the abundant mineral phases, such as quartz, feldspar, illite, kaolinite and carbonates. Mass transfer on a scale of more than a few metres is limited by the porewater flux that can be obtained during meteoric water-flow and compaction-driven flow. The low solubility of aluminium is a major control on illite precipitation. There is at present no evidence of potassium-rich brines, and most of the Zechstein salt has low potassium concentrations and cannot be a major source of potassium for illitization.

Porewater analyses and thermodynamic calculations suggest that the porewater in the North Sea Basin is in most cases supersaturated with respect to illite and undersaturated with respect to K-feldspar at temperatures below 120°C. At higher temperatures, illitization occurs faster and the porewater composition is close to the kaolinite–illite boundary. This coincides with an increased rate of K-feldspar dissolution.

In the Gulf Coast Basin, potassium concentrations close to the boundary between the stability field of kaolinite and illite are found in the presently subsiding basin. In sandstones that are not subsiding and in limestones, the porewater generally has potassium concentrations far above illite saturation, suggesting that the aluminium concentration in the porewater is the limiting factor for illite growth.

The concentration of potassium in the North Sea and Gulf Coast Basins seems to be controlled by the rate of K-feldspar dissolution versus the rate of illite precipitation. This suggests that the aluminium concentrations in the porewater are low and that the precipitation of illite is mostly controlled by the dissolution of aluminium-bearing precursor minerals such as smectite and kaolinite.

The primary mineral compositions at the time of deposition, i.e. the K-feldspar and smectite content and early diagenetic kaolinite, are the most important factors controlling precipitation of illite at depth.

This research has been supported by VISTA, a research cooperation between the Norwegian Academy of Science and Letters and Den Norske Oljeselskap (Statoil). Support from Norwegian Research Council (NAVF) is also appreciated. Review by M. R. Giles and an anonymous reviewer is very much appreciated.

References

AAGAARD, P., EGEBERG, P. K., SAIGAL, G. C., MORAD, S. & BJØRLYKKE, K. 1990. Diagenetic albitization of detrital K-feldspars in Jurassic; Lower Cretaceous and Tertiary clastic reservoir rocks from offshore Norway. II Formation water chemistry and kinetic considerations. *Journal of Sedimentary Petrology*, **60**, 575–581.

——, JAHREN, J. S. & EGEBERG, P. K. 1992. North Sea clastic diagenesis and formation water constraints. *In:* KHARAKA, Y. K. & MAEST, A. S. (eds) *Water–Rock Interaction Proceedings*. **WR7**, Park City, Utah, USA, 13–18 July, 1992.

BARTH, T. 1991. Organic and inorganic ions in water from

petroleum reservoirs, Norwegian Continental shelf. *Applied Geochemistry*, **6**, 1–15.

—— & BJØRLYKKE, K. 1993. Organic acids from source rock maturation: generation potentials, transport mechanisms and relevance for mineral diagenesis. *Applied Geochemistry*, **8**, 325–337.

BATH, A. H., MILDOWSKY, A. E. & STRONG, G. E. 1987. Fluid flow and diagenesis in the East Midland Triassic sandstone aquifer. *In*: GOFF, J. C. & WILLIAMS, B. P. J. (eds) *Fluid flow in Sedimentary Basins and Aquifers*. Geological Society, London, Special Publication, **34**, 127–140.

BJØRKUM, P. A. & GJELDSVIK, N. 1988. An isochemical model for formation of authigenic kaolinite, K-feldspar and illite in sediments. *Journal of Sedimentary Petrology*, **58**, 506–511.

BJØRLYKKE, K. 1984. Formation of secondary porosity. How important is it? *In*: MCDONALD, D. A. & SURDAM, R. C. (eds) *Clastic Diagenesis*. American Association of Petroleum Geologists, Memoir, **37**, 285–292.

—— 1993. Fluid flow in sedimentary basins. *Sedimentary Geology*, **86**, 137–158.

—— & EGEBERG, P. K. 1993. Quartz cementation in sedimentary basins. *American Association of Petroleum Geologists, Bulletin*, **77**, 1538–1548.

——, AAGAARD, P., DYPVIK, H., HASTINGS, D. S. & HARPER, A. S. 1986. Diagenesis and reservoir properties of Jurassic sandstones from the Haltenbanken area, offshore mid-Norway. *In*: SPENCER, A. M. *et al.* (eds) *Habitat of Hydrocarbons on the Norwegian Continental Shelf*. Norwegian Petroleum Society, Graham & Trotman, 275–286.

——, NEDKVITNE, T., RAMM, M. & SAIGAL, G. 1992. Diagenetic processes in the Brent Group (Middle Jurassic) reservoirs of the North Sea – an overview. *In*: MORTON, A. C., HASZELDINE, R. S., GILES, M. R. & BROWN, S. (eds) *Geology of the Brent Group*. Geological Society, London, Special Publication, **61**, 263–287.

BURLEY, S. D. & MACQUAKER, J. H. S. 1992. Authigenic clays; diagenetic sequences and conceptual diagenetic models in contrasting basin margin and basin center, North Sea Jurassic sandstones and mudstones. *In*: HOUSEKNECHT, D. W. & PITTMAN, E. D. (eds) *Origin, Diagenesis and Petrophysics of Clay Minerals in Sandstones*. SEPM, Special Publication, **47**, 81–110.

——, KANTOROWICZ, J. D. & WAUGH, B. 1985. Clastic diagenesis. *In*: BRENCHLEY, P. J. & WILLIAMS, B. P. J. (eds) *Sedimentology Recent and Applied Aspects*. Geological Society, London, Special Publication, **18**, 189–226.

CHESTER, R. 1990. *Marine Geochemistry*. Unwin Hyman, London.

DICKEY, P. A., COLLINS, A. G. & FAJARDO, I. 1972. Chemical composition of deep formation waters in southwestern Louisiana. *American Association of Petroleum Geologists, Bulletin*, **56**, 1530–1533.

DUTTON, S. P. & DIGGS, T. N. 1990. History of quartz cementation in the lower Travis Peak Formation, East Texas. *Journal of Sedimentary Petrology*, **60**, 191–202.

EGEBERG, P. K. & AAGAARD, P. 1989. Origin and evolution of formation waters from oil fields on the Norwegian Shelf. *Applied Geochemistry*, **4**, 131–142.

EHRENBERG, S. N. 1990. Relationship between diagenesis and reservoir quality in sandstones of the Garn Formation, Haltenbanken, Mid-Norwegian Continental Shelf. *American Association of Petroleum Geologists, Bulletin*, **74**, 1538–1559.

—— 1993. Preservation of anomalously high porosity in deeply buried sandstones by grain-coating chlorite: Examples from the Norwegian Continental Shelf. *American Association of Petroleum Geologists, Bulletin*, **77**, 1260–1286.

—— & BOASSEN, T. 1993. Factors controlling permeability variation in sandstones of the Garn Formation in Trestakk Field, Norwegian Continental Shelf. *Journal of Sedimentary Petrology*, **63**, 929–944.

—— & NADEAU, P. H. 1989. Formation of diagenetic illite in sandstones of the Garn Formation, Haltenbanken area, mid-Norwegian continental shelf. *Clay Minerals*, **24**, 233–253.

FISHER, J. B. & BOLES, J. R. 1990. Water–rock interaction in Tertiary sandstones, San Joaquin basin, California, U.S.A.: Diagenetic controls on water composition. *Chemical Geology*, **82**, 83–101.

GAUTIER, D. L., KHARAKA, Y. K. & SURDAM, R. C. 1985. *Relationship of organic matter and mineral diagenesis*. SEPM, Short Course No. **17**.

GILES, M. R. 1987. Mass transfer and problems of secondary porosity creation in deeply buried hydrocarbon reservoirs. *Marine and Petroleum Geology*, **4**, 188–201.

—— & DE BOER, R. B. 1990. Origin and significance of redistributional secondary porosity. *Marine and Petroleum Geology*, **7**, 378–397.

——, STEVENSON, S., MARTIN, S. V. & CANNON, S. J. C. 1992. The reservoir properties and diagenesis of the Brent Group: a regional perspective. *In*: MORTON, A. C., HASZELDINE, R. S., GILES, M. R. & BROWN, S. (eds) *Geology of the Brent Group*. Geological Society, London, Special Paper, **61**, 289–327.

GLASMANN, R. J. 1992. The fate of feldspar in Brent Group reservoirs, North Sea: a regional synthesis of diagenesis in shallow, intermediate, and deep burial environments. *In*: MORTON, A. C., HASZELDINE, R. S., GILES, M. R. & BROWN, S. (eds) *Geology of the Brent Group*. Geological Society, London, Special Publication, **61**, 329–350.

GLENNIE, K. W. 1990. *Introduction to the Petroleum Geology of the North Sea*. Blackwell, London.

GLUYAS, J. 1985. Reduction and prediction of sandstone reservoir potential, Jurassic, North Sea. *Royal Society of London, Philosophical Transactions*, **A315**, 187–202.

—— & COLEMAN, M. 1992. Material flux and porosity changes during sediment diagenesis. *Nature*, **356**, 52–54.

HAMILTON, P. J., GILES, M. R. & AINSWORTH, P. 1992. K–Ar dating of illites in Brent Group reservoirs: a regional perspective. *In*: MORTON, A. C., HASZELDINE, R. S., GILES, M. R. & BROWN, S. (eds) *Geology of the Brent Group*. Geological Society, London, Special Paper, **61**, 377–400.

HARRIS, N. C. 1989. Diagenetic quartzarenite and destruction of secondary porosity: An example from the Middle Jurassic Brent sandstone of northwest Europe. *Geology*, **17**, 361–364.

HARRISON, W. J. & SUMMA, L. L. 1991. Paleohydrology of the Gulf of Mexico Basin. *American Journal of Science*, **291**, 109–176.

HASZELDINE, R. S., BRINT, J. F., FALLICK, A. E., HAMILTON, P. J. & BROWN, S. 1992. Open and restricted hydrologies in Brent Group diagenesis: North Sea. *In*: MORTON, A. C., HASZELDINE, R. S., GILES, M. R. & BROWN, S. (eds) *Geology of the Brent Group*. Geological Society, London, Special Paper, **61**, 401–420.

HAYES, J. M. & BOLES, J. R. 1992. Volumetric relations between dissolved plagioclase and kaolinite in sandstones; Implications for aluminium mass transfer in the San Joaquin Basin, California. *In*: HOUSEKNECHT, D. W. & PITTMAN, E. D. (eds) *Origin, diagenesis and petrophysics of clay minerals in sandstones*. SEPM Special Publication, **47**, 111–123.

HERMANRUD, C., EGGEN, S. & LARSEN, R. M. 1991. Investigations of the thermal regime of the Horda Platform by basin modelling: implication for the hydrocarbon potential of the Stord basin, northern North Sea. *In*: SPENCER, A. M. (ed.) *Generation, accumulation and production of Europe's hydrocarbons*. Special Publication of the European Association of Petroleum Geoscientists, **1**, Oxford University Press, 65–73.

HOGG, A. J. C., SELLER, E. & JOURDAN, A. J. 1992. Cathodoluminescence of quartz cements in Brent Group Sandstones, Alwyn south, UK North Sea. *In*: MORTON, A. C., HASZELDINE, R. S., GILES, M. R. & BROWN, S. (eds) *Geology of the Brent Group*. Geological Society, London, Special Paper, **61**, 421–440.

HOWER, J. E., ESLINGER, E., HOWER, M. E. & PERRY, E. A. 1976. Mechanism of burial metamorphism of argillaceous sediments: I. Mineralogical and chemical evidence. *Geological Society of America, Bulletin*, **7**, 725–737.

HUTCHEON, I., SHEVALIER, M. & ABERCROMBIE, H. J. 1993. pH buffering by metastable mineral–fluid equilibria and evolution of carbon dioxide fugacity during burial diagenesis. *Geochimica et Cosmochimica Acta*, **57**, 1017–1027.

JESSEN, F. W. & ROLHAUSEN, F. W. 1944. Waters from the Frio Formation, Texas Gulf Coast. *AIME, Transactions*, **155**, 23–38.

JOHNSON, J. W., OELKERS, E. H. & HELGESON, H. C. 1992. SUPCRT92: A software package for calculating the standard modal thermodynamic properties of minerals, gases, aqueous species and reactions from 1–5000 bars and 0°C to 1000°C. *Computers & Geoscience*, **18**, 899–847.

JOURDAN, A., THOMAS, M., BREVART, O., ROBSON, P., SOMMER, F. & SULLIVAN, M. 1987. Diagenesis as the control of the Brent sandstone reservoir properties in the Greater Alwyn area (East Shetland Basin). *In*: BROOKS, J. & GLENNIE, K. (eds) *Petroleum Geology of North West Europe*. Graham & Trotman, 951–961.

KANTOROWICZ, J. D., EIGNER, M. R. P., LIVERA, S. E. & VAN SCHIJNDEL-GOESTER, F. S. 1992. Integration of petroleum into Brent Group reservoirs: some observations from the Gullfaks Field, Tampen Spur area, North Sea. *In*: MORTON, A. C., HASZELDINE, R. S., GILES, M. R. & BROWN, S. (eds) *Geology of the Brent Group*. Geological Society, London, Special Publication, **61**, 453–470.

KHARAKA, Y. K., CALLENDER, E. & CAROTHERS, W. W. 1977. Geochemistry of geopressured geothermal waters of the northern Gulf of Mexico basin, 1, Brazoria and Galveston Counties, Texas. *In*: PAQUET, H. & TARDY, Y. (eds) *Second International Symposium on Water-Rock Interaction*. Université Louis Pasteur, **II**, 32–41.

——, GUNTER, W., AGGARWAL, P. K., PERKINS, E. H. & DE BRAAL, J. D. 1988. SOLMINEQ 88: A computer programme for geochemical modelling of water–rock interactions. *U.S. Geological Survey Water Resources Investigations Report*, **88–4227**.

——, MAEST, A. S., CAROTHERS, W. W., LAW, L. M., LAMOTHE, P. J. & FRIES, T. L. 1987. Geochemistry of metal-rich brines from central Mississippi Salt Dome basin, U.S.A. *Applied Geochemistry*, **2**, 543–561.

KRAEMER, T. F. & KHARAKA, Y. K. 1986. Uranium geochemistry in geopressured-geothermal aquifers of the U.S. Gulf Coast. *Geochimica et Cosmochimica Acta*, **50**, 1233–1238.

LAND, L. S. & PREZBINDOWSKI, D. R. 1981. The origin and evolution of saline formation water, Lower Cretaceous carbonates, south-central Texas, U.S.A. *Journal of Hydrology*, **54**, 51–74.

LIEWIG, N., CLAUER, N. & SOMMER, F. 1987. Rb–Sr and K–Ar dating of clay diagenesis in Jurassic Sandstone Oil reservoir, North Sea. *American Association of Petroleum Geologists, Bulletin*, **71**, 1467–1474.

LØNØY, A., AKSELSEN, J. & RØNNING, K. 1986. Diagenesis of deeply buried sandstones from a deeply buried sandstone reservoir: Hild Field, Northern North Sea. *Clays and Clay Minerals*, **21**, 497–511.

MACPHERSON, G. L. 1992. Regional variations in formation water chemistry: major and minor elements, Frio Formation fluids, Texas. *American Association of Petroleum Geologists, Bulletin*, **76**, 740–757.

MORTON, A. C., HASZELDINE, R. C., GILES, M. R. & BROWN, S. (eds) 1992. *Geology of the Brent Group*. Geological Society, London, Special Publication, **61**.

NEDKVITNE, T., KARLSEN, D. A., BJØRLYKKE, K. & LARTER, S. R. 1993. The relationship between reservoir diagenetic evolution and petroleum emplacement in the Ula Field, North Sea. *Marine and Petroleum Geology*, **10**, 255–270.

ROBINSON, A. & GLUYAS, J. 1992. Duration of quartz cementation in sandstones, North Sea and Haltenbanken Basins. *Marine and Petroleum Geology*, **9**, 324–327.

SAIGAL, G. C., BJØRLYKKE, K. & LARTER, S. R. 1992. The effects of oil emplacements on diagenetic processes – Examples from the Fulmar Reservoir sandstones, Central North Sea. *American Association of Petroleum Geologists, Bulletin*, **76**, 1024–1033.

——, MORAD, S., BJØRLYKKE, K., EGEBERG, P. K. & AAGAARD, P. 1988. Daigenetic albitization of detrital K-feldspar in Jurassic, Lower Cretaceous, and Tertiary clastic reservoir rocks from offshore Norway, I. Textures and origin. *Journal of Sedimentary Petrology*, **58**, 1003–1013.

SCHMIDT, G. W. 1973. Interstitial water composition and geochemistry of deep Gulf Coast shales and sandstones. *American Association of Petroleum Geologists, Bulletin*, **57**, 321–337.

SCOTCHMAN, I. C., JOHNES, L. H. & MILLER, R. S. 1989. Clay minerals and oil migration in Brent group sandstones of NW Hutton Field, UK North Sea. *Clay Minerals*, **24**, 339–374.

SHARP, J. M., GALLOWAY, JR, W. E., LAND, L. S. ET AL. 1988. Diagenetic processes in Northwestern Gulf of Mexico sediments. *In*: CHILINGARIAN, G. V. & WOLF, W. H. (eds) *Diagenesis, II*. Developments in Sedimentology. Elsevier, Amsterdam, 43–113.

SMITH, J. T. & EHRENBERG, S. N. 1989. Correlation of carbon dioxide abundance with temperature in clastic hydrocarbon reservoirs: relationship to inorganic chemical equilibrium. *Marine and Petroleum Geology*, **6**, 129–135.

STOESSELL, R. K. & PITTMAN, E. D. 1990. Secondary porosity revisited: The chemistry of feldspar dissolution by carboxylic acid and anions. *American Association of Petroleum Geologists, Bulletin*, **74**, 1795–1805.

STORLI, A. 1980. *En undersøkelse av Zechstein-evaporittene i borhullene 17/4-1, 17/11-1 og 9/8-1*. Masters Thesis, University of Bergen, Norway.

SURDAM, R. C., BOESE, S. W. & CROSSEY, L. J. 1984. The chemistry of secondary porosity. *In*: MCDONALD, D. A. & SURDAM, R. C. (eds) *Clastic Diagenesis*. American Association of Petroleum Geologists, Memoir, **37**, 127–149.

——, CROSSLEY, L. J., HAGEN, E. S. & HEASLER, P. 1989. Organic–inorganic interaction and sandstone diagenesis. *American Association of Petroleum Geologists, Bulletin*, **73**, 1–23.

TRUESDELL, A. H. & JONES, B. F. 1974. WATEQ, a computer programme for calculating chemical equilibria of natural waters. *U.S. Geological Survey, Journal of Research*, **2**, 233–248.

WALDERHAUG, O. 1990. A fluid inclusion study of quartz cemented sandstones from offshore Mid-Norway – possible evidence for continued quartz cementation during oil emplacement. *Journal of Sedimentary Petrology*, **60**, 203–210.

Geochemical criteria for reservoir characterization

LEONID ANISSIMOV

Department of Hydrogeology, Saratov State University, Saratov 410601, Russia

Abstract: This paper shows how the basinal analysis of formation waters and petroleums may be used to deduce the general geological setting of the source-reservoir system. There are two extreme cases, principally depending on the relative volumes of the source and reservoir regions. If the source rock region dominates, a 'closed' system results. Otherwise, where the reservoir volume is large the system is 'open'. The 'reservoir' may be thought of as including the 'carrier bed' if appropriate.

In the 'closed' system, for example the carbonate reefs of the Western Canada Basin, water-soluble products of petroleum generation (e.g. benzene and H_2S) are preserved in the reservoir. There is insufficient water to remove these components by aqueous solution. The partial pressure of methane is expected to be the same in formation waters and petroleums, since they can equilibrate in a 'closed' system. In an 'open' system the volume of water in the 'reservoir' is large compared to the source. Water and petroleum expelled from the mature source rock kitchen cannot reach equilibrium with the vast volume of formation fluids. Regions which connect with local source kitchens are expected to experience alterations in salinity and dissolved organics. Concentrations of water-soluble components will therefore be reduced in petroleums.

Data from 'open' and 'closed' systems are presented from the Devonian of the Volga–Ural province. A reduction in water salnity below 3 km, near the source kitchen is attributed to expulsion of a low salinity water during oil generation. Related changes in aqueous benzene concentrations are also seen.

Porosity, permeability, oil saturation, gas saturation, water saturation and heterogeneity consitite the principal reservoir rock properties. The last parameter is the most difficult to describe mathematically and, therefore, it is considered under geological control. The role of heterogeneity is important and can be estimated by studying the reservoir as a whole, after drilling during the development of the field. Processes such as water and gas coning, pressure decline, oil and gas recovery and development of the field as the whole are influenced by this property. We can confirm this by considering the cases of three field examples: Orenburg (Russia), Hassi Messaoud (Algeria) and Lacq (France).

The gas–oil zone of the Orenburg Field is 520 m thick, yet water coning occured after some months of gas production. The productive oil and water zones of the reservoir are in communication which facilitates formation-water penetration. The oil rim cannot prevent water penetration. We can call this category of reservoirs, 'open'. Production problems of the Hassi Messaoud Field were caused by pressure decline. The Cambrian reservoir is represented by naturally fractured rocks (dual porosity model) with numerous fractures filled with clay, epigenetic minerals (calcite, anhydrite, barite, pyrite, quartz and bitumen. The formation waters are strong brines with salnity up to 310 g l^{-1}. Halite precipitation takes place in the reservoir and production lines during exploitation. Communication between different parts of reservoir is not perfect and we can call such categories of reservoir 'semi-closed'. Development of the Lacq gas Field begain in 1957. There is no remarkable formation water activity in spite of deep formation pressure decline. The field works as a closed system (Le Tran *et al.* 1973) and such reservoirs will be called 'closed'.

Heterogeneity characteristics may be assessed from the geological setting of a reservoir. Simple geological conditions generally produce 'open' reservoirs. Reef, deltaic and fluvial facies form semi-closed and closed reservoirs. But final appraisal of the connection between oil/gas and water zones is possible after studying all parts of the reservoir using well data. In this process, geochemical methods have a number of advantages if they play a pioneering role in reservoir characterization, at the initial stages of appraisal. This can be achieved if we can determine how geochemical characteristics relate to hydrocarbon/water ratios in reservoirs.

There are a number of studies where formation waters are considered as important hydrocarbon alteration processes (except biodegradation). Water washing is particularly effective for the C_{15-} fraction and hydrocarbons are removed from oils in the

From CUBITT, J. M. & ENGLAND, W. A. (eds), 1995, *The Geochemistry of Reservoirs*, Geological Society Special Publication No. 86, pp. 51–58.

sequence: aromatics, n-alkanes, then naphthenes. In the C_{15+} fraction, some aromatics and sulphur compounds especially dibenzothiophene are depleted (Lafarque & Barker 1988). The same mechanism causes the depletion of aromatic hydrocarbons by preferential removal from oils in the water-saturated system below the oil–water contact (Hillebrand & Leythaeuser 1992). In the present study, the author proposes models for the quantification of source and reservoir processes taking into account the mutual solubility of organic and inorganic compounds of oils, gases and formation waters.

Source processes model: compaction and generation

Clay compaction and sedimentary organic-matter maturation are accompanied by changes in volume and quality of fluid system components (Fig. 1). Gas volume (V_g) is determined by total organic carbon (TOC) and gas-generation rate (k_g). There are two zones where k_g is at a maximum: the biogenic and thermocatagenic zones (Tissot & Welte 1978; Vassoevich *et al.* 1979).

$$V_g = k_g \, [\mathrm{TOC}] \qquad (1)$$

Oil volume (V_o) is evaluated similary. Oil generation rate (k_o) is at a maximum in the temperature zone of 80 to 100°C.

$$V_o = k_o \, [\mathrm{TOC}] \qquad (2)$$

Water volume (V_w) is evaluated as clay porosity (n). Clay dewatering is considered primarily a mechanical compaction during which porosity decreases from 40% to 2% by the 3000 m depth (normal trend). At the next burial stage, water and hydrocarbon generation from organic matter and mobilization of interlayer water increase clay porosity to 4 to 6%. This level of porosity may be preserved in abnormal pressured zones. The state of water, oil and gas in the system is determined by mutual solubility. It depends on pressure, temperature, gas composition, etc. Hydrocarbon gas solubility in oil (a_o) is two orders of magnitude higher then the solubility in water (a_w). Gas absorption by oil ($a_o \times V_o$) and gas absorption by water ($a_w \times n$) determine the volume of dissolved gas and the phase state of the system. The equation of the system state is:

$$\frac{k_g \, [\mathrm{TOC}]}{a_o \, k_o \, [\mathrm{TOC}] + a_w \, n} = K \qquad (3)$$

$K > 1$ produces condition for gas-phase primary migration

$K < 1$ produces condition for oil and/or water-phase primary migration.

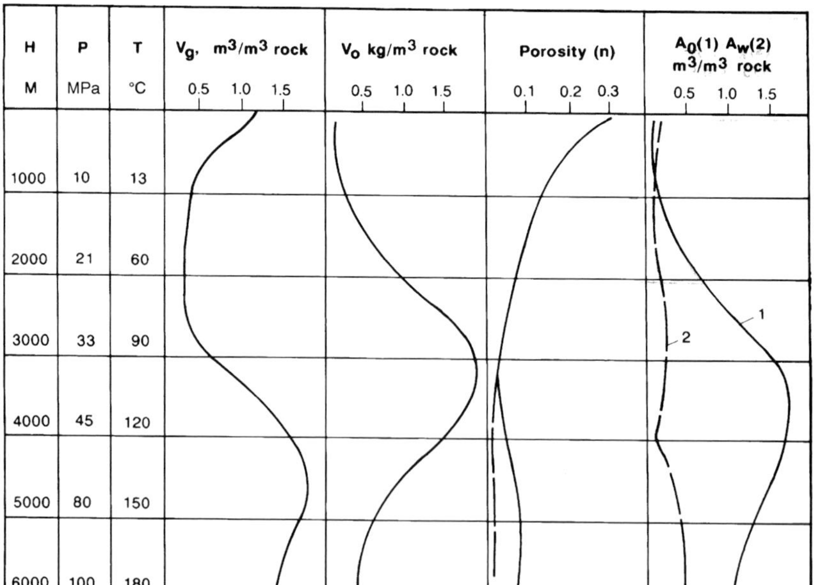

Fig. 1. Vertical variation of 'water–oil–gas' system parameters. V_g = gas generation volume; V_o = oil generation volume; A_o = gas absorption by oil; A_w = gas absorption by water; H = depth; P = pressure; T = temperature.

In-reservoir processes model: from dispersion to homogenization.

Three in-reservoir processes are considered: (a) the migration of fluids into the reservoir, (b) its subsequent homogenezation and (c) the accumulation of hydrocarbons. The mechanism and means of fluid expulsion depends on several factors, including matrix porosity and permeability, organic carbon content and distribution, the type of organic matter, oil, gas and source water generation and formation pressure distribution in the source rock-reservoir system. Commonly, the expulsion process begins after oil saturation reaches 20 to 30% of shale porosity, corresponding to the onset of the oil window. Such levels of hydrocarbon saturation occur when clay porosity decreases, as a result of dewatering, to 2 to 3%. This process is accompanied by peak oil and gas generation. Fig. 2 shows the dependence of expulsion (hydrocarbon plus source-rock water) on the differential pressure between source rocks and reservoir. This plot shows: (1) clay porosity increases due to generation of new fluids (maximal porosity); (2) average clay porosity of the overpressure zone and the related expulsion rate for the overpressured reservoir: Δn_1 (3) the trend for normal porosity related to the expulsion rate for a normally pressured reservoir: Δn_2. The next task is to distinguish between hydrocarbon and source-rock water fractions of expelled fluids. To assess the composition of hydrocarbons expelled from source rocks, we rely on experimental simulation of maturation processes. According to recent studies (Bechar *et al.* 1992), mass balances for Type II kerogen pyrolysis at a temperature of 350°C and a time of 24 hours are as follows: CO_2 = 53; H_2S = 24; H_2O = 48; C_1 = 13; C_2–C_5 = 32; C_6 = 366; residue = 466 (wt‰ of initial sample). Organic acids are an other pyrolysis product which can be recognized in formation waters of some basins (Fisher 1987; Eglington *et al.* 1987; Barth 1990; Davidson *et al.* 1991). All these products may be divided into water-soluble (source-rock water, organic acids), oil-soluble (hydrocarbon gases) and oil- and water-soluble (H_2S, CO_2). The disproportionate distribution of these components in the reservoir is demonstrated by Fig. 3 with the thickness of the arrows representing the portion of the component which is absorbed by oil or formation water. Carbon dioxide and hydrogen sulphide may be absorbed in the generation of new minerals resulting in a change to their concentrations in oil and formation water.

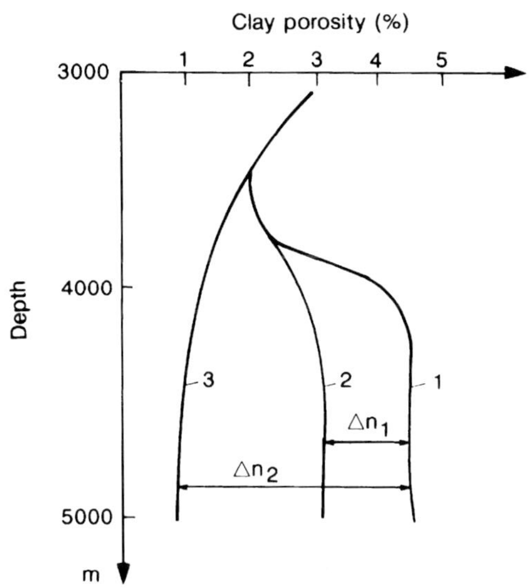

Fig. 2. Expulsion rate appraisal. 1, Porosity increase due to generation of pyrolysis products; 2, porosity of overpressure zone (expulsion rate for overpressure reservoir = Δn_1); 3, 'normal' porosity (expulsion rate for normal pressure reservoir = Δn_2).

The volume of expulsion water (V_s) is calculated as

$$V_s = h_s \times S \times \Delta n \tag{4}$$

where h_s is the thickness of the rocks for the hydrocarbon-bearing reservoir (half of the

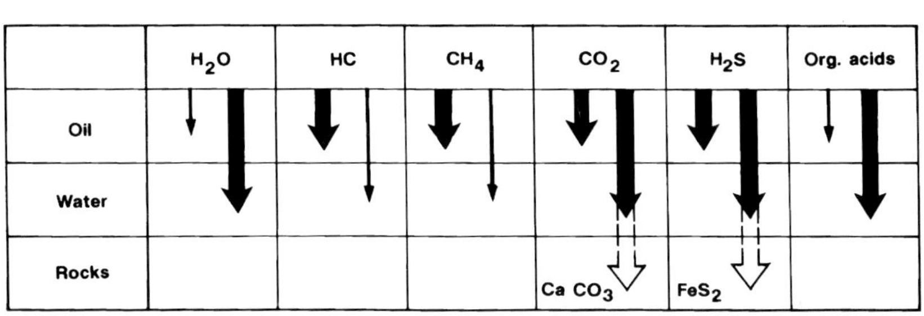

Fig. 3. Distribution of pyrolysis products in a reservoir.

thickness of the underlying and overlying clays).
$S`$ is the area
Δn is the expulsion rate for source-rock water.

The volume of expulsion water (V_s) may be greater than the volume of reservoir water (V_R), which we refer to as a displacement regime. The composition of reservoir water in this is similar to the composition of the source water. If $V_R > V_s$, the process is referred to as a mixing regime and in this case, the concentrations of soluble components in the source (C_s) and reservoir (C_R) waters may be represented by an equation of mass balance:

$$C_s \times V_s = C\,(V_s + V_R) \text{ or } C_R = C_s \frac{V_s}{V_s + V_R}. \quad (5)$$

This equation also may be expressed as:

$$C_R = C_s \frac{h_s \times \Delta n}{h_R \times n + h_s \times \Delta n} \quad (6)$$

the C_s/C_R ratio may then be considered as a reservoir characterization parameter.

In primary migration, oil, gas and water constitute a disperse system in equilibrium. After oil separation, however, the fates of the oil and source-rock water are different. Figure 4 shows the distribution of hydrocarbon gases in the course of maturation, expulsion and accumulation. There are three stages in the system development: (I) generation; (II) expulsion and (III) accumulation. At stages I and II, the system is dispersed and partial gas pressures in the oil and water phases are equal. During oil separation in the reservoir, however, gas dissipation from the oil phase is less than in the formation waters ($p_{oil} > P_{water}$).

The concentration ratio for methane and other hydrocarbon gases at equal partial pressures in oil and water is:

$$\frac{[CH_4]_{oil}}{[CH_4]_{water}} = n \times 10^2 \quad (7)$$

Under these circumstances, oil is the best adsorber of gas and the reservoir volume must be equal to $n \times 10^2$ of the injected oil volume to accomodate the same quantity of gas at equal partial pressures. The difference in partial pressures ($\Delta P`$) depends on ratio of expulsion and reservoir volumes. In the case of a small reservoir volume (closed reservoir), $P_{oil} = P_{water}$. In the case of vast (unlimited) reservoir P_{oil} is higher than P_{water}. ΔP can, therefore, be considered as an additional criterion for reservoir characterization. Secondary migration of oil creates geochemical conditions which can be grouped into three types.

(1) Oil and source waters fill the reservoir replacing formation waters completely. Equilibrium between oil and water is maintained.

(2) Oil and source water are expelled into a reservoir and are mixed with formation waters. Oil and gas pools are surrounded by waters with a high concentration of dissolved gases and organic components. Using a quantitative diffusion model, it is possible to assess the position of the pool more precisely.

(3) Oil and gas penetrates the formation (red beds, hydrodynamically active zones). They create qualitatively different zones between nearby pool waters and formation water.

Examples of reservoir characterization by fluid geochemistry

During exploration in the Volgograd region, hundreds of Palaeozoic formation waters were analysed for benzene and dissolved hydrocarbon gases. The depth of reservoirs differs from 500 to

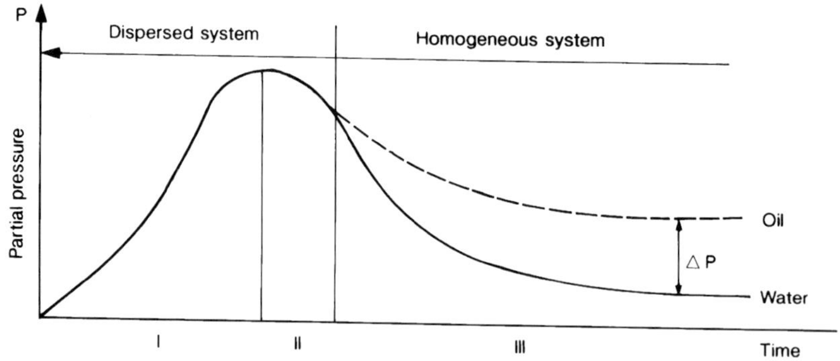

Fig. 4. Modelling of 'water–oil–gas' system in source and reservoir processes. P = partial pressure of dissolved gas.

5000 m and all hydrodynamic and maturation zones are present in the sedimentary section. The difference between source and reservoir waters can be evaluated by the histograms shown in Fig. 5. Line 1 is the assessment of C_{1+} and benzene concentration in source-rock water and line 2 is the same parameters in reservoir water. At the oil window the concentration difference is less due to active hydrocarbon generation and decrease in reservoir volume.

Demineralization (i.e. reduction in salinity) of formation waters begins at a depth greater than 3000 m. At this burial stage, active hydrocarbon generation is accompanied by low-salnity source water which decreases formation water salinity. At a depth to 3000 m, the formation-water salinity of terrigenous Devonian in the Volga–Ural Province is 230 to 260 g l^{-1}, but at the border of the Pericaspian depression in the Volgograd region, the salinity of formation waters decreases to 100–120 g l^{-1}. The zone of salinity minimum is situated near a palaeodepression, where clay sediments dominate over sandstones (Fig. 6). Low-salinity water characteristics include high concentrations of HCO_3, iodine and benzene relative to 'normal' salinity formation waters (Table 1). Figure 7 shows the salinity data of the terrigenous Devonian formation waters with reference to the ratio of thickness of the source and reservoir rocks. All the samples were taken from depths greater than 3000 m, mostly from abnormal pressure zones.

Quantitative appraisal of the expulsion rate was made on the base of formation-water salinity changes. Demineralization of formation water begins when h_R/h_s is equal to 0.1. This situation corresponds to the change of regime from displacement to mixing or $V_R = V_s$. If $V_R = h \times S \times n$ and $V_S = h_s \times S \times \Delta n$, the ratio $n/\Delta n$ is equal to 10. The porosity of Devonian sandstones is 10 to 15% and therefore, the expulsion rate of source-rock water is 1 to 1.5% of the rock volume. This is approximately equal to the expulsion rate of hydrocarbons.

Conclusions

Concentrations of organic and inorganic compounds in formation waters can serve as important characteristics of maturation processes as well as reservoirs. H_2S, CO_2, organic acids and some aromatic hydrocarbons are water-soluble and their

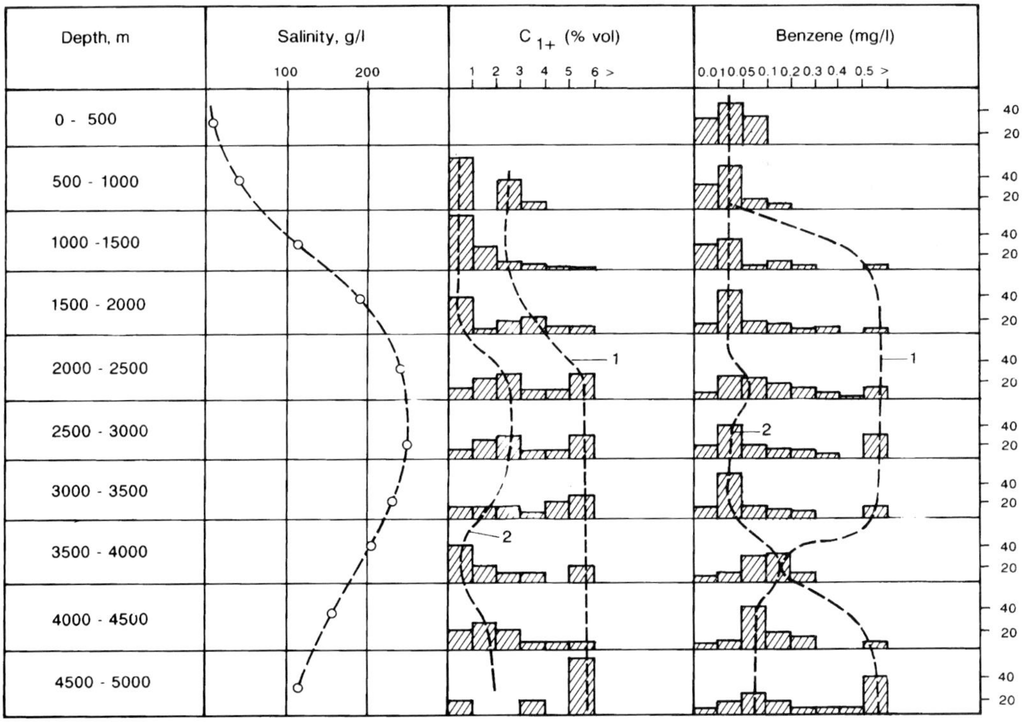

Fig. 5. Histograms showing the concentration of C_{1+} (% vol) and benzene (mg l^{-1}) dissolved in formation waters, Volograd region sedimentary section. Line 1 is source-rock water; line 2 is reservoir water.

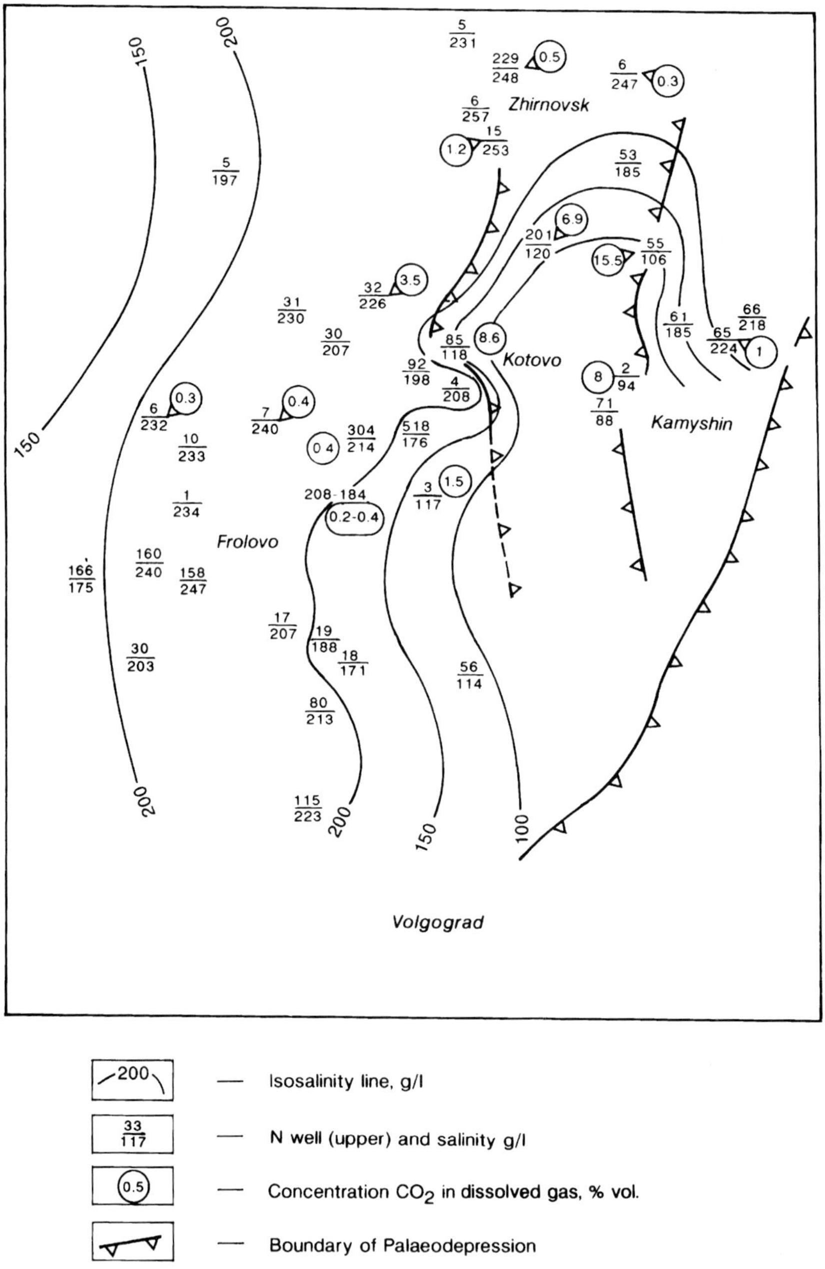

Fig. 6. Hydrochemical map of the Volgograd Region terrigenous Devonian deposits.

high concentration in oils, gases and formation waters may be observed only in closed (limited) reservoirs. Effective 'clean-up' of these components takes place in extended (unlimited) reservoirs, where there is vast volume of reservoir water. Sour gases of carbonate reefs in the Western Canada Basin and Lacq gas field are the best examples of water-soluble component preservation in isolated reservoirs.

In summary, reservoir geochemistry methods can be presented in the following operational framework.

Table 1. *Concentration of inorganic ions in Volgograd region formation waters*

No.	Locality	Well	Depth (m)	Cl	SO_4	HCO_3	Ca	Mg	Na + K	I	B	Total
'Normal' salinity formation waters												
1	Zhirnovsk	229	3133	4284	2.8	1.5	804	177	2329	8	1097	245 500
2	Nizhnaya Dobrinka	15	3210	4191	2.5	0	824	162	2225	—	—	237 200
3	Peskovatka	6	3004	4217	5.6	1.0	760	190	2331	7 5	1176	238 800
4	Berezovskaya	304	3240	3750	3.7	1.6	563	113	2409	8 8	801	214 300
5	Loznovskaya	80	3222	3757	0.7	0	675	107	2177	25	1258	213 400
6	Kudinovskaya	17	3083	3570	0.6	2.7	825	110	1704	15	1461	201 900
Low salinity formation waters												
7	Korobkovskaya	85	3860	2090	0.7	6.7	451	70	1049	47 2	571	118 000
8	East Umet	2	4808	1604	1.0	12.7	145	20	1238	154 5	442	94 000
9	South Umet	71	4818	1500	2.0	7.2	110	15	1261	160 0	373	88 000
10	East Umet	6	4873	1663	0.9	9.0	160	20	1315	145 0	489	97 000
11	Romanovskaya	3	3858	2046	0.6	2.4	262	43	1439	21 6	704	117 000
12	Petrov Val	5	4850	1360	9.0	5.3	75	15	1203	171	298	80 500

Major ion composition is given in millimolars; iodine, bromine and total salinity in mg 1^{-1}.

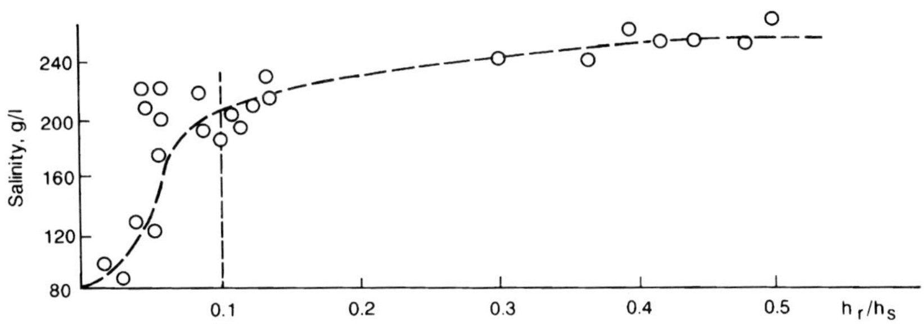

Fig. 7. Salinity plot for terrigenous Devonian formation waters; h_r/h_s = ratio of thickness of source to reservoir rocks.

(1) Formation of fluid parameter database for the basin. Maximum concentrations are interpreted to represent source-water parameters, whereas minimum concentrations would be considered indicative of reservoir-water parameters without significant source-fluid influence.

(2) Analysis of examples of formation fluid. Assessment of reservoir position in the range of model reservoirs.

(3) Modification of the assessment according to geological and hydrocarbon characteristics.

Geochemical analytical methodology for reservoir characterization can also be enhanced based on other parameters such as solid bitumen radioactivity and oil composition. These criteria as well as those mentioned previously are based on the different solubilities of gases, hydrocarbon and non-hydrocarbon compounds in oils and formation waters. Then, modelling of the three phase 'oil–gas–water' system increases the potential of employing geochemical methods for source and reservoir process characterization compared with binary 'gas–oil' system.

References

BARTH, T. 1991. Organic acids and inorganic ions in water from petroleum reservoirs, Norwegian continental shelf: a multivariante statistical analysis and comparison with American reservoir formation water. *Applied Geochemistry,* **6**, 1–15.

BECHAR, F., KRESSMAN, S., RUDKIEVICZ, J.L. & VANDENBROUCKE, M. 1992. Experimental Simulation in a confined system and kinetic modelling of kerogen and oil craking, *Organic Geochemistry,* **19**, 173–189.

DAVIDSON, R., ABERCROMBIE, H., BARKER, J., FOWLER, M. & SNOWDON, R. 1991. Water soluble organic acids from the hydrous pyrolysis of an immature, organic-rich shale from the Western Canada Sedimentary basin. *In*: MANNING, D. (ed.) *Organic Geochemistry. Advances and Application in Energy and the Natural Environment*. Manchester University Press, Manchester, 581–584.

EGLINTON, T., CURTIS, C. & ROWLANDS, S. 1987. Generation of water-soluble organic acids from kerogen during hydrous pyrolysis: implication for porosity development. *Mineralogical Magazine,* **51**, 495–503.

FISHER, J. 1987. Distribution and occurence of aliphatic acid anions in deep subsurface water. *Geochimica et Cosmochimica Acta,* **5**, 2459–2468.

HILLEBRAND, T. & LEYTHAEUSER, D. 1992. Reservoir geochemistry of Stockstadt oilfield compositional heterogeneities reflecting accumulation history and multiple source input. *Organic Geochemistry,* **19**, 119–131.

LAFARGUE, E. & BARKER, C. 1988. Effect of water washing on crude oil composition. *American Association of Petroleum Geologists, Bulletin,* **72**, 263–276.

LE TRAN, K., CONNAN, J. & VAN DER WEIDE, B. 1973. Problems relatif a la formation d'hydrocarbures et d'hydrogene sulfuré dans la basin Sud-Ouest Aquitain. *In: Advances in Organic Geochemistry.* Actes du 6e Congress International Paris, Technip, 762–789.

TISSOT, B.P. & WELTE, D.H. 1978. *Petroleum formation and occurence. A new approach to oil and gas exploration.* Springer Verlag, Berlin.

VASSOEVICH, N., KOZLOV, A. & LOPATIN, N. 1979. Gas generation on the different stages of lithogenesis. *Vestnik of Moscow University,* **1**, 36–43 [in Russian].

Compositional heterogeneities in oilfield formation waters: identifying them, using them

P. C. SMALLEY[1], T. A. DODD[2], I. L. STOCKDEN[1], A. RÅHEIM[3] & E. W. MEARNS[4]

[1]*BP Exploration, Chertsey Road, Sunbury-on-Thames, Middlesex TW16 7LN, UK*

[2]*BP Exploration, Farburn Industrial Estate, Dyce, Aberdeen AB2 0PB, UK*

[3]*Institute for Energy Technology, PO Box 40, 2007 Kjeller, Norway*

[4]*Isotopic Analytical Services Ltd, Campus 3, Aberdeen Science and Technology Park, Bridge of Don, Aberdeen AB22 8GW, UK*

Abstract: Two new techniques are now available for gaining information on formation water composition from core samples: core centrifugation, in which preserved core samples, complete with their preserved fluid content, are ultracentrifuged to extract the oil and water; and residual salt analysis (RSA), where salts are redissolved from unpreserved dry core and parameters such as $^{87}Sr/^{86}Sr$ are measured in the leachate. These methods are described and details given on the quality control measures that are crucial in order to avoid artefacts of core contamination by drilling fluids. Such methods have allowed formation-water compositional data to be obtained at the sub-metre scale, in some cases revealing significant small-scale variations (gradients and steps) in formation-water chemistry.

Resistivity (related to salinity) variations are important for interpretation of water saturation from electrical resistivity logs, which in turn affects the estimate of the total amount of oil contained in the field. In the Machar chalk-reservoired oilfield, formation-water data derived by RSA and core centrifugation both indicate variations in salinity (TDS 140 000–220 000 ppm) that significantly affect resistivity. In some oilfields, assumption of a single resistivity value for the whole field could thus lead to errors in petroleum reserves estimation.

Step changes in formation-water composition can be used as indicators of reservoir fluid compartmentalization, important for field development strategy. Examples are given of cases where RSA variations have been used to distinguish between laterally restricted and laterally extensive shales, both in the oil and water legs. Analysis of water compositional variation forms a valuable addition to the range of tools available for assessing reservoir compartmentalization during reservoir appraisal. This is because natural variations in fluid compositions provide quasi-dynamic data (i.e. how fluids have moved slowly during geological time), from which inferences can be made about more rapid fluid-flow patterns under dynamic (oil production) conditions.

Evidence is growing that formation waters in oilfields are sometimes heterogeneous in their bulk composition (Coleman 1992; Smalley & Oxtoby 1992). Even where there may not be noticeable variations in water bulk chemistry, subtle heterogeneities may be detected by the use of sensitive natural isotope tracers (Gibbons 1991; Smalley *et al.* 1992). This paper deals with new high-resolution water sampling techniques, because in many cases it is only through these that it is possible to identify variations in water composition. The paper also discusses the ways in which such water compositional information can be applied, because this type of data has important implications for many oilfield activities: petrophysics, production chemistry, facilities engineering, reservoir geology and so forth. Two cases are examined in detail: quantification of formation-water resistivity for interpretation of electrical well logs and reserves estimation, and the use of formation-water variations to assess reservoir compartmentalization.

Formation water sampling

Conventional methods

In order to identify formation-water variation, water must be sampled at a scale smaller than the scale of water heterogeneity. Within a well, this may require sampling at the metre scale, which cannot normally be achieved with conventional water sampling techniques.

The overwhelming majority of formation-water samples analysed by oil companies and published in the literature belong to one of the three main types: produced waters, drill stem tests or downhole samples.

From CUBITT, J. M. & ENGLAND, W. A. (eds), 1995, *The Geochemistry of Reservoirs*, Geological Society Special Publication No. 86, pp. 59–69.

(1) Produced waters are those that have been sampled at or after the well head, from the producing wellstream. These waters are coproduced with petroleum. Such samples have the advantage that they can be of high quality, relatively free of contamination from drilling or completion fluids. Later in field life, however, where water injection has been used as a secondary recovery method, produced waters can include a component of injected water. More importantly for the study of spatial heterogeneities of formation water, it is usually difficult to determine with any confidence from exactly where in the formation the water was derived. So, while produced waters are of some use for identifying large-scale fieldwide variations in water composition, they are of limited use for describing within-well variations.

(2) Drill stem test (DST) samples are those that have been collected from fluids that have flowed to surface with the well under test conditions. Their limitations are similar to those of produced waters, in that it is not usually possible to determine from exactly where the water has derived within the perforated interval. Further, it is unusual for DSTs to be carried out at multiple depths. They are thus of limited use for identifying within-well variations.

(3) Downhole samples are most often taken by the repeat formation tester (RFT) or similar tools. This wireline tool samples water directly from a single point down the well. The main limitations are sample quality (drilling fluid contamination can be a problem) and the fact that, mainly because of cost, RFT samples are not usually taken from multiple depths in a reservoir.

Clearly, if fine-scale variation in formation water is to be examined, water sampling techniques are needed that have the spatial resolution of the RFT tool, but have sufficiently low cost to allow detailed sampling at many different depths in a well.

Water extraction from core

The extraction of water directly from newly retrieved core samples is an attractive possibility, as it allows sampling to any degree of spatial resolution required. The main problem with this is that the water in the core can be contaminated by the drilling fluids in the well. Several factors affect the degree of core infiltration by drilling fluids: mud type, the pressure of the mud compared to that of the formation ('drilling overbalance'), the type of coring bit used, core permeability, including the presence of fractures in the core, and the length of time that the core has been in contact with the drilling fluid (hence the top few centimetres of a core may be excessively contaminated because it has been in contact with the drilling fluid for several hours while the previous bit or core is being retrieved from the well). Even if precautions are taken against these problems, for example using a low-invasion coring bit and avoiding core tops and highly permeable core, there will still be some drilling fluid invasion, especially towards the core edges. This can be minimized by drilling a plug for analysis along the axis of the core centre (the part of the core most likely to contain pristine formation water). This is ideally performed on the rig directly after core retrieval, and no water coolant should be used during plugging. The plug is preserved by wrapping it in plastic film, and taken onshore for analysis. The plug is then placed in a specially adapted ultracentrifuge, and the formation fluids spun out. The oil and water are collected in a tube and both are available for analysis.

A further precaution against contamination of the core by drilling fluid is to tag the drilling mud. A tracer (deuterated water is often used) is added periodically to the drilling mud, and mud deuterium to hydrogen (D/H) ratios are monitored during coring to make sure that they are much higher than the natural values in the formation water. D/H isotopic ratios are also measured in the water samples centrifuged from the core. Knowing the D/H values for the mud and for uncontaminated formation water, it is possible to calculate the degree of core invasion by drilling fluid in each sample. A dataset for a North Sea oilfield is shown in Fig. 1. Here, the degree of core invasion is strongly related to permeability, as might be expected, with all samples of more than 100 mD having more than 50% contamination. Never-

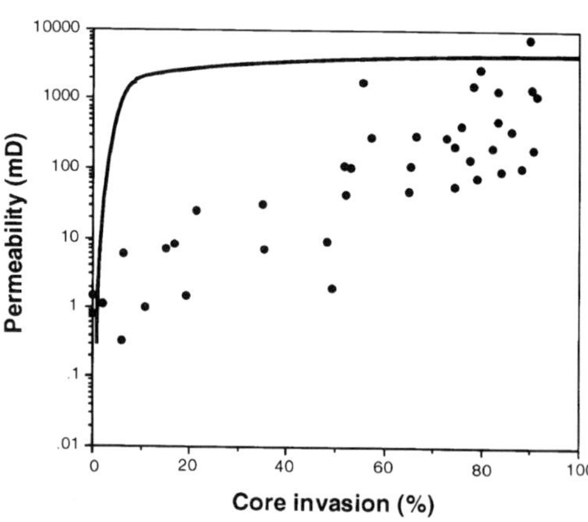

Fig 1. Degree of core contamination by drilling fluid (measured using a deuterium-tagged mud) compared to core permeability in a single well from a North Sea oilfield. The curve predicts the percentage change in measured water $^{87}Sr/^{86}Sr$ in the same system, for a formation water Sr content of 500 ppm and a drilling mud filtrate Sr content of 1 ppm.

theless, the D/H ratio of each sample can be used to correct the water composition for drilling fluid contamination (cf. Coleman *et al.* 1990) or, in extreme cases, highly contaminated samples can be screened out.

Residual salt analysis (RSA)

Another technique that is being used increasingly to obtain information on formation-water composition is residual salt analysis (RSA; Smalley *et al.* 1992; Smalley & England 1992; Gibbons 1991). This method can be used on old, dry, unpreserved core as well as on new preserved core. RSA makes use of the fact that when an unpreserved core is placed in storage, the formation water it contained will gradually evaporate, leaving behind residual salts precipitated in the pore spaces. In the laboratory, these residual salts can be redissolved by leaching the disaggregated rock with ultrapure water. This leachate now contains much of the material that was previously dissolved in the formation water. This leaching procedure is necessarily gentle, as it is important not to dissolve any of the minerals present in the rock. Because it is not always possible to completely redissolve all of the residual salts and it is also difficult to estimate the original amount of water in the sample, chemical analysis of the leachate will usually only give qualitative information concerning the original formation-water chemistry.

However, there is one parameter that can be measured on the leachate that is not affected by residual salt precipitation or incomplete residual salt redissolution: the $^{87}Sr/^{86}Sr$ isotopic ratio. This isotope ratio is an important natural tracer present in all formation waters and sensitive to water–rock interaction processes (Stueber *et al.* 1984; Smalley *et al.* 1989). The $^{87}Sr/^{86}Sr$ composition can be measured precisely (± 0.00001, 1σ) using mass spectrometry. Different minerals in a sandstone may have different $^{87}Sr/^{86}Sr$ compositions, e.g. carbonate (≤ 0.715), plagioclase (< 0.710), K-feldspar (> 0.730) and mica ($\gg 0.800$). Clearly, dissolution of any of these minerals into a formation water with original seawater composition ($^{87}Sr/^{86}Sr = 0.7068$–0.7092, 8 ppm) could have a major impact on the formation water $^{87}Sr/^{86}Sr$, even though its bulk composition may not be affected dramatically. The $^{87}Sr/^{86}Sr$ ratio can thus be used to identify subtle variations in formation-water composition.

Contamination of the core by drilling fluid is the main uncertainty in using RSA to study formation-water variations. The factors controlling the degree of contamination are the same as discussed previously with regard to water extraction from core, and similar precautions should be taken. Only the central parts of cores, away from core fractures, should be analysed.

Unfortunately, with RSA it is not possible to use a deuterium tracer in the mud to quantify the degree of mud infiltration. It is not possible, therefore, to correct RSA analyses for contamination. However, alternative approaches have been devised to identify and screen out contaminated samples. Within a single well, most of the factors affecting mud infiltration will be fairly constant (e.g. mud type and pressure, bit type), but permeability may be variable. In many wells it is thus observed that contamination of the core by drilling fluid is proportional to the permeability of the core (Fig. 1). Contaminated RSA samples may similarly be identified by plotting RSA $^{87}Sr/^{86}Sr$ against permeability (Fig. 2). This shows that, in this dataset, samples with permeabilities below about 300 mD have low $^{87}Sr/^{86}Sr$ values with a relatively narrow range, representative of the real natural variation in the formation water. Above 300 mD, however, samples have higher and more variable $^{87}Sr/^{86}Sr$ values, reaching towards the known $^{87}Sr/^{86}Sr$ value of the drilling fluid (~0.710). From this relation, an empirical cut-off of 300 mD is established, above which RSA analyses are suspect.

An important observation is that, while the relation between bulk water contamination and core permeability is linear (Fig. 1), the relation between RSA $^{87}Sr/^{86}Sr$ and permeability is a convex-upward curve (Fig. 2). This is explained by the mixing relations between formation-water Sr and mud filtrate Sr. Many North Sea fields have formation waters with Sr concentrations of several

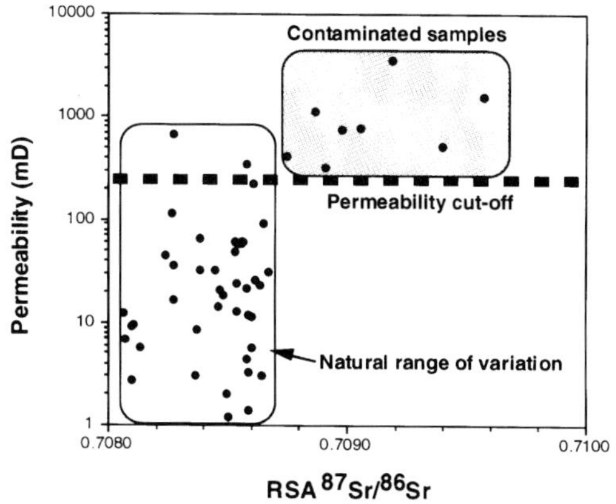

Fig. 2. RSA $^{87}Sr/^{86}Sr$ values of rocks from a North Sea well with varying core permeability. Low permeability samples have reliable $^{87}Sr/^{86}Sr$ values, while high permeability samples have $^{87}Sr/^{86}Sr$ values trending towards that of drilling fluid (cf. Fig. 1).

hundreds to thousands of ppm (Egeberg & Aagaard 1989), while all of the drilling fluid filtrates we have analysed to date have Sr contents of 1 to 10 ppm or less. This results in $^{87}Sr/^{86}Sr$ mixing lines between formation water and mud filtrate that are not linear; a modelled curve of $^{87}Sr/^{86}Sr$ contamination is shown on Fig. 1, and this is a similar shape to the pattern of RSA data in Fig. 2. This shows that even when there is over 80% bulk contamination of the formation water, the $^{87}Sr/^{86}Sr$ ratio of the water is modified by less than 5% (Fig. 1). RSA Sr isotope analyses are thus relatively resistant to drilling fluid contamination.

Another way of assessing the degree of contamination of an individual sample is to perform several RSA $^{87}Sr/^{86}Sr$ measurements at different places between the core centre and edge in the same core piece. The $^{87}Sr/^{86}Sr$ transect obtained is indicative of the degree to which drilling fluids have penetrated the core. In a moderately contaminated, but still usable, sample there is a clear plateau of RSA $^{87}Sr/^{86}Sr$ composition in the core centre (Fig. 3), indicating that contamination has not penetrated to the core centre. More extensive contamination frequently leads to a RSA $^{87}Sr/^{86}Sr$ profile where there is no distinct $^{87}Sr/^{86}Sr$ plateau in the core centre (Fig. 3). In such cases, it is

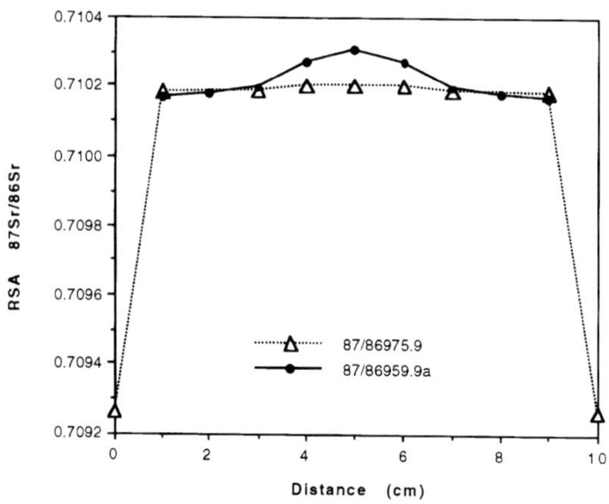

Fig. 3. RSA $^{87}Sr/^{86}Sr$ transects through two core samples from the Gypsy subsurface research site, Oklahoma, USA. A single radial transect was measured in each case (i.e. 0–5 cm); the mirror image (6–10 cm) was added for visual effect. The analysis at 0 cm is mud cake scraped from the edge of the core; it probably reflects drilling mud filtrate values. Though both samples have much higher RSA values than drilling fluid, the sample from 959.9 m depth has no clear plateau $^{87}Sr/^{86}Sr$ value in the core centre and may give unreliable data. The 975.9 m sample has a good plateau value of *c*. 0.7102, which is probably a reliable estimate of formation-water composition.

not possible to determine whether or not the $^{87}Sr/^{86}Sr$ value in the centre of the core has been contaminated.

In summary, it is possible to obtain information on formation-water composition from core. The main problem, drilling fluid contamination, can be minimized by avoiding highly permeable or obviously invaded samples. Quality control methods using mud tracers such as deuterium, cross-plots of composition against permeability, and core transects can all be used to identify and screen out contaminated samples.

Formation water resistivity

The amount of oil in a reservoir structure is, simply put, equal to:

$$\text{trap volume} \times \text{porosity} \times \text{oil saturation}$$

where oil saturation ($S_o = 1 -$ water saturation, S_w) is the proportion of the porosity filled by oil. S_w is thus a key parameter in reserves estimation. S_w is usually determined by exploiting the fact that oil has a higher resistivity than formation water. The resistivity of the formation can be measured by wireline electrical logs and, if the resistivity of the formation water (R_w) is known, S_w can be calculated throughout each logged well. Accurate estimation of S_w is vital; assumption of the wrong R_w value can lead to over- or underestimation of the hydrocarbon reserves. Assumption of a single R_w value in a system where R_w is variable could lead to errors in determination of where the oil is, something which is important for equity determinations (determining how hydrocarbon reserves should be apportioned in a field that spans more than one oil license area or international boundaries).

A common method of R_w determination is to take a sample of water from the aquifer underlying an oilfield, using RFT or DST sampling. There are two problems with this: (1) the aquifer water may not reflect the R_w of the oil-leg formation water; (2) the assumption of a single R_w may not be correct. This can lead to both types of errors described previously. The following sections deal with novel methods for complementing conventional techniques.

Water variations in Machar Field

Machar Field waters extracted from core

The Machar Field (UKCS Block 23/26a) consists of a condensed raft of reservoir rocks (mainly chalk) developed over a Zechstein salt diapir by a combination of salt movement and extensional faulting. In one well (23/26a-13), preserved chalk core samples were available, i.e. core that had been

Table 1. *Analytical results for waters centrifuged from preserved core, well 23/26a-13*

Sample depth (m)	$^{87}Sr/^{86}Sr$*	Na	K	Mg	Ca	Sr	Ba	Cl	SO_4	TDS†	R_w‡ (Ωm)
1889.10	0.708 08	53 600	400	483	1610	125	7	84 600	1050	141 875	0.0628
1934.25	0.708 06	53 800	1290	1090	5850	285	7	96 500	1870	160 629	0.0560
1977.10	0.707 92	69 900	710	791	3830	226	9	118 100	1430	194 996	0.0505
2016.10	0.707 96	79 600	785	663	1470	118	9	128 000	1660	212 305	0.0480
2016.25	0.707 94	65 000	1110	940	5280	218	13	115 500	2280	190 341	0.0517

All concentrations in ppm
* = ± 0.00003
† Total dissolved solids
‡ Measured at 25°C

encased in wax upon retrieval from the well in order to preserve the fluids in the core. For five of these samples it was possible to extract the water from plugs of the core by ultracentrifugation. Additional extractions were unsuccessful owing to problems with core fracturing and low matrix permeabilities. The extracted waters were analysed for ion concentrations using standard ICP-AES and ion chromatography techniques, and for Sr isotopes by standard mass spectrometric methods. The results are presented in Table 1. Core contamination by drilling fluids is negligible in these samples owing to the low matrix permeability of the chalk.

Over the 130 m sampled interval, the concentrations of individual species increase with depth (e.g. Cl from *c*. 85 000 to *c*. 130 000 ppm), total salinity increases from *c*. 140 000 to *c*. 220 000 ppm, giving rise to an apparent salinity (hence also R_w) gradient (Fig. 4). Over the same interval, $^{87}Sr/^{86}Sr$ decreases slightly from 0.7081 to 0.7079. There is a relation between formation water R_w and $^{87}Sr/^{86}Sr$ (Fig. 5). The R_w variation is not large, but

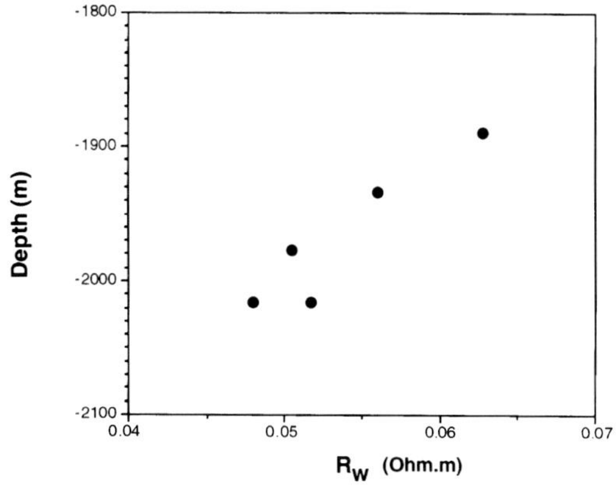

Fig. 4. Water resistivity gradient in well 23/26a-13 identified by waters centrifuged from preserved core.

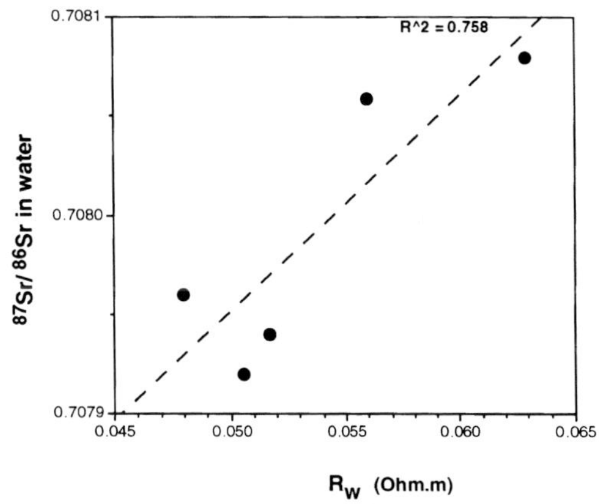

Fig. 5. Relationship between water resistivity and $^{87}Sr/^{86}Sr$ in waters centrifuged from preserved core in well 23/26a-13.

in fact the analysed samples only come from part of the chalk reservoir succession. Does R_w continue to fall with increasing depth?

Machar formation waters reconstructed from residual salts

Residual salts were leached from coarsely crushed chalk samples using a procedure similar to that described in Smalley *et al.* (1991). The leachates were analysed for Sr isotopes to a precision (1σ) of ± 0.00001. This allowed formation-water $^{87}Sr/^{86}Sr$ data to be determined on a much larger sample set (Table 2), as unpreserved core samples were more freely available than suitable preserved samples. The RSA data confirm the depthwise fall in formation water $^{87}Sr/^{86}Sr$ that was identified by the centrifuged water samples. However, the RSA data also indicate that below a depth of *c*. 2050 m,

Table 2. *RSA data for Machar well 23/26a-13, and water resistivities calculated from the relation in Fig. 5*

Depth (m)	$^{87}Sr/^{86}Sr$	R_w
1889.60	0.708 05	0.057
1905.80	0.708 05	0.057
1908.75	0.708 08	0.060
1920.73	0.707 96	0.051
1933.91	0.707 90	0.047
1951.75	0.707 86	0.044
1977.60	0.707 91	0.048
1982.85	0.707 84	0.043
2006.45	0.707 91	0.048
2015.73	0.707 84	0.043
2046.55	0.707 93	0.049
2070.05	0.707 96	0.051
2081.46	0.707 95	0.051
2081.48	0.707 96	0.051
2098.75	0.707 97	0.052

the formation water $^{87}Sr/^{86}Sr$ begins to rise again. Using the apparent relation between R_w and $^{87}Sr/^{86}Sr$ (Fig. 5), it is possible to convert the RSA data to R_w and thus determine how R_w may vary with depth (Fig. 6). The RSA-derived R_w data confirm the decrease in R_w with depth in the upper part of the reservoir (Fig. 6). However, in the lower part of the reservoir R_w does not continue to fall with increasing depth. Rather, below *c.* 1950 m, R_w begins to level out, and at greater depths (>2020 m) R_w begins to rise again (Fig. 5).

It is thought that transport of solutes through water in the oil leg by diffusion is so slow as to be insignificant. The cause of the variation is thus likely to be a pattern 'frozen in' during filling of the

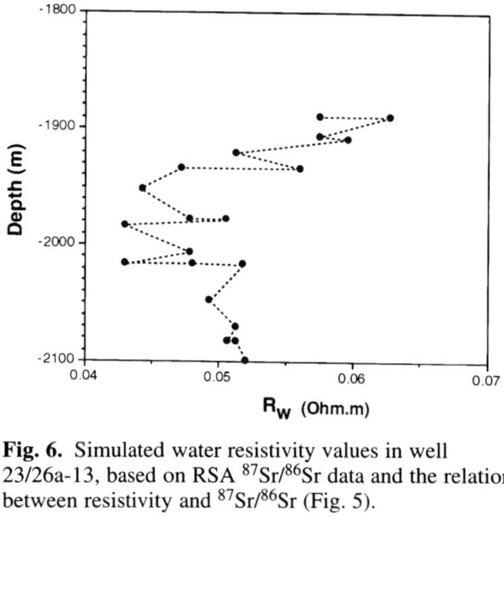

Fig. 6. Simulated water resistivity values in well 23/26a-13, based on RSA $^{87}Sr/^{86}Sr$ data and the relation between resistivity and $^{87}Sr/^{86}Sr$ (Fig. 5).

chalk with oil. The data indicate that salinity in the formation water was varying substantially during oil charging of the field. The fact that similar variations have been observed in other North Sea chalk oilfields (e.g. Ekofisk; Smalley *et al.* 1991) and sandstone reservoirs (Coleman 1992) could indicate that this situation is not unusual. In the case of the Machar Field, variable degrees of interaction with underlying salt deposits is a likely cause.

In this particular case, because of the low R_w values, the S_w calculations are not sensitive to R_w. However, this case study does demonstrate that significant variations in R_w can occur even on a relatively small scale, and this possibility should be taken into account when calculating water saturations from resistivity logs. These data also show that it is practical to map out water chemical variations by a combination of water centrifugation from preserved core, where it is available, and $^{87}Sr/^{86}Sr$ RSA where preserved core is unavailable.

Reservoir compartmentalization

A fundamental question during the development of an oilfield is whether it will behave during production as a single large tank of oil or as a number of smaller separate tanks. The degree to which a reservoir is compartmentalized controls the number and position of wells needed to recover oil in an optimal manner. Because compartmentalization will affect the type of facilities needed to develop the field, as well as the economics of the project, it is crucial to quantify compartmentalization and to identify the features that are acting as barriers to fluid flow as early as possible during field appraisal.

Compartmentalization is notoriously difficult to assess during field appraisal, as it relates to how the reservoir fluids will flow during production, but obviously no production data are available at the field appraisal stage. The problem is that dynamic data are required to assess compartmentalization definitively, while mainly static data are available during appraisal. One approach that is becoming more common is to use variations in reservoir fluid composition to derive quasi-dynamic information. This means using the way in which reservoir fluids have mixed through geological time to extrapolate how fluids would flow on a production time-scale.

A considerable amount of work has been done in some oil companies on using compositional variations in oil inherited from the filling history (England 1990; Smalley & England, 1992). Oil compositional variations are common in oilfields as a relic of the filling history. However, such variations should mix through geological time due to density-driven convection or diffusion. This mixing process can be modelled (England *et all.*

1987; England *et al.* 1995), and the degree to which oils 'should have mixed, yet have not' is an indicator of reservoir compartmentalization.

The same approach can also be used with water compositions. Compositional variations in water do occur at the reservoir and even within-well scale, as demonstrated by the Machar example discussed earlier. These differences can reveal information about the plumbing of the reservoir. In detail, however, the interpretation of water compositional variations depends on whether the samples come from within the oil leg or from the underlying aquifer.

Identifying fluid flow barriers in the aquifer

The identification of laterally extensive barriers to vertical fluid flow can be important, as it could affect exactly where water is injected into the formation for pressure support. Water compositional variations can occur in the water leg by local water–rock interaction (e.g. mineral dissolution) or by in-flow of new water. Figure 7 illustrates a case from a North Sea shallow marine sandstone reservoir where RSA has been used to reveal variations in water composition. The high net-to-gross succession has two prominent shales in the *c.* 20 m sampled interval. In the upper part of the sampled interval, RSA $^{87}Sr/^{86}Sr$ values are constant, indicating that the formation water has a homogeneous composition (Fig. 7). This is despite the presence of a shale, which has no effect on water composition. There is, however, a significant jump in RSA $^{87}Sr/^{86}Sr$ across the lowest shale (Fig. 7). This shale separates waters of different compositions and must thus represent a barrier to fluid movement.

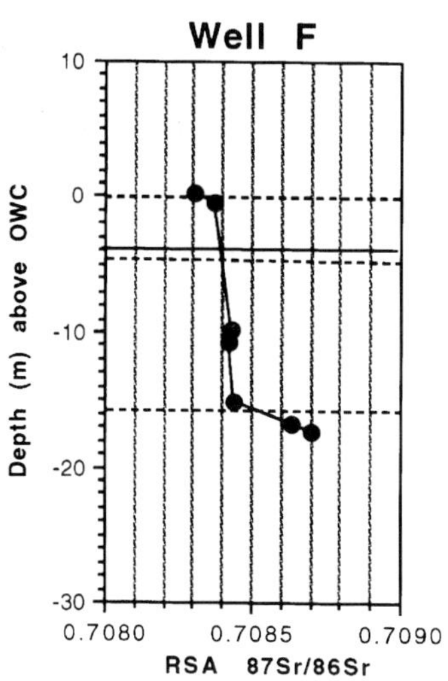

Fig. 7. Identification of a shale barrier to fluid movement in a North Sea reservoir aquifer using RSA $^{87}Sr/^{86}Sr$ data. Dashed lines are shales.

While the existence of a stepped RSA pattern (Fig. 7) indicates qualitatively that the shale is a laterally extensive fluid flow barrier, more quantitative constraints can be gained by considering the rate at which water would mix around such a barrier. It is unlikely (though not impossible) that the differences in water composition would be large enough to cause significant density differences. Density-driven convection is thus unlikely to be an important mixing process; diffusion would probably be more dominant. Diffusion will operate at all times in formation waters; if any fluid flow occurs in addition, this could only increase mixing rates. Diffusion thus represents the slowest mixing scenario.

Although shales, due to their low permeability, can form effective fluid flow barriers, it is not immediately obvious that they would be barriers to solute diffusion. Diffusional mixing times for Sr can be estimated using the relation:

$$T_{eq} = 0.1(L^2)/D_{eff} \tag{1}$$

where L is the diffusion distance and D_{eff} is the effective diffusion coefficient of Sr in the formation water. The effective diffusion coefficient is:

$$D_{eff} = (D_t\theta)/R \tag{2}$$

In this expression, D_t is the tracer diffusion coefficient for the solute in question, θ is the tortuosity of the porous medium and R is a retardation factor:

$$R = 1 + ((100\text{-}n)/n)\rho K_D \tag{3}$$

where n is the effective porosity (%), ρ is the density of the solid matrix and K_D is the distribution coefficient that describes the sorption of the solute onto the solid surfaces. It is the low porosity, high tortuosity and, most importantly, the high sorption characteristics of shales compared to sandstones that make them barriers which inhibit the diffusion of solutes. Our preferred D_{eff} for Sr isotopes in sandstones (at 100°C) is 8.3×10^{-11} m^2 s^{-1}, over *c.* 1250 times greater than that for shales (6.6×10^{-14} m^2 s^{-1}).

A quantitative 1-D computer code was used to model the diffusion of Sr through and around shales (performed by W.E. Falck at the British Geological Survey), aimed at estimating homogenization times for mixing of $^{87}Sr/^{86}Sr$ ratios between two sandstones with an initial $^{87}Sr/^{86}Sr$ difference of 0.01. Based on numerous modelling runs with different sand–shale scenarios, two diagrams were compiled to estimate quickly the times for $^{87}Sr/^{86}Sr$ homogenization through (Fig. 8) and around (Fig. 9)

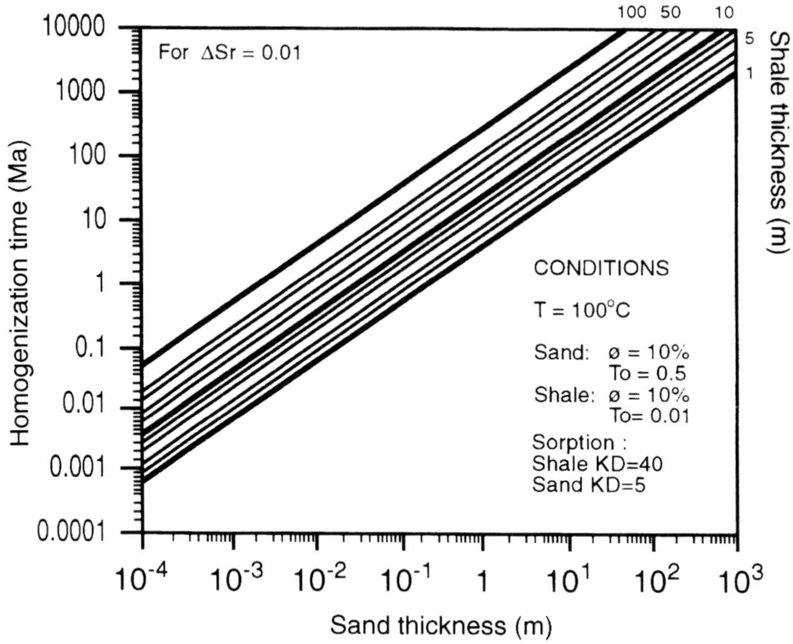

Fig. 8. Diagram for calculating time taken for formation waters in two sandstones to mix through an intervening shale. A line is drawn vertically from the mean sandstone thickness on the *x*-axis to the line for the appropriate shale thickness. A horizontal from here to the *y*-axis gives the mixing time (for Sr isotopes with an initial $^{87}Sr/^{86}Sr$ difference of 0.01).

shales of different thickness and lateral extent. These show that dissolved Sr in two 50 m thick sands would homogenize in *c*. 3 Ma in the absence of an intervening barrier (Fig. 8), while the presence of a 5 m thick shale between the sands would increase homogenization time to *c*. 50 Ma (Fig. 9).

The Sr mixing rates (Fig. 9) may be used to help interpret the dataset shown in Fig. 7. Figure 9 shows that, at our preferred sand D_{eff}, the $^{87}Sr/^{86}Sr$ of dissolved Sr on either side of the barrier shale would have homogenized in *c*. 1000 years in the absence of a barrier. The size of the barrier can be assessed only if something is known about the length of time that has elapsed since the step in $^{87}Sr/^{86}Sr$ arose. Some constraints are obvious, for example, the origin of the $^{87}Sr/^{86}Sr$ step must be younger than the age of the reservoir. In the typical Mesozoic reservoir of the North Sea this is not an especially useful constraint, though age of the reservoir may be a tight constraint in young reservoir systems (e.g. Pliocene, Gulf of Mexico).

Another constraint on the time elapsed since the heterogeneity arose can be gained by examination of the part of the well that has a uniform $^{87}Sr/^{86}Sr$ composition. About 15 m of core has a homogeneous $^{87}Sr/^{86}Sr$ ratio (Fig. 7). Assuming that this portion of formation water once contained $^{87}Sr/^{86}Sr$ variations that have now homogenized through diffusion, Fig. 9 can be used to determine the minimum time available for water mixing *c*. 300 000 years. An upper limit of time available for mixing is estimated from Fig. 8, by calculating how long it would take for the waters above and below the shale to homogenize by diffusion through the shale. For a sand thickness of *c*. 15 m and a shale thickness of 1 m, this yields *c*. 50 Ma. Plugging this time range of 300 000 to 50 Ma into Fig. 9 gives the range of possible minimum distances from the well to the edge of the shale: 15 to 200 m (minimum shale diameter 30 to 400 m).

In practice, a useful technique is to carry out similar work to that shown in Fig. 7 for several wells in the field under investigation. Extensive barriers, such as fieldwide shales, will then show up with a similar RSA signature in each well. The shale barrier shown in Fig. 7, which represents a major regional flooding surface, was identified in another well more than 2 km away.

Identifying fluid flow barriers in the oil leg

During oil charging of a reservoir, water saturation decreases until it becomes a discontinuous phase.

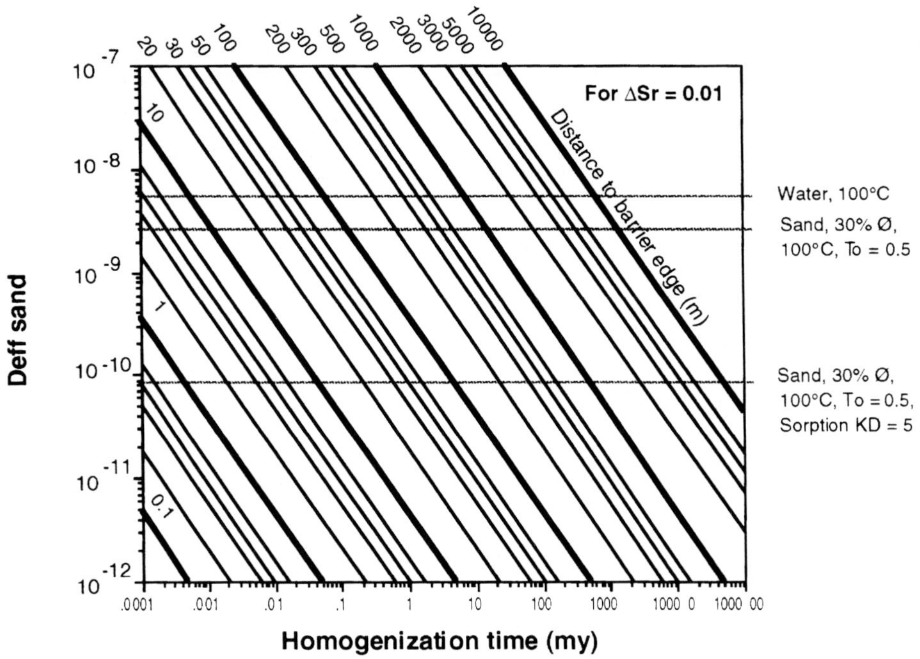

Fig. 9. Relations between diffusional mixing time (for $^{87}Sr/^{86}Sr$ in water) and distance to barrier edge for different effective diffusion coefficients.

At this stage, diffusion pathways through the remaining water films are so tortuous as to effectively bring a halt to diffusive mixing of solutes, 'fossilizing' the water that is trapped. Consequently, variations in formation-water composition in the oil leg cannot be interpreted using the diffusion modelling approach described previously.

Where variations in formation-water composition do exist in the oil leg (Fig. 4), these are most likely due to variations in formation-water composition with time during progressive filling of that segment of the reservoir with oil. Formation-water evolution trends with depth in the oil leg thus record the movement of oil–water contacts and may be used to reconstruct the filling history of the reservoir. This in turn enables the identification of barriers that are of sufficient size to have influenced the filling history. On its way to the crest of the reservoir, oil may pond underneath, or be trapped above, such intra-reservoir seals, temporarily leading to different oil–water contacts (OWCs) above and below the seal. The positions of such OWCs can be identified using water RSA analyses, providing a means of predicting fluid flow barriers.

The example shown in Fig. 10 is an RSA study of a well in a North Sea reservoir sandstone. The data are all from the oil leg. In the lower part of the well, RSA $^{87}Sr/^{86}Sr$ compositions are extremely uniform, indicating that the formation water has a homogeneous composition. This is despite the fact that there are two shales in this interval (Fig. 10). At *c.* 180 m above OWC, there is a sharp step in $^{87}Sr/^{86}Sr$ across a shale. Above this, there is once

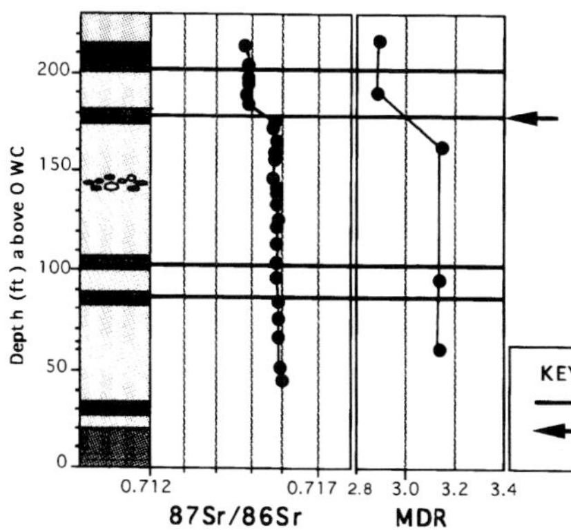

Fig. 10. RSA $^{87}Sr/^{86}Sr$ and core extract MDR (= 4-methyldibenzothiophene/1-methyldibenzothiophene) oil maturity parameter data in a North Sea oil well.A barrier is evident at 18 m above OWC.

again a constant water composition. The data thus indicate that the shale at 180 m above OWC is a barrier separating waters of two distinctly different compositions. In simple situations such as this, the approach of using cluster analysis of RSA $^{87}Sr/^{86}Sr$ data to identify flow units and the barriers separating them may give accurate results (Stølum & Smalley 1992).

Although the RSA pattern in Fig. 10 looks similar to that in Fig. 7, the causes must be different, as in Fig. 10 the waters are 'fossilized' within oil. Either there were two separate water compositions above and below the shale at the time when oil entered the reservoir and 'froze' the water compositions, or the two reservoir zones above and below the shale were filled with oil at slightly different times and a component of the water variation is due to the change in water composition with time. In either model, the shale must be a laterally extensive barrier to vertical fluid mixing. In the first model, it must form a barrier to the mixing of two water compositions in the same way as the shale barrier in Fig. 7. In the second model, the shale must be sufficiently laterally extensive to act as an intra-reservoir seal, influencing the oil-filling history of the reservoir by controlling the position of a temporary oil–water contact.

In the present case, the second model is favoured by the composition of the oil itself. Solvent extracts of core samples from the same well were analysed by gas chromatography–mass spectrometry for various organic parameters, including oil maturity indicators. Figure 10 shows how the MDR maturity parameter (4-methyldibenzothiophene/1-methyldibenzothiophene ratio) varies down the well. This reveals a distinct step change in oil composition at the same place at which the water composition changes. This supports the idea that the shale acted as a barrier during reservoir filling and that it has subsequently acted as a barrier to oil mixing. Integration with other data for this particular oilfield indicates that the upper part of the reservoir filled with oil first, and that the upper reservoir zone continued to fill downflank of the sampled well for some time before oil was finally emplaced beneath the barrier shale into the lower part of the reservoir. Between-well correlation indicates this shale to be continuous over *c.* 5 km.

Figure 11 illustrates another type of Sr RSA pattern that is more typical of lower net-to-gross with numerous barriers to vertical fluid movement, and more downflank reservoir well positions. This example comes from a sandstone reservoir of fluvial origin, containing floodplain shales, several of which are intersected by the sampled well (Fig. 11). These shales were deposited in a meandering river system and are probably several hundred metres in lateral extent.

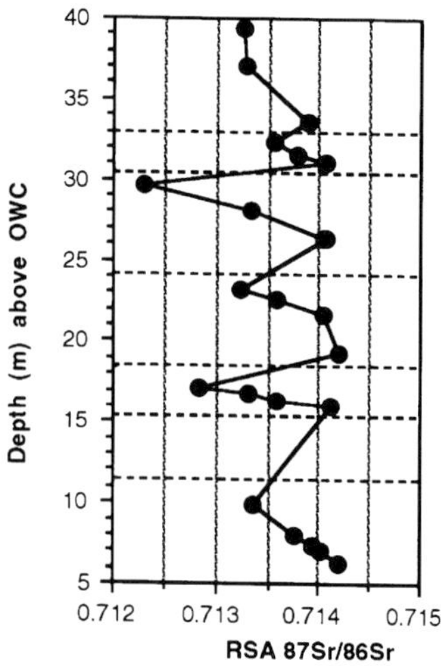

Fig. 11. RSA $^{87}Sr/^{86}Sr$ data for the North Sea showing a high degree of vertical compartmentalization by laterally extensive shales. Dashed lines denote the positions of shales.

The overall pattern of Fig. 11 is one of much more variation than in Figs 7 and 10. This in itself indicates a greater degree of vertical compartmentalization. In detail, RSA $^{87}Sr/^{86}Sr$ rises with depth beneath each shale until the next shale is reached and then there is an abrupt step back to lower $^{87}Sr/^{86}Sr$ values (Fig. 11). This 'saw-tooth' pattern is best interpreted as the result of gradually increasing formation water $^{87}Sr/^{86}Sr$ during filling of each sand layer from top to bottom with oil. The sand compartment between each shale has filled separately, each with its own temporary oil–water contact during oil charging. This indicates that the shales are sufficiently laterally extensive to have acted as an intra-reservoir seal, sustaining its own oil–water contact during filling.

Conclusions

Two new additions to the battery of water sampling techniques (water centrifugation from core and residual salt analysis of core) allow waters to be sampled from cored oil and gas reservoirs in great detail. These have allowed the detection of water compositional variations at the metre scale.

The identification and quantification of water compositional variations can be put to use in two main ways.

(1) Detailed mapping of salinity variations, illustrated here with an example from the Machar Field (central North Sea), provides a detailed understanding of water resistivity variations. This is important for the interpretation of electrical

resistivity logs to determine the oil saturation of the reservoir and thus the total amount of oil in place.

(2) Differences in formation-water composition on either side of potential fluid flow barriers in a reservoir identify those features having the greatest barrier effects. This is so both in the oil leg and water leg. Quantification of barrier importance has been attempted by diffusion modelling in the water leg and by interpretation of detailed filling history (from the 'fossilized' formation water variations) in the oil leg. This is important for modelling how fluids are likely to move through the reservoir during petroleum production.

We thank BP Exploration for permission to publish this work. Much of the Sr diffusion modelling was carried out by E. Falck in the Fluid Processes Group of the British Geological Survey.

References

COLEMAN, M. L. 1992. Water composition variation within one formation. *In*: KHARAKA, Y. K. & MAEST, A. S. (eds) *Water–Rock Interaction*. Balkema, Rotterdam, 1109–1112.

——, JONES, M. R. O. & COX, M. A. 1990. Analysis of formation water samples from core. *In*: BULLER, A. T., BERGE, E., HJELMELAND, O., KLEPPE, J., TORSAETER, O. & AASEN, J. O. (eds) *North Sea Oil and Gas Reservoirs-II*. Graham & Trotman, London, 165–171.

EGEBERG, P. K. & AAGAARD, P. 1989. Origin and evolution of formation waters from oil fields on the Norwegian Shelf. *Applied Geochemistry*, **4**, 131–142.

ENGLAND, W. A. 1990. The organic geochemistry of reservoirs. *Organic Geochemistry*, **16**, 415–425.

——, MACKENZIE, A. S., MANN, D. M. & QUIGLEY, T. M. 1987. The movement and entrapment of petroleum fluids in the subsurface. *Journal of the Geological Society, London*, **144**, 327–347.

——, MUGGERIDGE, A. H., CLIFFORD, P. J. & TANG, Z. 1995. Modelling density-driven mixing rates in petroleum reservoirs on geological time-scales, with application to the detection of barriers in the Forties Field (UKCS). *This volume*.

GIBBONS, K. A. 1991. Use of variations in strontium isotope ratios for mapping barriers: An example from the Troll Field, Norwegian Continental Shelf. *In*: *Proceedings of the Sixth European Symposium on Improved Oil Recovery*. Extended Abstracts, Norwegian Petroleum Directorate, **1**, 205–211.

SMALLEY, P. C. & ENGLAND, W. A. 1992. *Assessing reservoir compartmentalization during field appraisal: How geochemistry can help*. Society of Petroleum Engineers, Paper No. **25005**.

—— & OXTOBY, N. H. 1992. Spatial and temporal variations in formation water composition during diagenesis and petroleum charging of a chalk oilfield. *In*: KHARAKA, Y. K. & MAEST, A. S. (eds) *Water–Rock Interaction*. Balkema, Rotterdam, 1201–1204.

——, LØNØY, A. & RÅHEIM A. 1992. Spatial $^{87}Sr/^{86}Sr$ variations in formation water and calcite from the Ekofisk chalk oilfield: Implications for reservoir connectivity and fluid composition. *Applied Geochemistry*, **7**, 341–350.

——, RÅHEIM A., DICKSON, J. A. D. & EMERY, D. 1989. $^{87}Sr/^{86}Sr$ in waters from the Lincolnshire Limestone aquifer, England, and the potential of natural strontium isotopes as a tracer for a secondary recovery seawater injection process in oilfields. *Applied Geochemistry*, **3**, 591–600.

STØLUM, H. H. & SMALLEY, P. C. 1992. *A deterministic method for assessing reservoir communication based on strontium fingerprinting*. Society of Petroleum Engineers, Paper No. **25007**.

STUEBER, A. M., PUSHKAR, P. & HETHERINGTON, E. A. 1984. A strontium isotopic study of Smackover brines and associated solids. *Geochimica et Cosmochimica Acta*, **48**, 1637–1649.

Characterization of high molecular weight hydrocarbons ($>C_{40}$) in oils and reservoir rocks

R. P. PHILP[1], A. N. BISHOP[1], J.-C. DEL RIO[2] & J. ALLEN[1]

[1]*School of Geology and Geophysics, University of Oklahoma, Norman, OK 73019, USA*

[2]*Instituto de Recursors Naturales y Agrobiologia, C.S.I.C., Apdo. 1052, 41080 Seville, Spain*

Abstract: The presence of high molecular weight hydrocarbons in crude oils has been known for some time, particularly in view of their ability to precipitate and form barriers and blockages in pipelines. Recent advances in analytical technology have led to the development of capillary columns for gas chromatography capable of withstanding high temperatures (350 to 450°C), which allow the characterization of hydrocarbons with carbon numbers as great as C_{100} and above. Consequently, the individual high molecular weight hydrocarbon compounds in oils and waxes can now be identified. In the present paper, results are presented from high temperature gas chromatographic analyses of naturally occurring waxes. These clearly show the presence of significant concentrations of hydrocarbons with carbon numbers in excess of C_{40} in many of these samples. In order to assist in the quantification of these high molecular weight hydrocarbons, we have also assessed the potential of thermal microdistillation coupled with a field ionization detector, to obtain more accurate molecular weight distributions in wax samples. This is based on the concept that high temperature gas chromatography has an upper carbon number limit in the range C_{100} to C_{120}. Thermal distillation may provide the ability to exceed that limit and determine the presence of compounds above C_{120}. More detailed characterization of the wax fractions on a molecular level has the potential to provide clues as to the composition and origin of the residual components left in a petroleum reservoir at the end of its production history. A better understanding of the nature of this material will lead to improved secondary and tertiary recovery mechanisms, and other enhanced oil recovery procedures.

Petroleum geochemistry has developed rapidly over the past two or three decades (Peters & Moldowan 1993) and until recently most of the emphasis was directed towards exploration problems. However, the pace of this development has now inevitably slowed, resulting in a tendency to apply geochemistry to areas that are concerned more with the exploitation and production of fossil fuels, which has led to the development of reservoir geochemistry. Typically, two-thirds of the oil in place in a reservoir is not produced, even following secondary and tertiary recovery techniques. A number of processes are responsible for poor production, including reservoir heterogeneity, the formation of petroleum-derived barriers, either during accumulation or production (e.g. tar mats, asphaltenes, waxes), interaction of petroleum with surrounding minerals, low viscosity of waxy oils and petroleum biodegradation. An understanding of the processes by which oil entered the reservoir, the nature of inorganic/organic interactions and how petroleum compositions vary with respect to time, should lead to an improvement in methods available for the recovery of these residual oils.

A reservoir, in the context of petroleum, is simply defined as a rock containing petroleum (North 1985). Consequently, reservoir geochemistry can be considered as the study of the geochemical processes that take place in a porous sediment, following the commencement of petroleum accumulation. In a number of petroleum-producing regions, the precipitation of solids from the oil, either during accumulation or production, is a major problem. The composition of the precipitated solids ranges from high molecular weight hydrocarbons to bitumens and asphaltenes (Carnahan 1989). Use of the term 'reservoir bitumens' to describe petroleum-derived solids should not be encouraged, as this will simply lead to confusion with the bitumen of organic extracts and native bitumens such as the gilsonite veins of the Uinta Basin.

Precipitation may occur either in the reservoir or in the production string, as far up as the storage tanks. As a result, barriers may form within the reservoir and/or blockages develop in the pipelines themselves. By infilling the pores, the solids may restrict the pore throats and change the wettability characteristics of the mineral surfaces, which in turn will influence reservoir-quality predictions in play assessments and basin evaluations. Consequently, petroleum-derived solids are as significant as carbonate cements, silicate cements or authigenic clays, when it comes to the characterization and evaluation of petroleum reservoir systems (Lomando 1992). Finally the presence of

From CUBITT, J. M. & ENGLAND, W. A. (eds), 1995, *The Geochemistry of Reservoirs*, Geological Society Special Publication No. 86, pp. 71–85.

these materials can also cause reservoir damage through the migration of fines leading to reduction of porosity.

Waxes consisting primarily of high molecular weight hydrocarbons (i.e. > C_{40}), often referred to as paraffins or paraffin waxes, are a major problem in some fields. The factors controlling the deposition of such waxes have been reviewed by Carnahan (1989). Wax deposition is related to changes in the supercritical character of petroleum fluids during accumulation and production. The temperature in deeper reservoirs can exceed the critical temperature of the low molecular weight petroleum constituents, for example, methane or ethane. Hence these compounds may act as supercritical solvents for the high molecular weight hydrocarbons. Petroleum production inevitably results in a loss of reservoir pressure, which results in a reduction in the carrying capacity of the supercritical solvent system. This leads to the precipitation and deposition of the high molecular weight hydrocarbons. Deposition in the production facilities is a result of the drop in temperature of the petroleum after it has left the reservoir (Carnahan 1989).

A number of approaches have been utilized to remediate the effects of wax deposition (Tuttle 1983; Ashford *et al.* 1990), including hot oiling, wireline cutting and chemical treatments to prevent formation of wax deposits. In higher latitudes and beneath the sea, insulation may be necessary to prevent deposits forming in the well and transfer pipelines (Ashford *et al.* 1990). However, the application of such techniques clearly results in additional production costs, the actual amounts being linked to the wax content of the petroleum (Tuttle 1983).

Analytical developments

Despite the analytical advances made in the past few decades, the study of higher molecular weight (> C_{40}) hydrocarbons has been largely overlooked for two main reasons. The first has been the lack of appropriate techniques, because traditional gas chromatographic methods are limited to species with sufficient volatility. The second, particularly in the case of oils, is that high molecular weight hydrocarbons are often not present, or occur in low concentrations, in the oils collected at the well head. Rather, these compounds may precipitate due to their low solubility and mobility at surface temperatures and pressures.

With the recent development of columns suitable for high temperature gas chromatography (HTGC) (Lipsky & Duffy 1986*a,b*), and the introduction of supercritical fluid chromatography (SFC) (Hawthorne & Miller 1987, 1989; Smith *et al.* 1987), it has been possible to extend the carbon number range of the compounds that can be analysed and identified. Gas chromatographic capillary columns have been produced (made from fused silica, aluminium and stainless steel) containing thin films of thermostable, bonded and crosslinked stationary phases which can be maintained isothermally at 400 to 425°C and temperature programmed from 425 to 440°C. Lipsky & Duffy (1986*a,b*) first described the use of fused-silica columns for HTGC, showing that they were capable of analysing alkanes with carbon numbers up to C_{100}.

Capillary SFC can also be used for the separation of compounds that lack sufficient volatility to be separated by GC, and the use of SFC to separate high molecular weight compounds has been demonstrated (Hawthorne & Miller 1987, 1989). SFC is now recognized as an important method for analytical separation, because the physico-chemical properties of supercritical fluids impart significant chromatographic advantages for the analysis of thermally labile and higher molecular weight compounds. In addition, the use of small-diameter capillary columns (25–100 μm i.d.) for SFC, which have negligible pressure drops, permits exploitation of the full potential of pressure programming methods to obtain high-resolution separation, approaching that achievable with capillary gas chromatography.

The coupling of SFC or HTGC with mass spectrometry (MS) offers the potential to provide high-resolution separation of high molecular weight compounds with selective detection (Blum *et al.* 1990; Olesik 1991), and are thus the appropriate techniques required to deal with the characterization of high molecular weight biomarkers. HTGC–field ionization mass spectrometry (FIMS) has also been successfully applied to the analysis of porphyrins and complex mixtures of high molecular weight saturated and aromatic hydrocarbons (Gallegos *et al.* 1991; Carlson *et al.* 1993). The FIMS method permits the identification of molecular weights (and thus ring number) of high molecular weight hydrocarbons which exhibit no molecular ions in mass spectrometric studies employing other ionization methods. Hawthorne & Miller (1987) used SFC–MS to analyse several commercial waxes, including a microcrystalline wax, consisting of alkanes ranging up to C_{75}, and a synthetic Fischer–Tropsch wax, with hydrocarbons up to C_{100}. The potential of SFC–MS will continue to grow as SFC techniques are extended by the introduction of new mobile and stationary phases and as improved mass spectrometric detectors for higher molecular weight compounds (> 1000) become available.

Practical examples of high molecular weight hydrocarbon determinations

Experimental

A bitumen sample, ozocerite, from the Uinta Basin, Utah, a microcrystalline wax (Micro 195) and a synthetic wax obtained via a Fischer–Tropsch synthesis (FT H-1) were used as standards for this study along with various wax deposits and waxy oils collected from the Anadarko Basin, Oklahoma and the Uinta Basin, Utah. Where necessary, the saturate fractions of the oils were isolated by alumina column chromatography and elution with cyclohexane. Concentration of the high molecular weight hydrocarbon (HMWHC) fraction was also achieved by precipitation prior to asphaltene isolation by addition of hot tetrahydrofuran (THF) followed by cooling overnight at 0°C. The HMWHC precipitate could then be removed by filtration. Asphaltenes were isolated from the oils by precipitation with excess *n*-pentane and subsequent centrifugation followed by reprecipitation with *n*-pentane to purify the asphaltenes.

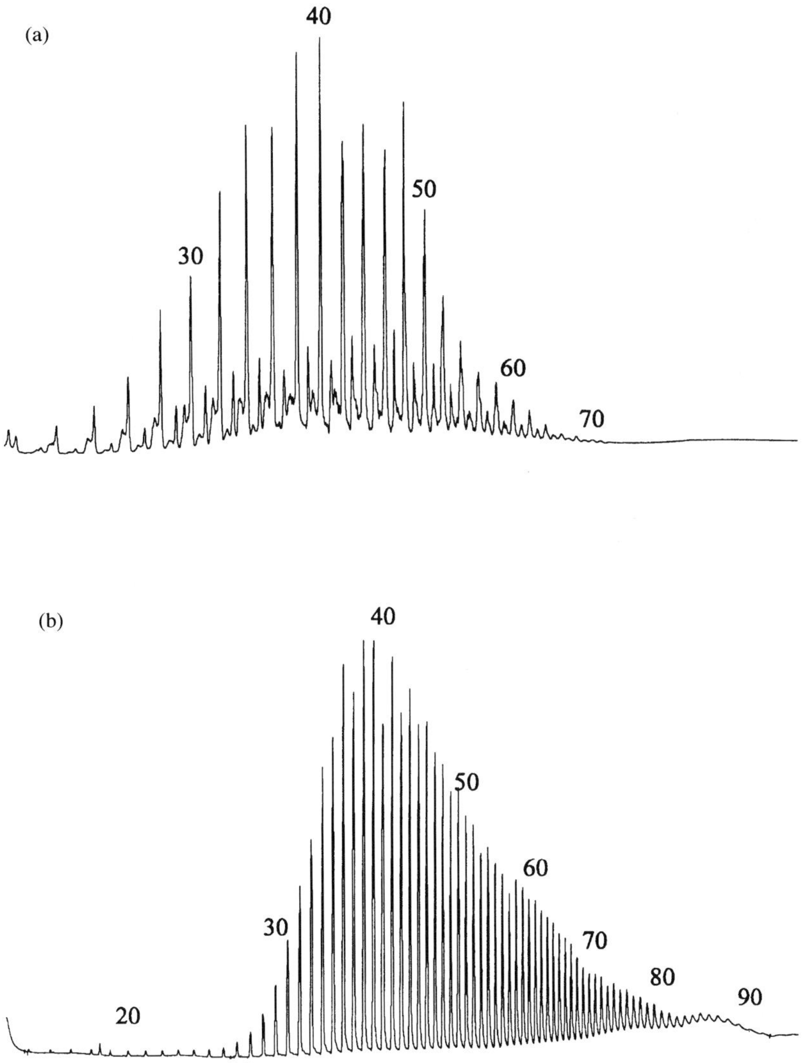

Fig. 1. Gas chromatograms obtained from analysis of two standard wax samples, **(a)** Microwax 195 and **(b)** Fischer–Tropsch wax sample, by high temperature gas chromatography.

High temperature gas chromatography analyses were performed using a Carlo Erba gas chromatograph equipped with a short 6 m aluminium-clad capillary column coated with HT-5 liquid film. The conditions were on-column injector and flame ionization detector at 400°C; column heated from 60 to 440°C at a rate of 8°C min^{-1} with 50 min final hold time. The samples were dissolved in warm *p*-xylene before injection.

Pyrolysis gas chromatography analyses of the asphaltenes were performed with a CDS Pyroprobe. The pyrolysis interface was held at 350°C and interfaced to a Varian 3300 GC. The gas chromatograph was equipped with an aluminium-coated column (6 m or 25 m) and temperature programmed from 0 to 400°C at 8°C min^{-1}.

Thermal distillation experiments of standards, wax samples and HMWHC concentrates were undertaken using a Pyran Level I pyrolysis system. The pyrocell was temperature programmed from 100 to 600°C at a rate of 10°C min^{-1} with 50 min hold time.

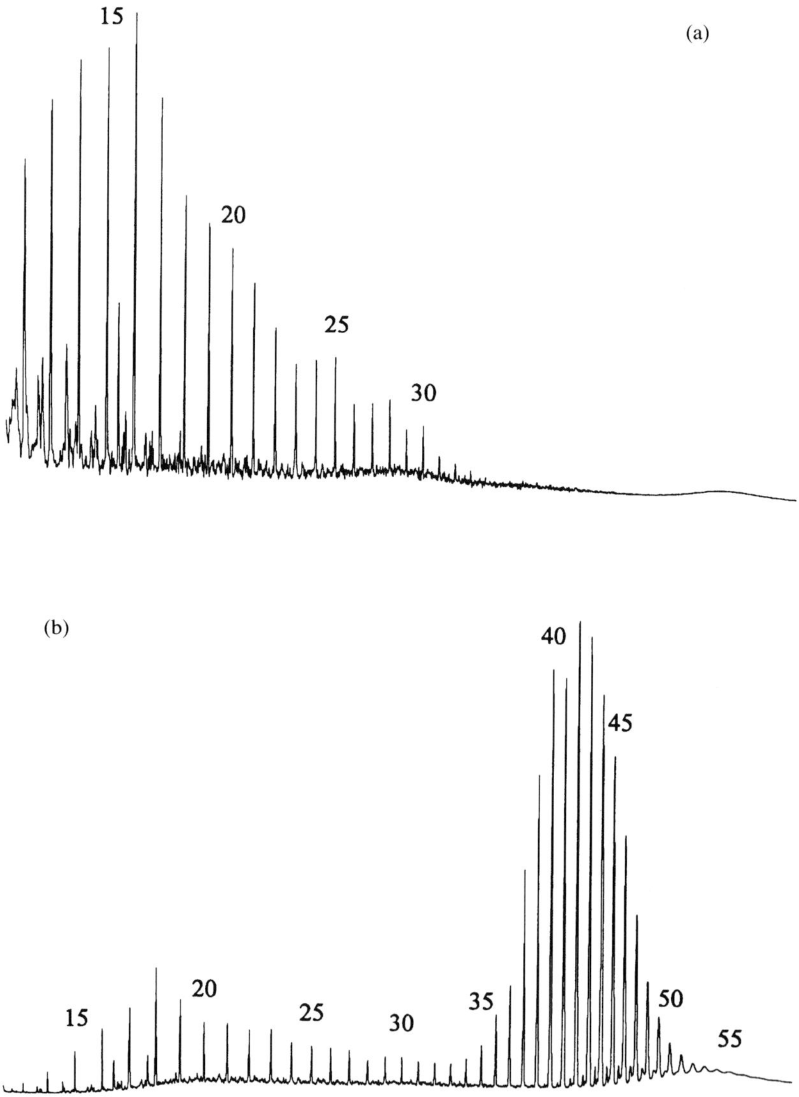

Fig. 2. Chromatograms showing a comparison between the *n*-alkane distribution in **(a)** an oil and **(b)** associated wax collected from the same well.

Characterization of natural and synthetic waxes

Chromatograms of the commercial waxes obtained from the analyses by HTGC using short aluminium-coated capillary columns produced the chromatograms shown in Fig. 1. The Micro 195 wax showed a distribution of *n*-alkanes up to C_{70}, with a maximum around C_{40} and a marked predominance of even over odd carbon number (Fig. 1a). The FT H-1 wax (Fig. 1b) showed an *n*-alkane distribution from C_{20} to C_{75} without any pronounced even/odd predominance as expected from a synthetic product of Fischer–Tropsch synthesis. Similar distributions of hydrocarbons were obtained by SFC, although the FT H-1 showed hydrocarbons extending up to C_{100} (Hawthorne & Miller 1987).

The saturate fractions of a number of oils and their associated waxes isolated from the oil-well drill-stem pipes have also been analysed by HTGC (Del Rio *et al.* 1992; Del Rio & Philp 1992*a*). A relatively high concentration of high molcular weight hydrocarbons (up to at least C_{60}) was observed in waxes but not in oils collected at the well-head (Fig. 2). The waxes dominated by *n*-alkanes also contain complex mixtures of branched/cyclic hydrocarbons, mainly alkylcyclohexanes and mid-chain methyl alkanes, probably formed by condensation reactions of fatty acids

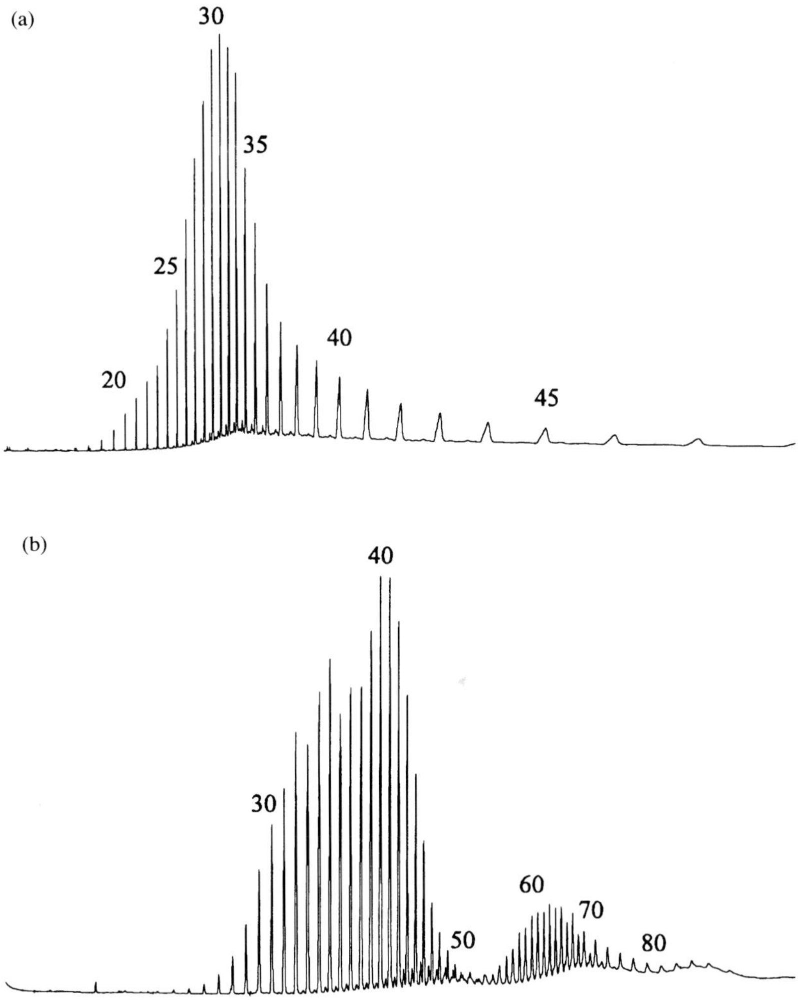

Fig. 3. Comparison between the hydrocarbons of ozocerite analysed on **(a)** a conventional GC column and **(b)** a higher temperature GC column.

(Del Rio & Philp 1992*b*). Similar results have also been reported by other authors (Carlson *et al.* 1993) who identified series of co-eluting monocyclic and acyclic saturated hydrocarbons extending to the C_{60} molecular weight range in a variety of crude oils. These cyclic compounds were observed to show an odd carbon number preference for marine-sourced oils and an even number preference for lacustrine-sourced oils. The oils produced a fairly typical *n*-alkane distribution ranging up to C_{35}. The presence of high concentrations of these hydrocarbons in wax deposits collected from drill-stem pipes was not altogether surprising, but the results demonstrate that high molecular weight hydrocarbons are absent, or present in low concentrations, from oils collected at the well-head. These components precipitate out in the drill-stem pipes due to changes in temperatures and pressures or remain in the reservoir rocks due to their low mobility.

Ozocerite, a fossil bitumen known from previous studies to be waxy, when analysed by conventional GC showed an *n*-alkane distribution maximizing around C_{35}–C_{40} (Fig. 3a). Analysis of the total saturate fraction by HTGC showed a distribution of *n*-alkanes ranging up to C_{75}, with a maximum at C_{32} and another at around C_{42} and no even/odd predominance (Fig. 3b). The saturate fraction was dominated by *n*-alkanes and relatively small amounts of branched/cyclic hydrocarbons, which after molecular sieving were found to be concentrated in the C_{50}–C_{75} *n*-alkane region. Electron impact mass spectra for both the total saturate and branched/cyclic fractions from the ozocerite, obtained using the direct insertion probe, showed the presence of molecular ions of hydrocarbons ranging up to C_{80} in the spectrum of the total saturate hydrocarbons (Del Rio *et al.* 1992). Note that in the higher molecular weight region of the chromatogram (i.e. $> C_{65}$) the alkanes at every second carbon atom are absent, which may be a product of the biosynthetic pathway responsible in their formation.

In one of the few reported applications of both HTGC and SFC for the analysis of HMWHC, Thomson & Rynaski (1992) reported a use for the simulated distillation of wax samples. The object of their study was to compare the degree of variation

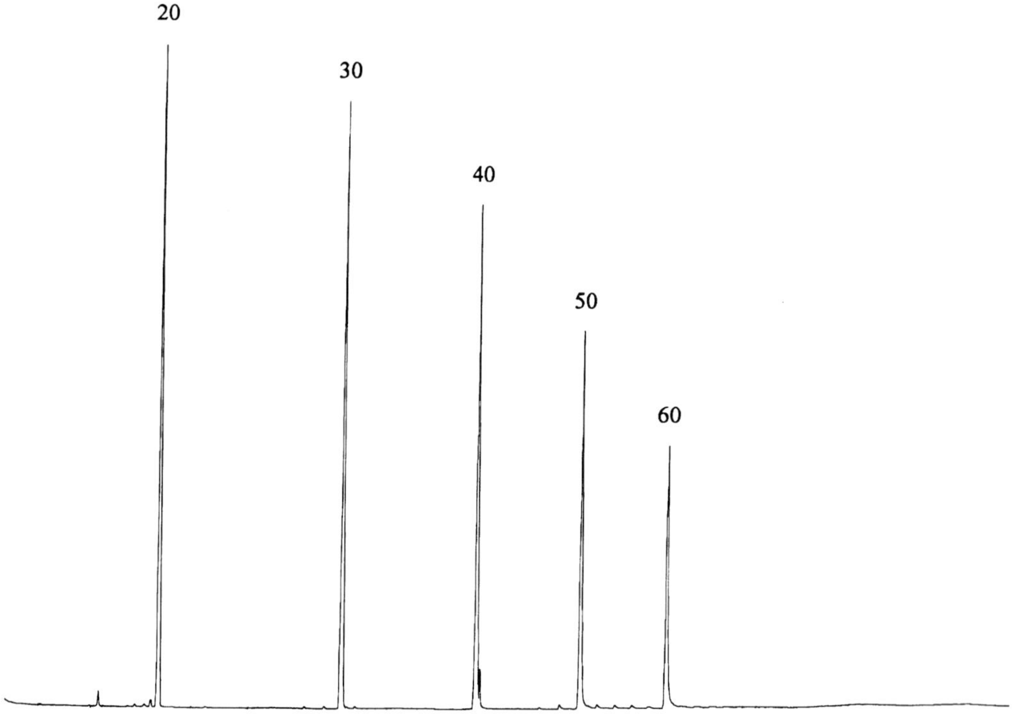

Fig. 4. Chromatograms obtained from analysis of a mixture containing equal concentrations of five *n*-alkanes. Note the decrease in response with increasing carbon number and decrease in retention-time differences with increasing carbon number.

in simulated distillation boiling range distributions from HTGC and SFC. The wax samples used in the study were obtained from the distillation and dewaxing of sludge from underground oil storage caverns. Previous studies have suggested that there are problems with both the HTGC and SFC approach for this type of analysis. Problems include thermal cracking at temperatures above 370°C as observed during thermogravimetric analysis, although in our own HTGC studies we have not observed this to be a problem. Sample introduction in SFC, particularly of these high molecular weight samples, has also caused problems in terms of compound discrimination, although this has now been rectified through the use of heated sample introduction units. Analyses of standard compounds and comparisons between the two techniques clearly showed that it was possible to get compounds up to C_{100} through both types of analytical systems. Results from the two techniques showed little difference and the authors concluded that, for the hydrocarbons at least, the preferred method will depend as much on instrument availability as anything else.

A key problem in the HTGC of HMWHCs is the bias of response towards the shorter-chain homologues. This is demonstrated in Fig. 4 where a number of *n*-alkanes of equal concentration are shown analysed by HTGC. Clearly, the apparent concentration decreases with increasing carbon number, such that the response of C_{60} is approximately one-third that of C_{30}. A number of factors are responsible for this discrimination, including the poor solubility of the HMWHCs in the solvent and the difficulty in eluting such high boiling compounds even at 440°C. The relationship between *n*-alkane carbon number, boiling point, and melting point is shown in Fig. 5. The boiling point of the *n*-alkanes begins to level off at approximately C_{50}, although it should be mentioned that these values were determined in a static atmosphere, whereas in HTGC a constant flow of He is present which will affect the boiling behaviour of the *n*-alkanes. Thus it is uncertain at which GC oven temperature the boiling point of the high molecular weight *n*-alkanes will be equivalent. The overall implication of these processes is that the concentrations of the HMWHCs are likely to be underestimated in HTGC. Consequently, the presence of a small peak at C_{100} may actually represent the largest *n*-alkane component in that sample.

High molecular weight hydrocarbons in asphaltene pyrolysates

The asphaltenes isolated from a number of waxy oils were characterized by flash pyrolysis–gas chromatography using an aluminium-coated fused-silica high temperature capillary column. (*Note*: It is important to emphasize here that the asphaltene fractions were prepared in the classical sense by precipitation with pentane and purification by reprecipitation with additional *n*-pentane. It is entirely possible that during this isolation process, HMWHCs also precipitate due to lack of solubility

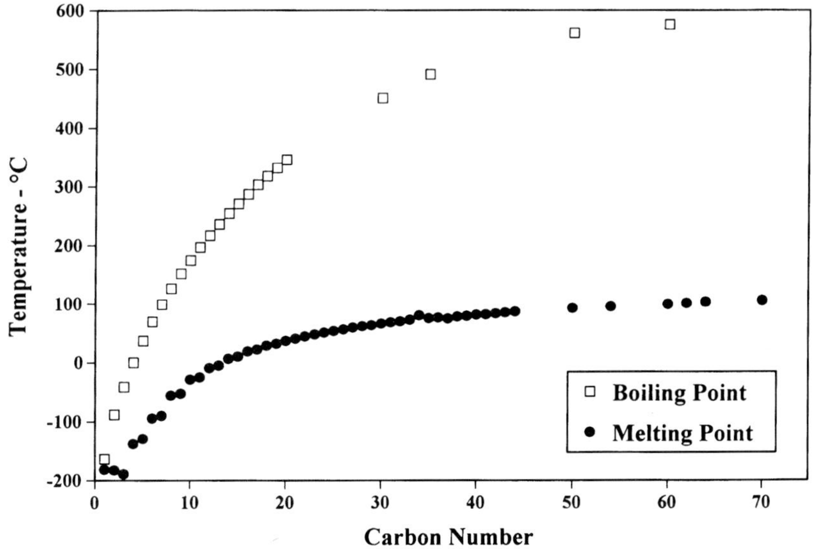

Fig. 5. Relationship between the *n*-alkane carbon number, boiling point and melting point.

in pentane. Despite not being chemically bound within the asphaltene structure, these hydrocarbons are by definition still a part of the so-called asphaltene fraction.) The results of these analyses showed that the pyrolysis products, for these particular asphaltenes, were dominated by *n*-alkanes extending to C_{60}, with a predominance of the higher molecular weight hydrocarbons (Fig. 6). As with the extractable material, these higher molecular weight compounds have been overlooked in the past owing to temperature limitations with the chromatographic columns. The question surrounding these compounds produced by pyrolysis is whether they are actually part of the asphaltene structure or simply trapped in the asphaltenes during the isolation process. Attempts to clarify this situation have led to the use of RuO_4, a mild and selective oxidizing reagent which preferentially destroys aromatic rings leaving the aliphatic and alicyclic portion intact, to degrade the asphaltenes. The carboxylic acid function in the aliphatic and acyclic structures produced marks the

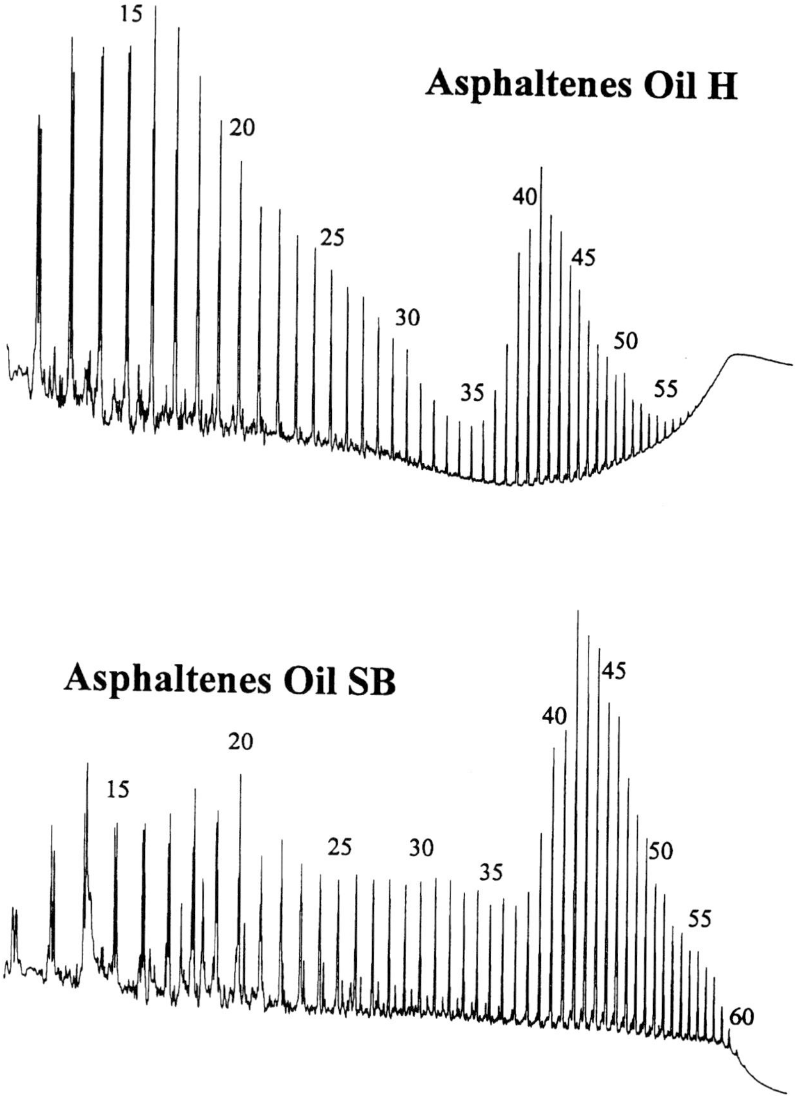

Fig. 6. Asphaltene pyrolysis products from oils containing no apparent high molecular weight hydrocarbons components.

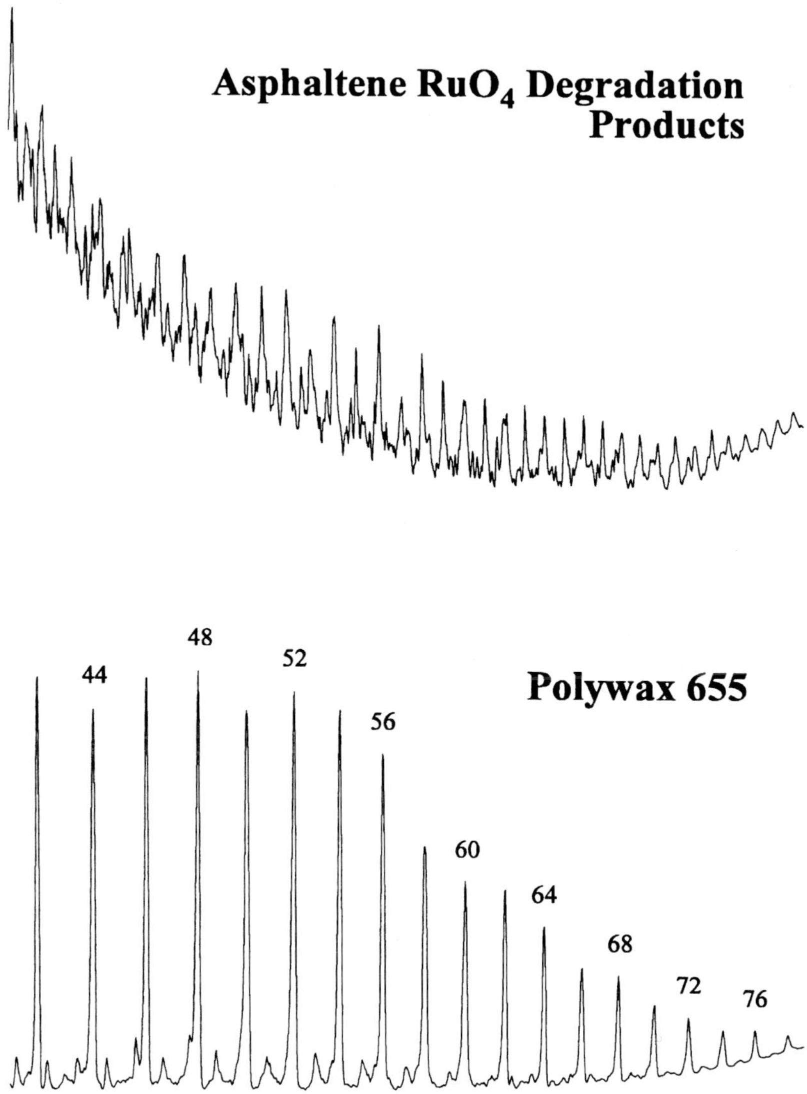

Fig. 7. Carboxylic acid oxidation products obtained from RuO_4 oxidation of an isolated asphaltene fraction compared with a standard wax sample in order to obtain the carbon distribution of degradation products.

point of attachment in the asphaltene structure. The products obtained in this way consist of carboxylic acids up to C_{75} (Fig. 7), supporting the idea that these long-chain compounds were indeed part of the asphaltene structure.

Alternative methods for the analysis of high molecular weight hydrocarbons

In an effort to extend the molecular weight range of compounds that may be determined in the crude oils or rock samples, a number of alternative approaches have been investigated for their applicability to this problem. In particular, the use of thermal distillation of samples directly into a flame ionization detector with no GC separation has been utilized to analyse both standards and samples with varying degrees of success. In the initial approach, a number of standard *n*-alkanes were analysed in this manner and it was noted that good separation could be obtained under the conditions being utilized as shown in Fig. 8. With

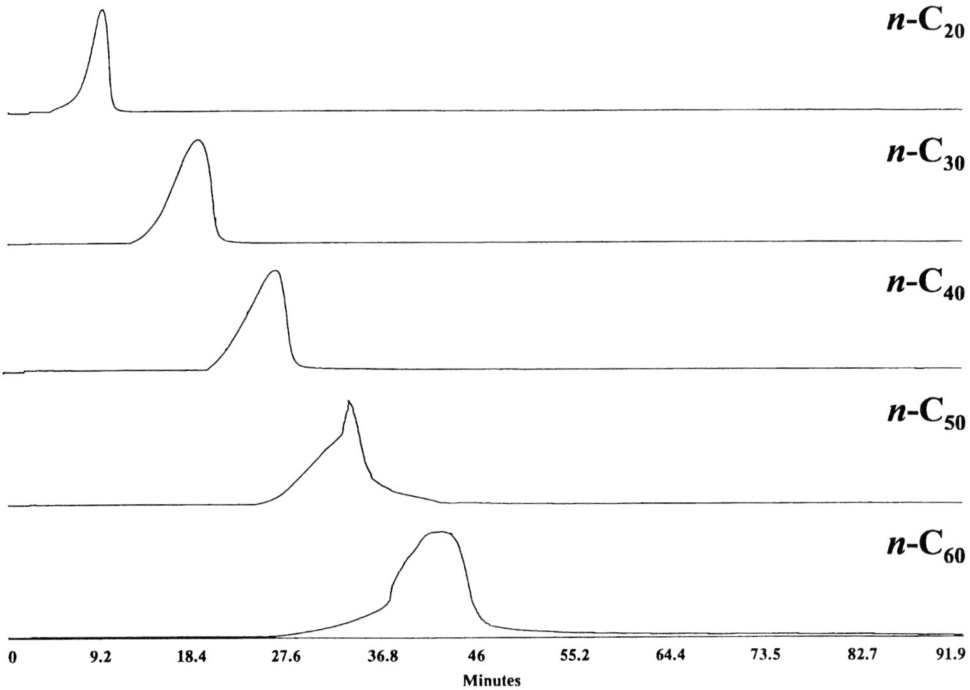

Fig. 8. Analyses of standard hydrocarbons by thermal distillation coupled directly to a flame ionization detector with no GC separation.

increasing carbon number, the amount of separation decreases because differences in the boiling points of these compounds decrease with increasing molecular weight, as shown in Fig. 5 and discussed previously. Following the initial success in separating the individual standards, the samples of the standard polywaxes were analysed in the same manner and it can be seen from Fig. 9 that the three standards can be differentiated on the basis of their molecular weights. Furthermore, these waxes can be used as a means of calibrating the molecular weight of unknown waxes isolated from the oils or extracts. The chromatogram for the ozocerite (Fig. 9d) has a trimodal distribution, and when compared with standards, displays hydrocarbon molecular weights ranging up to C_{80}, which agrees well with the gas chromatogram shown previously in Fig. 3. Other waxes analysed in this way showed that when used in conjunction with calibration standards, a good indication of the molecular weight distribution can be obtained.

In many of the oil samples analysed, concentrations of the high molecular weight fractions are relatively low because many of these HMWHCs have precipitated from solution during the production process or have remained in the reservoir. Therefore, if only oil samples are available for analysis, it is desirable to have a method of concentrating whatever alkanes are present in the samples. One approach involves taking advantage of the solubility differences of the HMWHC. Addition of warm THF to the waxy oil followed by cooling will lead to precipitation of the HMWHC which can then be removed by filtration. An example of this approach is shown in Fig. 10 where one of the waxy crudes from the Uinta Basin has been treated in this way. The chromatograms of the original oil and the HMWHC concentrate are shown. The use of this approach provides a rapid means of concentrating these compounds, enabling their characterization and analysis.

Origin of the high molecular weight hydrocarbons

In discussing the source of these compounds, the presence of natural precursors must be included, because polyisoprenyl alcohols up to C_{100} have been reported in several organisms (Hemming 1983; Lehle & Tanner 1983; Chojnacki & Vogtman 1984; Chojnacki *et al.* 1987; Swiezewska & Chojnacki 1988, 1989) and bacterial carotenoid skeletons containing up to C_{50} are well known

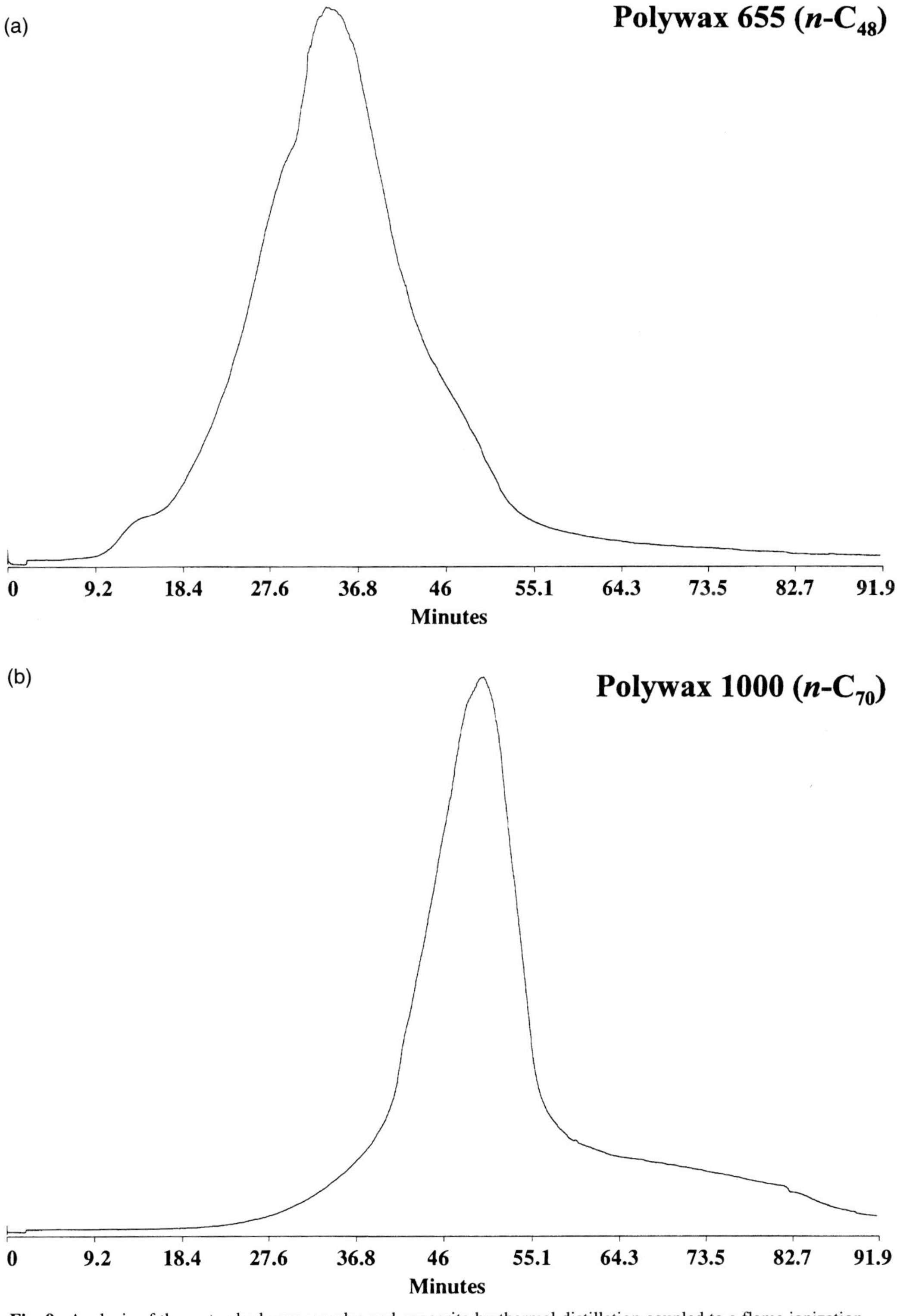

Fig. 9. Analysis of three standard wax samples and ozocerite by thermal distillation coupled to a flame ionization detector, showing variations resulting from differences in molecular weight distributions.

Fig. 9 continued overleaf.

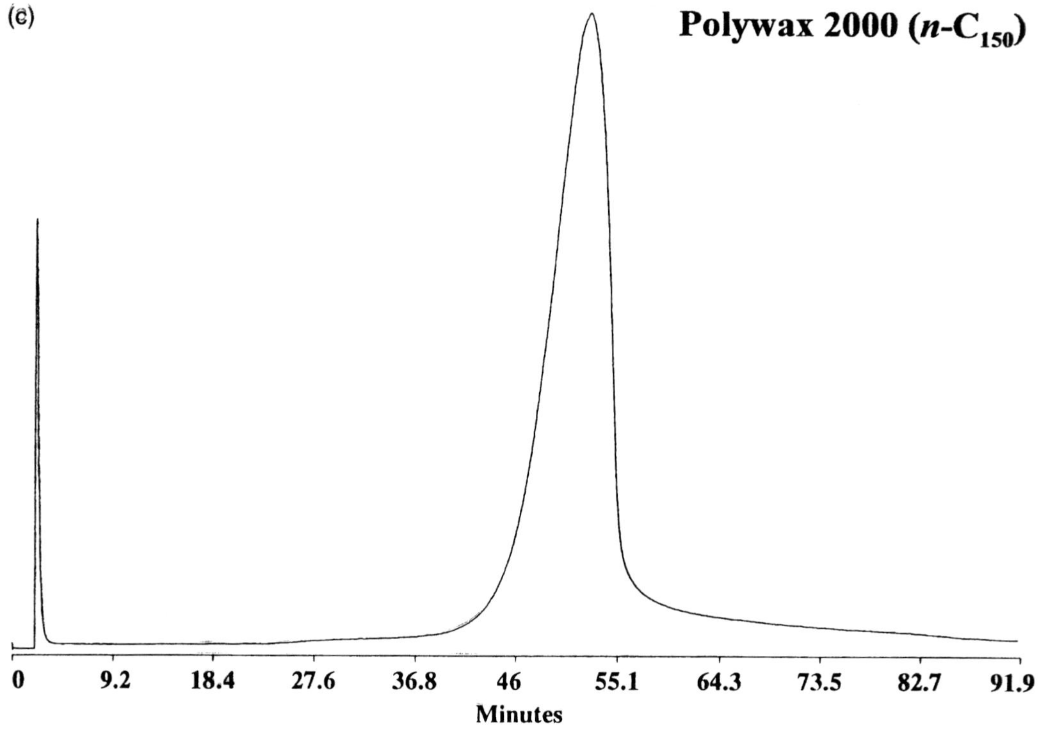

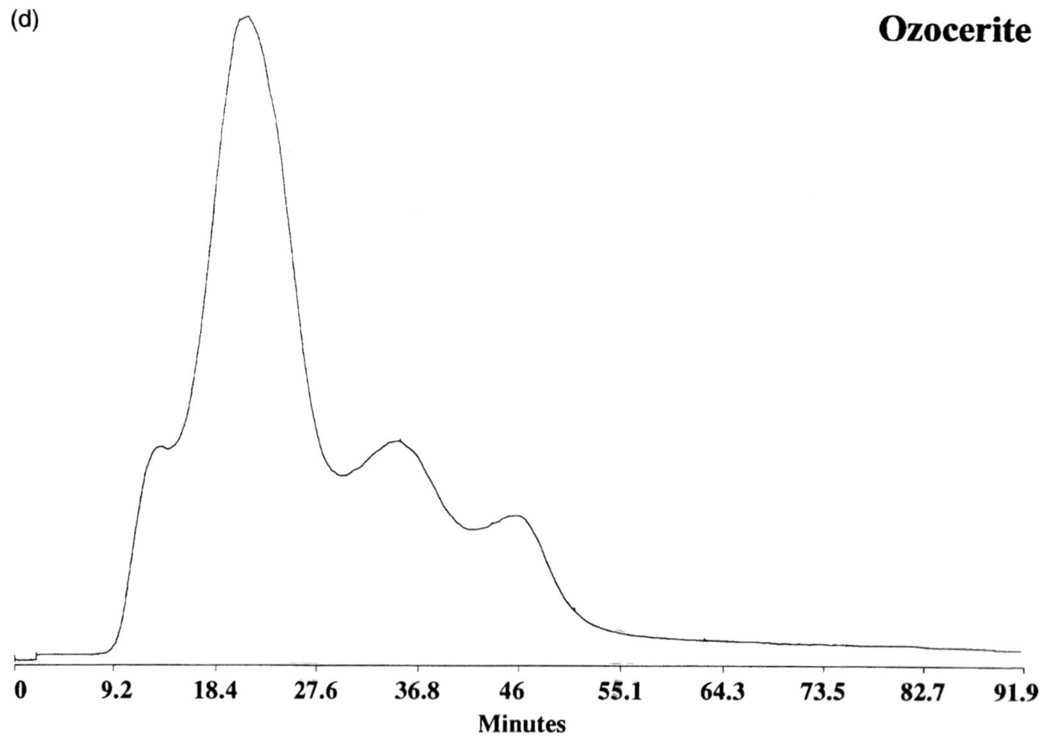

Fig. 9. continued

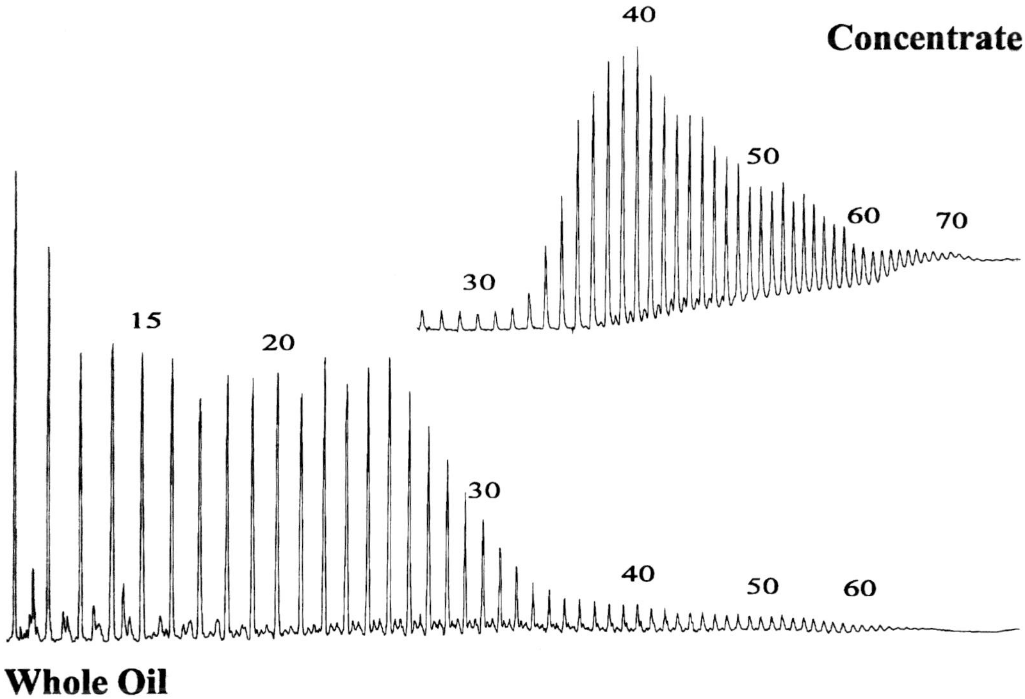

Fig. 10. Chromatograms showing the effects of concentrating the high molecular weight hydrocarbons in a yellow crude from the Bluebell Field by the addition of tetrahydrofuran to the saturate hydrocarbon fraction.

(Liaeen-Jensen 1990). Although detailed structures have not yet been determined, it is possible that some of the high molecular weight compounds in the oils could be formed by intramolecular cyclization of high molecular weight poly-isoprenoid alcohols. Moldowan *et al.* (1983) proposed that cyclization of solanesol (a C_{45} polyprenol) could lead to the extended series of tricyclic terpenoids up to C_{45}, and Albaiges (1980) also proposed polyprenols as precursors of isoprenoids extending up to C_{45}. These alcohols may be converted into high molecular weight isoprenoid alkanes just as the low molecular weight analogues (i.e. phytol) are converted into hydrocarbons by attack at the alcohol moiety, followed by reduction and thermal cracking.

Another possibility is that the high molecular weight hydrocarbons are di- and trimerization products of lower molecular weight precursors (Del Rio & Philp 1992*b*). Previous papers strongly support this hypothesis. For instance, De Leeuw *et al.* (1980) obtained di- and trimerization products of phytol after heating it in the presence of clay minerals. Rubinstein & Strausz (1979) reported the formation of dimerization products of fatty acids when heating them in the presence of clay minerals. More recently, Van Aarssen *et al.* (1992) discovered some C_{45} hydrocarbons, thought to be trimeric cadinanes, in crude oils and sediments from South East Asia, suggesting that sesquiterpenes may dimerize or oligomerize under appropriate conditions by an abiotic process.

Conclusions

High molecular weight hydrocarbons have been found to be concentrated in the wax deposits that frequently occur in drill-stem pipes of producing oil wells. Pyrolysis of the asphaltenes isolated from the oils suggests that thermal breakdown under reservoir conditions may be one mechanism leading to the formation of these high molecular weight hydrocarbons. The presence of high molecular weight hydrocarbons has been observed in the saturated fraction of ozocerite, a fossil bitumen. Cyclic structures ranging up to C_{80} could be observed in the ozocerite, although no detailed structures were confirmed for these compounds.

It is suggested that the study of high molecular weight hydrocarbons has been overlooked in the past for two main reasons: the first is the lack of appropriate analytical techniques; the second is that, particularly in the case of produced oils, the higher molecular weight components are often absent in the oil collected at the well-head. Rather, these components precipitate out in the drill-stem pipes or remain in the reservoir rocks due to their low mobility. In view of the results presented herein it is proposed that continued study of these high molecular weight fractions will provide additional insights into the origin and types of organic source materials responsible for various types of oils. Major advances can be expected in this area over the next few years, largely because of improvements in analytical techniques that will make it easier to identify the high molecular weight compounds on a molecular level.

Traditional gas chromatographic methods are limited to species with sufficient volatility. High temperature gas chromatography and supercritical fluid chromatography coupled with mass spectrometry offer the potential to provide high-resolution separation of high molcular weight compounds with selective mass spectrometric analysis. This could be the appropriate technique to analyse these new high molecular weight compounds, which may be potential biomarkers, although instrumentation and interfaces are still being developed. The potential of HTGC–MS and SFC–MS will continue to grow as HTGC and SFC techniques are extended by the introduction of new mobile and stationary phases, new approaches for the interfaces are developed and as improved mass spectrometric detectors continue to become available.

Finally, the importance of developing techniques to analyse for these high molecular weight compounds in the context of reservoir geochemistry cannot be overemphasized. These high molecular weight hydrocarbons will, in certain cases, be responsible for significant amounts of the residual components left in a reservoir. A better understanding of the nature of this fraction has the potential to improve current secondary and tertiary recovery processes and other enhanced oil recovery methods.

References

ALBAIGES, J. 1980. Identification and geochemical significance of long chain acyclic isoprenoid hydrocarbons in crude oils. *In*: DOUGLAS, A. G. & MAXWELL, J. R. (eds) *Advances in Organic Geochemistry 1979*. Pergamon Press, Oxford, 19–28.

ASHFORD, J. D., BLOUNT, C. G., MARCOU, J. A., RALPH, J. M. & SKAALURE, E. W. 1990. *Annular packer fluids for paraffin control: model study and successful application*. Society of Petroleum Engineers, Paper No. **19730**.

BLUM, W., RAMSTEIN, P. & EGLINTON, G. 1990. Coupling of high temperature glass capillary columns to a mass spectrometer. GC/MS analysis of metalloporphyrins from Julia Creek oil shale samples. *Journal of High Resolution Chromatography*, **13**, 85–94.

CARLSON, R. M. K., TEERMAN, S. C., MOLDOWAN, J. M. *et al.* 1993. High temperature gas chromatography of high wax oils. *Proceedings of the 20th Annual Convention of the Indonesian Petroleum Association*, Indonesian Petroleum Association, Jakarta, 483–507.

CARNAHAN, N. F. 1989. Paraffin deposition in petroleum production. *Journal of Petroleum Technology*, **41**, 1024–1025.

CHOJNACKI, T. & VOGTMAN, T. 1984. The occurrence and seasonal distribution of C_{50}–C_{60} polyprenols and of C_{100}– and similar long chain polyprenols in leaves of plants. *Acta Biochimica Polonica*, **31**, 115–126.

——, SWIEZEWSKA, E. & VOGTMAN, T. 1987. Polyprenols from plants – structural analogues of mammalian dolichols. *Chemica Scripta*, **27**, 209–214.

DE LEEUW, J. W., SIMONEIT, B. R. T., BOON, J. J. *et al.* 1980. Phytol derived compounds in the geosphere. *In*: DOUGLAS, A. G. & MAXWELL, J. R. (eds) *Advances in Organic Geochemistry 1979*. Pergamon Press, Oxford, 61–79.

DEL RIO, J.-C. & PHILP, R. P. 1992a. High molecular weight hydrocarbons: A new frontier in organic geochemistry. *Trends in Analytical Chemistry*, **11**, 187–193.

—— & —— 1992b. Oligomerization of fatty acids as a possible source for high molecular weight hydrocarbons and sulphur-containing compounds in sediments. *Organic Geochemistry*, **18**, 869–880.

——, —— & ALLEN, J. 1992. Nature and geochemistry of high molecular weight hydrocarbons (above C_{40}) in oils and solid bitumens. *Organic Geochemistry*, **18**, 541–553.

GALLEGOS, E. J., FETZER, J. C., CARLSON, R. M. & PENA, M. M. 1991. High-temperature GC/MS characterization of porphyrins and high molecular weight saturated hydrocarbons. *Energy and Fuels*, **5**, 376–381.

HAWTHORNE, S. B. & MILLER, D. J. 1987. Analysis of commercial waxes using capillary supercritical fluid chromatography–mass spectrometry. *Journal of Chromatography*, **388**, 397–409.

—— & —— 1989. Supercritical fluid chromatography–mass spectrometry of polycyclic aromatic hydrocarbons with a simple capillary direct interface. *Journal of Chromatography*, **468**, 115–125.

HEMMING, F. W. 1983. The biosynthesis of polyisoprenoid chains. *Transactions of the Biochemical Society*, **11**, 497–504.

LEHLE, L. & TANNER, W. 1983. Polyprenoid-linked sugars and glycoprotein synthesis in plants. *Transactions of the Biochemical Society*, **11**, 568–574.

LIAEEN-JENSEN, S. 1990. Marine carotenoids – selected topics. *New Journal of Chemistry*, **14**, 747–759.

LIPSKY, S. R. & DUFFY, M. L. 1986a. High temperature gas chromatography: the development of new aluminum clad flexible fused silica glass capillary columns coated with thermostable nonpolar phases. Part 1. *Journal of High Resolution Chromatography, Chromatography Communications*, **9**, 376–382.

—— & —— 1986b. High temperature gas chromatography: the development of new aluminum clad flexible fused silica glass capillary columns coated with thermostable nonpolar phases. Part 2. *Journal of High Resolution Chromatography, Chromatography Communications*, **9**, 725–730.

LOMANDO, A. J. 1992. The influence of solid reservoir bitumen on reservoir quality. *American Association of Petroleum Geologists, Bulletin*, **76**, 1137–1152.

MOLDOWAN, J. M., SEIFERT, W. K. & GALLEGOS, E. J. 1983. Identification of extended series of tricyclic terpanes in petroleum. *Geochimica et Cosmochimica Acta*, **47**, 1531–1534.

NORTH, F. K. 1985. *Petroleum Geology*. Allen and Unwin, Boston.

OLESIK, S. V. 1991. Recent advances in supercritical fluid chromatography/mass spectrometry. *Journal of High Resolution Chromatography*, **14**, 5–9.

PETERS, K. E. & MOLDOWAN, J. M. 1993. *The Biomarker Guide: Interpreting Molecular Fossils in Petroleum and Ancient Sediments*. Prentice Hall, Englewood Cliffs, New Jersey.

RUBINSTEIN, I. & STRAUSZ, O. P. 1979. Geochemistry of the thiourea adduct fraction from an Alberta petroleum. *Geochimica et Cosmochimica Acta*, **43**, 1387–1392.

SMITH, R. D., KALINOSKI, H. T. & UDSETH, H. R. 1987. Fundamentals and properties of supercritical fluid chromatography–mass spectrometry. *Mass Spectrometry Reviews 1987*, **6**, 445–496.

SWIEZEWSKA, E. & CHOJNACKI, T. 1988. Long chain polyprenols in gymnosperm plants. *Acta Biochimica Polonica*, **35**, 131–147.

—— & —— 1989. The occurrence of unique, long-chain polyprenols in the leaves of *Potentilla* species. *Acta Biochimica Polonica*, **36**, 143–158.

THOMSON, J. S. & RYNASKI, A. F. 1992. Simulated distillation of wax samples using super critical fluid and high temperature gas chromatography. *Journal of High Resolution Chromatography, Chromatography Communications*, **15**, 227–234.

TUTTLE, R. N. 1983. High-pour-point and asphaltic crude oils and condensates. *Journal of Petroleum Technology*, **35**, 1192–1197.

VAN AARSSEN, B. G. K., HESSELS, J. K. C., ABBINK, O. A. & DE LEEUW, J. W. 1992. The occurrence of polycyclic sesqui-, tri- and oligoterpenoids derived from a resinous polymeric cadinene in crude oils from South East Asia. *Geochimica et Cosmochimica Acta*, **56**, 1231–1246.

Overview of the geochemistry of some tar mats from the North Sea and USA: implications for tar-mat origin

A. WILHELMS[1] & S. R. LARTER[2]

[1] *Department of Geology, University of Oslo, PO Box 1047, 0316 Oslo, Norway*

[2] *Newcastle Research Group in Fossil Fuels & Environmental Geochemistry, University of Newcastle, Newcastle upon Tyne NE1 7RU, UK*

Abstract: Tar mats and oil legs from several marine source-rock derived oil fields (Ula and Oseberg, North Sea and an unidentified North American example) have been geochemically characterized. Tar mats were derived from the same petroleum system as the oil-leg oils and are found in the most porous and permeable portions of the reservoir. It is possible to eliminate some potential tar-mat formation mechanisms using the data from case histories and subsurface asphaltene solubility characteristics. Adsorption of asphaltenes on reservoir clays and light to moderate levels of biodegradation can probably be eliminated as feasible tar-mat formation mechanisms. Thermal degradation of oil in the carrier and reservoir causing asphaltene precipitation may be a major contributing factor to formation of tar mats in the Ula Field, whereas increased gas solution in the oil leg may have contributed to deasphaltation in the Oseberg Field. Based on tentative mass-balance calculations, the extreme pressure dependence of asphaltene solubility in subsurface crude oils and the occurrence of tar mats in zones of elevated permeability, it is suggested that tar-mat formation is initiated during petroleum migration. While tentative, this model may provide a more generic initiation mechanism than others in the literature.

Tar mats can best be defined as zones of asphaltene-rich petroleum with sharp compositional contacts to the overlying higher gravity petroleum column (Jones & Speers 1976; Dahl & Speers 1986; Wilhelms & Larter 1994*a*). Tar mats are typically close to geological discontinuities and are enriched in asphaltenes up to concentrations of 20 to 60 wt% of the C_{15+} fraction of the petroleum (Figs 1 & 2). It has been suggested that some tar mats can give indications of the filling history of a reservoir (Larter *et al.* 1990). For reservoir exploitation purposes, a predictive model of tar-mat areal extent would be beneficial both for field development and for secondary and enhanced oil recovery procedures. This is because tar mats dominantly represent non-producible oil in place. They may also represent low permeability intra-reservoir flow barriers, and petroleum reservoirs with tar mats often have production deasphaltation problems associated with them. A knowledge of spatial tar-mat distribution is important in equity discussions for a correct estimate of the producible reserves and in the effective placement of injection wells.

A variety of tar-mat formation processes have been discussed in the literature, including biodegradation (Connan 1984), in-reservoir oil mixing (Larter *et al.* 1990), adsorption of asphaltenes onto clay minerals, increased gas solution in oil legs (Dahl & Speers 1986), diffusion-mediated gravity segregation of oil columns (Schulte 1980), pressure reduction during reservoir inversion (Hirschberg 1984) and processes related to thermal convection in oil columns (Montel & Gouel 1985).

Here, we review previously proposed tar-mat formation models in the light of case studies that we have recently reported (Oseberg and Ula Fields, North Sea, and an unidentified North American example; Wilhelms and Larter 1994*a*), suggest an alternative tar-mat formation model and discuss future research routes for the study of tar mats and polar petroleum species in general.

Asphaltene structure and environment

The characteristics of marine petroleum derived asphaltenes (MPDA) are listed in Table 1. To describe the compounds present in subsurface crude oils that give rise to the laboratory product 'asphaltene', the term asphaltene-precursor-entity (APE) has been proposed (Wilhelms 1992; Wilhelms & Larter 1994*b*). Owing to asphaltene aggregation phenomena in solution, the molecular weight and other characteristics of petroleum APEs are not the same as for laboratory-precipitated asphaltenes.

The APE hypothesis suggests that APEs are kept in solution in subsurface oils by a process similar to steric stabilization in polymer solutions (cf. Hunter

From CUBITT, J. M. & ENGLAND, W. A. (eds), 1995, *The Geochemistry of Reservoirs*, Geological Society Special Publication No. 86, pp. 87–101.

Table 1. *Characteristics of marine petroleum derived asphaltenes (MPDAs)*

(1) Typical values for atomic ratios are: H/C 1.12 (range 1.05–1.21; average 1.12; extreme 0.89); O/C range 0.008–0.028. MPDA have H/C and O/C ratios similar to those of mature marine source rock kerogens, albeit lower S/C and N/C ratios.
(2) MPDA behave pyrolytically in a similar way to kerogens, with rather uniform *n*-hydrocarbon pyrolysate yields in the range of mature marine kerogens.
(3) Modal molecular weights are 1000 daltons or less.
(4) Nuclear magnetic resonance spectroscopy indicates black-oil asphaltenes aromaticities around 40 ± 10% and alkylation with abundant *n*-alkyl groups with average chain lengths of *c*. 10 to 17 carbon atoms. All asphaltene subfractions are alkylated.
(5) Functional group distribution: MPDAs are slightly depleted in heteroatom content compared to kerogens. MPDA clearly consist of mixed polarity assemblages with the following broad characteristics:
  (a) *c*. 40 to 50% of heteroatoms are in neutral groups (e.g. ether, thioether, pyrrole);
  (b) *c*. 50% of oxygen is acidic, probably phenolic;
  (c) *c*. 30% of nitrogen is basic (i.e. pyridinic) the rest being neutral (i.e. pyrrolic);
  (d) published data suggest that *c*. 60% of sulphur is thiophenic.

1989), this being related to the abundant alkyl side chains of MPDA (Wilhelms & Larter 1994*b*). They are considered to be a broad range of alkylated compounds (probably initially in true solution in crude oils) and a 'polymer solution' in a maltene solvent. Owing to changes in the solvent properties of the oil, the APEs start to aggregate, probably form lyophilic micelles, and ultimately precipitate as asphaltenes.

The factors controlling the solubility of APEs in crude oils appear to be pressure, temperature, oil composition and last, but not least, the characteristics of the asphaltenes themselves (Billheimer *et al.* 1949; Briant 1963; Lhioreau *et al.* 1967; Schulte 1980; Hirschberg & Hermans 1984; Burke *et al.* 1990; Wilhelms & Larter 1994*b*). A detailed discussion of these factors is reported elsewhere (Wilhelms & Larter 1994*b*), therefore only the salient features are given here.

Pressure and temperature controls on asphaltene solubility have been thermodynamically modelled (Bunger *et al.* 1979; Ali & Al-Ghannam 1981; Hirschberg & Hermans 1984; Hirschberg *et al.* 1984; Leontaritis & Mansoori 1987; Park & Mansoori 1988) based on the Flory–Huggins lattice model of polymer solution (Flory 1953), laboratory pressure–volume–temperature (PVT) experiments or polydisperse polymer solution theory (Scott & Magat 1945). These models allow a limited quantitative assessment of the factors affecting asphaltene solubility in crude oils.

The 'Hirschberg' model (Hirschberg 1984; Hirschberg & Hermans 1984) indicates that the volume fraction of asphaltenes soluble in a crude oil is given by:

$$f_{VA} = \exp\{V_A/V_L[1 - V_L/V_A - V_L/RT(\delta_A - \delta_L)^2]\} \quad (1)$$

where f_{VA} = volume fraction of asphaltenes soluble in the crude oil; V = molar volume of asphaltenes or liquid ($m^3\ kmol^{-1}$) under subsurface conditions; δ = solubility factor of asphaltenes or liquid ($MPa^{0.5}$); R = universal gas constant (8.3145 kJ $kmol^{-1}$); T = absolute temperature (K); and the subscripts A and L refer to asphaltene and crude oil, respectively.

Assuming asphaltene properties to be pressure-independent (Hirschberg *et al.* 1984) and using typical solubility factors for petroleums and asphaltenes (Hirschberg *et al.* 1984; Wilhelms & Larter 1994*b*), it can be seen that the principal factors governing the volume fraction of asphaltenes soluble in a crude oil are crude-oil (solvent) solubility factor (δ) related to bulk crude-oil composition, temperature and molar volumes (V_i) of solvent (crude oil) and asphaltenes. At pressures higher than the bubble-point pressure of the system, an increase in pressure will increase δ_L and decrease V_L (liquid compression and density increase), thus enhancing asphaltene solubility. The pressure dependency of the asphaltene solubility and δ_L are related to the density increase of the oil, which accounts for more 'solvent' molecules per unit volume. At pressures lower than the bubble-point pressure with a free gas cap present, an increase in pressure will result in an increase of dissolved gas in the crude, thus lowering δ_L (reducing the oil solvent properties) and decreasing asphaltene solubility (mixing of more paraffin-rich crude could thus also lower δ_L). The minimum asphaltene solubility for a given liquid petroleum will be reached at the bubble-point pressure; asphaltene solubility will then increase again with increasing pressure providing that gas is not added.

This means that a pressure reduction in a petroleum column at pressures lower than the bubble-point pressure generally should slightly increase the asphaltene solubility as gas exsolves. The effect of temperature on asphaltene solubility is less easy to predict because both V_L and δ_L are altered. However, this contribution is likely to be small when compared to the influence of pressure on changes in asphaltene solubility. Whatever the theoretical basis for the Hirschberg model, its calibrated nature does serve to illustrate the general processes involved in controlling asphaltene solubility in crude oils. The dependence of APE solubility on pressure in subsurface crude oils is consistent with APEs being initially in true solution in crude oils, pressure changes causing a change in the overall solubility factor of the crude oil and thus APE aggregation/asphaltene precipitation (i.e. crude oil is a maltene solvent containing sterically stabilized polymer in solution; Wilhelms & Larter 1994*b*).

A migrating petroleum charge moving along P–T reduction paths in a basin would be expected to show a decrease of APE (asphaltene) solubility with migration. Mature marine source rocks are known to contain significant quantities of asphaltenes (Leyhaeuser *et al.* 1988), and appear to expel asphaltene-rich petroleum (Tissot & Pelet 1971; Leyhaeuser *et al.* 1988; Wilhelms *et al.* 1990), the asphaltene content of which decreases rapidly with increasing distance from the source rock (Tissot & Pelet 1971). This might be taken as a tentative indication that petroleum charges during primary migration are at saturation solubility with respect to APEs.

The PVT changes during secondary migration (probably dominantly pressure and temperature reduction) result in an unstable petroleum system, which tries during secondary migration to approach a stable state which may involve asphaltene precipitation among other phase-related compositional fractionations (England *et al.* 1987; Larter & Mills 1991). It might also be speculated, based on this concept, that high pressured petroleum kitchens may lead to expulsion of petroleum richer in polar compounds (resins and asphaltenes) compared to lower pressured source kitchens. In-reservoir or migration-related processes involving addition of gas to undersaturated oil legs would also change the solubility factor of the oil and result in asphaltene precipitation. Conversely, removal of saturated hydrocarbons by petroleum biodegradation would increase the oil solvent properties for crude oils (at least under mild to moderate levels of degradation), thus stabilizing the APE/asphaltene solution.

Occurrence and setting of tar mats

Tar mats have been reported from both carbonate and clastic reservoirs (Lomando 1992; Wilhelms & Larter 1994*a*). The tar mats described briefly here have been discussed in detail elsewhere (Wilhelms 1992; Wilhelms & Larter 1994*a*). Oseberg and Ula Fields, both North Sea examples, and a North American field were studied by comparing oil-leg and tar-mat petroleum composition using standard bulk and molecular geochemical methods (Wilhelms & Larter 1994*a*) that employ the reservoir geochemical protocol developed at the University of Oslo (Karlsen & Larter 1989; Larter *et al.* 1990; Larter & Aplin 1995).

All the tar mats are from reservoirs containing oils sourced from mature marine oil-prone source-rock systems (oil-leg petroleum API gravity > 25°). Figures 1 & 2 show logs for the Oseberg and Ula Field case studies with tar-mat horizons indicated.

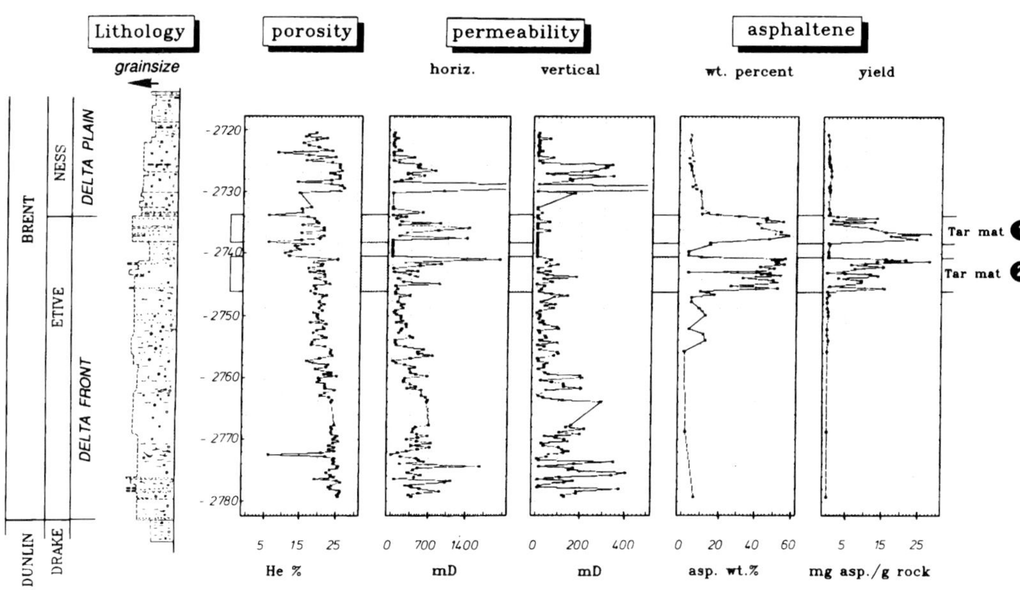

Fig. 1. Oseberg Field (North Sea) logs of petroleum reservoir zone illustrating the geological setting, rock properties and organic geochemical signal of the tar mats.

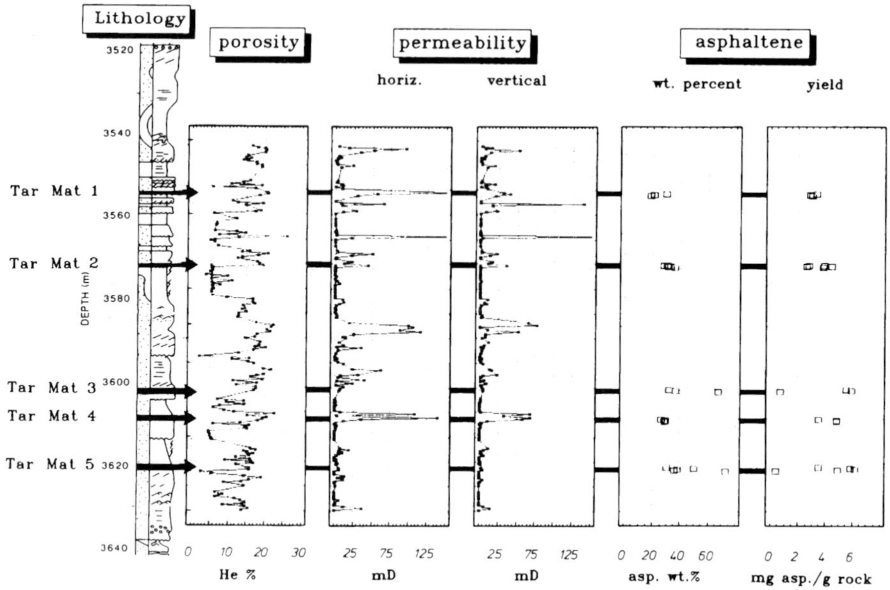

Fig. 2. Ula Field (North Sea) logs of petroleum reservoir zone illustrating the geological setting, rock properties and organic geochemical signal of the tar mats.

Basic oil-field data, ranges and average values of some organic geochemical compositional data for the reservoir rock extracts and related drill-stem test oils are listed in Tables 2 & 3.

The geological history of the Oseberg Field (blocks 30/6 and 30/9, Norwegian sector North Sea) has been described previously (Torvund & Nipen 1987; Wilhelms & Larter 1994*a*). The tar mats have been previously reported to be present in the southern part of the Oseberg Field (Dahl & Speers 1985) and were reported from two wells (Dahl & Speers 1986). The tar mats in well 30/9-1 are present in the Middle Jurassic Brent Group (Etive Fm) and are situated in high porosity (*c.* 20 to 25%) and high horizontal permeability (> 300 mD) delta front sandstones close to the oil–water contact (OWC), bordered by zones of low vertical and horizontal permeability (Fig. 1). They are overlain by an oil leg of variable continuous thickness (< 100 m) in the coastal plain deposits of the Ness Fm. Two tar mats are observed (4 m and 6 m thick respectively) and are defined on the basis of sharp kicks in the asphaltene content and asphaltene extract yield logs (Fig. 1; Table 3).

The Ula Field (block 7/12 Norwegian North Sea) has been described elsewhere (Larter *et al.* 1990). The tar mats studied (well 7/12-6) are observed in the lower Triassic Skagerrak Formation, at the base of fining-upwards channel sand sequences (permeability ranging from < 10 mD to *c.* 150 mD and porosity from 5% to 20%; Fig. 2). The tar-mat core samples again have higher porosity and horizontal permeability compared to the samples from the oil leg (Wilhelms & Larter 1994*a*). Here the thickness of the tar mats ranges between 0.4 and 0.9 m for each of the five tar mats (Fig. 2). The average asphaltene contents of the tar mats are

Table 2. *Basic oil field and produced oil data for tar mat-containing petroleum reservoirs*

Oil field	Formation	GOR (scf/bbl) (std m^3/std m^3)	API (deg)	Crude oil S content (%)	Reservoir type
Oseberg N. Sea	Etive & Ness Fm (M. Jurassic)	721–888 (130–160)	34.9	*c.* 0.3	Fluvial sandstone
Ula N. Sea	Skagerrak Fm (U. Triassic)	487 (88)	39.2	*c.* 0.1	Fluvial sandstone
N. America	–	680–830 (122–150)	24–28	0.8–2.2	Fluvial/marine sandstones

Table 3. *Summary table of gross compositional data (grouped after sample type, range and average) of core extracts from the case studies*

Oseberg Field case study (sample depth *c.* 2750 m)

Type of sample	Absolute yield (mg/g rock)					Relative composition					PYGC
	SAT	ARO	RES	ASP	EOM	% SAT	% ARO	% RES	% ASP	Wt % ASP	C_8–C_{30}
oil leg	0.4–7.4	0.0–3.2	0.4–1.9	0.1–1.0	1.2–11.0	31–73	0–29	9–47	5–20	3–10	12.8–28.3
(16)	(3)	(1.2)	(1.6)	(0.5)	(5.8)	(48)	(20)	(22)	(9)	(5)	(22.7)
tar mat-1	0.6–6.7	0.0–0.7	0.7–11.8	1.6–28	3.6–43.4	11–32	0–19	6–43	43–80	30–58	14.2–32.4
(9)	(3.9)	(0.1)	(4.4)	(14)	(22.7)	(19)	(2)	(19)	(60)	(47)	(25.1)
tar mat-2	0.8–5.6	0.0–3.0	0.5–6.5	2.9–27.9	4.9–46.0	13–22	0–12	7–24	56–73	33–56	17.6–34.5
(17)	(2.8)	(1.0)	(1.8)	(12)	(17.7)	(16)	(6)	(10)	(68)	(49)	(28.2)
lean zone	0.0–0.4	0.0–0.6	0.1–0.8	0.1–0.7	0.4–2.0	0–20	0–32	20–71	8–64	0–18	n.d.
(10)	(0.1)	(0.2)	(0.4)	(0.4)	(1.1)	(7)	(12)	(42)	(37)	(5)	
water leg	0.0–0.4	0.0–0.2	0.1–3.7	0.1–0.6	0.2–3.8	0–47	0–18	27–97	3–73	1–17	n.d.
(13)	(0.1)	(0.01)	(0.5)	(0.2)	(0.8)	(13)	(1)	(54)	(31)	(8)	
prod. oil						50	29	20	1		35.2

Ula Field case study (sample depth *c.* 3580 m)

Type of sample	Absolute yield (mg/g rock)					Relative composition				PYGC
	SAT	ARO	RES	ASP	EOM	% SAT	% ARO	% RES	% ASP	C_8–C_{30}
oil leg #1	(5.3)	(1.1)	(1.0)	(0.2)	(7.6)	(70)	(14)	(13)	(3)	
oil leg #2	(3.4)	(0.7)	(1.1)	(0.3)	(5.5)	(55)	(13)	(20)	(5)	
tar mats	4.2–9.4	1.0–2.6	0.4–1.3	2.6–6.2	8.3–17.7	38–58	9–16	4–10	18–50	14.5–27.0
(17)	(6.7)	(1.8)	(0.9)	(4.2)	(13.6)	(49)	(13)	(7)	(31)	(20.5)
lean zone	0.0–0.1	0.1	0.1	0.4–0.7	0.6–1.0	8–14	5–10	8–13	67–73	0.3–3.6
(2)	(0.1)			(0.6)	(0.8)	(11)	(8)	(11)	(70)	
DST-1						70	22	5	3	27.4
DST-2						69	24	5	2	27.7

North American case study

Type of sample	Absolute yield (mg/g rock)					Relative composition				PYGC
	SAT	ARO	RES	ASP	EOM	% SAT	% ARO	% RES	% ASP	C_8–C_{30}
oil leg	2.2–5.5	1.4–3.5	1.3–5.6	0.2–1.1	5.4–14.3	36–45	15–26	24–40	2–9	23.4–35.6
(6)	(4.2)	(2.1)	(3.9)	(0.6)	(10.8)	(39)	(20)	(35)	(6)	(27.6)
tar mat	2.6–8.3	1.7–8.7	1.5–8.2	1.8–17.6	7.6–42.8	12–34	16–22	15–40	14–50	13.6–30.5
(7)	(5.1)	(4.7)	(5.4)	(10.4)	(25.6)	(22)	(19)	(22)	(37)	(22.1)
water leg	1.6–1.7	0.1–0.7	1.8–2.0	0.2–0.3	4.4–4.5	36–38	13–16	41–44	4–7	19.8–23.8
(2)	(1.7)	(0.7)	(1.9)	(0.3)	(4.5)	(37)	(15)	(43)	(6)	(21.8)
prod. oil						44	28	26	2	22.9

EOM, extractable organic matter (mg/g rock); SAT, % SAT, saturated hydrocarbons (mg/g rock and percent of EOM) (TLC-FID); ARO, % ARO, aromatic hydrocarbons (mg/g rock and percent of EOM) (TLC-FID); RES, % RES, resins (mg/g rock and percent of EOM) (TLC/FID); ASP, % ASP, asphaltenes (mg/g rock and percent of EOM) (TLC/FID); Wt % ASP, asphaltene content measured by weight; PYGC C_8–C_{30}, flash pyrolysis *n*-hydrocarbon yield (μg/mg asphaltene) in C_8 to C_{30} carbon number range; prod. oil, produced oil; DST-1, drill-stem test oil from the Triassic reservoir (Ula Field); DST-2, drill-stem test oil from the Jurassic reservoir (Ula Field). No. of samples given in brackets below sample types.
#1 Ula Fm reservoir average of 34 samples (Karlsen unpublished data).
#2 Skagerrak Fm reservoir average of 24 samples (Karlsen unpublished data).

c. 30% of extract yield (Fig. 2; Table 3). The tar-mat extracts of the Ula Field are characterized by higher relative saturated hydrocarbon contents than those from the Oseberg Field (Table 3), which probably reflects the more paraffinic character and higher maturity of the Ula Field oil.

Based on gas-chromatographic and biomarker fingerprint analyses of the saturated and aromatic hydrocarbon fractions of core extracts of the tar mat and respective oil-leg intervals, it was concluded that the tar mats and oil-leg petroleums were from a common source, no sign of biodegradation being evident in any of the case studies (Wilhelms & Larter 1994*a*). While it has been assumed that tar mats are related to the overlying oil columns, to our knowledge this is the first time that this has been demonstrated.

Figure 3 shows a plot of carbon number versus *n*-hydrocarbon yield (sum of *n*-alkene and *n*-alkane in μg/mg asphaltene) for normal hydrocarbons generated during flash pyrolysis of oil, oil-leg extract and tar-mat extract asphaltenes from the Oseberg and Ula Fields. The oil-derived asphaltenes generated the highest *n*-hydrocarbon yields followed by tar-mat and oil-leg asphaltenes (Fig. 3). The higher abundance of *n*-hydrocarbon-generating moieties in the oil asphaltenes may be related to the mode of stabilization of the APEs/asphaltenes in the crude (steric stabilization, *vide supra*), i.e. the oil asphaltenes are better stabilized in the oil owing to their more abundant alkyl side chains compared to tar-mat asphaltenes, and thus have not precipitated. The same holds true for the *n*-hydrocarbon yields of the Ula asphaltenes, albeit with lower absolute yields (Fig. 3); this is probably related to the degradation of the Ula asphaltenes at the higher temperature of this reservoir (*c.* 140°C; Larter *et al.* 1990).

It appears therefore that the tar mats studied are situated in 'host rocks' of relatively high porosity and permeability (Wilhelms & Larter 1994*a*) and are characterized by consistently elevated relative and absolute asphaltene yields.

Evaluation of tar mat formation mechanisms

A wide range of processes leading to tar-mat formation have been considered in the literature, i.e. asphaltene precipitation (deasphalting) from the oil due to: segregation of asphaltenes as particulates or by diffusion in solution along gravitational potential gradients to the base of the oil column (Schulte 1980; Dahl & Speers 1986; Wilhelms & Larter 1994*b*); oil-leg biodegradation (Connan

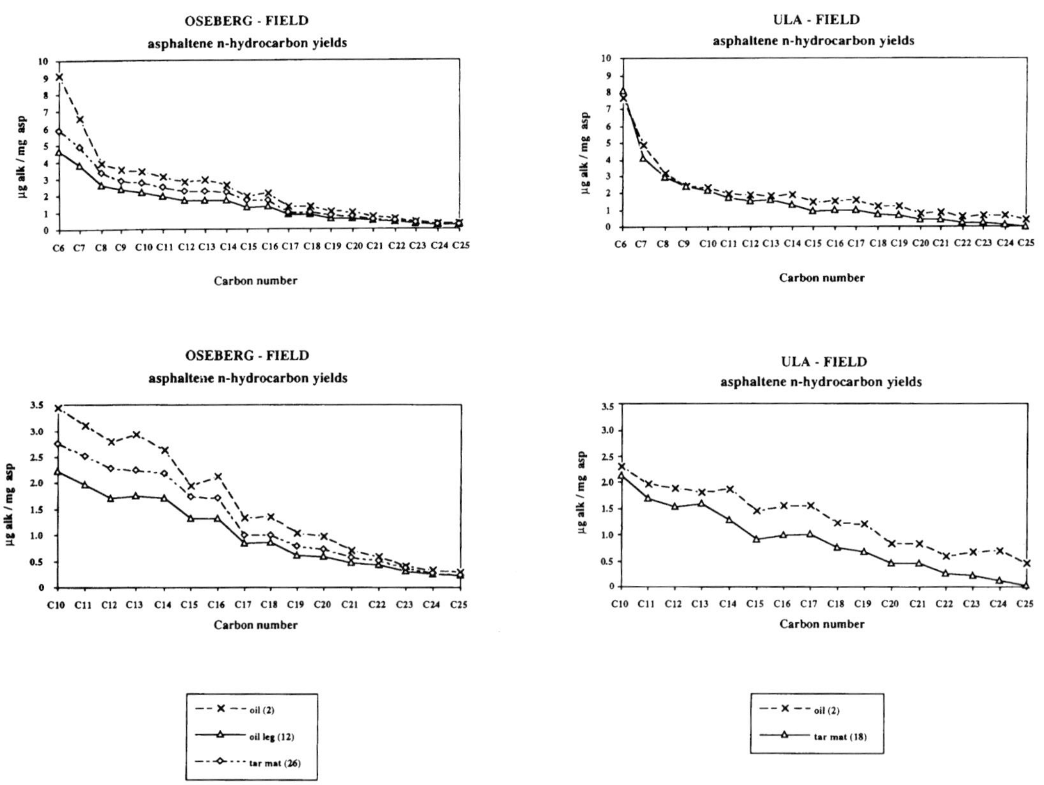

Fig. 3. Asphaltene flash-pyrolysis *n*-hydrocarbon yields (in μg/mg asphaltene) versus carbon number for oil, oil leg and tar-mat asphaltenes for Oseberg Field and Ula Field.

1984); gas injection ('deasphalting', *senso* Rogers *et al.* 1972); reservoir-pressure reduction due to uplift (Hirschberg 1984); adsorption of asphaltenes/resins onto reservoir clays (Larter *et al.* 1990); in-reservoir maturation of petroleum (Rogers *et al.* 1972; Larter *et al.* 1990); in-reservoir mixing of different maturity petroleums (Larter *et al.* 1990); convective processes in the oil column (Montel & Gouel 1985); and petroleum migration-related processes (Wilhelms & Larter 1994*b*).

Table 4 shows definitions of relevant terms used by the current authors and gives relevant references.

In order to understand the enrichment of asphaltenes in tar mats, one needs an estimate of the amounts of asphaltenes usually present in 'normal' oil legs. This is complicated in most situations because the asphaltene contents of produced oils are invariably lower than those of solvent extracts of reservoir cores. The enrichment of reservoir core extracts in asphaltenes compared to produced oils has been attributed to preferential adsorption of the asphaltenes onto mineral surfaces (Crocker & Marchin 1988; Dubey & Waxman 1989). Using published asphaltene adsorption data (Collins & Melrose 1983; Dubey & Waxman, 1989), it appears that *c.* 0.5 kg asphaltenes/tonne rock may typically be immobilized (presumably adsorbed) within an idealized sandstone reservoir (Wilhelms & Larter 1994*b*). Considering a reservoir volume of 1 km^3, approximately 0.7×10^6 to 1.2×10^6 tonnes of asphaltenes may be present in this reservoir volume as adsorbed or otherwise immobilized material.

The presence of such large amounts of asphaltenes 'fixed' in the reservoir, perhaps in part through adsorption, is important both in terms of mass-balance arguments and also as a constraint on mechanisms of tar-mat formation involving settling of asphaltene particles through the reservoir (see later).

Another important point to consider is the amount of asphaltenes present in the tar mats. Using average asphaltene yields from the case studies (Table 3) and oil-leg heights (*c.* 80 m Oseberg Field, 35 m Ula Field), it can be seen that tar mats are significantly enriched in asphaltenes (up to 25 times). These asphaltene enrichments probably require exchange of several oil-column

Table 4. *Terms relating to tar-mat formation as used by the current authors*

Term	Definition
Deasphalting	The process of asphaltene precipitation from a crude oil, mechanism non-specific (Dahl & Speers 1985).
Gas deasphaltation	Asphaltene precipitation caused by increased solution gas content of an oil leg (gas injection; Milner *et al.* 1977; Hirschberg 1984; Dahl & Speers 1985).
Gravitational segregation	Compositional gradients induced by diffusive movement of species in a petroleum column as a result of a gravitational field. Molecules move by diffusion trying to reach equilibrium between their chemical and gravitational potential. Less dense species tend to concentrate at the top of the reservoir, denser species at the base of the reservoir (Schulte 1980; Hirschberg 1984).
Biodegradation	Compositional change in a petroleum due to microbial attack, usually involving the consumption (removal) of *n*-alkanes and other petroleum components by aerobic bacteria (Connan 1984).
In-reservoir/ Carrier maturation	Thermal alteration of a petroleum or petroleum components in the reservoir or carrier system. Certain crude-oil components such as asphaltenes will have similar degradation kinetics to source-rock kerogen and can be expected to decompose significantly under geological conditions at temperatures *c.* 150°C. Such processes, and ultimately even thermal destruction of other crude oil components, will precipitate both solvent-soluble and insoluble phases as hydrogen balance must be maintained (Rogers *et al.* 1972; Milner *et al.* 1977; Larter *et al.* 1990).
Oil mixing	Petroleum reservoirs act as integrators of the petroleum charges filling a trap. With time, the petroleum composition will tend to be higher gravity, higher GOR and more paraffinic as source maturation proceeds (Larter *et al.* 1990; Leythaeuser & Rückheim 1991). More paraffinic petroleums have a lower capacity to dissolve asphaltenes, so precipitation may result. This is a less extreme case of the gas deasphaltation process.
Thermal convection	Density-driven overturning of a petroleum column due to temperature-gradient induced density differences (Montel & Gouel 1985).

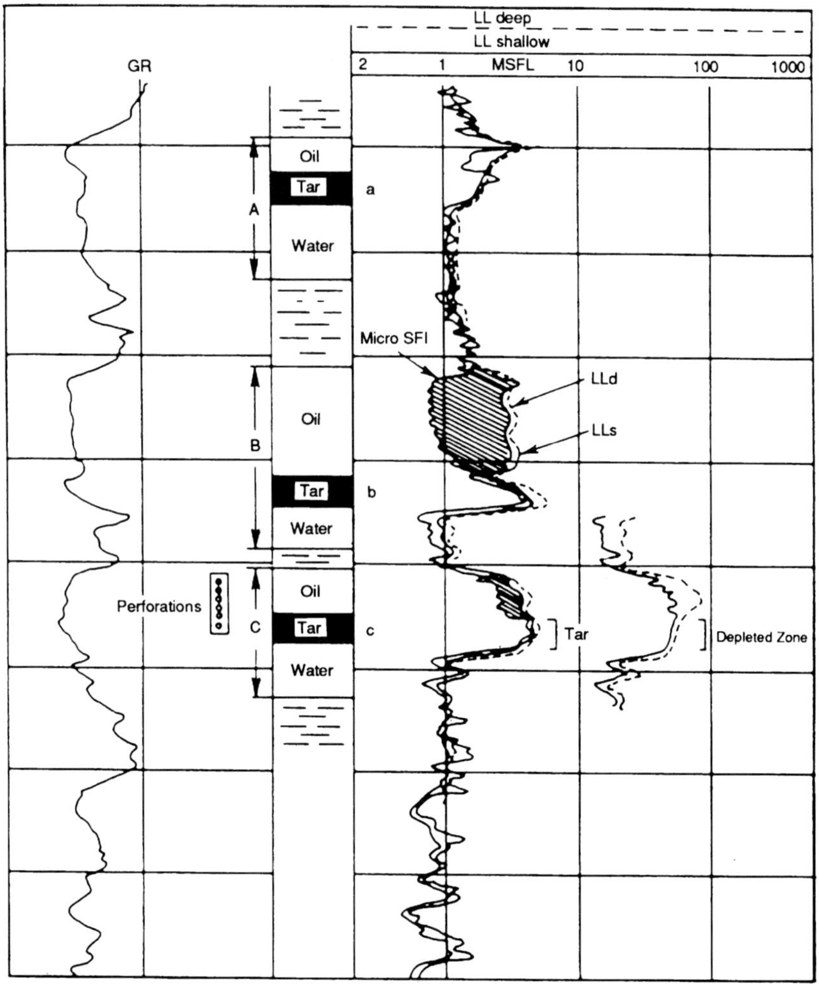

Fig. 4. Well log detected tar mats (modified after Dickey 1979, p. 213).

charges to account for the asphaltenes present (Wilhelms & Larter 1994*b*). We have previously estimated that around four and six volumes of current oil leg would be required to source the Ula and Oseberg tar mats, respectively. Figure 4 shows three log-detected staggered tar mats from an unidentified field. Here the ratio of oil-leg thickness to tar-mat thickness is two to three, and high initial asphaltene contents in the source oil would be needed to account for such features unless several oil volumes passing the site of tar-mat deposition were involved.

These general mass-balance assessments suggest that, in some cases, multiple volumes of oil were probably needed to account for the amounts of asphaltenes observed in the tar mats. This type of assessment puts constraints on some of those processes proposed to lead to tar-mat formation that derive the asphaltenes dominantly from the static overlying oil column (i.e. gravitational segregation, biodegradation and in-reservoir gas injection in a static oil column). Many of these mechanisms would only be viable if tar-mat formation occurred during field charging, when multiple oil volumes would be available near field-fill points.

Classification of tar-mat formation mechanisms

The processes invoked for the enrichment of asphaltenes from oil legs into tar mats can be subdivided as follows:

(a) accumulation/enrichment of precipitated asphaltenes by settling onto a physical barrier in a reservoir (OWC, sedimentological barrier);
(b) transport of APEs or weakly aggregated asphaltenes in solution by diffusion along potential gradients (chemical, thermal or gravitational) to the locus of APE/asphaltene enrichment.

In either case, the tar-mat forming asphaltenes might be contributed directly from in-place oil-column derived asphaltenes, from passage of several oil volumes past a site of asphaltene precipitation, from supply of precipitating asphaltenes from oil in the carrier system or from generation of asphaltenes in place by chemical reactions within the oil body or intra-reservoir source rocks.

Tar-mat formation mechanisms have previously been discussed in detail (Wilhelms & Larter 1994*b*). Here we simply group them as either potentially inviable or viable mechanisms. Let us recall that a viable tar-mat formation mechanism must account for:

(a) the occurrence of tar mats at geological boundaries (OWC and sedimentological boundaries; see Figs 1, 2 & 4);
(b) the occurrence of tar mats in zones of higher horizontal permeability and higher porosity (Lomando 1992; Wilhelms 1992; Wilhelms & Larter 1994*b*) than the adjacent reservoir and the sharp compositional contacts with the overlying oil column;
(c) the enrichment of asphaltenes in tar mats and the coeval nature of tar mats and oil legs, i.e. they are from the same source rock.

Additionally, one has to remember that any generic mechanism resulting in deasphaltation may lead to asphaltene-rich oil accumulations at a sedimentological (i.e. low permeability) barrier or phase (i.e. OWC) boundary simply due to physical constraints. However, such boundaries need not be predominantly involved in the precipitation mechanism itself, therefore, accumulation at an OWC need not imply any involvement of water in tar-mat formation.

Tar-mat formation mechanisms considered inviable for the study area

The settling of precipitated asphaltenes vertically through an oil column is not likely to lead to tar-mat formation (Wilhelms & Larter 1994*b*), as any minor permeability contrasts, such as silt or claystone layers, would be expected to lead to local 'micromat' formation. However, movement of APE/asphaltenes in solution along chemical or gravitational field gradients does seem plausible.

Asphaltene adsorption onto mineral surfaces has been widely noted (Clementz 1976; Czarnecka & Gillott 1980; Crocker & Marchin 1988; Dubey & Waxman 1989) and the amounts of asphaltenes adsorbed can be considerable. Apparently this is not, however, a major route to large-scale tar-mat formation as it appears that tar mats are associated with high, rather than low quality reservoir sections, i.e. all tar mats we have seen occur in the most porous and permeable, relatively clean reservoir sandstones (see Lomando 1992). Similarly, tar mats would be expected to be ubiquitous phenomena in more shaly reservoirs. We conclude that tar mats are not related to shaly sections, although clay layers may act as the physical barrier to downward asphaltene settling.

It has also been suggested, based on laboratory experiments and simple numerical modelling, that water-washing is probably not a viable tar-mat formation mechanism (Lafargue & Barker 1988). While biodegradation is capable of producing asphaltic oils (Rogers *et al.* 1972; Horsfield *et al.* 1991), initial to moderate stages of biodegradation are judged to be an unlikely mechanism for tar-mat formation in the present cases. This is based on the absence of signs of biodegradation in the three case studies, the lack of tar mats in at least one biodegraded North Sea petroleum reservoir (Horstad *et al.* 1990) and physico-chemical considerations. From a physico-chemical point of view, biodegradation of a petroleum, i.e. starting with preferred bacterial removal of *n*-alkanes, would be expected to enhance asphaltene solubility (Wilhelms & Larter 1994*b*) by increasing crude oil solubility parameters, i.e. enriching the oil in polar and aromatic species.

'In-carrier' mixing of two oil charges of different maturities, and thus composition, has also been suggested to have caused tar-mat formation in the Ula Field (Larter *et al.* 1990). The major difficulty with this mechanism stems from the difficulty of physically stirring large volumes of two oils. It is more likely that the more mature oil displaces the other oil with incomplete mixing (Leythaeuser & Rückheim 1989). Thus, in-carrier mixing seems unlikely to be involved in tar-mat formation, especially when one considers the oil volumes required.

Convection-related processes are not likely to be involved in tar-mat formation simply because there is no evidence that convection is a significant phenomenon in liquid petroleum columns. England *et al.* (1987, 1995) presented field and theoretical indications against the occurrence of thermal convection in petroleum reservoirs. Horstad *et al.* (1990) demonstrated that the Gullfaks Field petroleum column was not convective, and the persistence of lateral and vertical compositional fluctuations in most reservoirs is indicative of restricted overturning of petroleum columns.

Potentially viable tar-mat formation mechanisms

Diffusional transport of APE/asphaltenes in solution from the oil leg to the site of tar-mat formation plays a role in tar-mat formation, but, as described later, a purely diffusional process may not be entirely possible. Using diffusion coefficients for asphaltenes (Baltus & Anderson 1984), APE/asphaltene concentration equilibration in oil columns can be estimated to occur on scales of *c.* 100 to 150 m over petroleum-column history time-scales of *c.* tens of Ma (England *et al.* 1987). Thus, it is feasible that flocculating APEs in solution in oil could reach the site of tar-mat formation by diffusion through an oil column (hindered of course by adsorptive constraints). APE transport from an *in situ* oil column to the tar mat must occur primarily by diffusion along chemical and gravitational potential gradients. Increase in APE concentration near the base of an oil column may, with the correct oil composition, exceed local solubility maxima and asphaltenes may precipitate. This may also explain how tar mats form under laterally extensive permeability barriers (cf. Oseberg Field).

We consider it likely that should tar mats arise from precipitation from a bulk oil column in-reservoir, the actual phase change (i.e. from solution to solid phase) must take place near the tar mat itself, transport of flocculating APEs to the tar-mat site taking place in solution by diffusion. However, it seems unlikely that the rate of diffusion alone is adequate to move the total asphaltene mass required, and some advective component may be needed.

Gas solution in an oil leg (an extreme situation of petroleum mixing) has been discussed in the previous section. Provided the gas injection does not lead to 'catastrophic' asphaltene precipitation in the body of the reservoir (this did happen in the early history of the oil industry where enhanced oil recovery treatments using wet-gas injection reduced at least one field to a 'tarball'), there is abundant circumstantial and physico-chemical evidence that this is a viable locally active tar-mat formation mechanism (Rogers *et al.* 1972; Wilhelms & Larter 1994*b*).

Maturation of oil in place has been proposed as a mechanism causing tar-mat (Larter *et al.* 1990) and pyrobitumen formation during advanced stages of in-reservoir oil cracking (Rogers *et al.* 1974). None of the case studies examined here shows evidence for extensive oil cracking (Wilhelms & Larter 1994*b*); however, in the Ula Field signs of thermal degradation of the petroleum are found (Wilhelms & Larter 1994*b*), e.g. the *n*-hydrocarbon yield upon flash pyrolysis of the Ula asphaltenes is lower compared to Oseberg asphaltenes (Fig. 3), while Rock Eval T_{max} values are higher and the N/C, H/C data also show low values (Wilhelms & Larter 1994*a*). The *n*-hydrocarbon yields suggest reduced concentrations of *n*-alkyl groups on the asphaltenes, which might result in lowered asphaltene solubility in the Ula case (cf. APE hypothesis). This tentatively suggests that in-carrier/in-reservoir maturation may contribute to tar-mat formation as a destabilizing factor in the petroleum system.

It is possible that processes occurring during petroleum migration also may initiate APE/asphaltene precipitation (Wilhelms & Larter 1994*b*). During petroleum migration (both primary and secondary), pressure and temperature reduction occurs in the petroleum system, leading to reduced APE solubility in petroleum (cf. Hirschberg model), resulting in APE aggregation probably followed by separation of an asphaltene-rich material in the petroleum (Fig. 5). Residual asphaltene-rich petroleum in proposed secondary migration pathways has been observed (Miles 1990) and seems to support the hypothesis. Hypothetically, APE/asphaltene aggregates may be carried along during secondary migration as long as the dip of the carrier bed is bigger than a certain threshhold value, or, in contrast, left behind in the carrier system as the dip approaches zero, resulting in a decrease of buoyancy-driven migrational flux (Wilhelms & Larter 1994*b*).

Considering the tentative mass-balance arguments and the situation of tar mats in zones of elevated porosity, we tentatively propose that tar mats may form, in part, in feeder channels i.e. the carrier system for the main oil charge towards the trap. This model accounts for the multiple current oil-leg volumes needed to explain observed amounts of APE/asphaltenes in the tar mats, by influx of APE/asphaltenes or asphaltic-rich sludge or multiple oil volumes to the site of tar-mat formation.

This model is consistent with and potentially supported by the occurrence of tar mats in reservoir zones of increased horizontal permeability and high porosity, the amounts of asphaltenes observed and the occurrence of the tar mats both close to the OWC (Oseberg Field) and in the oil leg (Ula Field). However, several inconsistencies exist. It is not clear why some tar mats (e.g. Prudhoe Bay, Oseberg) are situated close to the OWC, which might imply that the APE/asphaltenes have settled some vertical distance. At least in Oseberg, how-

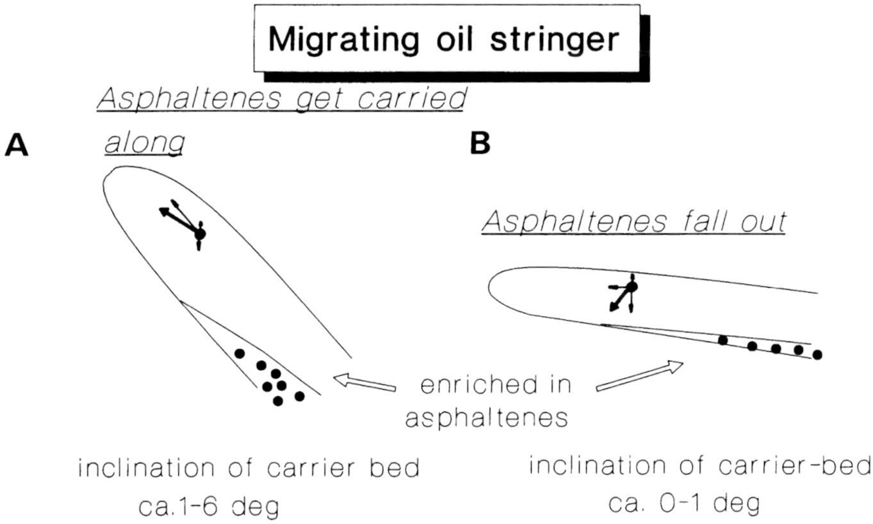

Fig. 5. **(a)** Schematic figure illustrating the movements of asphaltene particles in a migrating oil stringer in an inclined carrier. **(b)** Schematic figure illustrating the movement of asphaltene particles in an almost horizontal carrier.

ever, the OWC is in line with the present-day spill point of the structure (Dahl & Yükler 1991). Perhaps the biggest problem with the model is in explaining why tar mats are not ubiquitous, which is probably the more interesting question to ask. Additionally, this model still needs a diffusional/gravitational component in order to transport the APE/asphaltene aggregates either in dissolved or aggregated form to the base of the carrier/feeder system.

Conclusions, status quo and outlook

The geochemical signature indicates that in all the examples studied, tar mats are related to oil in the reservoir and were emplaced during or after reservoir filling. Based on these data and the solubility characteristics of APEs/asphaltenes in subsurface crude oils, this leads to the following conclusions.

(1) It is currently not possible to assign, conclusively, origins to the tar mats. However, formation mechanisms can be classified according to their feasibility and we can eliminate some mechanisms as being unlikely to contribute to tar-mat formation in a general context. Table 5 summarizes the general mechanisms proposed and correlates them with reported data. Two mechanisms seem to be improbable as tar-mat forming processes. (a) Gravitational fallout of flocculated particulate asphaltenes from a free oil column, and asphaltene adsorption onto clays can probably be eliminated as viable tar-mat formation mechanisms, otherwise small 'micromats' would be everywhere in the reservoir. (b) The compositional dependence of the solvent properties of crude oils strongly suggests that biodegradation of petroleum is likely to stabilize asphaltene solution in crude oil, and we are confident that light and moderate levels of biodegradation are most unlikely to be causes of tar-mat formation.

(2) A mass-balance problem seems to exist in some instances relating tar mats to current oil columns. This and the occurrence of tar mats, at least locally, under horizontal barriers, suggests that the tar mats are related in some instances to reservoir-filling processes (i.e. in feeder channels) or that APE diffusion in solution around barriers (perhaps along gravitationally induced compositional gradients) has occurred to the site of asphaltene precipitation. Gas injection into a migrating or stationary petroleum charge could probably form a tar mat below a barrier if the asphaltenes were transported to the site of ultimate precipitation by diffusion (Oseberg Field). Similarly, thermal alteration of the oil at an early stage, probably in the carrier (Ula Field), may have contributed to tar-mat formation. Multiple oil volumes must have been needed, however, and therefore the early stages of asphaltene flocculation most likely took place in the carrier system.

(3) While accurate mass balances on tar mat–oil systems are difficult to perform, those we have carried out, plus the strong pressure dependence of asphaltene solubility in crude oils, clearly suggest that there is room to consider migration-related tar-

Table 5. *Summary of requirements that a mechanism for tar-mat formation must fulfil in order to explain the characteristics of tar mats*

	Requirement*							
Mechanism	1	2	3	4	5	6	7	8
Gravitational fallout of flocculated asphaltenes from free oil column	×	×	×	–?	–?	❍	❍	–
Gravitational segregation of APEs	×	×	×	–	–	–	❍	–
Biodegradation-induced asphaltene precipitation	❍	×	–	–	–	–	–	❍
Gas injection-induced asphaltene precipitation	–	×	×	–	–	–	–	–
Pressure reduction (uplight) precipitation	×	–	–	–	–	–	❍	–
Adsorption onto clays, asphaltene precipitation	–	–	–	–	❍	❍	–	❍
Maturation-induced asphaltene precipitation	–	–	×	–	–	–	–	–
Oil mixing-induced asphaltene precipitation	–	×	×	–	–	×	×	–
Migration-controlled asphaltene precipitation and accumulation	×	×	–	–	×	×	×	–

*

1 Occurrence of tar mat both at base of oil column or mid-oil column.
2 Relative enrichment of tar-mat extract with asphaltenes (30–50 wt%).
3 Molecular weight differences between oil-column and tar-mat asphaltenes.
4 Yield of asphaltenes observed in tar mats.
5 Sharp compositional contact to oil column.
6 Occurrence in high permeability and porosity zones.
7 Bordered above by low permeability zones.
8 Observed composition of tar mats.

×, Definitely met.
–, Possibly met.
❍, Not possible.

mat precipitation mechanisms. Asphaltic residues in migrational pathways plus simple models do suggest that pressure reduction during secondary migration of petroleum and influx of asphaltic-rich petroleum from the carrier may be a possible additional mechanism for tar-mat formation.

Obviously, much further study is needed before we are capable of modelling tar-mat occurrence. In recent years, reservoir geochemistry has come of age and demonstrated its use, with interesting applications beginning to emerge (England *et al.* 1995; Larter & Aplin 1995). Further, as discussed previously, enormous quantities of asphaltenes and other polar species are adsorbed onto mineral surfaces in petroleum reservoirs. Integration of organic geochemistry with reservoir geology and sedimentology/diagenesis studies may help us to gain insight into the factors controlling composition and quantity of polar species adsorbed. Field wide reservoir geochemical studies using standard analytical techniques for saturated and aromatic hydrocarbon fraction show great potential, but resin and asphaltene fractions remain poorly characterized despite their crucial roles in many geological processes such as tar-mat formation, reservoir-surface wetting, crude oil physical properties etc. (Larter & Aplin 1995).

Following the work of Yamamoto (1992), nitrogen compound analyses are starting to be used in reservoir geochemical studies (Li *et al.* 1992; Li *et al.* 1995) but the methods remain significantly mired in gas chromatography-based analytical technology for routine work.

It is clear that in order to fully explain tar-mat occurrence, more detailed examination of asphaltene and resin characteristics at the molecular level is required, with a real need for methods capable of characterizing functionality and molecular size on a routine basis. Similarly, our understanding of the compositional phase behaviour of tar-mat prone oil charges is general and poor (e.g. the PVT influence of C_{15+} compounds lacks precision), and the understanding of organic–rock interaction lacks sophistication.

It is likely that features such as tar mats may form as a result of subtle changes in petroleum environment, and our current understanding is perhaps inadequate for realistic predictive models. Our experience of small tar mats in petroleum reservoirs (such as those in the Ula Field) continues to grow and, in some cases, they do enhance existing barriers or baffles in petroleum reservoirs and are thus relevant to understanding petroleum production problems. Detection of these features is a developing art. Classic reservoir geochemical screening for small tar mats by Iatroscan (Karlsen & Larter 1989, 1991) or Rock-Eval (Wilhelms 1992; Ghenima 1993; Wilhelms & Larter 1994*a*) or other core-based methods (Wilhelms *et al.* 1994) is now viable and recommended. Improved log-detection methods can be expected to be developed (Desbrandes 1982; Herron 1987; Wilhelms *et al.* 1994).

The status quo appears to be that while we are capable of readily finding and identifying tar mats and characterizing them, several areas exist where new knowledge is needed. With analytical techniques now capable of detecting small tar mats ('micromats'), we hope that new approaches and technologies will be applied to these phenomena.

This study is part of a collaborative German–Norwegian research programme financed by VISTA. A.W. acknowledges a NTNF fellowship and financial support from IFP during post-doctoral studies. We thank BP Norway for financial support of the Organic Geochemistry Project at the University of Oslo, Department of Geology, and Norsk Hydro, BP Norway, BP UK and Unocal for providing sample material.

References

ALI, L. H. & AL-GHANNAM, K. A. 1981. Investigation into asphaltenes in heavy crude oils. I. Effect of temperature on precipitation by alkane solvents. *Fuel*, **60**, 1043–1046.

BALTUS, R. E. & ANDERSON, J. L. 1984. Comparison of g.p.c. elution characteristics and diffusion coefficients of asphaltenes. *Fuel*, **63**, 530–535.

BILLHEIMER, J. S., YUNDT, C. G., SAGE, B. H. & LACEY, W. N. 1949. Multiple condensed phases in the methane–decane–tetralin–bitumen system. *Transactions of the American Institute of Mining and Metallurgical Engineers*, **186**, 265–270.

BRIANT, J. 1963. Sur quelques facteurs influençant la formation de certains dépots (paraffins, asphaltènes) dans les installations de production. *Revue de l'Institut Français du Pétrole*, **18**, 1–16.

BUNGER, J. W., THOMAS, K. P. & DORRENCE, S. M. 1979. Compound types and properties of Utah and Athabasca tar sand bitumens. *Fuel*, **58**, 183–195.

BURKE, N. E., HOBBS, R. E. & KASHOU, S. F. 1990. Measurement and modeling of asphaltene precipitation. *Journal of Petroleum Technology*, **42**, 1440–1446.

CLEMENTZ, D. M. 1976. Interaction of petroleum heavy ends with montmorillonite. *Clays and Clay Minerals*, **24**, 312–319.

COLLINS, S. H. & MELROSE, J. C. 1983. *Adsorption of asphaltenes and water on reservoir rock minerals*. Society of Petroleum Engineers, Paper No. **11800**, 249–256.

CONNAN, J. 1984. Biodegradation of crude oils in reservoirs. *In*: BROOKS, J. & WELTE, D. H. (eds) *Advances in Petroleum Geochemistry*. Academic Press, London, 299–335.

CROCKER, M. E. & MARCHIN, L. M. 1988. Wettability and adsorption characteristics of crude-oil asphaltene and polar fractions. *Journal of Petroleum Technology*, **40**, 470–474.

CZARNECKA, E. & GILLOTT, J. E. 1980. Formation and characterization of clay complexes with bitumen from Athabasca oil sand. *Clays and Clay Minerals*, **28**, 197–203.

DAHL, B. & SPEERS, G. C. 1985. Organic geochemistry of the Oseberg Field (I). *In*: THOMAS, B. M. *et al.* (eds) *Petroleum Geochemistry in Exploration of the Norwegian Shelf*. Norwegian Petroleum Society, Graham & Trotman, London, 185–195.

—— & —— 1986. Geochemical characterization of a tar mat in the Oseberg Field Norwegian sector, North Sea. *Organic Geochemistry*, **10**, 547–558.

—— & YÜKLER, A. 1991. The role of petroleum geochemistry in basin modeling of the Oseberg Area, North Sea. *In*: MERRILL, R. K. (ed.) *Source and Migration Processes and Evaluation Techniques*. Treatise of Petroleum Geology, Handbook of Petroleum Geology, American Association of Petroleum Geologists, Tulsa, Oklahoma, 65–85.

DESBRANDES, R. 1982. *Diagraphies dans les sondages*. Editions Technip, Paris.

DICKEY, P. A. 1979. *Petroleum Development Geology*. The Petroleum Publishing Company, Tulsa, Oklahoma.

DUBEY, S. T. & WAXMAN, M. H. 1989. *Asphaltene adsorption and desorption from mineral surfaces*. Society of Petroleum Engineers, Paper No. **18462**, 51–62.

ENGLAND, W. A., MACKENZIE, A. S., MANN, D. M. & QUIGLEY, T. M. 1987. The movement and entrapment of petroleum fluids in the subsurface. *Journal of the Geological Society, London*, **144**, 327–347.

——, MUGGERIDGE, A. H., CLIFFORD, P. J. & TANG, Z. 1995. Modelling density-driven mixing rates in petroleum reservoirs on geological time-scales, with application to the detection of barriers in the Forties Field (UKCS). *This volume*.

FLORY, P. J. 1953. *Principles of Polymer Chemistry*. Cornell University Press, Ithaca, NY.

GHENIMA, R. 1993. *Etude des roches-meres paleozoiques du bassin de Ghedames. Modelisation des hydrocarbures migres et application a l'etude du gisement d'el Borma*. Ph.D. thesis, University of Orleans, France.

HERRON, M. M. 1987. Future applications of elemental concentrations from geophysical logging. *Nuclear Geophysics*, **1**, 197–211.

HIRSCHBERG, A. 1984. *The role of asphaltenes in compositional grading of a reservoir's fluid column*. Society of Petroleum Engineers, Paper No. **13171**.

—— & HERMANS, L. 1984. Asphaltene phase behaviour: A molecular thermodynamic model. *In*: *Characterisations des huiles lourdes et des residues pétroliers*. Édition Technip, Paris, 492–497.

——, DEJONG, L. N. J., SCHIPPER, B. A. & MEIJER, J. G. 1984. *Influence of temperature and pressure on asphaltene flocculation*. Society of Petroleum Engineers Paper No. **11202**, 283–293.

HORSFIELD, B., HECKERS, J., LEYTHAEUSER, D., LITTKE, R. & MANN, U. 1991. A study of the Holzener Asphaltkalk, Northern Germany: Observations regarding the distribution, composition and origin of organic matter in an exhumed petroleum reservoir. *Marine and Petroleum Geology*, **8**, 198–211.

HORSTAD, I., LARTER, S. R., DYPVIK, H., AARGAARD, P., BJØRNVIK, A. M., JOHANSEN, P. E. & ERIKSEN, S. 1990. Degradation and maturity controls on oil field petroleum column heterogeneity in the Gullfaks field, Norwegian North Sea. *Organic Geochemistry*, **16**, 497–510.

HUNTER, R. J. 1989. *Foundations of Colloid Science, Vol. 1*. Oxford Science Publications, New York.

JONES, H. P. & SPEERS, R. G. 1976. Permo-Triassic reservoir of Prudhoe Bay field. North Slope, Alaska. *American Association of Petroleum Geologists Memoir*, **24**, 23–50.

KARLSEN, D. & LARTER, S. R. 1989. A rapid correlation method for petroleum population mapping within individual petroleum reservoirs: Applications to petroleum reservoir description. *In*: COLLINSON, J. D. (ed.) *Correlation in Hydrocarbon Exploration*. Norwegian Petroleum Society, Graham & Trotman, London, 77–85.

—— & —— 1991. Analysis of petroleum fractions by TLC-FID: Applications to petroleum reservoir description. *Organic Geochemistry*, **17**, 603–617.

LAFARGUE, E. & BARKER, C. 1988. Effect of water washing on crude oil composition. *American Association of Petroleum Geologists, Bulletin*, **72**, 263–276.

LARTER, S. R. & APLIN, A. C. 1995. Reservoir geochemistry: methods, applications and opportunities. *This volume*.

—— & MILLS, N. 1991. Phase-controlled molecular fractionations in migrating petroleum charges. *In*: ENGLAND, W. A. & FLEET, A. J. (eds) *Petroleum Migration*. Geological Society, London, Special Publication, **59**, 137–147.

——, BJØRLYKKE, K. O., KARLSEN, D. A. ET AL., 1990. Determination of petroleum accumulation histories: Examples from the Ula field, Central Graben, Norwegian North Sea. *In*: BULLER, A. (ed.) *North Sea Oil and Gas Reservoirs*. Graham & Trotman, London, 319–330.

LEONTARITIS, K. J. & MANSOORI, G. A. 1987. *Asphaltene flocculation during production and processing: A thermodynamic colloidal model*. Society of Petroleum Engineers, Paper No. **16258**.

LEYTHAEUSER, D. & RÜCKHEIM, J. 1989. Heterogeneity of oil composition within a reservoir as a reflection of accumulation history. *Geochimica et Cosmochimica Acta*, **53**, 2119–2123.

——, SCHAEFER, R. G. & RADKE, M. 1988. Geochemical effects of primary migration of petroleum in Kimmeridge source rocks from Brae field area, North Sea. I: Gross composition of C_{15+}-soluble organic matter and molecular composition of C_{15+}-saturated hydrocarbons. *Geochimica et Cosmochimica Acta*, **52**, 701–713.

LHIOREAU, C., BRIANT, J. & TINDY, R. 1967. Influence de la pression sur la flocculation des asphaltènes. *Revue de l'Institut Français du Pétrole*, **22**, 797–806.

LI, M., LARTER, S. R., STODDART, D. & BJORØY, M. 1992. Liquid separation schemes for pyrrole and pyridine nitrogen aromatic heterocycle fractions from crude oils suitable for rapid characterisation of geochemical samples. *Analytical Chemistry*, **64**, 1337–1344.

——, ——, —— & —— 1995. Fractionation of pyrrolic nitrogen compounds in petroleum during migration: derivation of migration-related geochemical parameters. *This volume*.

LOMANDO, A. J. 1992. The influence of solid reservoir bitumen on reservoir quality. *American Association of Petroleum Geologists, Bulletin*, **76**, 1137–1152.

MILES, J. A. 1990. Secondary migration routes in the Brent sandstones of the Viking Graben and East Shetland basin: Evidence from oil residues and subsurface pressure data. *American Association of Petroleum Geologists, Bulletin*, **74**, 1718–1735.

MILNER, C. W. D., ROGERS, M. A. & EVANS, C. R. 1977. Petroleum transformations in reservoirs. *Journal of Geochemical Exploration*, **7**, 101–153.

MONTEL, F. & GOUEL, P. L. 1985. *Prediction of compositional grading in a reservoir fluid column*. Society of Petroleum Engineers, Paper No. **14410**.

PARK, S. J. & MANSOORI, G. A. 1988. Aggregation and deposition of heavy organics in petroleum crudes. *Energy Sources*, **10**, 109–125.

ROGERS, M. A., BAILEY, N. J. L., EVANS, C. R. & MCALARY, J. D. 1972. An explanatory and predictive model for the alteration of crude oils in reservoirs in the Western Canada Basin. *In*: *Mineral fuels – Combustibles Mineraux*. Section 5, 24th International Geological Congress, Montreal, Canada, 48–55.

——, MCALARY, J. D. & BAILEY, N. J. L. 1974. Significance of reservoir bitumens to thermal-maturation studies, Western Canada Basin. *American Association of Petroleum Geologists, Bulletin*, **58**, 1806–1824.

SCHULTE, A. M. 1980. *Compositional variations within a hydrocarbon column due to gravity*. Society of Petroleum Engineers, Paper No. **9235**.

SCOTT, L. & MAGAT, M. 1945. The thermodynamics of high-polymer solutions: I. The free energy of mixing of solvents and polymers of heterogeneous distribution. *Journal of Chemical Physics*, **13**, 172–177.

TISSOT, B. & PELET, R. 1971. Nouvelles données sur les mécanismes de genèse et de migration du pétrole. Simulation mathématique et application à la prospection. *Proceedings of the 8th World Petroleum Congress, volume* **2**, 35–46.

TORVUND, T. & NIPEN, O. 1987. The Oseberg field: Potential for increased oil recovery by gas injection in a high permeability environment. *In*: KLEPPE, J., BERG, E. W., BULLER, A. T., HJELMELAND, O. & TORSAETER, O. (eds) *North Sea Oil and Gas Reservoirs*. Graham & Trotman, London, 61–74.

WILHELMS, A. 1992. *An investigation into the factors influencing tar mat formation in petroleum reservoirs*. PhD thesis, University of Oslo, Norway.

—— & LARTER, S. R. 1994*a*. On the origin of tar mats in petroleum reservoirs: Part I. Introduction and case studies. *Marine and Petroleum Geology*, **11**, 418–441.

—— & —— 1994*b*. On the origin of tar mats in petroleum reservoirs: Part II. Formation mechanisms for tar mats. *Marine and Petroleum Geology*, **11**, 442–456.

——, CARPENTIER, B. & HUC, A. Y. 1994. New methods to detect tar mats in petroleum reservoirs. *Journal of Petroleum Science & Engineering*, **12**, 147–155.

——, LARTER, S. R., LEYTHAEUSER, D. & DYPVIK, H. 1990. Recognition and quantification of the effects of primary migration in a Jurassic clastic source-rock from the Norwegian continental shelf. *Organic Geochemistry*, **16**, 103–113.

YAMAMOTO, M. 1992. Fractionation of azarenes during oil migration. *Organic Geochemistry*, **19**, 389–402.

Fractionation of pyrrolic nitrogen compounds in petroleum during migration: derivation of migration-related geochemical parameters

MAOWEN LI[1,3], S. R. LARTER[1], D. STODDART[1] & M. BJORØY[2]

[1] *Fossil Fuels & Environmental Geochemistry (Postgraduate Institute): NRG, Drummond Building, University of Newcastle upon Tyne, Newcastle upon Tyne, NE1 7RU, UK*

[2] *Geolab Nor, Hornebergveien 5, PO Box 1581, 7002 Trondheim, Norway*

[3] *Present address: Geological Survey of Canada, Institute of Sedimentary & Petroleum Geology, 3303 33rd St, N.W. Calgary, Alberta, Canada*

Abstract: Analysis of a range of crude oils and source rocks appears to indicate that primary and secondary oil migration has a strong impact on the distribution of pyrrolic nitrogen compounds in reservoired petroleum. The migration effects controlling the composition of pyrrolic nitrogen compounds recognized tentatively in this study include (a) the relative enrichments of alkyl carbazoles to alkylbenzocarbazoles, (b) nitrogen-shielded isomers to nitrogen-exposed isomers, and (c) higher homologues to the lower homologous species. A suite of molecular migration parameters based on chemical compositions of pyrrolic nitrogen compounds in crude oils have been proposed. If substantiated in more case history studies, these parameters are potentially useful in petroleum population mapping.

While hydrocarbon compositions are quickly equilibrated vertically over a few tens of metres in reservoired petroleum columns (England 1990), there is substantial evidence that the same is not true for polar species such as resins and asphaltenes (Stoddart *et al.* 1995). Large heterogeneities in polar compound concentration also exist in the form of tar mats (sharply defined zones of asphaltene-rich petroleum, 1 m to tens of metres thick, usually occurring in light paraffinic oil reservoirs) (Larter *et al.* 1991). In addition to these macroscopic compositional features, the work of Horstad *et al.* (1990) suggests that certain polar species are adsorbed onto reservoir surfaces, resulting in increased concentrations of polar compounds in reservoir core extracts as compared to bulk drill-stem test (DST) oil samples produced from the same depth intervals.

Organic nitrogen contents in most oils range from 0.1 to 2.0%. Although present in such small quantities, organic nitrogen compounds have been studied for many years, largely because of their role as a health hazard and involvement in catalyst poisoning and sludge formation in distillate products (Schmitter & Arpino 1985; Jokuty & Gray 1991). Organic nitrogen in crude oils is present mainly as aromatic cycles with a predominance of neutral pyrrolic structures over basic pyridinic forms. For example, Richter *et al.* (1952) reported that crude oils have typically 25 to 30% of the nitrogen in basic form. Most of the nitrogen in crude oils is associated with high boiling-point petroleum fractions, principally asphaltenes (Tissot & Welte 1984), and recent work (Wilhelms *et al.* 1992) suggests that approximately one-third of asphaltene nitrogen is typically in alkylated pyridine, azanaphthalene and azaphenanthrene forms, the remainder being pyrrolic in nature. The majority of published geochemical studies of organic nitrogen compounds in sediments, fossil fuels and environmental samples have focused on the structural determination of these compounds. A great deal of such work was done in the 1930s and 1940s at the University of Texas (Perrin & Bailey 1933; Schenck & Bailey 1941*a,b*; Shive *et al.* 1942), in the 1960s at the Union Oil Company of California (Snyder 1970) and in the early 1980s at the Laboratoire de Chimie Analytique Physique of Ecole Polytechnique in France (Schmitter & Arpino 1983; Schmitter *et al.* 1982, 1983; Dorban *et al.* 1982, 1984*a,b*; Ignatiadis *et al.* 1985). For more comprehensive literature reviews see Schmitter & Arpino (1985), Yamamoto *et al.* (1991) and Yamamoto (1992) on basic nitrogen compounds; Frolov *et al.* (1989) on pyrrolic nitrogen compounds, and Bakel (1990), Bakel & Philp (1990) and Li *et al.* (1992) on organic nitrogen compounds in general. An important property of the aromatic nitrogen heterocycles is their capacity for interactions with reservoir

From CUBITT, J. M. & ENGLAND, W. A. (eds), 1995, *The Geochemistry of Reservoirs*, Geological Society Special Publication No. 86, pp. 103–123

surfaces or formation water via hydrogen bonding (pyrrolic nitrogen compounds), and ionic bonding or hydrogen bonding (basic pyridinic nitrogen compounds) (Larter & Aplin, this volume).

Polar compound interactions with reservoir surfaces have clear implications both for practical concerns, such as reservoir production, and for the fundamental study of petroleum migration processes and reservoir filling mechanisms. For example, the strongly adsorbing organic nitrogen and oxygen compounds have been suggested to play a significant role in influencing reservoir wettability (Mitchell *et al.* 1990). In the study of a series of crude oils from the Sarukawa Oilfield (southeastern Japan), Yamamoto (1992) observed systematic variations in the molecular compositions of alkylbenzoquinolines from source rocks to oils and with increasing migration distance. He suggested that distributions of basic nitrogen compounds can be used to estimate the degree of fractionation of crude oils during petroleum migration.

In the present study, a wide range of crude oils and source rock extracts have been analysed for organic nitrogen compounds, aimed at (1) determining the general molecular distributions of organic nitrogen compounds in a range of crude oils, (2) studying the possible effects of organic source inputs, depositional environment, thermal maturation and oil migration, and (3) on the basis of the study proposing a series of organic nitrogen molecular parameters relevant to petroleum reservoir filling. We discuss in this paper only pyrrolic nitrogen fractions. The results of our work on pyridinic nitrogen fractions will be published later. Though standing on its own, this paper should be read together with Stoddart *et al.* (1995).

Samples

Two oils and six shales from Liaohe Basin, China, two oils and 11 shales from Los Angeles Basin, California and 18 oils from a variety of other fields were included in this study (Tables 1 & 2).

Samples from the western depression of the Liaohe Basin, northeast China

These samples consist of two oils (Os4 and Os3) and six shales. The geology of the western depression of the Liaohe Basin has been described previously (Huang *et al.* 1984), and the regional petroleum geochemical investigation on the major source rock unit, the Shahejie Formation (Es), can be found in Yang *et al.* (1982) and Huang *et al.* (1984). The Os4 oil was produced from the Es4 reservoir (Fig. 1a) which is vertically interbedded with and laterally in direct contact with mature Es4 source rocks, whilst the Os3 oil was produced from the shallower Es3 reservoir which is interbedded with the immature Es3 mudstones. Lu *et al.* (1990) reported that the distributions of acyclic isoprenoid

Table 1. *Description of crude oil samples used in this study*

Code	Basin or field	Age	Reservoir Fm unit	Source rock type
W153	Nanyang	Oligocene	–	Inland freshwater lacustrine
S452	Nanyang	Oligocene	–	Inland freshwater lacustrine
L49	Ordos	Triassic	Yanchang	Inland freshwater lacustrine
F135	Songliao	Cretaceous	Putaohua	Paralic freshwater lacustrine
P15	Shengli	Eocene	No. 4 Shahejie	Paralic freshwater lacustrine
Y67	Shengli	Oligocene	No. 3 Shahejie	Paralic freshwater lacustrine
Y18	Shengli	Eocene	No. 4 Shahejie	Paralic freshwater lacustrine
Os3	Liaohe	Oligocene	No. 3 Shahejie	Paralic freshwater lacustrine
Os4	Liaohe	Eocene	No. 4 Shahejie	Paralic freshwater lacustrine
W32	Jianghan	Oligocene	No. 3 Qianjiang	Inland brackish–brine lacustrine
G33	Jianghan	Oligocene	No. 3 Qianjiang	Inland brackish–brine lacustrine
W203	Zhongyan	Paleocene	–	Inland brackish–brine lacustrine
H79	Chaidam	Miocene	U. Ganchaigou–	Inland brackish lacustrine
CA1	Chaidam	Jurassic	Dameigou	Inland swampy lake
PDR1	Los Angeles	Miocene	Lower oil zone	Marine
PDR2	Los Angeles	Pliocene	Upper oil zone	Marine
A1	North Sea	Danian	–	Marine
A2	North Sea	Danian	–	Marine
A3	North Sea	Danian	–	Marine
B1	North Sea	Maastrichtian	–	Marine
C2	North Sea	Maastrichtian	–	Marine
C2	North Sea	Maastrichtian	–	Marine

Table 2. *Descriptions and basic geochemical data for the shale samples used in this study**

Code	Age	Formation	Depth (m)	TOC (%)	HI†	T_{max} (°C)	$\beta\beta(\alpha\alpha+\beta\beta)$‡	S/(S + R)§
Liaohe shales								
8611	Eocene	Shahejie	1279.0	3.8	412	414	0.15	0.23
8613	Eocene	Shahejie	3601.0	2.2	288	430	0.39	0.45
8614	Eocene	Shahejie	2305.5	2.8	207	416	0.16	0.22
8615	Eocene	Shahejie	2789.2	4.5	449	422	0.20	0.21
8616	Eocene	Shahejie	2928.5	2.1	169	434	0.63	0.57
8617	Eocene	Shahejie	3122.0	2.8	285	427	0.38	0.46
Californian shales								
PDR1	U. Miocene	Middle E	3103.7–3104.0	7.5	302	427	0.24	0.21
PDR2	U. Miocene	Lower E	3164.6–3164.9	–	–	–	0.05	0.17
PDR4	U. Miocene	Upper E	1757.0–1758.8	7.5	405	414	0.14	0.21
PDR5	Pliocenee	Repetto	1307.9–1405.5	1.9	52	415	0.50	0.48
PDR6	U. Miocene	Middle E	1440.9–1450.3	9.9	416	393	0.24	0.24
PDR7	U. Miocene	Middle E	1956.1–1966.1	10.1	454	412	0.14	0.17
PDR8	U. Miocene	A	1693.3–1755.8	1.4	88	420	0.10	0.12
PDR9	U. Miocene	Lower E	3163.6–3164.9	6.4	185	419	0.43	0.28
PDR10	Pliocene	Repetto	984.5–1085.7	3.5	80	410	0.05	0.03
PDR11	Pliocene	Repetto	1161.0–1236.3	3.9	86	424	0.41	0.26
PDR12	U. Miocene	Lower E	1994.8–2010.7	16.4	467	399	0.46	0.47

* Each set of samples was taken from different wells
† Hydrogen index (mg HC/g TOC)
‡ $\beta\beta(\alpha\alpha+\beta\beta)$ ratio for C_{29} steranes
§ 20S/(20S + 20R) ratio for C_{29} steranes

alkanes (i.e. predominance of phytane over pristane), steranes and terpanes for oil from both the Es3 reservoir (Os3) and the Es4 reservoir (Os4) have a good match with the Es4 source rocks. However, they are different from the Es3 source rocks, suggesting that both oils were sourced from the Eocene Es4 source rocks, with the Os4 oil being sourced locally and the Os3 oil having migrated a longer distance to the shallower depth. According to Lu *et al.* (1990), other parameters, such as the distributions of *n*-alkanes, methylphenanthrenes and aromatic steranes, are also consistent with oils derived mainly from the Es4 source rocks. The Es4 source rock sequence in the sampling area belongs to a special lithological section composed mainly of carbonate, dolomite and oil shale which contain type II kerogens consisting of mixtures of lacustrine algal remains and terrestrial plant debris. Six shale samples from different burial depths were also collected from the Es sequence.

Samples from Los Angeles Basin, California

A generalized structural section for the samples collected is shown in Fig. 1b. Eleven shale samples covering the stratigraphic interval from the top of the Lower Pliocene to the bottom of the Upper Miocene were sampled. One of the shales (sample PDR2, Table 2) contains 'tarball' (migrated solid bitumen in the shale fractures). Two crude oils produced from unrelated reservoirs (J. R. Fox, pers. comm.) were also studied. The major oil source in this area was the black nodular marine shale (Miocene Division E) containing organic matter derived mainly from algal material (Walker *et al.* 1983, J. R. Fox, pers. comm.).

Other oil samples

Eighteen other crude oils from a variety of sources were also included in this study. None of these oils shows any evidence of biodegradation. On the basis of the geological relationships and previous studies of oil–source rock correlation (e.g. Huang & Li 1982; Wang *et al.* 1988; Fu *et al.* 1986, 1991), the depositional environments of source rocks for these oils can be divided into four groups.

Group A. Paralic lake. Crude oils belonging to this group include F135 (Daqin Oilfield), P15, Y67 and Y18 (Shengli Oilfield), deriving from either mainly lacustrine algal (type I) or mixed algal/terrestrial type II organic matter (Yang *et al.* 1982; Shi *et al.* 1982).

Group B. Inland freshwater lake. High wax oils derived mainly from type III kerogens of higher plant material from the Ordos Basin (L49) and Nanyang Oilfield (W153 & S452) are examples in this group (Huang & Li 1982; Huang *et al.* 1984).

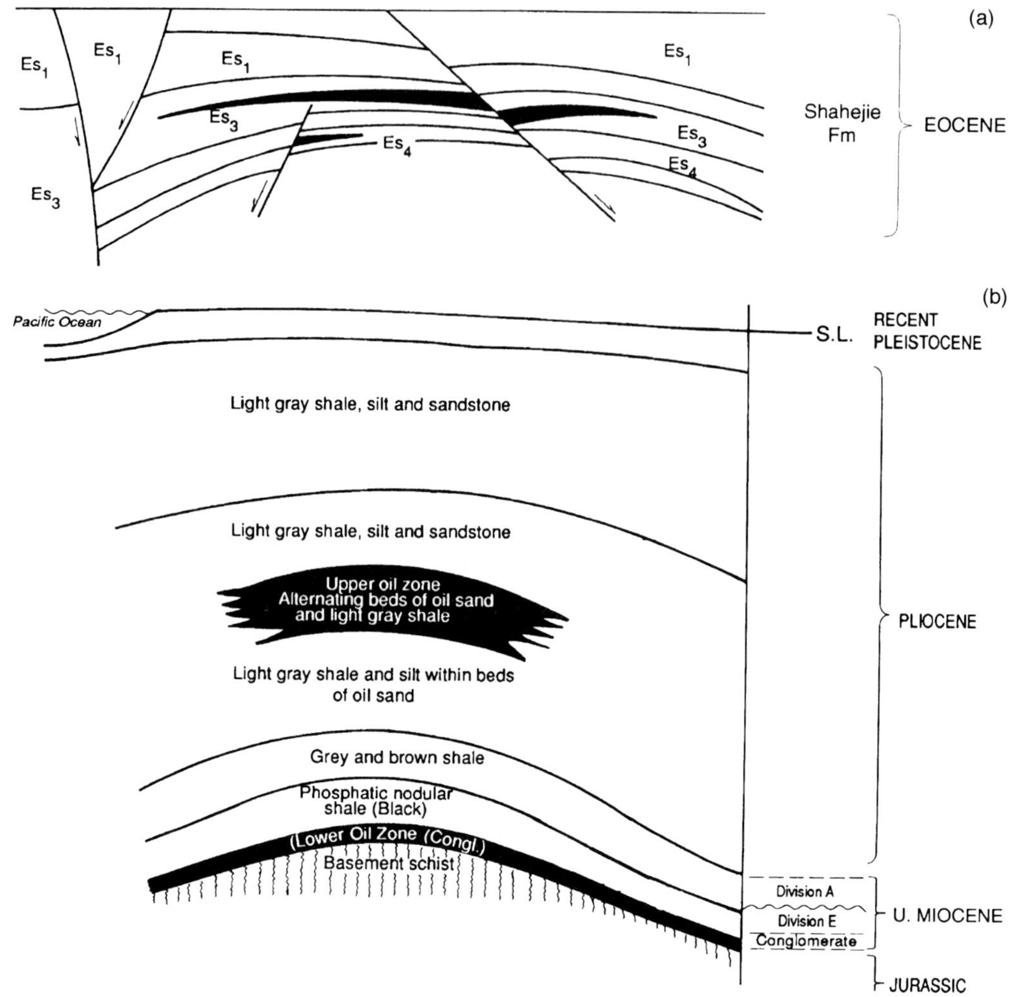

Fig. 1. Structural cross-sections of **(a)** a Liaohe oilfield and **(b)** a Californian oilfield.

Group C. Inland brackish–brine water lake. Well known examples are oils sourced from the Eocene Qianjiang Formation in the Jianghan Basin (W32 & G33) (Fu *et al.* 1986; Sinninghe Damste *et al.* 1987) and the Miocene Upper Ganchaigou Formation in Chaidam Basin (H79) (Huang *et al.* 1990, 1991).

Group D. Marine. These were taken from a carbonate chalk field in the Central Graben of the North Sea, sourced from Upper Jurassic Mandal Formation source rocks containing abundant type II organic matter with marine algae as the dominant organic precursors (Hughes *et al.* 1985; Stoddart *et al.* 1995).

Analytical methods

Solvent extraction of core samples

It has been shown that the choice of organic solvent mixtures is crucial in determining the yields and distributions of organic nitrogen species obtained during the extraction of core samples, especially shales (Yamamoto 1992). A number of solvents or solvent mixtures commonly used in organic geochemical studies were tested and a mixture of dichloromethane/methanol (93 : 7, Soxhlet, 72 h) was selected in this study. Higher percentages of methanol in the solvent mixture gave higher yields of the total polar compounds, but had no detectable effect on yields of pyrrolic and pyridinic nitrogen species.

Isolation of nitrogen heterocycles and other fractions from crude oils and rock extracts

An analytical flow chart is presented in Fig. 2. A review of detailed procedures and instrumentation for rapid isolation and characterization of undistorted nitrogen fractions from crude oils and shale extracts has been reported previously (Li *et al.* 1992).

Owing to its ease of use and large sample capacity, the two-step liquid column (LC) chromatographic separation method reported by Later *et al.* (1981) is used here, with slight modifications, in routine isolation of nitrogen-fractions. A deasphalting step was introduced prior to the LC steps, which was found necessary for better recovery of nitrogen fractions that can be characterized with gas chromatography-related analytical techniques. The fractions of aliphatic hydrocarbons, aromatic hydrocarbons and nitrogen heterocycles were separated from deasphaltened oil or rock extract (100 to 300 mg) using a 25 cm × 7 mm i.d. column packed with 6 g of neutral alumina, eluting sequentially with *n*-hexane (50 ml), toluene (50 ml) and chloroform/ethanol (98 : 2, 70 ml). For the samples containing abundant fluorenones, cleaner fractions of nitrogen heterocycles can be obtained by subjecting these fractions to a repeated alumina separation step. Separation of the nitrogen heterocycles into pyrrolic nitrogen (possibly amine) and pyridinic nitrogen fractions was then achieved using a silicic acid column (2 g) eluting sequentially with *n*-hexane/toluene (1 : 1, 50 ml), toluene (30 ml) and toluene/anhydrous diethyl ether (1 : 1, 50 ml).

Characterization of pyrrolic nitrogen fractions

This was achieved using three different techniques: gas chromatography (GC)–nitrogen-specific detection, gas chromatography–mass spectrometry (GC-MS) and direct insertion probe mass spectrometry (PMS).

The gas chromatograph used for the characterization of pyrrolic nitrogen fractions was an HP 5890 Series II equipped with a flame ionization detector (FID) and a nitrogen phosphorus detector (NPD). A SGE fused-silica capillary column (25 m × 0.33 mm i.d.) coated with BPX5 (film thickness 0.25 μm) was used with helium as carrier gas. The pyrrolic nitrogen fractions (in dichloromethane) were injected at 40°C and the oven was subsequently programmed to 150°C within 2 min and then to 310°C at 4°C min^{-1}. Identification of carbazole and its C_1 and C_2 alkyl derivatives was made by co-injection with appropriate authentic standard compounds (Li *et al.* 1992).

GC–MS of the pyrrolic nitrogen fractions was performed using an HP 5970 mass selective detector (MSD), with GC conditions similar to those described previously except that a 25 m × 0.22 mm i.d. HP-5 column was used. The mass spectrometer was operated in full scan and selective-ion monitoring (SIM) electron-impact modes with an electron energy of -70 eV. Isomeric compound ratios were calculated using the peak intensities in the corresponding molecular-ion chromatograms.

PMS of the pyrrolic nitrogen fractions was obtained with a VG TS250 mass spectrometer.

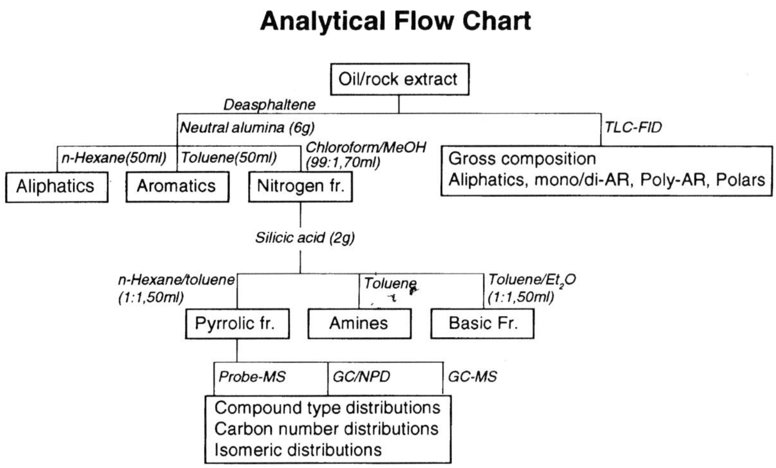

Fig. 2. Analytical routines used in this study.

Table 3. *Geochemical parameters of the crude oil samples investigated in this study*

Sample code	Gross composition (%)			C_{29} Steranes				Methylphenanthrene index*		4-MDBT/ 1-MDBT
	Saturates	Aromatics	Polar	S/(S + R)	ββ(αα + ββ)	S/(S + R)	Pr/Ph	1	2	
Non-marine oils										
W153	57.6	21.5	20.9	0.37	0.28	0.55	1.00	0.70	0.68	3.23
S452	54.0	18.1	27.9	0.48	0.31	0.59	0.61	0.95	1.05	3.57
L49	69.2	20.8	10.0	0.53	0.49	0.59	0.98	0.84	0.89	4.17
F135	50.7	32.5	16.8	0.52	0.66	0.60	1.25	0.42	0.39	4.76
Y67	57.8	27.2	15.0	0.51	0.47	0.59	1.30	0.74	0.73	3.45
Y18	43.5	18.7	37.8	0.17	0.07	0.55	0.63	1.18	1.14	2.45
Os4	59.5	25.7	14.8	0.50	0.60	0.44	0.89	0.89	0.84	3.13
Os3	22.8	19.3	57.9	0.35	0.29	0.55	0.38	1.22	0.98	2.17
W32	26.2	34.8	39.0	0.28	0.23	0.52	0.42	0.85	0.88	2.38
G33	18.9	29.2	51.9	0.24	0.22	0.47	0.31	0.88	0.98	0.78
H79	40.8	18.8	40.4	0.42	0.28	0.57	0.38	0.99	1.03	3.85
W203	73.3	13.7	13.0	0.56	0.61	0.56	0.63	0.59	0.62	5.00
CA1	81.0	10.5	8.5	0.56	0.41	0.45	0.80	0.91	0.90	3.70
Marine oils										
PDR1	6.3	38.6	55.1	0.44	0.47	0.61	0.95	–	–	–
PDR2	18.9	9.4	71.7	0.32	0.43	0.57	1.31	–	–	–
A1	59.5	24.2	15.8	0.38	0.49	0.60	1.42	0.51	0.58	3.57
A2	57.4	22.9	19.7	0.38	0.46	0.60	1.41	0.52	0.55	2.70
A3	57.2	24.9	17.9	0.34	0.46	0.60	1.41	0.53	0.54	2.78
B1	58.5	25.2	15.8	0.37	0.46	0.60	1.32	0.52	0.54	3.13
C1	56.4	22.1	21.5	0.38	0.48	0.60	1.35	0.52	0.54	2.70
C2	57.5	24.6	17.9	0.37	0.46	0.60	1.37	0.51	0.54	2.78

* For definitions see Radke *et al.* (1982)

Table 4. *C_3/C_2 alkylcarbazole ratios and group type compositions of the pyrrolic nitrogen fractions in the oils and source rocks from Liaohe and California*

Sample	C3/C2	CA*	Group type composition (%)		
			Carbazoles	Benzo-	Dibenzo-
Liaohe oils					
Os4	0.46	0.27	21	57	22
Os3	1.78	0.54	46	41	13
Liaohe shales					
8611	0.87	0.11	9	65	26
8613	0.93	0.17	14	70	16
8614	1.12	0.27	21	57	22
8615	0.60	0.28	23	61	16
8616	0.80	0.39	27	42	31
8617	0.89	0.15	11	59	30
Californian oils					
PDR1	0.90	0.53	52	46	2
PDR2	0.95	0.51	50	48	2
Californian shales					
PDR1	0.77	0.10	8	71	21
PDR2	0.65	0.15	13	61	26
PDR4	0.88	0.12	9	70	21
PDR5	0.74	0.17	14	60	26
PDR6	0.79	0.21	19	72	9
PDR7	0.81	0.25	23	68	9
PDR8	0.78	0.24	20	67	13
PDR9	0.67	0.19	17	70	13
PDR10	0.81	0.22	28	53	19
PDR11	0.74	0.23	19	64	17
PDR12	0.88	0.26	22	65	13

* CA = alkylcarbazoles/(alkylcarbazoles + alkylbenzocarbazoles)

A mass range of m/z = 50 to 650 was scanned at a resolution of approximately 1000. Samples were introduced into the ion source *via* the direct insertion probe. Mass spectra were obtained at an ionizing voltage of −20 eV as the probe and source were heated from 150 to 350°C. In the integrated mode, all spectra were summed (signal averaging) to obtain a representative spectrum of the average composition of the sample. Assuming constant ionization efficiency (although it was certainly not true), relative intensities of molecular ions in the mass spectra were used without correction in the determination of molecular weight distributions for pyrrolic nitrogen compounds. In cases where fractions were found to contain impurities, contributing ions with the same m/z as the molecular ions of pyrrolic nitrogen compounds, GC–MS data were consulted to deduce the contributions from impurities. For instance, the ion at m/z 167 is the molecular ion of carbazole, but the same ion is also the major fragment ion of the phthalate contaminant. In this situation, the relative intensities of carbazole and phthalate in the m/z 167 mass fragmentogram were used to estimate the relative contributions of the two components to the total intensity of the m/z 167 ion in the probe mass spectrum of the pyrrolic nitrogen fraction. Relative abundances of three pyrrolic nitrogen group types, i.e. alkylcarbazoles, alkylbenzocarbazoles and alkyldibenzocarbazoles, were calculated using the sum of the parent ring molecules and their C_1 to C_{10} alkyl derivatives in each group.

Characterization of aliphatic and aromatic hydrocarbon fractions

Gas chromatographic analyses of the aliphatic and aromatic hydrocarbon fractions were performed on two separate Carlo Erba Mega Series 5150 gas chromatographs connected with a VG Multichrom laboratory data system. One gas chromatograph was fitted with an OV-1 fused silica column

Table 5. *C_3/C_2 alkylcarbazole ratios and group type compositions of the pyrrolic nitrogen fractions in a range of Chinese non-marine oils and North Sea oils*

			Group type composition (%)		
Sample	C3/C2	CA*	Carbazoles	Benzo-	Dibenzo-
Non-marine oils					
W153	0.44	0.42	48	34	18
S452	0.70	0.45	41	50	9
L49	1.60	0.55	34	49	17
F135	1.21	0.52	39	46	16
P15	1.20	0.48	15	65	20
Y67	0.98	0.64	59	33	8
Y18	1.29	0.66	63	33	4
W32	1.78	0.58	67	30	3
G33	1.97	0.60	65	31	4
H79	0.50	0.39	49	48	3
W203	0.46	0.33	34	32	33
North Sea oils					
CA1	0.80	0.80	79	20	1
A1	1.44	0.53	49	44	7
A2	2.10	0.73	67	25	8
A3	1.20	0.74	71	22	7
B1	1.36	0.72	66	26	8
C1	1.41	0.70	64	27	9
C2	1.25	0.63	57	34	9

* CA = alkylcarbazoles/(alkylcarbazoles + alkylbenzocarbazoles)

(30 m × 0.32 mm i.d.) for analysis of aliphatic hydrocarbons, and another with a DB-5 column (30 m × 0.32 mm i.d., film thickness 0.2 μm) for aromatic hydrocarbons. Samples were injected onto the column via a cold on-column injector. The oven was maintained at 50°C for 2 min and then programmed to 300°C at 4°C min^{-1}, holding this final temperature for 20 min. GC–MS analyses were performed on the same GC–MS system as used for analysis of pyrrolic nitrogen fractions, except that an HP-1 fused silica capillary column (25 m × 0.22 mm i.d.) was used for analysis of aliphatic hydrocarbons. Splitless injection was used. Molecular parameters were calculated using

Table 6. *Isomeric compositions of C_1 to C_3 alkylcarbazoles in the Liaohe crude oils and shales studied**

	Methylcarbazoles			Dimethylcarbazoles						Trimethylcarbazoles					
Sample	1-	3- + 2-	4-	1,8-	1,4-	3,5-	G1	G2	G3	A	B	C	G1	G2	G3
Crude oils															
Os4	2	69	29	1	4	95	<1	15	85	2	7	91	2	32	66
Os3	35	44	21	41	33	26	15	47	38	40	34	26	36	38	26
Shales															
8611	0	61	39	0	4	96	0	3	97	0	4	96	0	3	97
8613	1	49	50	1	4	95	1	4	95	1	5	94	1	4	96
8614	4	43	53	1	6	93	1	6	94	0	4	96	1	5	94
8615	0	67	33	0	0	100	0	3	97	0	2	98	0	5	95
8616	4	43	53	1	5	94	1	5	94	0	4	96	0	6	94
8617	1	48	51	1	5	94	1	6	93	1	5	94	<1	5	94

* Ternary relative percentages: for methylcarbazoles the sum of 1-, 3- + 2- and 4-isomers is 100%; for dimethylcarbazoles, the sum of 1,8-, 1,4- and 3,5-isomers is 100%, and so is the sum of G1, G2 and G3; for trimethylcarbazoles, the sum of compounds A, B and C is 100%, and so is the sum of G1, G2 and G3

the peak heights of appropriate compounds in the following mass fragmentograms: *m/z* 217 for steranes, 191 for triterpanes, 178 + 192 for methylphenanthrene indices (MPI, Radke *et al.* 1982), and 198 for 4-methyldibenzothiophene (MDBT)/1-MDBT ratios.

Bulk chemical analyses

The total organic carbon (TOC) contents of powdered rocks were determined using a LECO carbon analyser. Corrections for carbonate contents were made by pretreatment of the samples with hydrochloric acid. A petroleum potential survey (hydrogen index and pyrolysis T_{max} of the shales) was conducted using a LECO THA-200 thermolytic hydrocarbon analyser. Group type compositions of crude oils (i.e. percentage aliphatic hydrocarbons, aromatic hydrocarbons and polars) were determined using a standard Iatroscan–FID technique (Karlsen & Larter 1991).

Results and discussion

Basic geochemical data for the samples used in this study are given in Tables 1 to 3. Molecular weight distributions of the pyrrolic nitrogen fractions and isomeric compositions of the C_1 to C_3 alkylcarbazoles in these samples are presented in Tables 4 to 7.

Distributions of pyrrolic nitrogen compounds in crude oils and rock extracts

Typical chromatograms of the pyrrolic nitrogen fractions obtained in this study are shown in Fig. 3. An example of the probe mass spectra of the pyrrolic nitrogen fractions is given in Fig. 4.

The pyrrolic nitrogen fractions obtained from all oils and shales comprise carbazole, benzocarbazole and dibenzocarbazole and their alkyl derivatives. Alkylpyrrole and alkylindole derivatives are either absent or present in only minor abundance. Of the five possible methylcarbazole isomers, the four non-N-substituted isomers are present in the samples analysed. Chromatographic co-injections with standard compounds suggest that, of the C_2-alkylcarbazoles, only one isomer has an ethyl substituent (at position 1) and the rest are dimethylcarbazoles with methyl substituents at either one aromatic ring or at both aromatic rings of the

Table 7. *Isomeric compositions of C_1 to C_3 alkylcarbazoles in a range of crude oils studied**

	Methylcarbazoles			Dimethylcarbazoles						Trimethylcarbazoles					
Sample	1-	3- + 2-	4-	1,8-	1,4-	3,5-	G1	G2	G3	A	B	C	G1	G2	G3
Non-marine oils															
W153	27	47	26	26	30	44	5	55	40	20	14	66	8	54	38
S452	3	72	25	3	7	90	<1	23	77	5	19	76	2	36	62
L49	42	41	17	32	46	22	10	66	24	18	56	26	10	65	25
F135	34	35	31	40	31	29	10	60	30	26	42	32	17	56	27
P15	30	46	24	16	50	34	4	58	38	8	50	42	6	58	36
Y67	42	42	16	38	41	21	13	62	25	39	41	20	25	50	25
Y18	30	54	16	28	45	27	7	47	26	37	40	23	28	46	25
W32	38	41	21	37	36	27	12	56	32	36	36	29	35	39	26
G33	54	31	15	44	38	18	20	55	25	46	36	17	44	39	17
H79	18	53	29	13	24	63	3	44	53	9	9	82	8	49	43
W203	10	53	37	6	14	80	1	25	73	3	4	93	7	41	52
CA1	44	38	18	45	37	18	17	62	21	53	33	14	42	41	17
Marine oils															
PDR1	39	42	19	27	43	30	10	61	29	17	49	35	21	46	33
PDR2	36	43	21	25	42	33	9	60	31	15	49	37	19	45	36
A1	51	36	13	45	33	22	20	56	24	57	25	18	45	39	16
A2	54	31	15	49	32	19	20	64	16	64	22	14	48	39	13
A3	52	33	15	48	31	21	19	63	18	63	21	16	47	39	14
B1	49	35	16	46	35	19	18	54	28	60	23	17	40	41	19
C1	50	36	14	49	37	14	20	57	23	66	24	10	42	38	20
C2	55	29	16	45	38	16	22	50	29	57	21	22	41	39	20

* See footnote to Table 6.

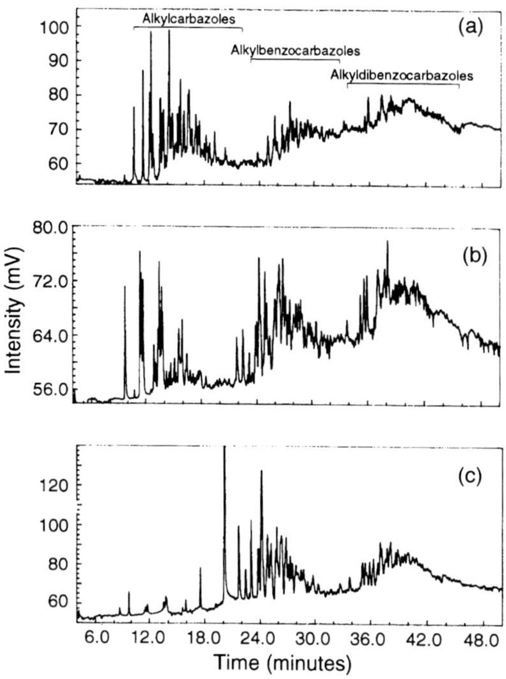

Fig. 3. Nitrogen-specific detection–gas chromatograms of the pyrrolic nitrogen fractions from a Liaohe oilfield: **(a)** Es3 oil; **(b)** Es4 oil; **(c)** Es4 source rock.

carbazole nucleus. Three benzocarbazole isomers were identified, i.e. benzo[a]carbazole, benzo[c]-carbazole and, in much lower abundance, benzo[b]carbazole. Other derivatives of carbazole, benzocarbazole and dibenzocarbazole could not be further characterized owing to the lack of reference compounds.

Examination of many gas chromatograms indicates that, although the range and type of pyrrolic nitrogen compounds detected are more or less comparable across a wide range of oil samples, the relative distributions of these compounds vary significantly. The following variations in the distributions of the pyrrolic nitrogen fractions have been observed to occur.

Relative distributions of alkyl-, alkylbenzo- and alkyldibenzocarbazoles. A ternary plot depicting the relative compositions of alkylcarbazoles, alkylbenzocarbazoles and alkyldibenzocarbazoles is employed to illustrate the pyrrolic nitrogen group-type variations (Fig. 5). It is apparent from Fig. 5 that the major difference among the samples is in the relative abundances of alkylcarbazoles versus alkylbenzocarbazoles, with dibenzocarbazoles being consistently the least abundant components. All the source-rock samples are enriched in alkylbenzocarbazoles whilst crude oils contain relatively higher percentages of alkylcarbazoles.

Isomeric distributions of alkylcarbazoles. According to the presence or absence of alkyl substituent(s) at 1 and 8 positions, the structures of 9-unsubstituted alkylcarbazoles can be categorized into three groups ($n \geq 1$):

Group 1 (G1). Pyrrolic N–H-shielded: R1, R2=C_nH_{2n+1}; e.g. 1,8-dimethylcarbazole

Group 2 (G2). Pyrrolic N–H partially shielded:

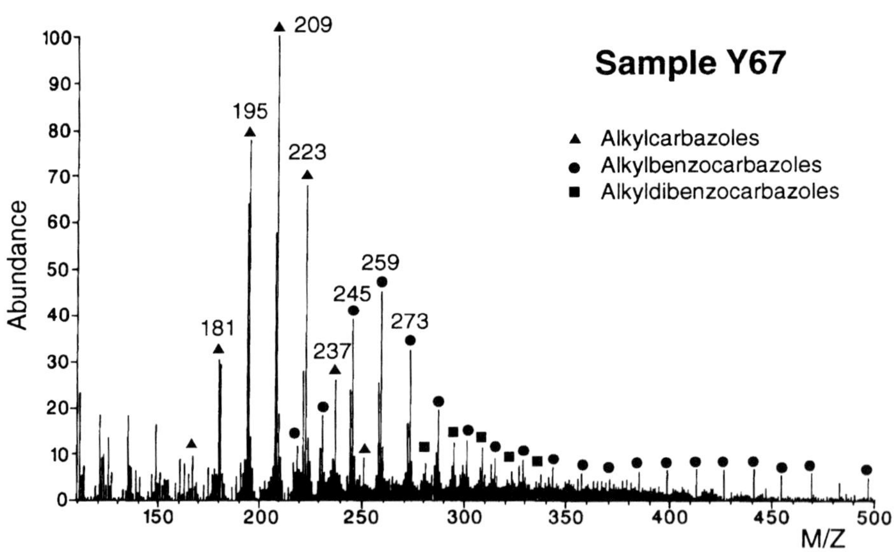

Fig. 4. Probe mass spectrum of the pyrrolic nitrogen fraction obtained from the Shengli oil (sample Y67).

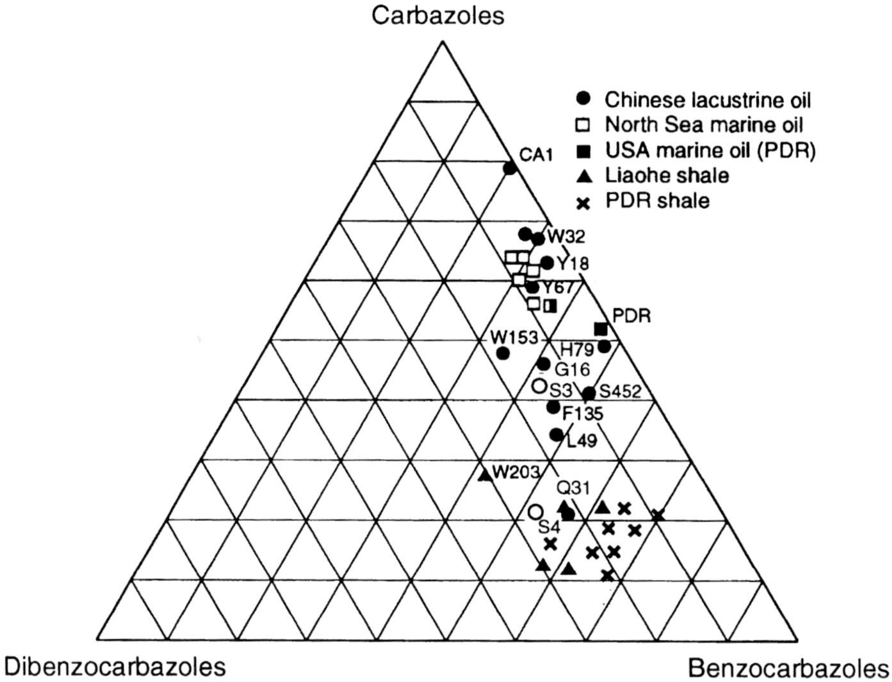

Fig. 5. Group-type distributions of the pyrrolic nitrogen compounds in the samples investigated. See text for descriptions of individual compounds included in each group type.

R1=H, R2=C_nH_{2n+1}; e.g. 1-methyl- and 1,4-dimethylcarbazole

Group 3 (G3). Pyrrolic N–H-exposed: R1=R2=H; e.g. 3,4- and 3,5-dimethylcarbazole.

Ternary plots reflecting the isomeric variations among the crude oils and shale samples are presented in Fig. 6. Mass chromatograms of C_1 to C_3 alkylcarbazoles from the oils in the Liaohe Field and Shengli Field are given in Figs 7 & 8, respectively, as examples. All the shale samples display similar isomeric distributions in which Group 3 (pyrrolic N-H-exposed isomers) predominates over the other isomers for C_2 and C_3 substituted carbazoles. For methylcarbazoles, the 4-methyl isomer dominates, with the 1-methyl isomer being negligible in all the Liaohe shale samples analysed here. We have also had cases where the 1-methylcarbazole is the most abundant methylcarbazole isomer, e.g. in North Sea source-rock samples (Stoddart *et al.* 1995). Nevertheless, the 1-/4-methylcarbazole ratios are generally much lower in source rocks than in crude oils.

Pseudo-homologous distributions of alkylcarbazoles. The relative abundances of the lower molecular weight pseudohomologues of alkylcarbazoles (C_1 to C_3) in all the samples were examined. It was found that samples enriched in pyrrolic N–H-shielded isomers at any carbon number contain higher proportions of the higher molecular weight homologues than those with more nitrogen-exposed ones. This trend is clearly indicated in Fig. 9.

Good correlations have been observed between the variations described above (Fig. 10). For example, the proportions of pyrrolic N–H-exposed isomers (Group 3) correlate well for C_1, C_2 and C_3 alkylcarbazoles; similarly the proportions of Group 3 compounds correlate inversely with the ratios of alkylcarbazoles/(alkylcarbazoles + alkylbenzocarbazoles). This suggests that similar geochemical processes may be responsible for these variations.

Geochemical controls on pyrrolic nitrogen distributions in reservoired petroleum

The crude oils and source rock samples analysed in this study have different origins and geochemical backgrounds. A detailed investigation of the nitrogen compound compositional variations can

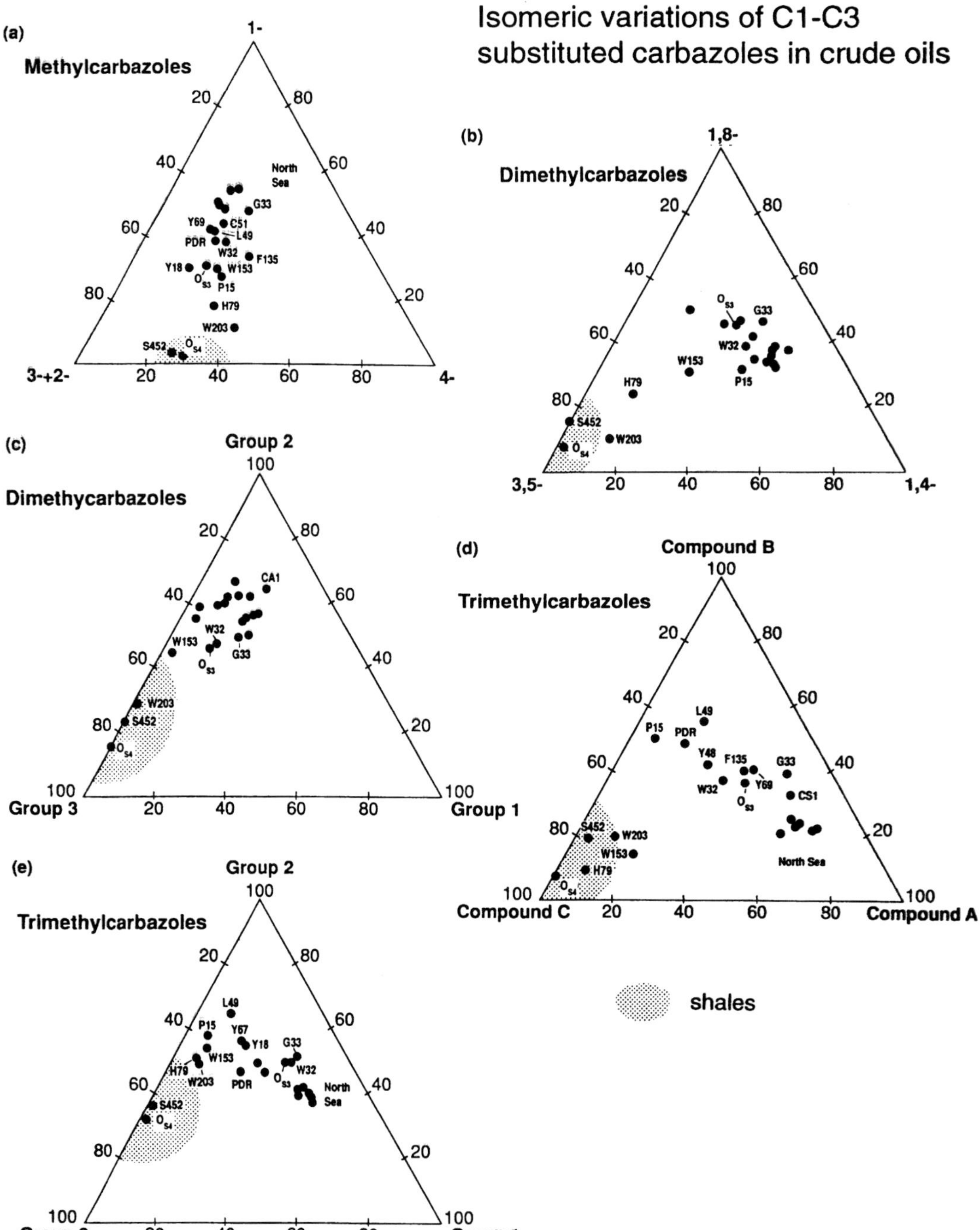

Fig. 6. Isomeric distributions of the C_1 to C_3 alkylated carbazoles in the samples investigated. See text for definitions of Groups 1, 2 and 3 and Figs 7 & 8 for individual compounds included in each group.

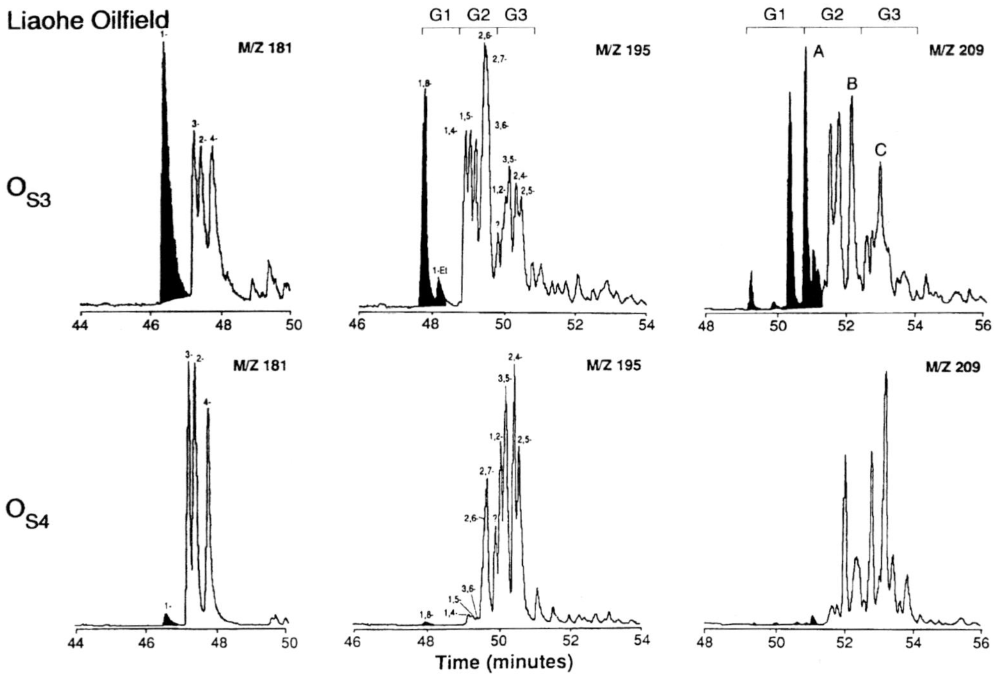

Fig. 7. Mass chromatograms m/z 181, 195 and 209 showing the isomeric distributions of the C_1 to C_3 alkylated carbazoles from the two Liaohe crude oils. G1, G2 and G3 indicate compounds included in each group.

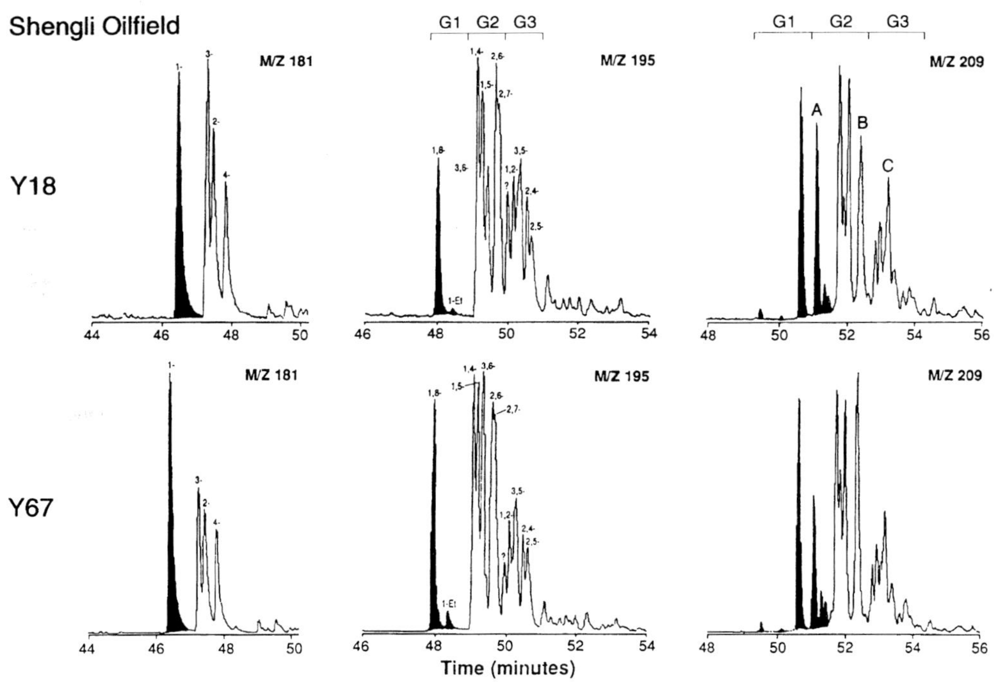

Fig. 8. Mass chromatograms m/z 181, 195 and 209 showing the isomeric distributions of the C_1 to C_3 alkylated carbazoles from the two Shengli crude oils.

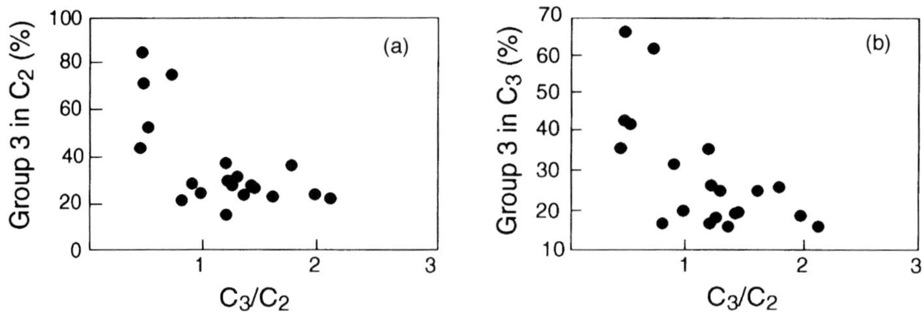

Fig. 9. Plots showing correlations of C_3/C_2 alkylcarbazole ratios with percentages of Group 3, pyrrolic N–H-exposed isomers in **(a)** C_2 and **(b)** C_3 alkylcarbazoles.

provide significant insight into the major (i.e. first order) geochemical controls on the distributions of pyrrolic nitrogen compounds in reservoired petroleums.

Organic source inputs and pyrrolic nitrogen precursors. Our knowledge of the origin of non-porphyrinic pyrrolic nitrogen compounds in the geosphere is limited. The chemical structures of these compounds may either be directly inherited from biological precursors or represent diagenetic artefacts through intensive reworking of nitrogen-containing sedimentary organic matter or incorporation of inorganic nitrogen species into organic carbon skeletons.

After intensive studies of nitrogen species in Californian oils, Snyder (1965) proposed that many of the alkylcarbazoles and alkylbenzocarbazoles in petroleum may be derived from alkaloids with an indole nucleus, often distributed in higher plants (Kapil 1971; Hesse 1974) and also in blue-green algae (Cardelina *et al.* 1979). If this hypothesis holds true, the following features of alkaloid pyrrolic nitrogen distributions should be observed for petroleum pyrrolic nitrogen distributions: (a) absence of linear anneleted benzo[b]carbazole structure, and (b) predominance of alkyl substituents on a single aromatic ring of the carbazole and benzocarbazole backbones.

Data presented by Dorban *et al.* (1984*b*) and in this study clearly show the presence of benzo[b]carbazole in all the samples analysed, although in lower abundance than its angular counterparts. Carbazole derivatives with substituents on both aromatic rings of the carbazole backbone, such as 1,8-dimethylcarbazole, are also significant and indeed dominant components in most oil samples studied. Furthermore, the crude oils and source-rock samples included here have different origins and geochemical characteristics. Despite the large variations in their relative distributions, the range and group type of pyrrolic nitrogen species encountered are more or less similar. In particular, the pyrrolic nitrogen compositions of the high-wax oils derived from higher plant material in the Ordos Basin and Nanyang Oilfield do not differ significantly from those of other oils of lacustrine and marine origins. These facts appear to suggest that plant alkaloids are not the major source of pyrrolic nitrogen compounds in petroleum.

Alternative, diagenetic formation mechanisms may have to be considered. This could involve the condensation of low molecular weight nitrogen species such as hydroxylamines with compounds such as 1,4-diketones (Dorban *et al.* 1984*b*) or polyunsaturated lipids as proposed for sulphur heterocycles (de Graaf *et al.* 1992). A contribution from other nitrogen-containing sedimentary matter, such as proteins and plant pigments, cannot be ruled out, and the similarities in alkylcarbazoles may well implicate a consistent facies-independent precursor, such as proteins or their nitrogen-cycle products, as a universal nitrogen source. Similarities in isomeric distributions between several classes of sulphur-, oxygen- and nitrogen-bearing aromatic heterocycles (i.e. alkyl derivatives of fluorenone, dibenzofuran, dibenzothiophene and carbazole) suggest common modes of origin for several of these important compound classes.

Thermal maturation. Two suites of source rock samples were selected in this study. In general, samples in each suite differ mainly in their levels of thermal maturity, with less variable organic source inputs and depositional environment. The maturity ranges of these samples are demonstrated using the sterane maturity parameters (Fig. 11), and are also reflected in the data of Rock-Eval T_{max} values (Table 2).

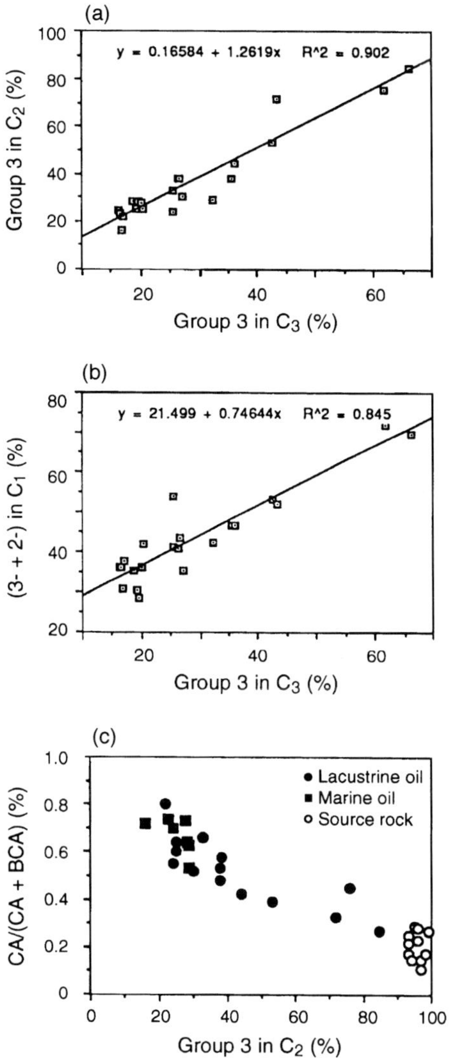

Fig. 10. Correlations of various pyrrolic nitrogen ratios in the samples investigated. Group 3 in C_2 (%): percentage of pyrrolic N–H-exposed isomers in C_2 alkylcarbazoles; Group 3 in C_3 (%): percentage of pyrrolic N–H-exposed isomers in C_3 alkylcarbazoles; (3-+2-) in C_1 (%): percentage of 3- and 2-isomers in the methylcarbazoles; CA/(CA + BCA) (%): relative percentage of alkylcarbazoles within the sum of alkylcarbazoles and alkylbenzocarbazoles.

We observed that shale samples in both suites are characterized by the predominance of alkylbenzocarbazoles over alkylcarbazoles and of pyrrolic N–H-exposed isomers over their pyrrolic N–H-shielded counterparts (Figs 5 & 6), even though their maturity levels are different based on sterane biomarkers. As shown in Fig. 12, no obvious correlations exist between maturity and the variations in pyrrolic nitrogen distributions. This may suggest that, in these cases, thermal maturation is unlikely to be a factor of primary importance in controlling the relative distributions of pyrrolic nitrogen compounds.

A geochemical study based on bulk chemical data and biomarker maturity parameters reveals that the apparent maturity variations in the Californian shale set are affected by oil stains (Li & Larter unpublished results), but this does not seriously affect our conclusion here.

Supporting evidence can also be obtained from the analyses of closely related crude oil samples. For the two Liaohe oils, the shallower one (Os3) is an example of a so-called 'immature oil' in which its apparent maturity is low, whilst the deeper one (Os4) represents a normal maturity oil (Table 3). In the more mature oil (Os4) there is a significant predominance of alkylbenzocarbazoles over alkylcarbazoles and of pyrrolic N–H-exposed isomers over the pyrrolic N–H-shielded ones, thus resulting in a much lower CA/(CA + BCA) ratio and a higher G3 abundance in the C_2 and C_3 alkylcarbazoles than in the immature oil (Os3) (Figs 5 to 7). In the case of two Shengli oils (Y67 and Y18), the reverse trend is observed with maturity (Table 3, Fig. 8). Oils Os4 and Y18 represent oils found near their respective source rocks (Lu *et al.* 1990; Shi *et al.* 1982).

It should be pointed out, however, that the maturity range of the shale samples included in this study is relatively narrow. In a survey of 192 crude oils generated from a broad spectrum of source rocks, Bakel (1990) observed a general trend of decrease in nitrogen content of oils with decreasing oil API gravity and increasing burial depth of the petroleum reservoir. To what extent thermal maturation affects the contents and distributions of the pyrrolic nitrogen compounds in crude oils remains to be seen.

Environment of deposition. The depositional environments of the source rocks for the crude oils selected in this study range from marine, freshwater and brackish–brine lacustrine to swampy lake. So far we have not found any characteristic pyrrolic nitrogen distributions specific to a given depositional environment using the compound types examined here. This suggests that depositional environment, though impacting strongly on the chemical characteristics of the sedimentary organic matter initially incorporated into sediments, may not play a major role in controlling pyrrolic nitrogen distributions in reservoired petroleum. We have not analysed a large enough suite of source rocks to define the

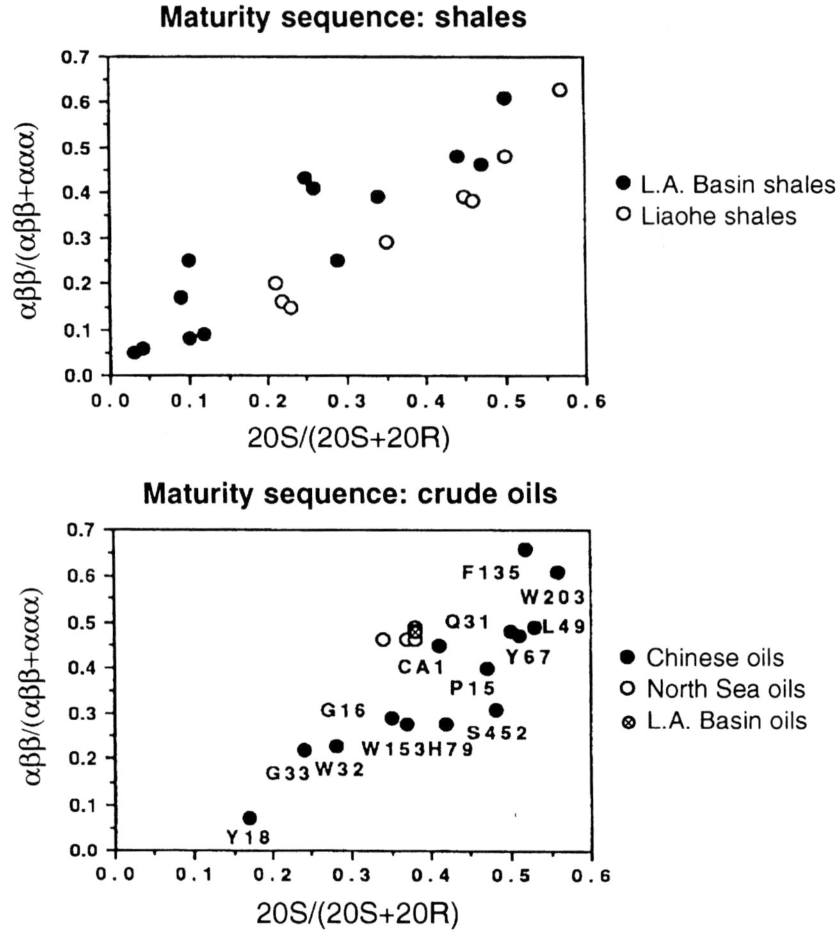

Fig. 11. Maturity sequences determined from the C_{29} sterane biomarker ratios for the shales and oil samples studied.

primary depositional controls on organic nitrogen compound distributions.

Primary oil migration. As described previously, striking differences are observed in the distributions of pyrrolic nitrogen compounds between crude oils and corresponding shale extracts. Shale extracts have significantly lower CA/(CA + BCA) ratios (Fig. 5), higher percentages of pyrrolic N–H-exposed isomers (Fig. 6) and lower C_3/C_2 ratios for alkylcarbazoles (Fig. 9), when compared to related crude oils.

In the previous sections, we have demonstrated that organic source inputs, depositional environments and thermal maturation appear not to be the prime controls responsible for the differences observed. Such differences may be explained by processes occurring during primary oil migration.

(1) Pyrrolic nitrogen compounds are known to interact with formation water and mineral surfaces, probably through hydrogen bonding between the pyrrolic N–H with hydroxy groups in the surrounding environment (Dorban *et al.* 1984*b*). Similarly, the solubility of 1,8-dimethylcarbazole is 15 to 20 times higher than its N–H-exposed isomers in petroleum phases (Frolov *et al.* 1989) and presumably immobile, polar organic phases. Indeed, partitioning of petroleum components into solid organic matter has been proposed as important for primary migration (Pepper 1991). Preliminary data from organic-rich source rocks do indicate some TOC control on nitrogen compound explusion (Chen *et al.* unpublished results). It is possible that crude oils are expelled at saturation solubility with regard to several crude oil components including asphaltenes (Wilhelms & Larter 1994). This could also apply to carbazoles. A weaker retention

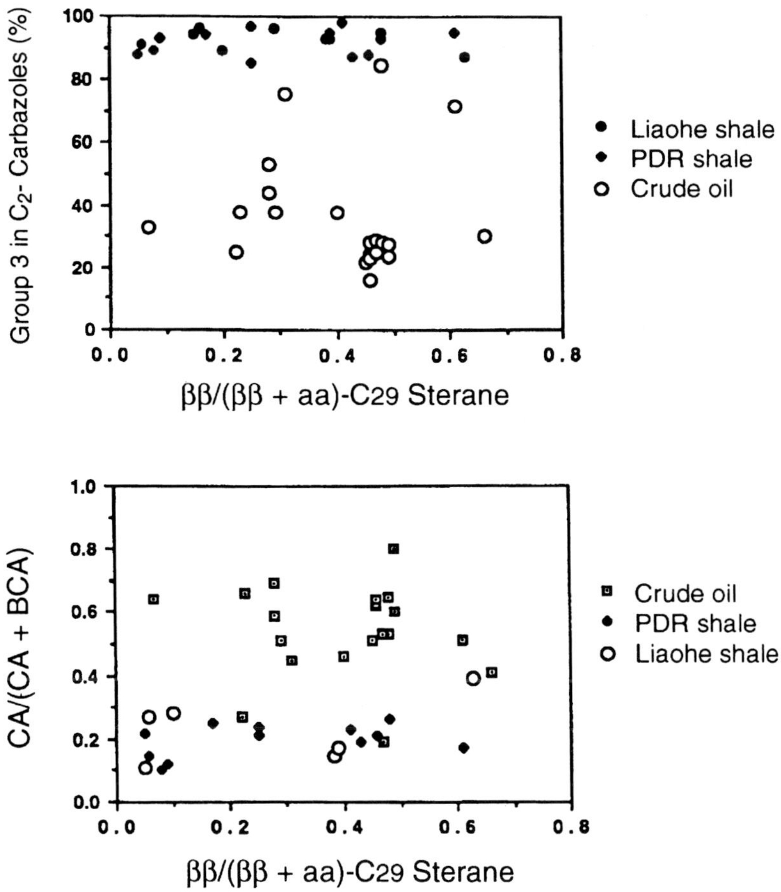

Fig. 12. Correlations of pyrrolic nitrogen compound ratios with sterane maturity parameters. See Fig. 10 for y-axis abbreviations.

and preferential movement of the more nitrogen-shielded isomers, such as 1,8-dimethylcarbazole, during migration can be expected compared with their other isomeric counterparts. This is probably related to the shielding effect of the alkyl substitution. Similar arguments have been suggested for fractionations observed in alkylquinolines between shales and oils (Yamamoto 1992), where stronger, ionic interactions between the quinoline derivatives and mineral surface are also possibly involved.

(2) The predominance of alkylcarbazoles over alkylbenzocarbazoles in crude oils is thought to result from a 'geochromatographic' effect during primary migration. It is known that increased aromatic ring number retards expulsion in aromatic hydrocarbons and heterocycles (Leythaeuser & Schaefer 1984). An interesting parallel can be seen in laboratory normal-phase high performance liquid chromatography experiments using silica or alumina stationary phases (Ruckmick & Hurtubise 1985; Li & Larter 1993).

(3) It is not straightforward to explain the relative enhancement of higher *pseudo*-homologues of alkylcarbazoles in crude oils. However, work conducted by Charlesworth (1986) has provided a significant clue towards the understanding of such fractionation effects. He observed that some nitrogen compounds are removed by irreversible adsorption on clay minerals. The major controls are the size of the non-polar portion of the nitrogen compounds and the nature of clay minerals. Laboratory experiments with model compounds (Li & Larter 1993) also showed that the relative mobility of alkylcarbazoles decreases as their molecular weights decrease. The increase in C_3/C_2 ratios in alkylcarbazoles in crude oils clearly implies that selective adsorption is important during primary migration.

The actual mechanisms involved are not fully understood, but may reflect adsorption processes between oil phases and mineral surfaces and may also involve partition between oil and organic kerogen network and immobile bitumens. In water-wet soils, it is generally expected that partition of hydrophobic components into bound organic phases dominates over mineral adsorption at TOC values as low as 0.1% (Lee *et al.* 1988). Clearly, simplistic models of mineral adsorption cannot be applied to explain our results.

Secondary oil migration. Data shown in Figs 4 & 5, comparing integrated spectra for a wide range of crude oils, present a similar empirical trend in the changes of pyrrolic nitrogen distributions during secondary oil migration to that in primary migration (comparing source rocks versus all oils). This trend is exemplified by the two crude oils taken from the Es3 and Es4 sequence in a Liaohe Field (Fig. 1a). The shallower oil (Os3) has migrated upwards for a significantly longer distance, whilst the deeper oil (Os4) has been reservoired close to its source bed. It is obvious from Figs 3 & 7 that the oil Os4 has retained more significant portions of pyrrolic N–H-exposed isomers and alkylbenzocarbazoles than the oil Os3.

This suggests that similar mechanisms operated during secondary migration. Prior to oil entering the reservoir, however, relatively porous reservoir beds are initially saturated with water, and hence the mineral surfaces are deactivated by water. Consequently, distribution coefficients of pyrrolic nitrogen compounds between the solid mineral/liquid petroleum phases in a typical petroleum carrier bed are at least two orders of magnitude lower than those measured in laboratory experiments using dry minerals (Li & Larter 1993). In such cases, the intensities of the interaction of organic nitrogen compounds in petroleum charge with mineral surface depend on their relative solubilities in hydrocarbons and interstitial water films. The solubilities of 1,8-, 1,4- and 3,5-dimethylcarbazole are reported to decrease in this sequence in non-polar organic solvents (Frolov *et al.* 1989), and a reverse trend can be expected for their solubilities in water, consistent with the increased strengths of the hydrogen-bonding interaction with water. The roles of solid organic matter and bitumens in the carrier bed may also be important, and may dominate in organic-rich carrier systems (TOC > 0.1%) (Lee *et al.* 1988). Comparisons of reservoir core extracts and reservoired oils suggest that fractionation between free petroleum and other phases continues in the reservoir (Stoddart *et al.* 1995) and that the process is slow. We suspect that the rate-limiting step to surface adsorption is, in part, slow transport through grain-boundary water films before the petroleum components can be adsorbed onto the mineral surface. This is why the processes are so difficult to mimic with wet rocks in the laboratory (Li & Larter 1993).

Adsorption isotherms obtained in an over-idealized artificial organic phase/mineral system (Li & Larter 1993) show that pyridinic nitrogen compounds adsorb in much greater quantities than pyrrolic nitrogen compounds. For example, for each gram of montmorillonite, 55 and 70 mg of 8- and 4-methylquinoline, respectively, are adsorbed, and only 130 μg of carbazole is adsorbed. The relative amounts adsorbed are approximately equivalent to monolayers where the methylquinolines are stacked in folded butterfly-wing orientation above the mineral surface, probably ionically bound in salt form. Molecular modelling shows that alkylcarbazoles must take the alternative configuration covering the surface as planar tiles with their maximum cross-section sub-parallel to the surface. While we have not yet proven the bonding spectroscopically in either case, the circumstantial evidence, based on the amounts adsorbed and orientation, suggests ionic bonds in the case of alkylquinolines and hydrogen bonds in the case of alkylcarbazoles.

Summary

Data presented in this study have demonstrated that primary and secondary migration may be the primary geochemical controls on the pyrrolic nitrogen distributions in reservoired petroleum.

The migration effects recognized tentatively concerning the composition of pyrrolic nitrogen compounds include the relative enrichments of alkylcarbazoles to alkylbenzocarbazoles, nitrogen-shielded isomers to nitrogen-exposed isomers and higher homologues to the lower homologous molecules. These effects appear to be supported by laboratory simulation experiments (Li & Larter 1993).

A detailed analysis of organic nitrogen compounds in general, and pyrrolic nitrogen compounds in particular, in crude oils may not only aid a better understanding of hydrocarbon migration, but also provide a useful tool in the study of petroleum reservoir filling.

We are currently working on the quantitative aspects of organic nitrogen speciation in a few well defined case histories. Quantitation of the compositional changes of nitrogen species during primary migration may help in assessing hydrocarbon expulsion efficiencies. Quantitation of the transport of nitrogen compounds to reservoir surfaces during secondary migration may help in determining source rock–reservoir distances and residual oil losses in the carrier beds. While we have reported only general arguments to describe possible mechanisms for fractionation of organic nitrogen compounds during migration, we have proposed a suite of currently uncalibrated (in an absolute sense) migration range parameters that can be applied to the internal mapping of petroleum transport within basins or within reservoirs (Stoddart *et al.* 1995).

In addition, caution should be taken when dealing with biodegraded petroleum reservoir systems. It has been shown previously by laboratory experiments (Fedorak & Westlake 1984) that mixed microbial populations are capable of degrading a wide variety of alkylcarbazoles in crude oils. To what extent the natural microbial degradation affects the distributions of pyrrolic nitrogen compounds in biodegraded oils is currently under investigation.

We thank Phillips Petroleum (Norway) for financial upport of the research and Drs John Fox, Bruce Bromley (UNOCAL) and Joe Senftle (ARCO) for support and Drs Archie Douglas and Martin Jones (NRG) for useful discussions.

References

BAKEL, A. J. 1990. *The Geochemistry of Nitrogen in Crude Oils*. PhD Dissertation, University of Oklahoma.

—— & PHILP, R. P. 1990. The distribution and quantitation of organonitrogen compounds in crude oils and rock pyrolysates. *In*: DURAND, B. & BEHAR, F. (eds) *Advances in Organic Geochemistry 1989*. Pergamon Press, London, 353–368.

CARDELLINA, J. K., KIRKUP, M. P., MOORE, R. E., MYNDERSE, J. S. & SIMMONS, C. J. 1979. Hyellazole and chlorohyellazole, two novel carbazoles from the blue-green alga *Hyella caespitosa* Born. et Flah. *Tetrahedron Letter*, **51**, 4915–4916.

CHARLESWORTH, J. M. 1986. Interaction of clay minerals with organic nitrogen compounds released by kerogen pyrolysis. *Geochimica et Cosmochimica Acta*, **50**, 1431–1435.

DE GRAAF, W., SINNINGHE DAMSTE, J. S. & DE LEEUW, J. W. 1992. Laboratory simulation of natural sulphurization: I. Formation of monomeric and oligomeric polysulphides by low temperature reactions of inorganic polysulphides with phytol and phytadienes. *Geochimica et Cosmochimica Acta*, **56**, 4321–4328.

DORBAN, M., IGNATIADIS, I., SCHMITTER, J. M., ARPINO, P., GUIOCHON, G., TOULHOAT, H. & HUC, A. 1984*a*. Identification of carbazoles and benzocarbazoles in a coker gas oil and influence of catalytic hydrotreatment on their distribution. *Fuel*, **63**, 565–570.

——, SCHMITTER, J. M., ARPINO, P. & GUIOCHON, G. 1982. Carbazoles et lastames du petrole: methode d'extraction et caracterisation. *Journal of Chromatography*, **246**, 255–269.

——, ——, GARRIGUES, P., IGNATIADIS, I., EDWARD, M., ARPINO, P. & GUIOCHON, G. 1984*b*. Distribution of carbazole derivatives in petroleum. *Organic Geochemistry*, **7**, 111–120.

ENGLAND, W. A. 1990. The organic geochemistry of petroleum reservoirs. *Organic Geochemistry*, **16**, 415–426.

FEDORAK, P. M. & WESTLAKE, D. W. S. 1984. Microbial degradation of alkylcarbazoles in Norman Wells crude oil. *Applied and Environmental Microbiology*, **47**, 858–862.

FROLOV, Y. B., SMIRNOV, M. B., VANYUKOVA, N. A. & SANIN, P. I. 1989. Carbazoles of crude oils. *Petroleum Chemistry USSR*, **29**, 87–102.

FU, J., SHENG, G., PENG, P., BRASSELL, S. C., EGLINTON, G. & JIANG, J. 1986. Peculiarities of salt lake sediments as potential source rocks in China. *In*: LEYTHAEUSER, D. & RULLKOTTER, J. (eds) *Advances in Organic Geochemistry 1985*. Pergamon Press, London, 119–126.

——, ——, ——, XU, J. & HUANG, Y. 1991. The major types of terrestrial source rocks and the peculiarities of biomarker composition of terrestrial oils in China. *Journal of Southeast Asian Earth Science*, **5**, 167–174.

HESSE, M. 1974. Indolalkaloids. *In*: BUDZIKIEWICZ, H. (ed.) *Progress in Mass Spectrometry*. Verlag Chemie, Weinheim, 165–189.

HORSTAD, I., LARTER, S. R., DYPVIK, H., AAGAARD, P., BJORNVIK, A. M., JOHANSEN, P. E. & ERIKSEN, S. 1990. Degradation and maturity controls on oil field petroleum column heterogeneity in the Gullfaks field, Norwegian North Sea. *In*: DURAND, B. & BEHAR, F. (eds) *Advances in Organic Geochemistry 1989*. Pergamon Press, London, 497–510.

HUANG, D. & LI, J. 1982. *Petroleum Generation in the Chinese Terrestrial Sedimentary Basins*. Petroleum Press, Beijing, 355 p. (in Chinese).

——, —— & ZHANG, D. 1990. Maturation sequence of continental crude oils in hydrocarbon basins in China and its significance. *Organic Geochemistry*, **16**, 521–530.

——, ——, ——, HUANG, X. & ZHOU, Z. 1991. Maturation sequence of Tertiary crude oils in the Chaidam Basin and its significance in petroleum resource assessment. *Journal of Southeast Asian Earth Science*, **5**, 359–366.

——, ——, ZHOU, Z., GU, X. & ZHANG, D. 1984. *Evolution and hydrocarbon generation from terrestrial organic matter*. Petroleum Press, Beijing, 228 p. (in Chinese).

HUGHES, W. B., HOLBA, A. G., MILLER, D. E. & RICHARDSON, J. S. 1985. Geochemistry of Greater Ekofisk crude oils. *In*: THOMAS, B. M. *et al.* (eds) *Petroleum Geochemistry in Exploration of the Norwegian Shelf*. Graham & Trotman, London, 75–92.

IGNATIADIS, I., SCHMITTER, J. M. & ARPINO, P. J. 1985.

Separation et identification par chromatographie en phase gazeuse et chromatographie en phase gazeuse-spectrometrie de masse de composes azotes d'une huile Lourde desasphaltee-evolution de leur distribution apres un hydrotraitement catalytique. *Journal of Chromatography*, **324**, 87–111.

JOKUTY, P. L. & GRAY, M. R. 1991. Resistant nitrogen compounds in hydrotreated gas oil from Athabasca bitumen. *Energy & Fuels*, **5**, 791–795.

KAPIL, R. S. 1971. The carbazole alkaloids. *In*: MANSKE, R. H. F. (ed.) *The Alkaloids, V. 13*. Academic Press, New York, 273–302.

KARLSEN, D. A. & LARTER, S. R. 1991. Analysis of petroleum fractions by TLC-FID: applications to petroleum reservoir description. *Organic Geochemistry*, **17**, 603–617.

LARTER, S. R. & APLIN, A. C. Reservoir geochemistry: methods, applications and opportunities. *This volume*.

——, BJORLYKKE, K. O., KARLSEN, D. A. ET AL. 1991. Determination of petroleum accumulation histories: examples from the Ula Field, Central Graben, Norwegian North Sea. *In*: BULLER, A. T., BERG, E., HJELMELAND, O., KLEPPE, J., TORSAETER, O. & AASEN, J. O. (eds) *North Sea Oil & Gas Reservoirs – II*. Graham & Trotman, London, 319–330.

LATER, D. W., LEE, M. L., BARTLE, K. D., KONG, R. C. & VASSILAROS, D. L. 1981. Chemical class separation and characterisation of organic compounds in synthetic fuels. *Analytical Chemistry*, **53**, 1612–1620.

LEE, L. S., RAO, P. S., BRUSSEAU, M. L. & OGWADA, R. A. 1988. Nonequilibrium sorption of organic contaminants during flow through columns of aquifer materials. *Environmental & Toxicological Chemistry*, **7**, 779–793.

LEYTHAEUSER, D. & SCHAEFER, R. G. 1984. Effects of hydrocarbon expulsion from shale source rocks of high maturity in Upper Carboniferous strata of the Ruhr area, Federal Republic of Germany. *Organic Geochemistry*, **6**, 671–681.

LI, M. & LARTER, S. R. 1993. Interactions of organic nitrogen species with mineral water and organic networks: Implications to petroleum geochemistry. *205th American Chemical Society Meeting*. Division of Geochemistry, American Chemical Society, Abstract Paper **89**.

——, ——, STODDART, D. & BJOROY, M. 1992. Practical liquid chromatographic separation schemes for pyrrolic and pyridinic nitrogen aromatic heterocycle fractions from crude oils suitable for rapid characterisation of geochemical samples. *Analytical Chemistry*, **64**, 1337–1344.

LU, S., HE, W. & HUANG, H. 1990. The geochemical characteristics of heavy oil and its recovery in Liaohe basin, China. *In*: DURAND, B. & BEHAR, F. (eds) *Advances in Organic Geochemistry 1989*. Pergamon Press, Oxford, 437–450.

MITCHELL, A. G., HAZELL, L. B. & WEBB, K. J. 1990. *Wettability determination: Pore surface analysis*. Society of Petroleum Engineers, Paper No. **2050**, 351–360.

PEPPER, A. S. 1991. Estimating the petroleum expulsion behaviour of source rocks: a novel quantitative approach. *In*: ENGLAND, W. A. & FLEET, A. J. (eds) *Petroleum migration*. Geological Society, London, Special Publication, **59**, 9–32.

PERRIN, T. S. & BAILEY, J. R. 1933. The nitrogen compounds in petroleum distillates. IV. Cumulative extraction of kero bases. The isolation of 2,4,8-trimethylquinoline among the kero bases. *Journal of American Chemical Society*, **55**, 4136–4140.

RADKE, M., WELTE, D. H. & WILLSCH, H. 1982. Geochemical study on a well in the Western Canada Basin: relation of the aromatic distribution pattern to maturity of organic matter. *Geochimica et Cosmochimica Acta*, **46**, 1–10.

RICHTER, P. P., CEASSER, P. D., MEISEL, S. L. & OFFENHAUSER, R. D. 1952. Distribution of nitrogen according to basicity. *Industrial & Engineering Chemistry*, **44**, 2601–2605.

RUCKMICK, S. C. & HURTUBISE, R. J. 1985. Liquid chromatographic systems for the separation of polycyclic aromatic hydrocarbon and nitrogen heterocycle compounds present in coal liquids. *Journal of Chromatography*, **321**, 343–352.

SCHENCK, L. M. & BAILEY, J. R. 1941*a*. The nitrogen compounds in petroleum distillates. XXI. Isolation and synthesis of 2,3,4-trimethyl-8-isopropylquinoline. *Journal of American Chemical Society*, **63**, 1364–1365.

—— & —— 1941*b*. The nitrogen compounds in petroleum distillates. XXII. Isolation and synthesis of 2,3-dimethyl-4,8-diethylquinoline (I) and 2,3-dimethyl-4-ethyl-8-n-propylquinoline (II). *Journal of American Chemical Society*, **63**, 1365–1367.

SCHMITTER, J. M. & ARPINO, P. J. 1983. Possible origin and fate of α-methylbenzo[h]quinolines from crude oils. *In*: BJOROY, M. (ed) *Advances in Organic Geochemistry 1981*. J. Wiley, Chichester, 808–812.

—— & —— 1985. Azaarenes in fuels. *Mass Spectrometric Review*, **4**, 87–121.

——, COLLIN, H., EXCOFFIER, J. L., ARPINO, P. & GUIOCHON, G. 1982. Identification of triaromatic nitrogen bases in crude oils. *Analytical Chemistry*, **54**, 769–772.

——, IGNATIADIS, I. & ARPINO, P. 1983. Distribution of diaromatic nitrogen bases in crude oils. *Geochimica et Cosmochimica Acta*, **47**, 1975–1984.

SHI, J., MACKENZIE, A. S., EGLINTON, G., GOWAR, A. P. & MAXWELL, J. R. 1982. Biological markers for petroleum and shales in Shengli Oilfield and their geochemical significance. *Geochimica*, **1**, 1–20 (in Chinese).

SHIVE, B., ROBERTS, S. M., MAHAN, R. I. & BAILEY, J. R. 1942. The nitrogen compounds in petroleum distillates. XXIII. The structure of a $C_{16}H_{25}N$ base from Californian petroleum. *Journal of American Chemical Society*, **64**, 909–912.

SINNINGHE DAMSTE, J. S., KOCK-VAN DALEN, A. C., DE LEEUW, J. W., SCHENCK, P. A., SHENG, G. Y. & BRASSELL, S. C. 1987. The identification of mono-, di- and trimethyl 2-methyl-2-(4,8,12-trimethyltridecyl) chromans and their occurrence in the geosphere. *Geochimica et Cosmochimica Acta*, **51**, 2393–2400.

SNYDER, L. R. 1965. Distribution of benzocarbazole isomers in petroleum as evidence for their biogenic origin. *Nature*, **205**, 277.

—— 1970. Petroleum nitrogen compounds and oxygen

compounds. *Accounts on Chemical Research*, **3**, 290–299.

STODDART, D. P., HALL, P. B., LARTER, S. R., BRASHER, J., LI, M. & BJORØY, M. 1995. The reservoir geochemistry of the Eldfisk Field, Norwegian North Sea. *This volume*.

TISSOT, B. P. & WELTE, D. H. 1984. *Petroleum Formation and Occurrence*. Springer Verlag, Berlin.

WALKER, A. L., MCCULLOH, T. H., PETERSON, N. F. & STEWARD, R. J. 1983. Anomalously low reflectance of vitrinite, in comparison with other petroleum source rock-maturation indices, from the Miocene Modelo Formation in the Los Angeles Basin, California. *In*: ISAACS, C. M. & GARRISON, R. E. (eds) *Petroleum Generation and Occurrence in the Miocene Monterey Formation, California*. Society of Economic Paleotologists & Mineralogists, L.A., 185–190.

WANG, T. G., FAN, P. & SWAIN, F. M. 1988. Geochemical characteristics of crude oils and source beds in different continental facies of four oil-bearing basins, China. *In*: FLEET, A. J., KELTS, K. & TALBOT, M. R. (eds) *Lacustrine Petroleum Source Rocks*. Blackwell Scientific Publications, Oxford, 309–325.

Wilhelms, A. & Larter, S. R. 1994. Origin of tar mats in petroleum reservoirs. Part II: Formation mechanisms for tar mats. *Marine and Petroleum Geology*, **11**(4), 442–456.

——, PATIENCE, R. L., LARTER, S. R. & JORGENSEN, S. 1992. Nitrogen functionality distributions in asphaltenes isolated from several oils from different source rock types. *Geochimica et Cosmochimica Acta*, **56**, 3745–3750.

YAMAMOTO, M. 1992. Fractionation of azarenes during oil migration. *Organic Geochemistry*, **19**, 389–402.

——, TAGUCHI, K. & SASAKI, K. 1991. Basic nitrogen compounds in bitumen and crude oils. *Chemical Geology*, **93**, 193–206.

YANG, S., LI, Q. & PIAO, M. 1982. Petroleum source beds of the Eocene Shahejie Formation in Western Depression, Liaohe Basin. *In*: HUANG, D. & LI, J. (eds) *Petroleum Generation in the Chinese Terrestrial Sedimentary Basins*. Petroleum Press, Beijing, 282–295 (in Chinese).

Diagenesis of the Rotliegend Sandstones in the V-Fields, southern North Sea: a fluid inclusion study

B. McNEIL[1,2], H. F. SHAW[1] & A. H. RANKIN[1,3]

[1]*Department of Geology, Imperial College, London SW7 2BP, UK*

[2]*Present address: Institute of Geosciences, University of Tsukuba, Ibaraki 305, Japan*

[3]*Present address: Department of Geology, University of Kingston, Kingston-on-Thames, UK*

Abstract: Illite, quartz, Fe-dolomite (with some dolomite), anhydrite and barite are the principal pore-filling cements in the analysed samples of the Rotliegend sandstones from the Vulcan, South Valiant and Vanguard Fields of the southern North Sea. Apart from illite, these cements also infill fractures cutting the Rotliegend sandstones, but in the fractures there is a predominance of anhydrite and barite whereas in the pore infills, quartz and Fe-dolomite predominate. The quartz, Fe-dolomite and sulphate cements have been studied using a variety of analytical techniques but with the emphasis on fluid inclusion microthermometry. The results indicate that fluids at temperatures of 70 to 125°C, with salinities of 17 to 24 eq. wt% NaCl and an overall composition of Na–Ca–Mg–K–Cl–SO_4, were involved in authigenic mineral formation in pore spaces and fractures. Hydrocarbons were detected in the fluid inclusions hosted in the cements and indicate that cementation occurred in the presence predominantly of methane and only rarely liquid hydrocarbons. Relating the microthermometry homogenization/formation temperatures to the burial history, quartz cementation occurred mainly in the late Permian to late Cretaceous, Fe-dolomite during the Jurassic to mid-Cretaceous, and anhydrite and barite mainly in the mid-to late Cretaceous.

The area of study is located in the UK sector of the southern North Sea, in the region known as the Sole Pit High–Inde Shelf (Fig.1). All samples were taken from the Rotliegend sandstone horizons in the Conoco-operated V-Fields, i.e. the Vulcan, South Valiant and Vanguard fields (Fig. 1). The Rotliegend sandstones, which form the main gas reservoirs in the southern North Sea, overlie the gas-prone source rocks in the Carboniferous Coal Measures (Cornford 1990) and are sealed by the Zechstein evaporites and carbonates (Fig. 2).

All samples chosen for analysis contained fracture-filling and pore-filling quartz, anhydrite, barite and Fe-dolomite cements. The aim of this study was to better define the diagenetic environments in which these minerals formed and, in particular, the nature and sources of the fluids from which they precipitated. This was achieved principally through the analysis of the fluid inclusions hosted by these materials and the sulphur isotopes in anhydrite and minor barite cements.

Geological history

The depositional and diagenetic history of the Rotliegend sediments is well documented (Lahann *et al.* 1993; Guapp *et al.* 1993; Robinson *et al.* 1993; Purvis 1993; Abbotts 1991; Sullivan *et al.* 1990; Glennie & Provan 1990; Arthur *at al.* 1986; Pye & Krinsley 1986; Rossel 1982; Nagtegaal 1979; Marie 1975). The pattern of diagenetic characteristics shows some variation depending on the location and the geological history of the area concerned.

After deposition of the Rotliegend sandstones in the arid environments of the early Permian that affected much of Northern Europe, subsidence took place, resulting in a transgression of the Zechstein Sea over the Southern Permian Basin (Glennie 1990). Continual and rapid burial of the Rotliegend sediments during the Triassic and Jurassic resulted in a maximum burial depth of approximately 4000 m. Rapid burial also took the Carboniferous Coal Measures into a temperature regime capable of producing gas, i.e. 100 to 160°C (Glennie 1986) at a depth range of 4000 to 6000 m (Barnard & Cooper 1983). At their maximum burial depth, the Rotliegend sediments were at temperatures in excess of 125°C. The burial history curve for the Vulcan Field (Fig. 3) is similar to those for the Vanguard and South Valiant Fields and can be taken as indicative of the burial history for all three fields.

In the early Cretaceous, uplift of the Sole Pit Basin took place and continued to the late Cretaceous, when a further, minor period of subsidence occurred. Minor uplift affected the Sole Pit area again in the middle Tertiary but subsidence is taking place at present (Ziegler, P. A. 1975; Ziegler,

From CUBITT, J. M. & ENGLAND, W. A. (eds), 1995, *The Geochemistry of Reservoirs*, Geological Society Special Publication No. 86, pp. 125–139.

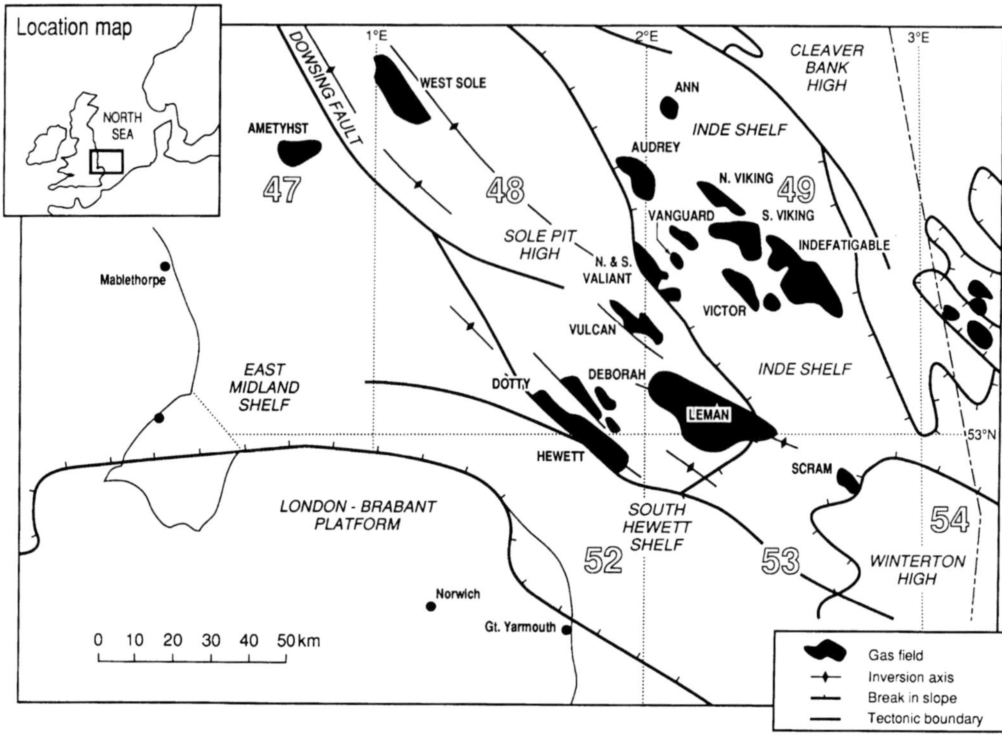

Fig. 1. The study area within the Sole Pit Region.

W. H. 1975). During these phases of uplift, the formation of the now cemented fractures occurred which cut the Rotliegend sandstones (Glennie & Provan 1990). It is the fractured sandstones in addition to the unfractured but cemented sandstones that were analysed in this study.

Techniques

Standard optical petrological techniques were used to examine thin sections of the sandstone and to define the paragenesis of the cement minerals. Additional information on cement petrology was provided by scanning electron microscopy (SEM) analyses of the sandstones employing both secondary and backscatter imaging modes. In both instances, standard analytical techniques were used.

Fluid inclusion microthermometry was performed using a Linkam THM 600 heating – freezing stage attached to a Leitz petrological microscope. The fluid inclusion wafers were produced using the process described by McNeil & Morris (1992). Freezing temperatures were obtained using a Linkam CS 196 cryogenic unit.

Ultraviolet analyses of fluid inclusions employed a mercury–xenon source lamp with an excitation filter at 365 ± 10 nm and barrier filter at 400 nm.

Fluid inclusions were studied by placing crystals in an immersion oil between two glass plates then crushing by hand under the microscope. Gas chromatography involved the use of a Perkin–Elmer 8310GC and a helium carrier gas. The method requires dry grinding of the minerals in sealed containers and the liberated gas was removed by a syringe for analysis by gas chromatography (McNeil 1992).

Sulphate samples for sulphur isotope analyses were prepared using the method of Coleman & Moore (1978). Sulphur isotope studies were performed at the NERC Isotope Facility (BGS Keyworth) using a VG SIRA 10 gas source mass spectrometer. All $\delta^{34}S$ values are reported relative to the Canyon Diablo troilite standard (CDT).

Sandstone petrography

All samples examined in this study are fine to medium-grained quartz arenites and sublitharenites (Pettijohn *et al.* 1972) deposited in aeolian and fluvial environments. The sandstones are cut by fractures that vary in size but are commonly less

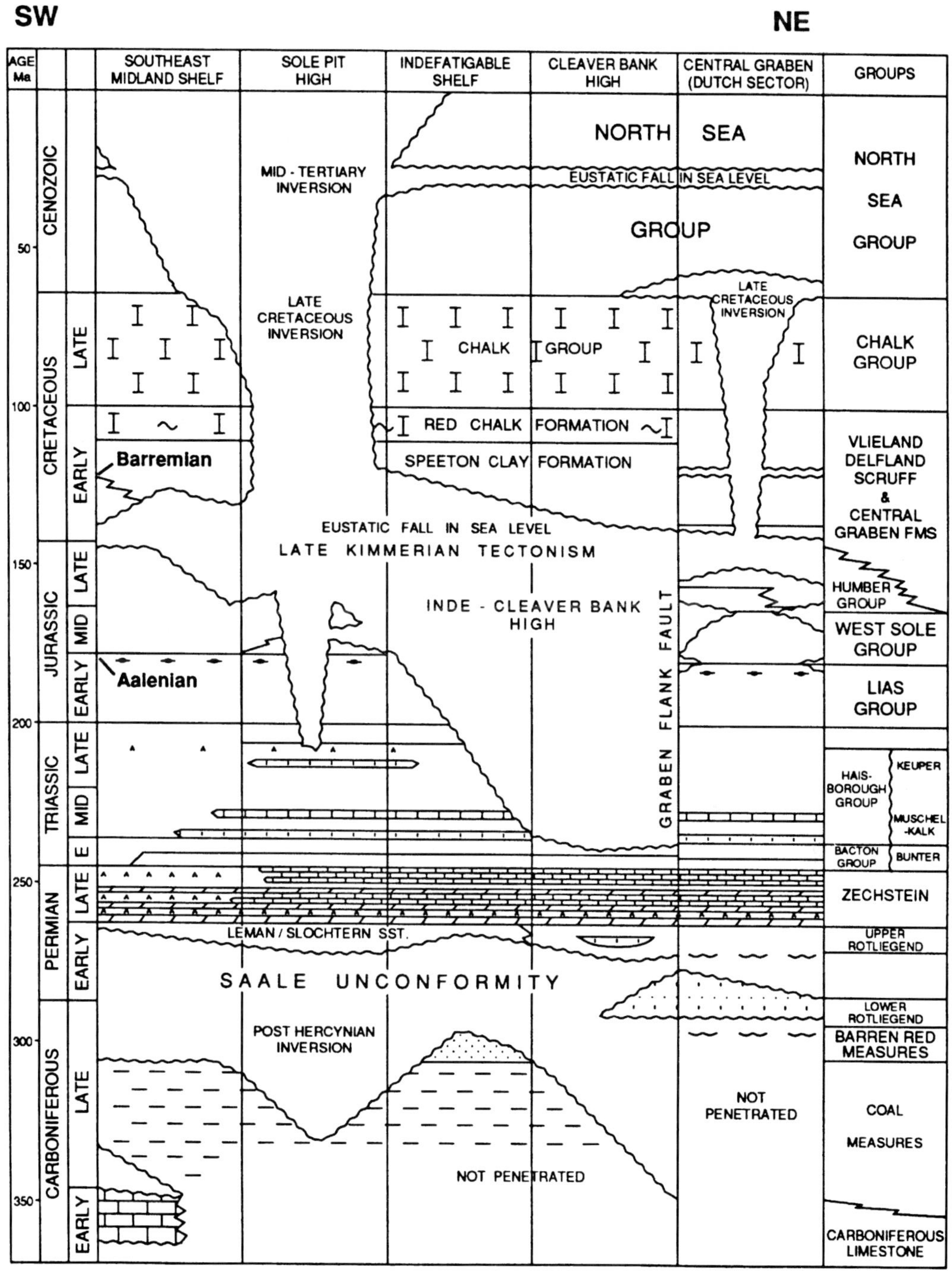

Fig. 2. Stratigraphy of the Southern Permian Basin (from Glennie & Provan 1990).

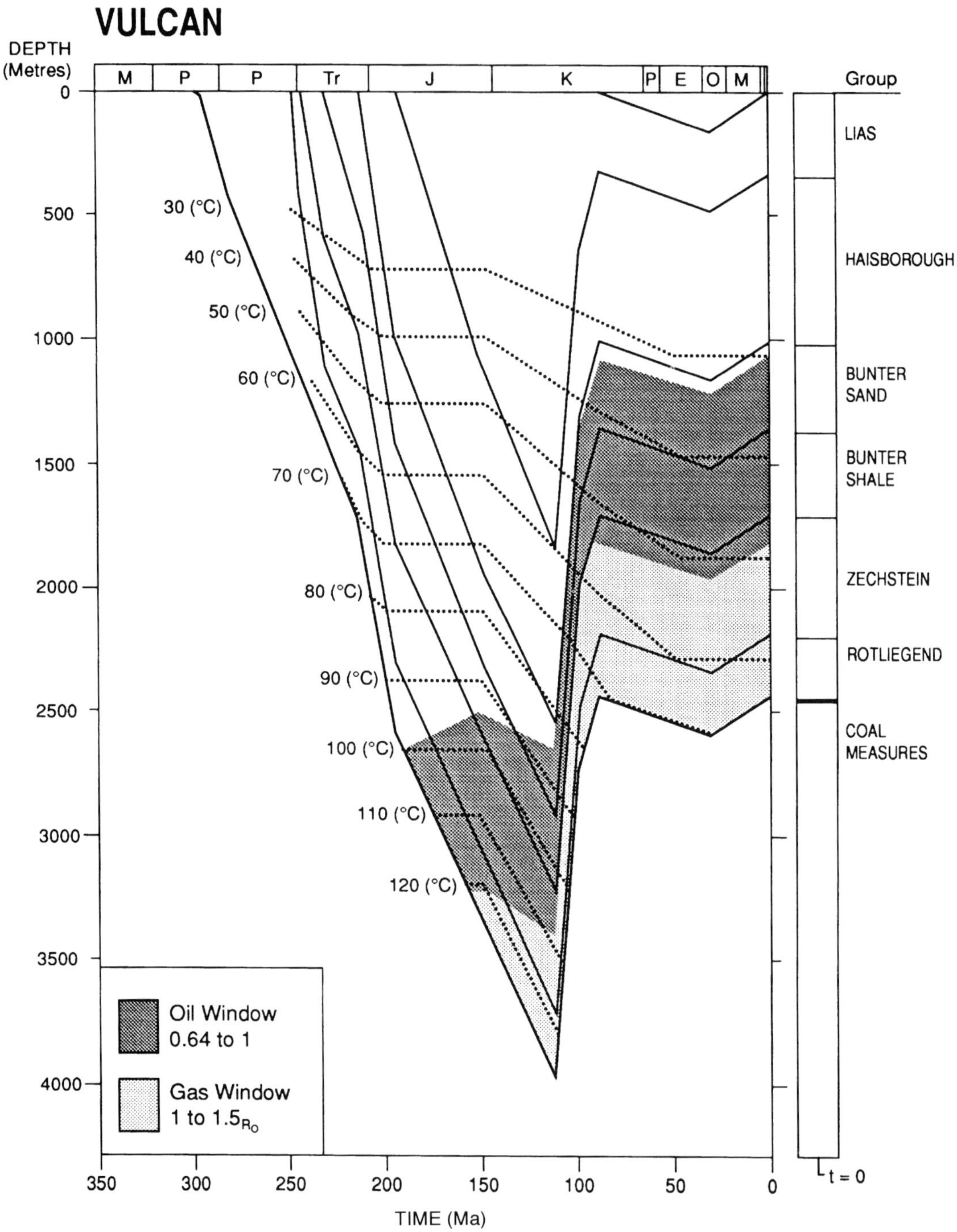

Fig. 3. Burial history curve for the Vulcan Field.

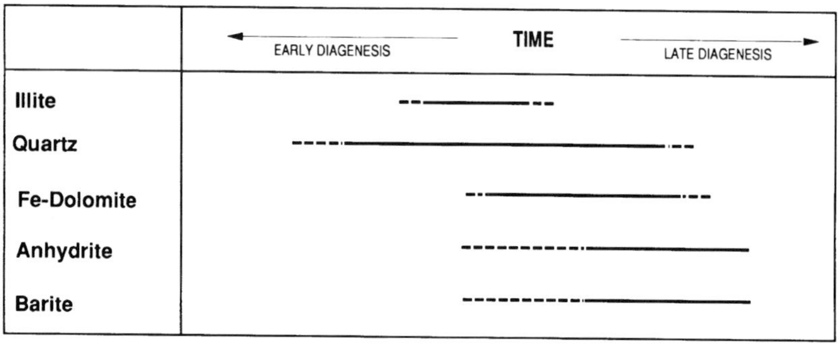

Fig. 4. Cement paragenesis for the Rotliegend Sandstone in the V-fields.

than 3 cm in width. Based on optical and SEM petrographic studies carried out at Imperial College and at Conoco (UK) Ltd, the major cements and their order of precipitation in the Rotliegend sandstones of the V-Fields have been defined (Figs 4, 5 & 6). The order of cementation is generally the same in the fractures as in the sandstone pore spaces except that no illite is found in the fractures. This suggests that as the fractures are uplift-related (Glennie & Provan 1990), the formation of cements in the Rotliegend sandstones has been influenced by the uplift events and their associated fracturing. This is supported by a greater degree of pore-fill cementation in the sandstones adjacent to the fractures, which was also observed by Guapp *et al.* (1993) and Purvis (1992).

In the sandstone pore spaces there is a greater dominance of Fe-dolomite and quartz relative to

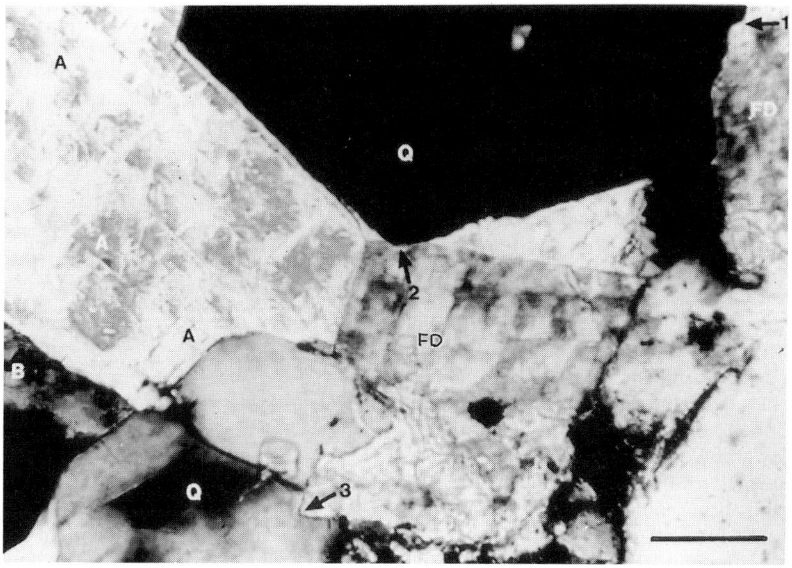

Fig. 5. Cement infilling pore spaces in Rotliegend sandstone. Prismatic quartz (Q) is present together with rhombic Fe-dolomite (FD), pore-filling anhydrite (A) and anhedral barite (B). The crystal boundary at arrow 1 suggests possible coprecipitation of quartz and Fe-dolomite but the crystal boundary at arrow 2 suggests formation of euhedral quartz before euhedral Fe-dolomite. The contact at arrow 3 points to precipitation of Fe-dolomite before quartz. An interpretation of quartz and Fe-dolomite coprecipitation is proposed with later anhydrite and barite cements. Scale bar = 0.2 mm.

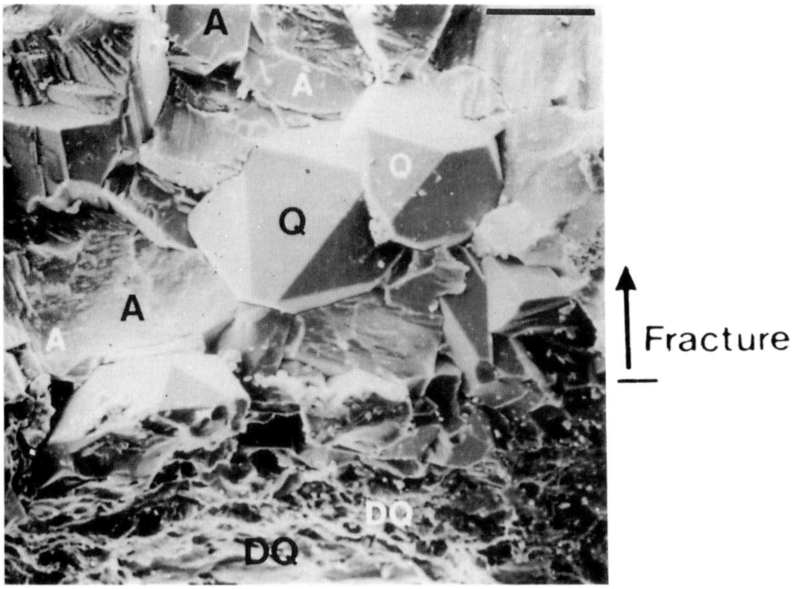

Fig. 6. Scanning electron micrograph of euhedral quartz (Q) and blocky and anhedral anhydrite (A). DQ = detrital quartz. The proportion of anhydrite increases with distance from the fracture walls, representing increased anhydrite precipitation with time during uplift. Scale bar = 60 μm.

anhydrite and barite (5:1) whereas in the fractures there is a dominance of sulphates (9:1) relative to quartz and Fe-dolomite. This suggests that increasing sulphate precipitation accompanied basin inversion and fracturing.

These clay and non-clay cements with a similar order of paragenesis have been reported in other studies of the Rotliegend sandstones in the North Sea (Sullivan *et al.* 1990; Lahann *et al.* 1993). There are some differences from these previous studies in that no kaolinite or chlorite cements were identified in this study. There was also little evidence of an early dolomite cement reported in many studies of the Rotliegend sandstones, apart from what appeared to be some resorbed relicts of dolomite earlier than the later ferroan dolomite cements described here. Glennie *et al.* (1978) and Pye & Krinsley (1986) have suggested that much of the early formed dolomite cements in the Rotliegend sandstones could have been dissolved and reprecipitated during the later Cretaceous inversion. Although Fe-dolomite is the dominant carbonate cement in the samples analysed, there is some evidence of associated non-ferroan dolomites that formed slightly earlier but at a similar time to the Fe-dolomite cements. It appears that the dolomite cements became increasingly Fe-rich with time.

Fluid inclusion microthermometry

Microthermometric analyses were made of aqueous fluid inclusions in pore filling and fracture-filling Fe-dolomite, quartz and anhydrite cements. Barite could not be analysed as it forms less than 1% of the cements in the sandstones. Fluid inclusions in Rotliegend cements show a variety of inclusion shapes, sizes (5–30 μm) and phase characteristics (mainly two-phase liquid and vapour, or mono-phase liquid inclusions but also some liquid–vapour–solid inclusions).

Precipitation temperatures

Heating experiments provide homogenization temperatures (T_h) when the vapour bubble is removed through the thermal expansion of the liquid. This is the minimum trapping temperature for each inclusion analysed and represents the minimum formation temperature of the host crystal. A pressure correction must be applied to this homogenization temperature to obtain true formation temperatures (T_f) (McNeil 1992; Shepherd *et al.* 1985; Roedder 1984).

Homogenization temperatures for Fe-dolomite, quartz and anhydrite aqueous fluid inclusions are presented in Table 1. It is reasonable to assume that

Table 1. *Aqueous fluid inclusion microthermometric results for Rotliegend Group cements*

	Homogenization temp.(°C)			First melting temp. (°C)			Eutectic temp. (°C)			Final ice melting temp. (°C)		
	No. of analyses	Range	Mean	No. of analyses	Range	Mean	No. of analyses	Range	Mean	No. of analyses	Range	Mean
Anhydrite												
Vulcan	161	56.5 to 370.0	181.0	204	–79.6 to –45.0	–63.8	51	–59.2 to –50.1	–50.7	210	–36.6 to –10.5	–21.8
S. Valiant	71	73.5 to 348.0	162.8	75	–76.3 to –52.8	–60.0	19	–57.0 to –42.8	–52.0	71	–24.4 to –11.6	–20.4
Vanguard	31	45.8 to 117.5	82.8	59	–73.3 to –53.2	–60.5	24	–54.9 to –45.6	–50.5	52	–37.1 to –12.7	–20.9
Quartz												
Vulcan	195	40.0 to 126.9	86.3	160	–76.5 to –42.0	–60.4	8	–55.8 to –43.0	–51.5	167	–34.4 to –10.6	–20.5
S. Valiant	141	65.9 to 115.7	85.8	56	–64.7 to –49.1	–55.6	2	–52.8 to –48.8	–50.8	65	–24.5 to –2.4	–16.9
Vanguard	10	76.2 to 113.2	99.6	2	–57.0 to –56.9	–57.0	–	–	–	3	–22.9 to –18.4	–19.7
Dolomite												
Vulcan	24	101.4 to 159.7	137.8	15	–73.1 to –50.7	–65.6	7	–56.0 to –50.7	–53.4	13	–34.3 to –17.6	–23.5

Initial data, uncorrected for pressure

the data for anahydrite are also representative of the conditions for barite precipitation, bearing in mind their apparent coprecipitation based on the petrographic evidence (Fig. 6). Alteration of inclusions after formation must not have taken place if a reliable interpretation of T_h is to be made (Osborne & Haszeldine 1993). Quartz-hosted fluid inclusions are here considered to be the most reliable for the following reasons.

The phenomenon of 'necking' of the inclusions was rarely observed and such inclusions were not analysed. The liquid/vapour ratios for the same inclusion population within individual crystals were routinely measured and found to be similar. Most of the quartz cements hosting the inclusions have not been subjected to substantial burial after their formation. The inclusions formed in quartz cements that precipitated at up to 4000 m depth and subsequently suffered conditions of decreasing pressure and temperature during later basin uplift (Fig. 3). There was no evidence of increased homogenization temperatures during initial or repeat heating runs which might have indicated stretching in any quartz-hosted inclusion. However, this point in itself is not evidence for the lack of stretching phenomena (Roedder 1984). A range of homogenization temperatures observed in a group of crystals within these samples showed reproducible and similar T_h values. In addition, most of the observed T_h values in this study do not even closely approach the maximum burial temperature of 125°C that affected this area (Fig. 3), and there is no observed correlation between larger inclusions and higher T_h values reported by Osborne & Haszeldine (1993). Published data by Ulrich & Bodnar (1988) and Robinson *et al.* (1992) also provide evidence for the maintenance of the integrity of fluid inclusions in quartz with overheating.

The temperature range displayed here by quartz-hosted fluid inclusions therefore represents the precipitation of quartz over a range of burial temperatures. This interpretation is supported by several previous studies (Guapp *et al.* 1993; Hancock 1978; Glennie *et al.* 1978) which indicated that quartz cementation in the Rotliegend Group sandstones occurred over a range of depths, from shallow to maximum burial depth, and throughout the period of basin uplift. Purvis (1992) concluded that quartz precipitation is limited to depths of more than 2000 m.

In most circumstances, a pressure correction is needed to derive formation temperatures from the measured homogenization temperatures. In this study of the quartz-hosted inclusions, the trapped fluids contain gas in both the liquid and vapour phase and may also have a high salinity (see later discussion). A pressure correction can be omitted if gas is pressure at saturation levels (Ravenhurst *et al.* 1989; Malley *et al.* 1986; Hanor 1980) and little gas is needed to achieve saturation in a high-salinity fluid (Haas 1978). In this case therefore, it has been assumed that the homogenization temperature is equivalent to the formation temperature. The results in Table 1 show that the mean temperature at which quartz precipitated can be related to the uplift of the Sole Pit region. The highest mean temperatures (99.6°C) were recorded from the Vanguard Field samples, which had the slowest rate of uplift, and lower formation temperatures were recorded for Vulcan and South Valiant Field samples (86.3°C and 85.8°C) where the rate of uplift was relatively fast. Studies of quartz-hosted fluid inclusions from the Leman Field, carried out as part of a wider study, recorded mean formation temperatures for quartz of 64.8°C. The Leman Field lies on the Sole Pit axis where the rate of uplift was most rapid. These temperatures are supported by Sullivan (1991), whose oxygen isotope and fluid inclusion data for Leman Field sandstones suggests that quartz formation commenced at temperatures of less than 60°C.

The interpretation of measured T_h values for anhydrite and Fe-dolomite is more difficult as the inclusions hosted in these minerals often display the effects of stretching and leakage. A wide range of liquid/vapour ratios were observed in the inclusions hosted in these minerals and repeated heating runs did not provide reproducible T_h values (Osborne & Haszeldine 1993), but careful observation of other characteristics of these inclusions can provide some useful information. First, some inclusions in anhydrite are monophase at room temperature (25°C) and the vapour bubbles that should have formed on cooling are not present, but a slight reduction in temperature to 15 to 20°C causes a vapour bubble to nucleate. These inclusions were then heated to determine the T_h value in the knowledge that stretching could not have altered the T_h value. All measurements for monophase anhydrite-hosted inclusions average 70 to 75°C.

Quartz also generally appears to have formed before anhydrite in the uplift precipitation sequence. The maximum measured T_h value for quartz was 127°C (Table 1) and places an upper limit of 127°C for anhydrite formation.

Finally, observation of the histograms of T_h values for dolomite and anhydrite shows bimodality (Fig. 7). The lower temperature range probably reflects a more realistic minimum value for trapping of inclusions (100 to 125°C for Fe-dolomite and 45 to 120°C for anhydrite), while the upper temperature range reflects the effects of stretching and leakage. Oxygen isotope values for carbonates and anhydrite cements on Leman Field

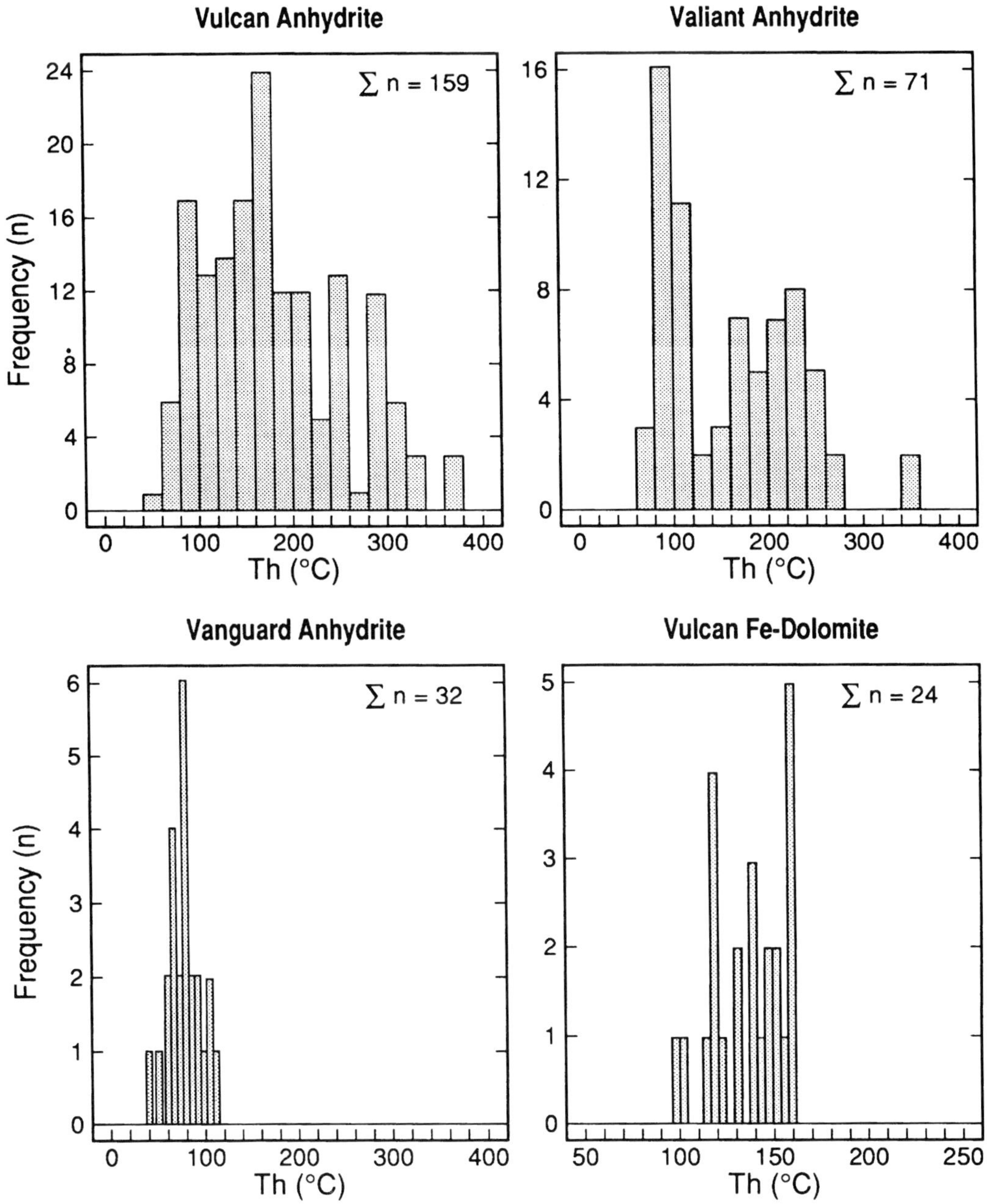

Fig. 7. Frequency histograms of homogenization temperatures (T_h) of inclusions in anhydrite and dolomite cements.

sandstones reported by Sullivan (1991) and Sullivan *et al.* (1994) suggested maximum temperatures of 120°C to 140°C for cement precipitation.

In summary, the microthermometry analyses suggests that Fe-dolomite precipitated at minimum temperatures of between 100°C and 125°C (Fig.7), quartz between 40°C and 127°C (mean values of 86°C and 100°C) and anhydrite between 70°C and 127°C (the latter probably skewed towards the lower end of this range, Fig. 7). This supports the precipitation sequence suggested from inter-

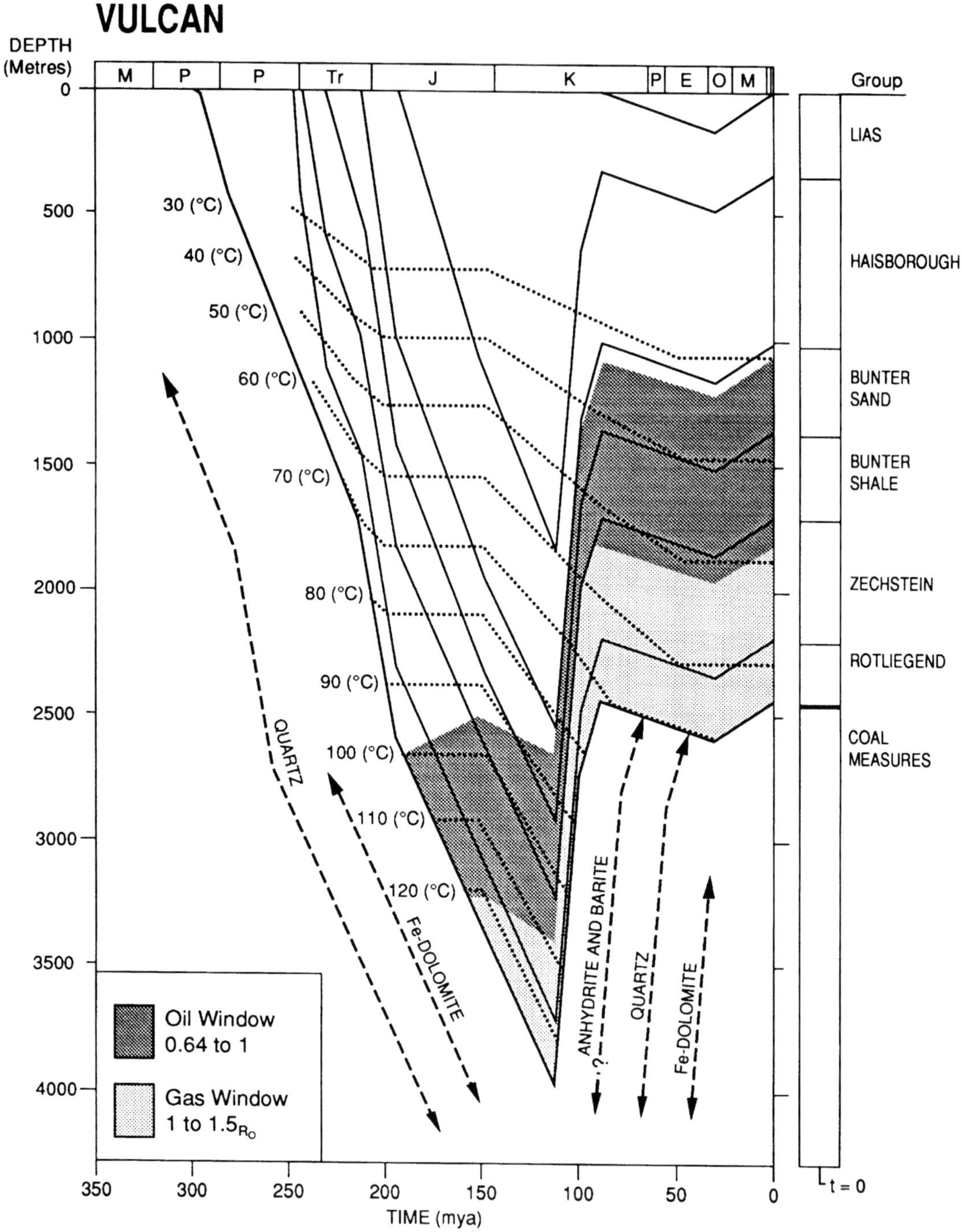

Fig. 8. The formation of the quartz, Fe-dolomite, anhydrite and barite cements in relation to the burial history curve of the Vulcan Field, considered generally representative of the V-Fields area.

pretations of the petrographic studies. Some low T_h values of between 40°C to 60°C (Table 1) were recorded for quartz and anhydrite-hosted inclusions, but these were only six measurements out of a total of 633 analyses.

Relating these formation temperatures to the burial curve allows the timing of the formation of these cements to be estimated and related to the basin uplift event in the early Cretaceous (Fig. 8).

Fluid chemistry

Chemical composition. Fluid inclusion analyses can be used to investigate the chemical nature of the fluid from which minerals precipitated. By freezing and melting the inclusion fluid and recording the temperatures at which phase changes take place, fluid compositions can be estimated. The fluid chemistry can be derived from eutectic and first melting temperatures while salinity is derived from final (ice) melting temperatures. The results of these measurements are listed in Table 1.

The first melting temperatures vary from –60 to –63.8°C for anhydrite, –55.6 to –60.4 °C for quartz and –65.5°C for Fe-dolomite. First melting temperatures below –35°C occur only in $CaCl_2$-bearing systems (Crawford *et al.* 1979) or $CaCl_2$-LiCl- or $FeCl_3$-enriched fluids (Burley *et al.* 1989). Initial melting temperatures as low as –60°C also suggest that the brines in the inclusions are enriched in $CaCl_2$ (Crawford *et al.* 1979). Davis *et al.* (1990) stated that phase changes at temperatures between –85°C and –60°C may be due to metastable $CaCl_2$ hydrate melting, and the melting and recrystallization of ice. In addition, first melting temperatures between –56°C and –47°C probably refer to the stable $CaCl_2$ hydrate reaction. Ravenhurst *et al.* (1989) and Burruss (1981) have proposed that low melting temperatures of around –65°C may indicate the presence of methane and/or carbon dioxide in addition to $CaCl_2$. Thus the melting temperatures of the inclusions in the Rotliegend cements in the range of –56°C to –47°C probably represent eutectic temperatures (whether they are first melting temperatures or not). Those below this range are due to metastable phase behaviour and also, possible, to the presence of CH_4 and/or CO_2.

The first melting temperatures and the eutectic temperatures of anhydrite, quartz and Fe-dolomite in this study differ markedly from the eutectic temperatures of –20.8°C for the system NaCl–H_2O and –22.9°C for the system NaCl–KCl–H_2O (Burley *et al.* 1989; McLimans 1987; Shepherd *et al.* 1985). The recorded temperatures are, however, much closer to the eutetic temperatures of between –49.5°C and –57°C for NaCl–$CaCl_2$ and NaCl–$MgCl_2$- bearing brines (Shepherd *et al.* 1985). Therefore, a complex NaCl–$CaCl_2$–$MgCl_2$–KCl–H_2O composition is indicated from the analyses of the Rotliegend fluid inclusions.

Salinity Final ice melting temperatures represent the removal of ice platelets, except in a small number of cases where clathrates persisted to temperatures above 10°C. The mean final ice melting temperature varied between –17°C and –24°C and, using the data of Shepherd *et al.* (1985), these correspond to salinities of 17 and 24 equivalent wt% NaCl for the Rotliegend fluids. The presence of gas in the fluid inclusions allows absorption of water during inclusion freezing to form clathrates. These clathrates were sometimes observed in this study and, as the clathrates melt after ice, the recorded ice melting temperatures may therefore have been lowered and the interpreted salinites higher than in reality. Collins (1979) stated that salinity estimates due to clathrate formation may be in error by up to 50%. Therefore the true salinity values may be at the lower end of the estimated 17–24 eq. wt% NaCl range and in some cases may be much lower (Table 1).

Hydrocarbons. Examination of the fluid inclusions under ultraviolet light occasionally revealed the presence of liquid hydrocarbons in some fluid inclusions. A bright blue fluorescence was observed, suggesting the existence of low carbon number, high API° gravity liquids (Bodnar 1990) dominated by aliphatic hydrocarbons (Khorasani 1987). Guapp *et al.* (1993) noted the presence of oil condensate in the Rotliegend sandstones of the North German Basin, and Lahann *et al.* (1993) recorded liquid hydrocarbons in Vanguard Field sandstones. The detection of clathrate solids during freezing of fluid inclusions in anhydrite and quartz also suggests the presence of methane (Rankin *et al.* 1990; Collins 1979). Microscopic examination of cement minerals during crushing showed that vapour bubbles expanded when a fracture intersected the inclusions, indicating the presence of a gas under pressure in the inclusions. Gas chromatography analysis of the released gas revealed the dominance of methane (Table 2). The overall composition of the gas is similar to that observed in the reservoirs of the Sole Pit area (Abbotts 1991) and suggests that gas was being transported to the reservoirs via fractures during uplift and was present in small amounts during cement precipitation.

A late migration of methane was suggested by Guapp *et al.* (1993) and was considered to be related to Zechstein brine movement. Purvis (1992) proposed that hydrocarbon gas migration halted

Table 2. *Gas chromatographic results (volume or mole %) for anhydrite fluid inclusions*

Sample	Methane	Ethane	Propane	Ethene
1	98.2	1.4	–	–
2	92.8	5.3	1.0	1.0
3	94.0	3.2	1.8	1.0
4	94.0	3.2	1.8	1.0
5	95.6	2.2	0.8	1.4
6	95.1	2.2	2.7	–
Average	95.0	2.9	1.6	1.1

diagenesis in the early Cretaceous which would generally coincide with uplift of the V-Fields area.

Isotopic studies

Analyses of the $\delta^{34}S$ values in the anhydrite and barite cement shows a narrow range of values (+10.7 to +7.7‰) for anhydrite and +8.5‰ for barite (Table 3). These values are similar to the $\delta^{34}S$ values for marine evaporites of Permo-Triassic age reported by Nielsen (1979) (Fig. 9) and suggest that the Permo-Triassic evaporites provided the most likely SO_4 source to form the anhydrite and barite cements in the Rotliegend sandstones. Purvis (1992) presented higher $\delta^{34}S$ values for samples immediately east of this study area but still suggested a predominantly Zechstein source for the anhydrite and barite.

The results indicate either that the sulphate minerals in the sandstones represent one precipitation event or that one source of sulphate was involved in several precipitation events. Pye & Krinsky (1986) proposed that all early sulphate was dissolved from the sandstones and that the present sulphate in the southern North Sea is the result of one late diagenetic event (also Lahann *et al.* 1993). These interpretations, together with the abundance of sulphate in pore spaces around fractures, generally support the conclusion that the anhydrite represents a single late diagenetic event.

Table 3. *Sulphur isotope results from Rotliegend sulphates (relative to Canyon Diablo Troilite standard)*

Sample	$\delta^{34}S$ (‰)
Vulcan	
RO7 9631 fracture-fill anhydrite	+ 8.5
UO1 7735 fracture-fill anhydrite	+ 7.7
UO1 7735 fracture-fill barite	+ 8.5
RO 8835 fracture-fill anhydrite	+ 9.5
UO1 7694 pore-fill anhydrite	+10.7
S Valiant	
TO3 9068 fracture-fill anhydrite	+ 7.8
Vanguard	
QO2 12588.3 fracture-fill anhydritee	+ 9.4
QO2 12552.5 fracture-fill anhydrite	+ 9.3
QO2 12636.8 fracture-fill anhydrite	+ 9.0
49/16-10 8556.8 pore-fill anhydrite	+10.1

Replicate analyses deviate by ±0.1‰

Conclusions

1. Fe-dolomite, quartz, anhydrite and barite are the dominant pore-and fracture-filling cements in the Rotliegend sandstones of the V-Fields. Illite also occurs as a pore-filling cement but is absent from the fractures. Quartz precipitation took place from shallow to deep burial and continued during inversion. Generally the Fe-dolomite cements are post-quartz but there is some evidence of co-precipitation or even pre-quartz precipitation. Anhydrite and barite (minor) formed in general during the uplift and are the dominant phases in the inversion fractures.

2. Quartz cements precipitated at temperatures between 40°C and 127°C (mean temperatures between 86°C and 100°C); Fe-dolomite precipitated at temperatures between 100°C and 125°C; anhydrite precipitated at temperatures between 70°C and 127°C (probably skewed towards the lower end of this range, Fig. 7). Homogenization temperatures can be regarded as being approximately equivalent to true formation temperatures in these samples. According to thin-section studies, quartz and Fe-dolomite cements were precipitated mainly in pore spaces in the Rotliegend sandstones, with relatively minor precipitation in sandstone fractures. In contrast, anhydrite and barite precipitated mainly in the fractures.

3. Relating the homogenization temperatures to the burial history curve (Fig. 8), quartz cementation took place mainly in the late Permian to late Cretaceous (250 Ma–120 Ma), Fe-dolomite precipitation during the Jurassic to mid-Cretaceous (200 Ma–120 Ma), and anhydrite and barite mainly during mid- to late Cretaceous.

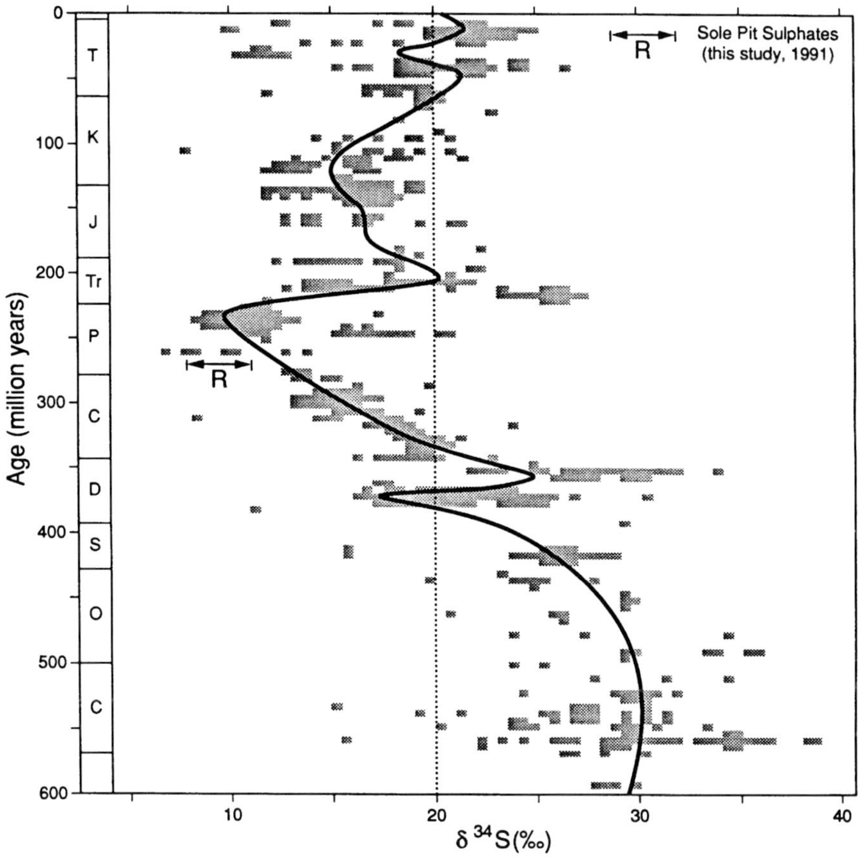

Fig. 9. Variations in $\delta^{34}S$ values for evaporitic sulphates over geological time (Nielsen 1979).

4. The lowest mean quartz T_h values occur in the Leman Field samples (65°C) on the axis of the Sole Pit High and these values increase in Vulcan (86°C) and South Valiant (86°C), and on the flanks of the anticline in the Vanguard Field (100°C). A correlation therefore appears to exist between uplift and quartz cementation.

5. Eutectic temperatures of between –49.5°C and –57.0°C suggest a fluid of complex composition, i.e. $NaCl–CaCl_2–MgCl_2–KCl–H_2O$. First melting temperatures of –65°C and less are due to metastable phase changes as a result of a relatively high calcium content and the presence of methane.

6. The salinity of fluids trapped in the inclusion varies between 17 and 24 eq.wt ‰ NaCl but these may be overestimates because of the presence of gas in the fluids. Consequently the lower value of 17 eq.wt% NaCl should be taken as a more realistic estimate of salinity.

7. A source of sulphate in the Zechstein, suggested from the sulphur isotope studies, supports the existence of a complex fluid. As anhydrite and barite are the main fracture-filling cements, it is envisaged that during inversions, fluids from the Zechstein, where high pore pressure existed due to gypsum dehydration (Purvis 1992), escaped along fractures into the Rotliegend.

8. The detection of a predominantly methane gas component suggests the presence of gas within the formation fluids during diagenesis but not in sufficient quantities to halt diagenetic processes, and that the fractures acted as migration pathways for hydrocarbon gases.

9. Relating the gas migration to the fracturing suggests that migration took place during the mid- to late Cretaceous inversion event.

Permission to publish this paper was kindly given by Conoco (UK) Ltd. The support provided by Conoco (UK) Ltd and the Geochem Group during the PhD research leading to this publication is gratefully acknowledged.

References

ABBOTTS, I. L. 1991. *United Kingdom Oil and Gas Fields, 25 Years Commemorative Volume.* Geological Society, London, Memoir, **14**, 573 p.

ARTHUR, T. J., PILLING, D., BUSH, D. & MACCHI, L. 1986. The Leman Sandstone Formation in UK Block 49/28: sedimentation, diagenesis and burial history. *In*: BROOKS, J. GOFF, J., & VAN HOORN B., (eds) *Habitat of Palaeozoic Gas in NW Europe.* Geological Society, London, Special Publication, **23**, 251–266.

BARNARD, P. C. & COOPER, B. S. 1983. A review of geochemical data related to the NW European gas province. *In*: BROOKS, J. (ed.) *Petroleum Geochemistry and Exploration of Europe.* Geological Society, London, Special Publication, **12**, 19–23.

BODNAR, R. J. 1990. Petroleum migration in the Miocene Monterey Formation, California, USA: constraints from fluid inclusion studies. *Mineralogical Magazine*, **54** (**2**), 295–304

BURLEY, S. D., MULLIS, J. & MATTER, A. 1989. Timing diagenesis in the Tartan Reservoir (UK North Sea): constraints from combined cathodoluminescence microscopy and fluid inclusion studies. *Marine and Petroleum Geology*, **6**, 98–120.

BURRUSS, R. C. 1981. Hydrocarbon fluid inclusions in studies of sedimentary diagenesis. *In*: HOLLISTER, H. S. & CRAWFORD, M. L. (eds) *Fluid inclusions: applications to Petrology.* Mineralogical Society of Canada, Short Course Handbook, **6**, 138–156.

COLEMAN, M. & MOORE, M. P. 1978. Direct reduction of sulphate to sulphur dioxide for isotopic analysis. *Analytical Chemistry*, **28**, 199–260.

COLLINS, P. L. F. 1979. Gas hydrates in CO_2-bearing fluid inclusions and the use of freezing data for estimation of salinity. *Economic Geology*, **74**, 1435–1444.

CORNFORD, C. 1990. Source rocks and hydrocarbons of the North Sea. *In*: GLENNIE, K.W. (ed.) *Introduction to the Petroleum Geology of the North Sea.* Blackwells Scientific Publications, Oxford, 294–361

CRAWFORD, M. L., KRAUS, D. W. & HOLLISTER, L. S. 1979. Petrologic and fluid inclusion study of calc-silicate rocks, Prince Rupert, British Columbia. *American Journal of Science,* **279** (10), 1135–1159.

DAVIS, D. W., LOWENSTEIN, T. K. & SPENCER, R. J. 1990. Melting behaviour of fluid inclusions in laboratory-grown halite crystals in the system NaCl-H_2O, NaCl-KC1-H_2O, NaCl-$MgCl_2$-H_2O, NaCl-$CaCl_2$-H_2O. *Geochimica et Cosmochimica Acta*, **54**, 519–601.

GUAPP, R., MATTER, A., PLATT, J., RAMSEYE, K. & WALZEBUCK, J. 1993. Diagenesis and fluid evolution of deeply buried Permian (Rotliegend) gas reservoirs, Northwest Germany. *American Association of Petroleum Geologists*, Bulletin, **77** (7), 1111–1128.

GLENNIE, K. W. 1986 Development of NW Europe's Southern Permian Gas Basin. *In*: BROOKS, J. & VAN HOORNE, B. (eds) *Habitat of Palaeozoic Gas in NW Europe.* Geological Society, London, Special Publication, **23**, 3–22.

—— 1990. Lower Permian-Rotliegend. *In*: GLENNIE, K. W. (ed.) *Introduction to the Petroleum Geology of the North Sea.* Blackwells Scientific Publications, Oxford, 120–152.

—— & PROVAN, D. M. J. 1990. Lower Permian Rotleigend reservoir of the Southern North Sea Gas Province. *In*: BROOKS, J. (ed.) *Classic Petroleum Provinces.* Geological Society, London, Special Publication, **50** 399–416.

——, MUDD, G. C. & NATEGAAL, P. J. C. 1978. Depositional environment and diagenesis of Permian Rotliegend sandstones in Leman Bank and Sole Pit Areas of the UK Southern North Sea. *Journal of the Geological Society, London,* **135**, 25–34.

HAAS, J. L. 1978. An empirical equation with the tables of smoothed solubilities of methane in water and aqueous sodium chloride solutions up to 25 weight percent, 360°C and 138 MPa. *U.S. Geological Survey Open-File Report*, **78-1004**.

HANCOCK, N. J. 1978. Possible causes of Rotliegend sandstone diagenesis in northern West Germany. *Journal of the Geological Society, London*, **135**, 35–40.

HANOR, J. S. 1980. Dissolved methane in sedimentary brines: potential effect on the PVT properties of fluid inclusions. *Economic Geology*, **75**, 603–617.

KHORANSANI, G.K. 1987. Novel development in fluorescence microscopy of complex organic mixtures: application in petroleum geochemistry. *Organic Geochemistry*, **11**, 157–168.

LAHANN, R. W., FERRIER, J. A. & CORRIGAN, S. 1993. Reservoir heterogeneity in the Vanguard Field. *In*: ASHTON, M. (ed.) *Advances in Reservoir Geology.* Geological Society, London, Special Publication, **69**, 33–56.

MCLIMANS, R. K. 1987. The application of fluid inclusions to migration of oil and diagenesis in petroleum reservoirs. *Applied Geochemistry*, **2**, 585–603.

MCNEIL, B. 1992. *Fluid processes during diagenesis of the Rotleigend sandstones of the Southern North Sea.* PhD thesis, University of London. 435 p.

—— & MORRIS, E. 1992. The preparation of double-polished fluid inclusion wafers from friable water-sensitive material. *Mineralogical Magazine*, **56**, 120–122.

MALLEY, P., JOURDAN, A. & WEBBER, F. 1986. Etude des inclusions fluides dans les nourrissages siliceux des gres reservoirs de Mer du Nord: une nouvelle lecture possible de l'histoire diagenetique du Brent de la region d'Alwyn. *Academy of Science, Paris*, **302**, 653–658.

MARIE, J. P. P. 1975. Rotliegendes stratigraphy and diagenesis. *In*: WOODLAND, A. W. (ed.) *Petroleum Geology and the Continental Shelf of NW Europe.* Institute of Petroleum/Applied Science Publications, 205–210.

NAGTEGAAL, P. J. C. 1979. Relationship of facies and reservoir quality in Rotliegend desert sandstones, Southern North Sea region. *Journal of Petroleum Geology*, **1**, 145–158.

NIELSEN, J. 1979. Sulfur Isotopes. *In*: JAEGER, E. & HUNZITER, J.C. (eds) *Lectures in Isotope Geology*, Springer-Verlag, Berlin, 283–312.

OSBORNE, M. & HASZELDINE, S. 1993. Evidence for resetting of fluid inclusion temperatures from quartz cements in oilfields. *Marine and Petroleum Geology*, **10**, 271–278.

PETTIJOHN, F. J., POTTER, P. E. & SIEVER, R. 1972. *Sand and Sandstones*. Springer-Verlag, Berlin, 618 p.

PURVIS, K. 1992. Lower Permian Rotliegend sandstone, Southern North Sea. A case study of sandstone diagenesis in evaporite associated sequences. *Sedimentary Geology*, **77**, 155–171.

PYE, K. & KRINSLEY, D. H. 1986. Diagenetic carbonate and evaporite minerals in Rotliegend aeolian sandstone of the southern North Sea: their nature and relationship to secondary porosity development. *Clay Minerals*, **21**, 443–457.

RANKIN, A. H., HODGE, B. L. & MOSER, M. 1990. Unusual oil-bearing inclusions in fluorite from Baluchistan, Pakistan. *Mineralogical Magazine*, **54**, 335–342.

RAVENHURST, C. E., REYNOLDS, P. H., ZENTILLI, M., KRUEGER, H. W. & BLENKINSOP, J. 1989. Formation of Carboniferous Pb–Zn and barite mineralisation from basin-derived fluids, Nova Scotia, Canada. *Economic Geology*, **84**, 1471–1488.

ROBINSON, A. G., COLEMAN, M. L. & GLUYAS, J. G. 1993. The age of illite cement growth Village Fields area, southern North Sea: evidence from K/Ar ages and $^{18}O/^{16}O$ ratios. *American Association of Petroleum Geologists, Bulletin*, **77**, 68–80.

——, GRANT, S. & OXTOBY, N. 1992. Evidence against natural deformation of fluid inclusions in diagenetic quartz. *Marine and Petroleum Geology*, **9**, 568–572.

ROEDDER, E. 1984. *Fluid inclusions*. Reviews in Mineralogy, **12**, Mineralogical Society of America, 644 p.

ROSSEL, N. C. 1982. Clay mineral diagenesis in Rotliegend aeolian sandstones of the Southern North Sea. *Clay Minerals*, **17**, 69–77.

SHEPHERD, T., RANKIN, A. H. & ALDERTON, D. H. M. 1985. *A Practical Guide to Fluid Inclusion Studies*. Blackie, Glasgow, 239 p.

SULLIVAN, M. D. 1991. *Diagenetic study of the Lower Permian Rotliegend Sandstone, Leman Field, Southern North Sea*. PhD thesis, University of Glasgow.

——, HASZELDINE, R.S., BOYCE, A. J., ROGERS, G. & FALLICK, A. E. 1994. Late anhydrite cements mark basin inversion: isotopic and formation waters evidence in Rotliegend Sandstone, North Sea. *Marine and Petroleum Geology*, **11**, 46–54.

——, HASZELDINE, R. S. & FALLICK, A. 1990. Linear coupling of carbon and strontium isotopes in Rotliegend Sandstone, North Sea: evidence for cross-formational fluid flow. *Geology*, **18**, 1215–1218.

ULRICH, M. R. & BODNAR, R. J. 1988. Systematics of stretching of fluid inclusions II: barite at 1 atm confining pressure. *Economic Geology*, **83**, 1037–1046.

ZIEGLER, P. A. 1975. North Sea history in the tectonic framework of NW Europe. *In*: WOODLAND, A. W. (ed.) *Petroleum Geology and the Continental Shelf of NW Europe*. Institute of Petroleum/Applied Science Publications, 131–148.

ZIEGLER, W. H. 1975. Outline of the geological history of the North Sea. *In*: WOODLAND, A. W. (ed.) *Petroleum Geology and the Continental Shelf of NW Europe*. Institute of Petroleum/Applied Science Publications, 165–187.

The filling and emptying of the Ula Oilfield: fluid inclusion constraints

NORMAN H. OXTOBY[1], ALAN W. MITCHELL[2,3] & JON G. GLUYAS[1]

[1] *BP Exploration Operating Company Ltd, BPX Technology Provision, Chertsey Road, Sunbury on Thames, Middlesex. TW16 7LN, UK*

[2] *BP Norway UA, Forusbeen 35, 4033 Forus, PO Box 197, Stavanger, Norway*

[3] *Present address: BP Exploration, Blackhill Road, Holton Heath Trading Park, Poole, Dorset. BH16 6LU, UK*

Abstract: The Norwegian sector Ula Field contains 430 million barrels of oil reserves trapped in a dip-closed Upper Jurassic sandstone reservoir and has multiple oil-water contacts (OWCs). Historically, the depth to the contacts has been the subject of some debate and development drilling has now proven the existence of at least six. The precise nature of some of these contacts remains uncertain because all were discovered after production start-up and determination of their origin is critical to future field development. The relationship of oil inclusion abundances to OWCs and rock properties (porosity, permeability) has been evaluated (a) to gain an insight into the filling history and mechanisms, and (b) to attempt to define pre-production OWCs. Oil inclusion abundances decrease with depth and towards present OWCs in Ula, but oil inclusions also occur below them. They indicate that Ula filled from the crest down, and that the filling was synchronous with cementation to at least 3500 metres subsea. Below this depth, cementation appears to have reduced permeability to such an extent prior to charge that the resulting filling geometry was irregular. Local abundance peaks within the oil column correlate with the tops of coarsening-upward cycles. Rare oil inclusions below the oil legs may represent either migration pathways updip from the source or invasion downdip from the main oil leg along higher permeability zones. Oil inclusions persist to greater depths in the east and south-east of the field which indicates that oil charged the field from these directions. A number of models explaining the variability of the OWCs have been ruled out by this work, thereby reducing the uncertainty in field development planning.

Sandstone reservoirs are often laterally and vertically heterogeneous because of lithological, diagenetic and tectonic factors (Gibbons & Irwin 1991; Reinson 1991; Ashton 1993; McMahon 1993). The distribution of producible oil and gas in these heterogeneous networks is complex and it is possible to leave substantial amounts of potentially producible petroleum behind during production if development does not take such heterogeneity into account. Understanding the distribution of oil and gas in relation to permeability heterogeneity is thus of major importance in effective development and production. To achieve this understanding, it is helpful to know how the reservoir filled, in addition to evaluating the porosity and permeability architecture and the fluid properties of an accumulation. For example, water-saturated zones found within structural closure during drilling could be the result of lithological, tectonic or diagenetic effects operating at any time before, during or after filling; proximity to source 'kitchens'; trap leakage; production, or any combination of these factors. If the reasons for such water-saturated zones are not understood ahead of drilling, planned producing wells and injectors may not be optimally located.

We report fluid inclusion data from the Ula Field which suggest that quantified petroleum inclusion abundances can be a useful tool for evaluating the filling mechanisms of petroleum accumulations, ultimately leading to more effective development strategies. This work validates studies by Nedkvitne *et al.* (1993) and extends them to cover a larger area of the Ula Field as well as attempting to provide a quantitative measure of petroleum inclusion abundance. We believe that the technique can augment the information on filling given by organic geochemistry (Hillebrand & Leythaeuser 1992; Trindade & Brassell 1992), strontium isotope analysis (Smalley *et al.* 1995) and numerical modelling (Burrus *et al.* 1991). Integration of all four approaches must be a powerful weapon in the armoury of the development geologist.

Geology of the Ula Field

The Ula Field is situated in Block 7/12 on the eastern margin of the Central Graben in the

From Cubitt, J. M. & England, W. A. (eds), 1995, *The Geochemistry of Reservoirs*, Geological Society Special Publication No. 86, pp. 141–157

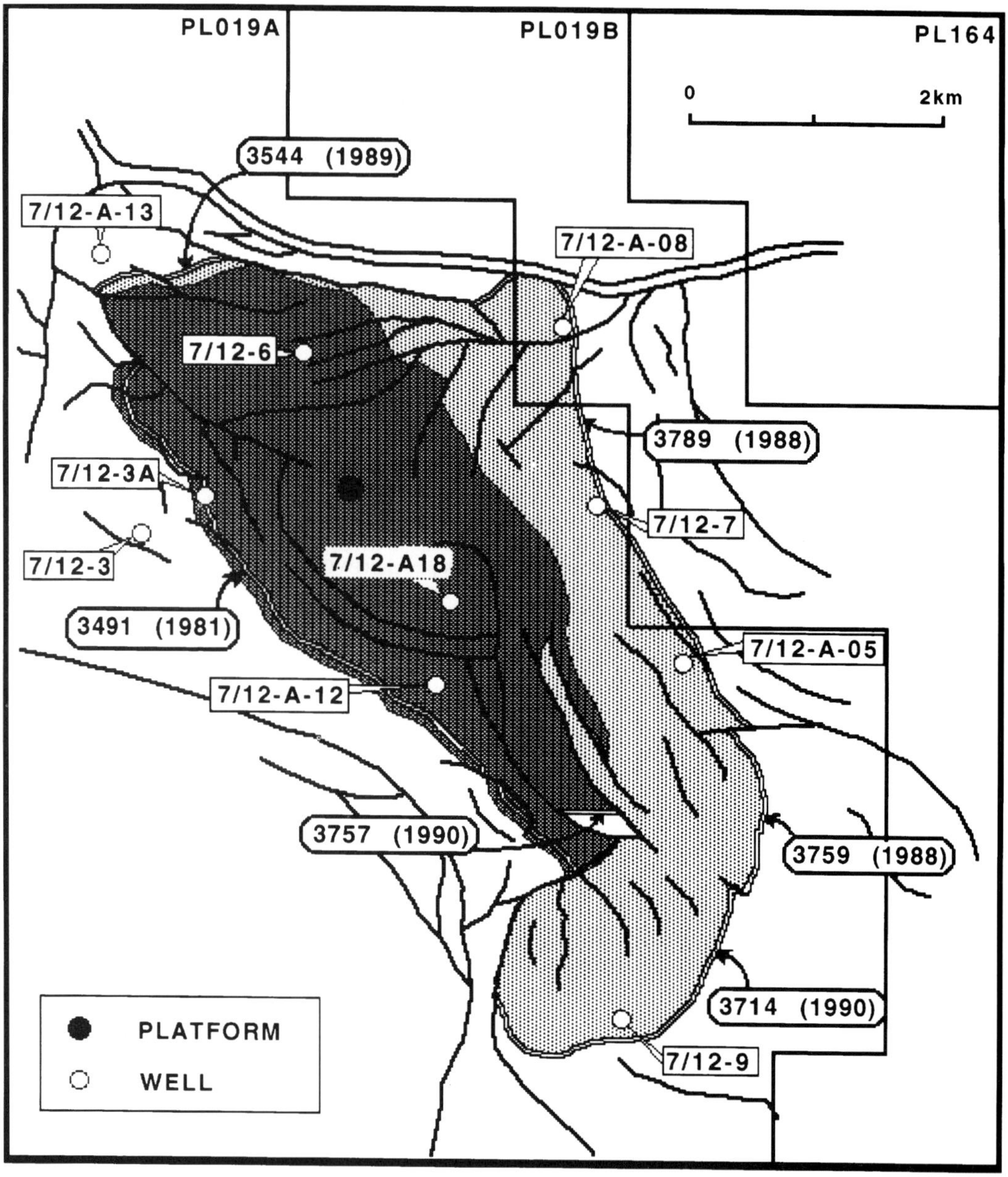

Fig. 1. Location map of Ula Field showing structure, oil–water contacts (depths in metres subsea (mss) and year of discovery in rounded boxes). Faults are indicated by single solid lines, oil–water contacts by parallel lines. The dark shading indicates the portion of the field above 3500 mss and the lighter shading the portion below 3500 mss.

Norwegian Continental Shelf. The trap is relatively simple: a salt-cored, largely dip-closed pericline, elongated NW–SE, with *c.* 500 m of vertical closure, dissected into two major and a number of minor blocks by normal faults (Fig.1). It developed as a result of Early Cretaceous, Late Cretaceous (Middle to Late Campanian) and Tertiary (Late Ypresian and Middle Oligocene) episodes of structural inversion along a base Zechstein fault (Stewart 1993). It contains *c.* 860 million barrels in place of undersaturated 39° API oil with an average gas/oil ratio of 490 scf/bbl. The source rock for the oil is the overlying Mandal Formation, of Portlandian age, which has supplied the trap from source volumes to the west and south. The main phase of petroleum charge occurred in response to rapid burial in the late Neogene (<10 Ma BP) and has continued to the present (Larter *et al.* 1990).

The main oil accumulation is in sandstones of the late Jurassic Ula Formation. These sandstones are predominantly fine- to medium-grained and relatively homogeneous, and are interpreted as a composite progradational–aggradational sandsheet developed on a shallow marine shelf between storm and fairweather wavebase. They vary in thickness from 84 m in well 7/12-4 to 166 m in well 7/12-3A and are thickest in the northwestern part of the field (Fig. 1). The Ula Formation is divided into five zones (Fig. 2). The basal part of the Ula Formation consists of a transgressive sandstone (zone 5) overlain by an open marine silt-mudstone (zone 4). The mudstone coarsens upwards into a progradational shoreface sequence of well-sorted and cross-bedded sandstones overlain by a retrogradational shelf sequence of small-scale coarsening and fining-upwards units. Most of the sandstones are thoroughly bioturbated. These sandstones form zone 3. A second phase of shelf progradation–aggradation produced a coarsening upwards sequence of very fine- to medium-grained sandstones (zone 2). The uppermost reservoir sands form a retrogradational shelf sequence of very fine-grained argillaceous sandstones (zone 1).

The fine-grained sandstones which form the lower part of each coarsening-upwards unit are rich in detrital clay (zones 2B and 3B). They have a

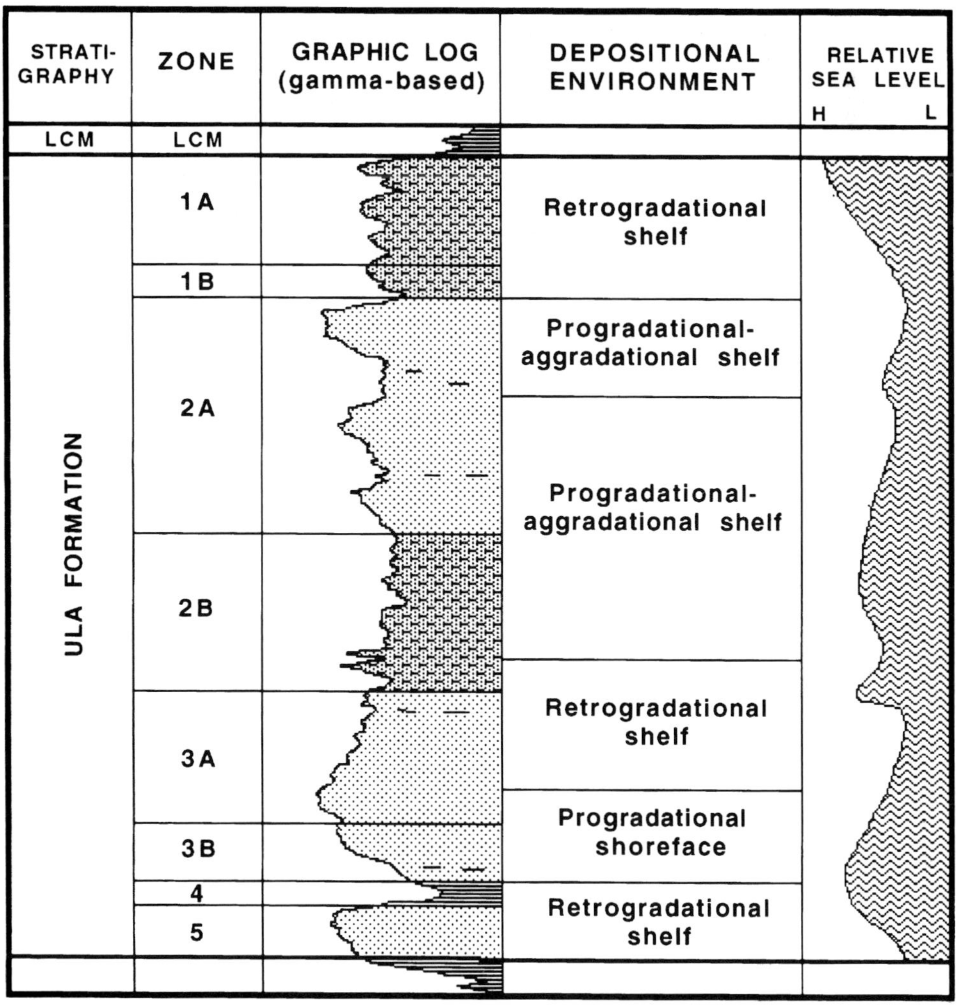

Fig. 2. Schematic stratigraphy and zonation of the Ula reservoir.

relatively low quartz cement content and low permeabilities (60–80 mD). The clean sandstones at the top of the coarsening-upwards units contain high proportions of quartz cement but nonetheless form the best reservoir sands with moderately good permeability. Zones 2A and 3A have average permeabilities in the region of 230 mD and 100 mD, respectively.

Six oil–water contacts (OWCs) have been found within the field area (Fig. 1). In the southwestern fault block, a free OWC at 3491 metres subsea (mss) has been postulated from logging and testing in wells 7/12-3A and 7/12-A12. In the northern fault panel, a free OWC has been established at 3544 mss in 7/12-A13A. In the northeast producible oil has been established at 3789 mss in well 7/12-7, 31 m above the mapped closing contour at *c*. 3820 mss. Well 7/12-A08, mapped within the same fault panel as 7/12-A13, has deeper oil but it is uncertain whether this is producible because the well was not tested. Some of these wells were drilled after production start-up and their oil legs may have been disturbed by production, but others appear to record pre-production differences in OWCs. There are two likely controls on variable original OWCs in the field: the normal fault system, and reservoir quality variation (Brown *et al.* 1992). To optimize production from the field it is important to evaluate where the original OWCs were, and what controls their distribution. Analysis of three dimensional seismic data, dynamic production data and residual oil geochemistry has helped in this evaluation (Larter *et al.* 1990; Brown *et al.* 1992), but pertinent data were also acquired from petroleum inclusions, trapped in detrital and diagenetic minerals.

Previous work with impact on this study.

Petroleum inclusions have been used as a tracer for petroleum migration since 1980 with the application of microthermometry, fluorescence microscopy, fluorescence spectroscopy and gas chromatography (see reviews in Pagel *et al.* 1986; McLimans 1987; Burruss 1991; Eadington *et al.* 1991). Petroleum inclusion abundances have not been routinely reported in studies of this kind. The pioneering work at the University of Oslo on the Ula Field (Nedkvitne *et al.* 1993; Karslen *et al.* 1993) showed what might be achieved with abundance data. They used qualitative assessments of petroleum inclusion abundances. as well as detailed geochemical analyses of inclusion oils, to identify the patterns and the timing of petroleum emplacement in the Ula Field. The abundance data were reported on a scale of 0 to 5, and the authors tentatively correlated abundance maxima with zones of good reservoir quality. The Oslo group's data suggested that a more detailed analysis of petroleum inclusion abundances might yield economically useful information. Accordingly, a simple method for quantifying petroleum inclusion abundances, based on point counting, was developed and applied to a larger sample set from the Ula Field (see appendix A for the methodology).

Most authors have assumed that petroleum inclusions are trapped on migration routes (e.g. Burrus *et al.* 1983; Bodnar 1990). Some undoubtedly are, but we consider that it is equally plausible for inclusions to be trapped in the transition zone in zones of active diagenesis during the development of OWCs. England *et al.* (1987) argue that migration routes form only a small proportion of carrier bed volume (<10%), while transition zones will sweep the entire reservoir volume during filling. This mechanism for petroleum inclusion trapping is implicit in the studies by Saigal *et al.* (1992) and Nedkvitne *et al.* (1993). Petroleum inclusion distribution should thus provide a partial 'view' of an accumulation at an earlier stage in its development. This differs from the evidence provided by the free reservoir oil in that different populations of inclusions may trap petroleum charges of different composition which have been mixed to a greater or lesser degree in the free reservoir. In other words, we can potentially use inclusion data to compare the present accumulation geometry with its geometry in the past, attempt to identify the pre-production OWCs and gain some insight into the mode of filling, migration routes and high permeability conduits. The link with transition zones is important for industrial applications because the latest inclusions to form should be related to pre-production OWCs and it may be possible to use oil inclusions to identify OWCs where production has destroyed the original configuration.

Earlier studies of the Ula reservoir focus on the inverse problem: how reservoir filling has affected reservoir quality in the Ula Field. Gluyas *et al.* (1990) have shown that an increase in quartz cement volume with depth is the most likely cause of declines in porosity and permeability with depth in the Ula Field. Near the crest of the structure there is minor cement and permeabilities are over 1 D, while on the flanks, quartz cementation reaches *c*. 20% of rock volume and permeabilities are tens of mD or lower. These intra-field gradients are larger than the regional gradients reported by Ehrenberg (1990), Gluyas *et al.* (1993) and Haszeldine & Osborne (1993). The relatively rapid increase in volume of cement from crest to flank could have been established prior to petroleum emplacement (by some unknown mechanism), but it seems more likely that the gradient is due to the inhibiting effect of the petroleum, progressively filling the trap from the top down, and progressively halting cementation. The identification of oil inclusions in quartz overgrowths in the studies by Nedkvitne *et al.* (1993) and Karlsen *et al.* (1993), and here, supports this inference. The inclusion and petrographic data suggest that the onset of quartz cementation and petroleum emplacement sourced from the Mandal Formation were virtually coincident (to within the time take to form 2 to 3% cement) at the crest. With numerical abundance data from petroleum inclusions, we can begin to discern some of the details of the interaction between oil emplacement and cementation.

Results

By inspecting doubly polished rock wafers under ultraviolet illumination, oil inclusions were identified in detrital quartz, diagenetic quartz overgrowths, carbonates and authigenic and detrital feldspars. Burruss (1991) gives details of the technique. Inclusions in quartz and carbonate were classified into primary (formed during mineral precipitation) or secondary (formed after mineral precipitation along healed microfractures in detrital quartz and within detrital feldspar vacuoles) types, as recommended by Roedder (1984). Inclusions in feldspar were not subdivided by host mineral because of the difficulty in identifying the type of feldspar grain or cement in the thick inclusion sections. All the inclusions observed were two-phase, containing relatively small vapour bubbles (<20% by volume at room temperature) in a fluorescent liquid, and no visible water. The phase proportions, and the homogenization data reported by Nedkvitne *et al.* (1993), show that an oil has been trapped. The fluorescence colours of the liquid phase generally shows little variation between samples, host-mineral phases or inclusion type (primary vs. secondary). A continuum of the blue-green to blue-white fluorescence colours was observed. No gas-condensate inclusions (i.e. those containing large vapour bubbles at room temperature and showing homogenization to vapour) or gas inclusions (i.e. all vapour at room temperature) have been observed to date.

The numerical abundance data for each sample are given in Table 1. Oil inclusions occurred in at least one sample in all the wells. Oil inclusions in the wells intersecting the oil leg showed major variations in total abundance ranging from zero to almost 3500 per scan. The maxima in each well are similarly highly variable: 58 in 7/12-3 to 3393 in 7/12-6 (Fig. 1). Crestal wells (7/12-6 and -A18) contain the highest numbers, but oil inclusions also occur in wells 7/12-A13 and 7/12-3, which lie wholly within the water leg today. We evaluated which, if any, geological parameters or rock properties correlated with oil inclusion abundances. The two most striking features of the data are, first, that different populations of inclusions in different host minerals and of different genetic type (secondary or primary) show the same overall trends in abundances, though absolute abundances differ. However, in low abundance samples, not all of the possible hosts contained the oil inclusions. In view of this, we used the total oil inclusion abundances for each sample in our correlations. The second feature is that there are fewer oil inclusions, including the subset of primaries in authigenic quartz, where there is more cement and vice versa. this result was unexpected. We expected that more inclusions would be trapped if there were more cement, given synchronous oil migration and cementation. An explanation for these phenomena is given later, after a discussion of the relationships of oil inclusion abundances to depth, reservoir architecture, OWCs and rock properties.

Relationship to depth

On a field scale, there is clearly a decrease in oil inclusion abundance with depth (Fig. 3). This may be viewed in two ways. Firstly, there may be two depth-related zones: (1) above *c.* 3500 mss where oil inclusions are abundant and decrease in abundance with depth; (2) below *c.* 3500 mss where oil inclusions are present but rare. Secondly, the decrease may be a power law or expotential decline in numbers. The apparent break of slope in inclusion numbers at 3500 mss could reflect the segment of an exponential decline where the slope becomes roughly asymptotic.

Further case studies should reveal whether oil inclusion trends are typically continuous or discontinuous, or, indeed, whether every field is different.

Relationships to reservoir architecture

A summary of the relationship to structure and field reservoir zonation is given in Table 1 and an overview of the fieldwide relationships at sample scale as given on Fig. 4.

Within individual wells, oil inclusions are most abundant near the top of the reservoir zone 2A. This reservoir zone boundary represents the transition from progradational sedimentation, and provides the most pronounced lithological contrast in the sampled sections (Fig. 2). Zone 2A is at varying elevations within different wells, suggesting that the control on inclusion abundance is not wholly structural, though overall abundances in this zone are greater in wells closer to the crest of the field.

Table 1. *Summary of relationship of structural position, reservoir zones and drilling data to oil inclusion abundances*

Well	Structural position	Reservoir zones sampled	Maximum numbers	Remarks
7/12-3	SW flank	1A, 1B	max 58 near base 1B, no primaries	Drilled below the west flank OWC at 3491 m TVDss. No primary oil inclusions were observed, though secondary oil inclusions in feldspar were identified in two samples.
7/12-3A	SW flank	1A, 2A	max 922 near top 2A	Tested water up to 3509 m TVDss. Logs suggest a possible OWC at 3491 m TVDss within zone 1A. The core is variably oil stained below this depth. Oil inclusions are abundant below the suggested OWC, and down to 3530 m TVDss. They occur in much lower abundances in the two deeper sample points.
7/12-A12	SW flank	2B, 3A	max 350 in mid 2B	Drilled at the same time as 7/12-7. From logs, the original contact was identified at 3491 m TVDss in zone 2A and a producing contact is currently placed at 3462 m TVDss. All the core recovered was from below these depths. Oil inclusions are present down to 3530 m TVDss.
7/12-A13	NW of NE flank	1A, 1B, 2A, 2B	max 98 near base 2B, no primaries	Drilled below the OWC. The OWC was defined in an updip sidetrack at 3544 m TVDss. Oil inclusions are in low abundance or absent.
7/12-A08	N of NE flank	2A, 2B, 3A	max 174 near top 2A	Drilled three years after production start-up. The well is expected to have the same contact as 7/12-7 (3789 m TVDss) but this is difficult to prove because it has extremely low poroperm and net sand is rare. An OWC has been tentatively identified at 3779 m TVDss in zone 2A but this would be a producing contact, not original. Inclusion data suggest a minimum in abundance at or slightly below the 3789 m TVDss contact.
7/12-7	E of NE flank	2A, 2B	max 384 near top 2B	Drilled two years after production start-up but log and core evidence suggests an original OWC at 3789 m TVDss in zone 2B. There appears to be a peak in oil inclusion abundances around the proposed OWC: samples above and below have lower abundances, but inclusions are not entirely absent.
7/12-9	SE of NE flank, 1A to 2A(2) condensed	2A	max 367 at top preserved 2A	Drilled four years after production start-up and has an OWC at 3710 m TVDss in zone 3A. This is believed to be an original contact. The abundance data are sparse, but suggest a decline towards the level of the OWC.
7/12-6	Crestal, NE area	1A, 2A, 2B, 3A	max 3393 near top 2A	Crestal, wholly within the oil column. Oil inclusion abundances are high.
7/12-A18	Crestal, NE area	1B, 2A, 2B, 3A	max 1782 near top 2A	A crestal well. It is thought to have an OWC at 3491 m TVDss but this could be interpreted as a producing contact, the original contact lying deeper. Oil inclusion abundances decline to low abundances at the suggested OWC.

OWC, oil–water contact
TVDss, true vertical depth subsea

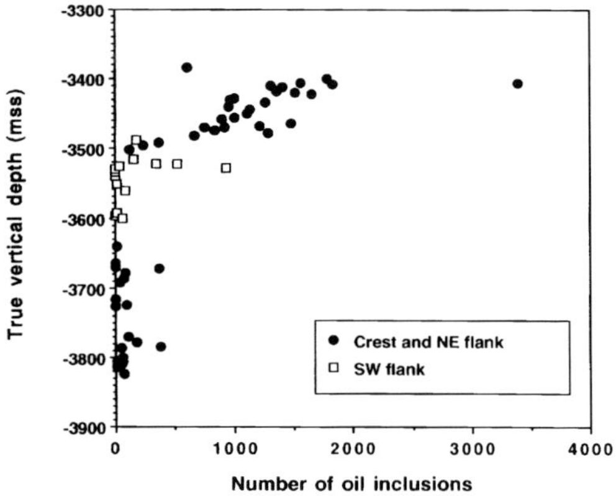

Fig. 3. Plot of oil inclusion numbers against depth in the Ula Field, differentiating samples from SW and NE flanks.

The maxima are not systematically related to the position of the present OWCs, suggesting that they reflect some pre-existing feature of the field, namely the configuration of oil distribution at some point during oil emplacement. In any case, it seems that oil is present throughout the system, both within and below the oil legs.

Relationship to oil–water contacts

The variability of the depth to the OWC in different sectors of the Ula Field is thought to be a primary geological feature. The initial geometry has been disturbed by production, such that wells drilled since production start-up now have OWCs which are higher than they were originally. Oil inclusions are not affected by production and their distributions may reflect original OWCs. The different classes of oil inclusions (primary, secondary) might also reveal palaeo-OWCs established during filling. This section discusses these ideas in the light of this study. The relationship of oil inclusion abundances to OWCs is summarized on Table 1.

For wells drilled wholly within the water leg or the oil leg (-3, -6 and -A13), oil inclusion abundances reflect these positions, being low (though never entirely absent) and high, respectively. Oil inclusion abundances appear to reflect the position of the original OWC in wells -9, -A08 and possibly -A18. In these cases, there is often a general decline in inclusion numbers towards the OWCs. In wells -3A, -7 and -A12, however, the trend of decline in numbers points to OWCs 20 to 40 m below the log/test OWCs, though only -3A is thought to have an OWC which is in doubt. It is suggested that two effects are combining to produce this situation: (1) oil inclusions are trapped in the zone of immovable oil (transition zone) below an OWC defined on the basis of producibility; (2) oil inclusions are trapped at deeper levels along migration routes. The inclusion data thus appears to reflect the sum of movable oil and immovable oil and may not be a guide to OWCs defined by the lower limit of movable oil. With this proviso, it may be possible to define a ‘pseudo-OWC’ from inclusion data by observing or extrapolating declining trends in abundances to zero abundance. Table 2 compares pseudo-OWCs interpreted from the declines in oil inclusion abundances with log/test derived OWCs.

Relationship to rock properties

The identification of local abundance maxima below zones of lithological contrasts led us to explore the relationship with rock properties. Linear regressions of inclusion numbers and rock properties were attempted, utilizing petrographic and poroperm data (Carruthers, R. A., Mitchell, A. W., Payne, D.F. & Slordal, J. personal communications). The correlations between total number of inclusions and petrographic data are relatively poor ($r^2 < 0.5$) and are not illustrated here. The scatter in the correlations using the petrographic sample and inclusion sample points (there are several metres mismatch in some cases). However, taking the nearest sample depths in the two datasets as equivalent, negative correlations, significant at the 95% confidence level, occur between inclusion numbers in progradational/aggradational zones 2 and 3 and (i) in the sum of the authigenic and ductile minerals and (ii) amount of quartz cement and detrital clay. Although sparse, data points from retrogradational zone 1 produce a separate trend paralleling the zone 2 and 3 data trend, but displaced from it.

A higher correlation coefficient characterizes the regression between oil inclusion abundances and permeability. In this case, depth correspondence in sample points was closer, there being a maximum of 15 cm mismatch. It is possible exact depth correspondence would produce a better correlation, but the present correlation is remarkably good, considering the potential counting errors.

An even better correlation was found between the logarithm of the average total number of inclusions per sample and the logarithm of arithmetic average permeability (Fig. 5). This translates to the power law relationship:

$$n = 15.89k^{0.62} \qquad (1)$$

$$(r^2 = 0.984)$$

where n is the average number of inclusions per sample per well and k is the average permeability per well (in mD). This relationship is expected to be

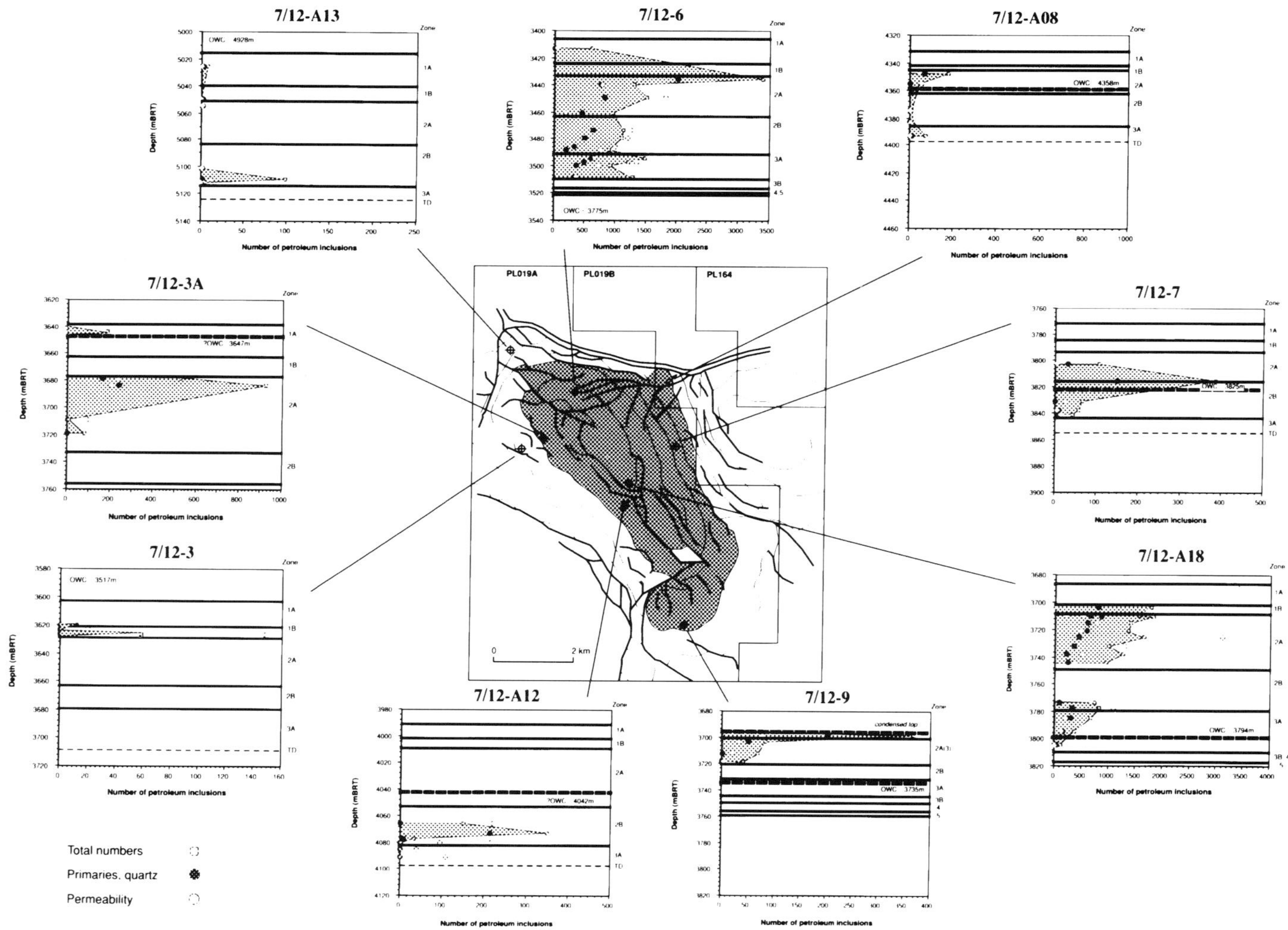

Fig. 4. Synoptic diagram illustrating the geographical and depth variation in oil inclusion numbers, and their relationship to geological features. Note different scales on horizontal axes of well panels.

Table 2. *Comparison of oil–water contacts estimated from log/test data and predicted from oil inclusion trends*

Well	Log/test OWC (m TVDss)	Inclusion pseudo-OWC (m TVDss)	Difference (m)
7/12-3	below OWC	below OWC	–
7/12-3A	?3491	*c.* 3530	*c.* 40
7/12-6	above OWC	above OWC	–
7/12-7	3789	>3808	*c.* 20
7/12-9	3710	*c.* 3710	*c.* 0
7/12-A08	?3789	*c.* 3790	*c.* 0
7/12-A12	3491	*c.* 3530	*c.* 40
7/12-A13	below OWC	below OWC	–
7/12-A18	?3491	*c.* 3495	*c.* 0

OWC, oil–water contact
TVDss, true vertical depth subsea

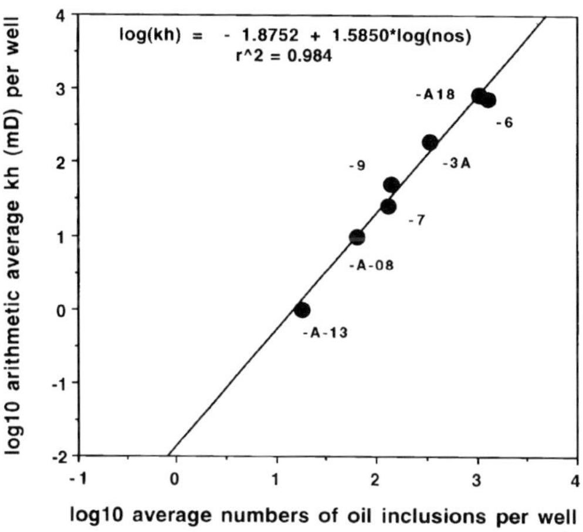

Fig. 5. Correlation of oil inclusion numbers and permeability, both parameters averaged by well.

unique to the Ula reservoir. Clearly, there seems to be a broad connection between inclusion numbers and permeability or permeability-influencing parameters.

Discussion

The trends and correlations described previously can be most simply explained if oil inclusions are assumed to form the periphery of invading petroleum stringers, with total inclusion numbers depending on the density of stringers which in turn depends on permeability (Fig. 6). The dependence of petroleum-stringer density on permeability is well known through computer modelling, and supported by visualization experiments. (Frette *et al.* 1992) and core tomography (Wang *et al.* 1984; Hove *et al.* 1985). Permeability is dominated by flow on an effective path of larger pore throats in a porous medium (Thompson 1991). Computer realizations of percolation mechanisms have been used to model the invasion of a fluid into a porous medium (Dias & Payatakes 1986; Petersen 1988; Kiriakides *et al.* 1991, 1993). A fluid is allowed to invade a lattice and the flow paths are determined by assigning a transmission probability to the lattice. Insofar as permeability is analogous to transmission probability, the results of modelling oriented percolation show that the amount of invaded rock increases with the transmission probability. These visualizations should be interpreted with care because they are strictly only applicable to (a) a geometrically random distribution of pore throat radii (a homogeneous medium), (b) an unchanging lattice (cementation and compaction are not occurring during percolation), and (c) a situation where the invading fluid is not drained out of the system (it is not connected to a higher permeability region).

However, the oil inclusion data closely approximate this model and suggest that permeability variation is the fundamental control on numbers of oil inclusions trapped. Variations in inclusion numbers reflect the accessibility of trapping sites to petroleum, and although permeability has an obvious role to play in this, other factors may play a part, three of which are discussed.

1. *Availability of trapping sites*: grain surface sites and intragranular sites (microfractures and vacuoles) must be present. Most primary oil inclusions in quartz cements are trapped at the interface between the detrital grain and the overgrowth. Secondary oil inclusions in quartz grains are trapped in microfractures. More oil inclusions are trapped in microfractures than at grain–overgrowth interfaces. The sites at which inclusions are

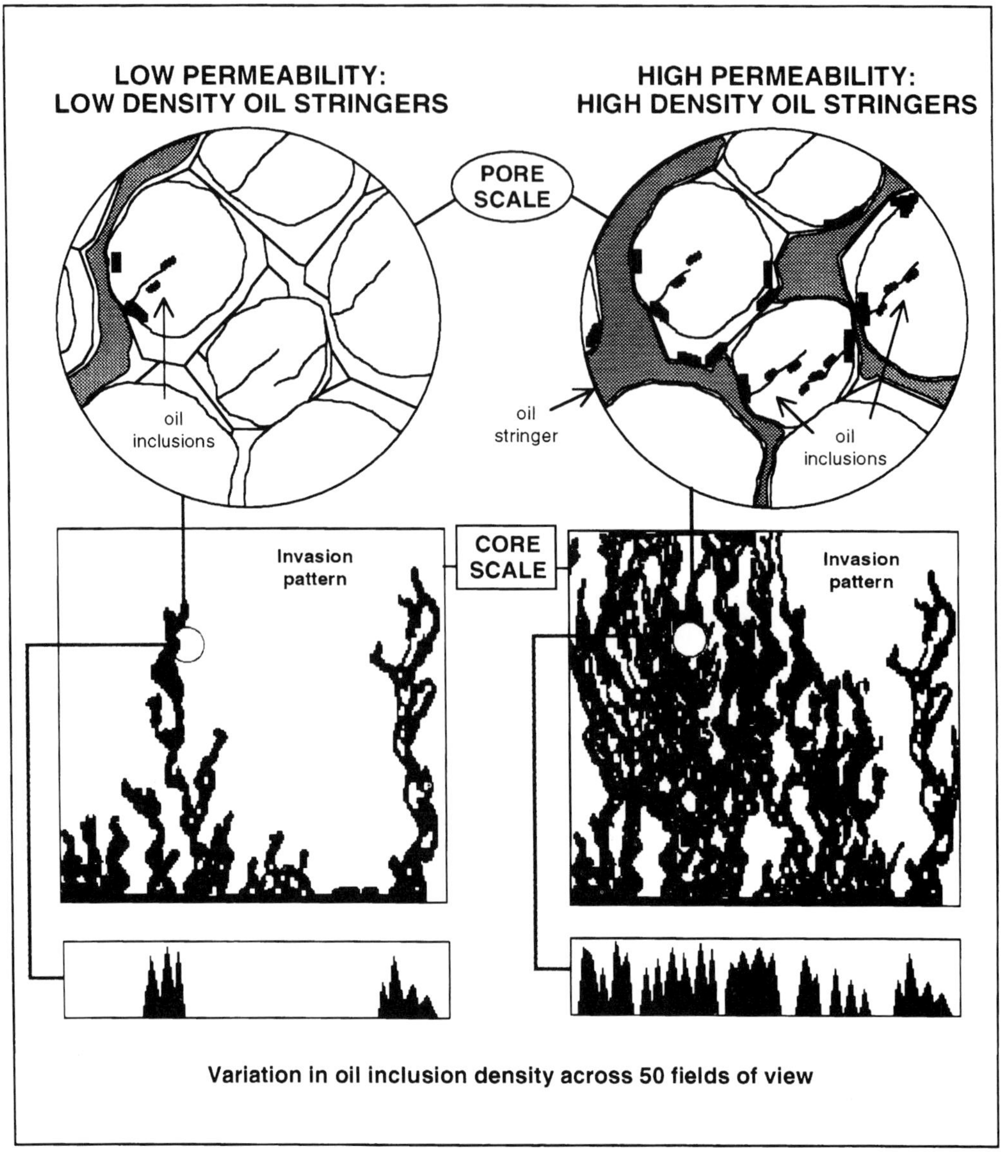

Fig. 6. Conceptual model for the entrapment of oil inclusions in sandstones, illustrating the effect of permeability on numbers trapped.

trapped might be expected to depend partly on the wetting properties of the surface with respect to the fluid being trapped. However, the difference between numbers trapped (secondary vs. primary in quartz) can simply be explained by the difference between the areal extent of healed microfracture surfaces and that of grain coverage by quartz cement, so we conclude that any dependency on wettability seems to be subordinate to other factors for oil inclusion trapping in quartz.

Individual feldspar grains can contain large numbers of oil inclusions (more than are trapped in quartz per unit volume). As overall abundance increases, the number of grains containing inclusions increases, but individual grains contain approximately the same number of inclusions. these may be located either at cement–grain interfaces, as with primary oil inclusions in authigenic quartz, or, more commonly, filling vacuoles in microporous grains. The most typical occurrence in a low abundance sample is one or two vacuolized grains containing large inclusion numbers. In the lowest abundance samples it is generally only feldspars that trap oil inclusions. These observations suggest that feldspars preferentially trap oil inclusions relative to quartz and this may be due to wettability differences. However, feldspar abundance and extent of vacuolization are roughly constant throughout the Ula Formation, which, with the foregoing, suggests that site availability is not the primary control on numbers trapped.

So far we have been concerned mainly with relative differences in oil inclusion abundances. The absolute numbers of oil inclusions trapped will probably depend on the number and distribution of potential sites on mineral surfaces, which in turn may be a function of surface topography, surface chemistry and fluid chemistry (Karlsen *et al.* 1993). Thus, absolute numbers of inclusions trapped may turn out to be field- or lithology-specific. Elucidating these aspects and the fundamental mechanisms enabling petroleum inclusion trapping will undoubtedly be an exciting topic for further research.

2. *Availability of petroleum*: Obviously, if no petroleum were present, no trapping could occur. Petroleum must have been present, but how much and for how long? Is there a relationship between the time-integrated amount of petroleum in contact with a mineral and the numbers of inclusions trapped? The time aspect cannot be addressed without some independent geochronometer, but there does appear to be a link with oil saturation. Wells 7/12-A12, -6 and -A18 lie wholly within the oil leg but only the crestal, high permeability, high oil-saturation wells -6 and -A18 have high abundance of oil inclusions, while samples from below oil legs have lower abundances that those from within oil legs. Insofar as oil saturations, as conventionally measured, are a spatial average of water-filled and oil- and water-filled pores, we believe that oil-stringer density variation is the fundamental reason for these effects.

As Nedkvitne *et al.* (1993) and Karlsen *et al.* (1993) have pointed out, a column of oil several metres high is required to force oil into small apertures, such as microfractures or vacuole entrances, in water-wet reservoirs. We do not envisage this to be a problem in Ula because the oil column is hundreds of metres high, sufficient to fill nanometre-scale apertures, and was probably tens of metres high throughout a large part of its filling history.

3. *Availability of sealing agents*: cementation must occur. Both primary and secondary inclusion types require authigenic growth for trapping. We might therefore expect more inclusions to be trapped where quartz cementation is greatest, at least in the oil leg. This is the complete opposite of what is actually observed. Fewer inclusions are trapped in sediments with more cement. the large numbers of oil inclusions in the crestal walls (virtually every microfracture, vacuole and overgrowth contains them) implies high oil saturations. It is commonly supposed that there is a lower limiting water saturation for diagenesis to proceed. Our data gives no indication of such a limit, though one may exist. However, if diagenesis occurs at high oil saturations, from relative permeability considerations, the mass transport mechanism must be largely diffusive. This does not preclude transport by mass advection in those parts of a reservoir with lower oil saturation.

The conclusion from this discussion is that there is no need to invoke complex variations on wettability or duration of charging or cementation to account for the inclusion abundances. Stringer density variation is adequate.

So what is the evidence for stringers of petroleum in the samples? On the scale of individual thin sections inclusions distribution was heterogeneous; patches of high oil-inclusion abundance alternated with patches of low abundance (Fig. 6). In general, as total numbers per sample declined, the distance between the high abundance patches increased. This, rather than a decline in peak numbers in the high abundance patches, was responsible for most of the variation between samples. Such features are considered to be direct evidence for the presence of petroleum as stringers in the rock during filling.

Some scatter in inclusion number-permeability correlation is expected at small scale (even in the absence of counting errors) because petroleum density at any given location will depend not only upon the local permeability, but also on the extent of viscous fingering and the larger-scale permeability structure of the rock which the petroleum has passed through (Elder 1978). For example, low stringer densities developed while crossing a low permeability zone and entering a high permeability zone are unlikely to develop high stringer densities near the interface. At large scale, however, such features may be smoothed out, which should be reflected in an improvement in correlation if averages are taken. This appears to be the situation for the Ula reservoir, where there are strong large-scale trends in rock properties (Gluyas *et al.* 1993). This conclusion cannot be extended to other reservoirs, especially where reservoir heterogeneity or compartmentalization dominates. However, in these cases, we speculate that analysis of oil inclusion abundances could provide additional data to assess their effects on reservoir filling.

We can use this model to account for the data

trends. In the crestal, high permeability case (7/12-6 and -A18), petroleum emplacement would show minor dispersion, i.e. in any given rock volume the distance between adjacent invading petroleum stringers would be relatively small. As a result, inclusions form closely spaced arrays, giving high abundances. In the flank (all other wells studied), lower permeabilities developed, in this case through cementation, would produce a more dispersed invasion pattern and in any given rock volume the distance between adjacent invading petroleum stringers would be relatively large. Inclusions thus form widely spaced arrays and abundances are low.

Figure 7 shows the number-permeability plot with data points classified according to whether they are in the water leg or oil leg (-3A and -A08 have been plotted separately because there is doubt over the original OWC in these wells). Only the points plotted on the permeability axis have evidently never seen petroleum (i.e. they are in a 'true' water leg), but note that not all have low permeabilities. This variation must have a purely lithological-diagenetic origin. The 'main' trend crossing the figure contains samples from both the oil and water legs (identified from logs). Water-leg samples generally exhibit low numbers and low permeabilities, and oil-leg samples high numbers and high permeabilities. The inclusion data show that this set of water-leg samples has, however, been in contact with petroleum. The implication of this is that because the definition of oil legs from

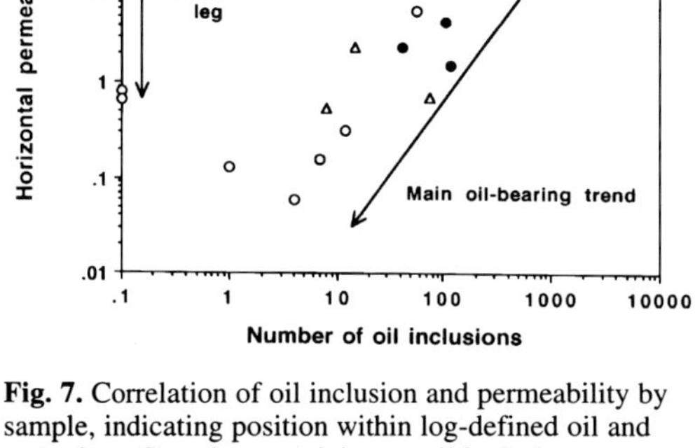

Fig. 7. Correlation of oil inclusion and permeability by sample, indicating position within log-defined oil and water legs. Samples containing no oil inclusions are plotted at 0.1 on the inclusion number scale. These all correspond to log-defined water legs and are interpreted as samples that have never been in contact with oil.

logging and testing is usually partly controlled by an arbitrary permeability limit, we would expect oil inclusions to persist to greater depths.

Permeability and depth are strongly correlated regardless of location (Gluyas *et al.* 1993). However, there appear to be significant differences between the bases of free oil legs in the southwestern and northeast areas of the field (Fig. 4) and this is supported by the distribution of oil inclusions. Filling direction may be the missing factor. Because of the buoyancy of petroleum, it is easier for a slug of petroleum to move upwards than downwards. Oil should collect at the termini of migration routes and the effect will be enhanced in low-permeability rocks, because of the inverse relation of entry pressure to pore throat radius ($P \propto 2\gamma/r$). Thus, it is easier to emplace oil upwards during migration than to cause the OWC to descend in rocks of similar pore throat radii. Based on these arguments, the data for Ula would suggest that the eastern flank is closer to an active source volume, as illustrated on Fig. 8. This filling direction is consistent with that inferred from maturity modelling (Larter *et al.* 1990). The inclusion distributions therefore support the contention that OWCs were variable across the Ula Field before being distributed by production.

This evidence tests the three models that Brown *et al.* (1992) proposed to account for the variable OWCs. It effectively precludes the model in which the variable OWCs were due to production. According to Brown *et al.* (1992) the remaining possibilities are: (1) there is compaction-driven water flow from the west and the crestal fault is sealing – the OWCs on each side of the fault are different because of differing aquifer pressures; (2) the reservoir is discontinuous to the west and oil charging was from the east. Brown *et al.* (1992) pointed out that there were problems with both models: flow rates associated with compaction were expected to be too low to cause the observed difference in east and west-flank OWCs in the first model, and there is only limited seismic evidence for reservoir deterioration to the west to support the second. The inclusion data suggest that crest to flank capillary pressure variation (as a result of synchronous petroleum emplacement and cementation) combined with charging from the east can account for the variable OWCs. Of the two models, the second is closest to this scenario.

From a field development standpoint, the most important findings are that pre-production OWCs appear to have been variable and that the inclusion data give some guide to their likely distribution across the field. The value of the inclusion data in development is that they have helped to reduce the uncertainty that an analysis of conventional geological data has produced.

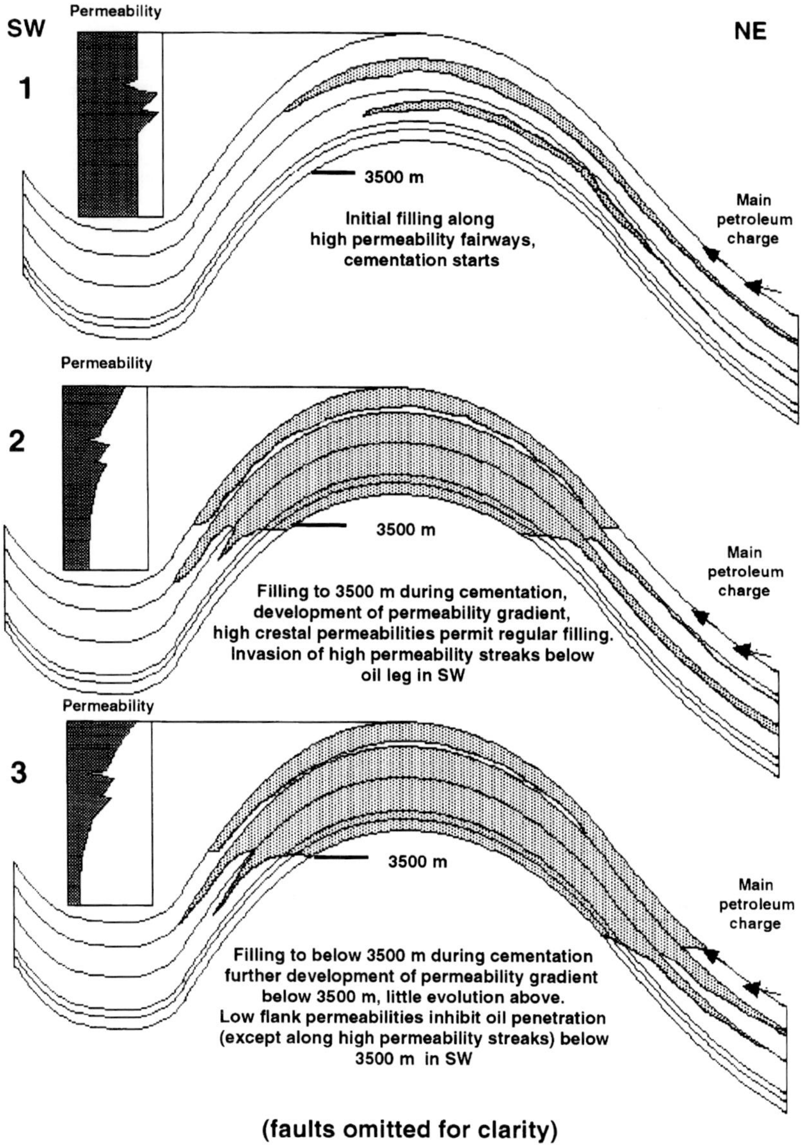

Fig. 8 Model for the sequential filling of the Ula Field, based on oil inclusion abundance data.

Conclusions

Oil inclusions in the Ula Formation reflect the distribution of unproducible (immovable) and producible (movable) oil. They may be used to identify oil–water contacts only if the oil leg is defined to include immovable oil. Distributions and numbers of oil inclusions appear to reflect the manner in which petroleum invades rocks of variable permeability, including non-producible migration routes below the present oil legs.

The inclusion data suggest that the Ula Field filled from crest downwards. This was synchronous with cementation to *c.* 3500 m, with the top of reservoir zone 2A being a major migration route. Below 3500 m, cementation appears to have reduced permeability to such an extent before petroleum entered that petroleum emplacement was irregular. However, oil inclusions persist to a greater depth in the east and southeast and it is inferred that this is the filling direction. This filling direction is consistent with that inferred from maturity modelling (Larter *et al.* 1990). This study therefore supports the contention that oil-water contacts were variable across the Ula Field before being disturbed by production. This finding rules out some of the models proposed to explain the variability in oil–water contacts (Brown *et al.* 1992) and helps to refine some others: future development strategy will take account of this.

Reservoir quality analysis of the Ula Formation (Gluyas *et al.* 1993) has suggested that synchronous cementation and petroleum emplacement is responsible for the systematic declines in porosity and permeability and the increase in quartz cement volume with depth. The oil inclusion data generally support this model, but also suggest that the patterns of filling and cementation were more complex than hitherto understood. The new data indicate that permeable zones below less permeable parasequence boundaries were often migration pathways and suggest that the presence of oil in these sandstones suppressed cementation. The inclusion data also suggest the mechanism by which the reservoir quality trends were produced.

The oil inclusion distributions suggest that in high permeability rocks, migrating petroleum occupies a large proportion of pore space as a dense network, while in low permeability rocks it is present as dispersed stringers. In the initial stages of reservoir filling, oil preferentially migrates along higher permeability units, but is concentrated in structurally higher positions because of buoyancy. The high concentration of petroleum stringers will ensure that little cementation can occur, but will result in large numbers of oil inclusions being trapped in whatever sites are available, i.e. feldspar grains, calcite cement, authigenic quartz and quartz grains in Ula. As cementation proceeds in other parts of the structure, permeability is reduced so that oil invasion is more dispersed. Fewer inclusions are trapped despite similar numbers of potential sites for trapping. This process leads to a correlative decline in numbers of oil inclusions with permeability. In a structure filling from crest downwards, there will thus be a power law or exponential decline in number of oil inclusions with depth.

Oil inclusions found below the oil legs, although rare, may represent either migration pathways updip from the source, or invasion downdip from the main oil leg along higher permeability zones.

The method presented here could be applied in other sedimentological or tectonic settings to elucidate filling mechanisms and pre-production oil distributions. It can also be used to investigate the relationship between reservoir quality and oil inclusion abundances for different relative timings of petroleum emplacement and cementation at different depths.

The authors would like to thank BP Norway UA and its partners Statoil, Svenska Petroleum A/S, Conoco Norway Inc. and A/S Pelican & Co., for permission to publish the data; their colleagues at BP for discussion; Andrew Hogg (BP) for suggestions for improving the format; Graham Brumby for sample preparation; and reviewers A. H. Rankin (Kingston University) and D. A. Karlsen (Oslo University) for their insights.

References

ASHTON, M. (ed.) 1993 *Advances in Reservoir Geology.* Geological Society, London, Special Publication, **69**.

BEATTIE, P. 1993. On the occurrence of apparent non-Henry's Law behaviour in experimental partitioning studies. *Geochimica et Cosmochimica Acta*, **57**, 47–55.

BODNAR, R. J. 1990. Petroleum migration in the Miocene Monterey Formation, California, USA: constraints from fluid inclusion studies, *Mineralogical Magazine*, **54**, 289–304

BROWN, A., MITCHELL, A. W., NILSSEN, I. R., STEWART, I. J. & SVELA, P. T. 1992. Ula field: relationship between structure and hydrocarbon distribution? *In*: LARSEN, R.M. *et al.* (eds) *Structural and Tectonic Modelling and its Applications to Petroleum Geology*. NPF, Special Publication, **1**, 409–420.

BURRUS, J., KUHFUSS, A., DOLGEZ, B. & UNGERER, P. 1991. Are numerical models useful in reconstructing the migration of hydrocarbons? A discussion based on the Northern Viking Graben. *In*: ENGLAND, W.A. & FLEET, A.J. (eds) *Petroleum Migration.* Geological Society, London, Special Publication, **59**, 89–110.

BURRUSS, R. C. 1981. Hydrocarbon fluid inclusions in studies of sedimentary diagenesis. *In*: HOLLISTER, H. S. & CRAWFORD, M. L. (eds) *Short course in Fluid Inclusions: Applications to petrology.* Mineralogical Society of Canada. Short Course Handbook, **6**, 138–156.

—— 1991. Practical aspects of fluorescence microscopy of petroleum fluid inclusions. *In:* BARKER, C. E. & KOPP, O. C. (eds) *Luminescence Microscopy: Quantitative and Qualitative aspects.* Society of Economic Paleontologists and Mineralogists, Short Course, **25**.

——, CERCONE, K. R. & HARRIS, P. M. 1983. Fluid inclusion petrography and tectonic-burial history of the Al Ali No2. well: evidence for the timing of diagenesis and oil migration, northern Oman foredeep. *Geology*, **11**, 567–570.

DIAS, M. M. & PAYATAKES, A. C. 1986. Network models for two-phase flow in porous media Part 1. Immiscible microdisplacement of non-wetting fluids. *Journal of Fluid Mechanics,* **164**, 305–336.

EADINGTON, P. J., HAMILTON P. J. & BAI, G. P. 1991. Fluid history analysis–a new concept for prospect evaluation. *Australian Petroleum Exploration Association, Journal,* **31**, 282–294.

EHRENBERG, S. N. 1990. Relationship between diagenesis and reservoir quality in sandstones of the Garn Formation, Haltenbanken, mid-Norwegian Continental Shelf. *American Association of Petroleum Geologists, Bulletin,* **74**, 1538–1558.

ELDER, J. 1978. *The Bowels of the Earth.* Open University Press, Milton Keynes, 222 p.

ENGLAND, W. A., MACKENZIE, A. S., MANN, D. M. & QUIGLEY, T.M. 1987. The movement and entrapment of petroleum fluids in the subsurface. *Journal of the Geological Society of London,* **144**, 327–348.

FRETTE, V., FEDER, J., JØSSANG, T. & MEAKIN, P. 1992. Gradient-driven migration in porous media: experiments and simulations. *In:* AASEN, J.O. *et al.* (eds) *North Sea Oil and Gas Reservoirs.* Trondhein, Norway, Nov, 30–Dec. 2, 1992, Abstracts.

GIBBONS, K. & IRWIN, H. 1991. Controls on calcite cemented horizons by sequence stratigraphy and hydrocarbon infilling of the reservoir, Troll field, Norwegian continental shelf. *American Association of Petroleum Geologists, Bulletin,* **75**, 580.

GLUYAS, J. G., LEONARD, A. J. & OXTOBY, N. H. 1990. Diagenesis and petroleum emplacement: the race for space – Ula Trend, North Sea. *13th International Sedimentological Congress.* Abstracts, International Association of Sedimentologists, Utrecht, 193.

——, ROBINSON, A. G., EMERY, D., GRANT, S. M. & OXTOBY, N. H. 1993. The link between petroleum emplacement and sandstone cementation. *In:* PARKER, J. R. (ed.) *Petroleum Geology of Northwest Europe.* Geological Society, London, 1395–1402.

HASZELDINE, R. S. & OSBORNE, M. 1993. Fluid inclusion temperatures in diagenetic quartz rest by burial: implications for oil field cementation. *In*: HORBURY, A. D. & ROBINSON, A. G. (eds) *Diagenesis and Basin Development.* 35–46

HILLEBRAND, T. & LEYTHAEUSER, D. 1992. Reservoir geochemistry of Stockstadt oilfield: compositional heterogeneties reflecting accumulation history and multiple source input. *Organic Geochemistry,* **19**, 119–131

HOVE, A., RINGEN, J. K. & READ, P. A. 1985.Visualization of laboratory corefloods with the aid of computerized tomography of X-rays. *Society of Petroleum Engineers, Paper No.* **13654**, 611–620.

KARLSEN, D. A., NEDKVITNE, T., LARTER, S. R. & BJØRLYKKE, K. 1993. Hydrocarbon composition of authigenic inclusions: application to elucidation of petroleum reservoir filling history. *Geochimica et Cosmochimica Acta,* **57**, 3641–3659.

KIRIAKIDIS, D. G., MITSOULIS, E. & NEALE, G. H. 1991. Linear displacement of a wetting fluid by an immiscible non-wetting fluid in a porous medium: a predictive algorithm. *Canadian Journal of Chemical Engineering*, **69**, 557–563.

——, —— & —— 1993. Computer simulations of immiscible displacement in a porous medium containing a region of a different wettability. *Journal of Canadian Petroleum Technology*, **32**, 21–25.

LARTER, S. R., BJØRLYKKE, K., KARLSEN, D. A. ET AL. 1990. Determination of petroleum accumulation histories: examples from the Ula field, Central Graben, Norweigen North Sea. *In:* BULLER, A. T., BERG, E., HJELMELAND, D., KLEPPE, J., TORSAETER, O. & AASEN, J. O. (eds) *North Sea Oils & Gas Reservoirs II.* Graham & Trotman, London, 319–330.

MCLIMANS, R. K. 1987. The application of fluid inclusions to migration of oil and diagenesis in petroleum reservoirs. *Applied Geochemistry,* **2**, 585–603.

MCMAHON, D. A. 1989. Secondary porosity in sandstones and diagenesis of adjacent shales, Oligocene South Texas. *Transactions of the Gulf Coast Association of Geological Societies,* **XXXIX**, 207–220.

NEDKVITNE, T., KARLSEN, D. A., BJØRLYKKE, K. & LARTER, S. R. 1993. The relationship between diagenetic evolution and petroleum emplacement in the Ula Field, North Sea. *Marine and Petroleum Geology*, **10**, 255–270.

PAGEL, M., WALGENWITZ, F. & DUBESSY, J. 1986. Fluid inclusions in oil and gas-bearing sedimentary formations. *In*: BURRUS, J. (ed.) *Thermal Modelling in Sedimentary Basins.* Editions Technip, Paris, 565–583.

PETERSEN, I. 1988. *The Mathematical Tourist.* W. H. Freeman & Co., London, 187–194.

REINSON, G.E. 1991. Depositional facies control of reservoir heterogeneity and performance, Cretaceous sandstone reservoirs, south-central Alberta Basin. *American Association of Petroleum Geologists, Bulletin,* **75**, 659.

ROEDDER, E. 1984. Fluid inclusions. *Reviews in Mineralogy,* **12**, 644.

SAIGAL, G. C., BJØRLYKKE, K. & LARTER, S. R. 1992. The effects of oil emplacement on diagenetic processes – examples from the Fulmar reservoir sandstones, central North Sea. *American Association of Petroleum Geologists, Bulletin,* **76**, 1024–1033.

SMALLEY, P. C., DODD, T. A., STOCKDEN, I. L., RAHEIM, A. & MEARNS, E.W. 1995. Compositional heterogeneities in oilfield formation waters: identifying them, using them. *This volume.*

STEWART, I. J. & SCHJERVERUD, K. 1993. Structural controls on the late Jurassic age shelf system, Ula Trend, Norwegian Sea. *In*: PARKER, J. R. (ed.) *Petroleum Geology of Northwest Europe.* Geological Society, London, 469–483.

THOMPSON, A.H. 1991. Fractals in rock physics. *Annual Review of Earth and Planetary Science*, **19**, 237–262.

TRINDADE, L. A. F. & BRASSELL, S. C. 1992. Geochemical assessment of petroleum migration phenomena on a regional scale: case studies from brazilian marginal basins. *Organic Geochemistry*, **19**, 13–27.

WANG, S. Y., ARYAL, S., CASTELLANA, F. S. & GRYTE, G. C. 1984. Reconstruction of oil saturation distribution histories during immiscible liquid–liquid displacement by computer-assisted tomography. *American Institute for Chemical Engineering, Journal,* **30**, 642–646.

Appendix A: methods

The sample set consisted of 63 core chips from nine wells whose locations are shown on Fig. 1. Care was taken to sample, as far as possible, parts of the core having comparable detrital textures and mineralogies. These were made into doubly polished wafer with porosity impregnated by a blue-dyed resin. Numerical abundance data were collected according to the procedures described later.

Each wafer was examined on a petrographic microscope under incident ultraviolet illumination from an HBO 100 Hg lamp with a 20× fluorescence objective and 10× eyepieces. This combination was found to maximize fluorescence intensity and thus optimize the chance of detecting oil inclusions. The field of view was a circle of 700 μm diameter (0.385 mm^2). All oil inclusions visible within the field of view were categorized according to mineral host, inclusion type and fluorescence colour, and counted. Oil inclusions in 50 fields of view were counted for each wafer in three to four transects of different parts of the wafer, throughout the thickness of the wafer. A Swift point counter was used to step between fields of view using a stage interval of three, which ensured that the fields of view in each transect did not overlap. Thus, the total area covered is 19.25 mm^2 and the total volume *c.* 1.5 mm^3. If a counting run failed to reveal any oil inclusions, the whole section was examined. However, if oil inclusions were identified in this general scan, their presence was recorded separately and does not feature in the numerical data (in the Ula dataset this does not occur). This procedure undoubtedly involves counting errors (Beattie 1993), depending on operator skill in identifying oil inclusions and the potential sources of error discussed later. However, we find that there are typically several orders of magnitude of variation in inclusion abundances within datasets and we were satisfied that the abundance data give us at least a semi-quantitative parameter which can be used in numerical correlation studies.

In preliminary tests, we have found that reproducibility for single operator counting inclusions in one sample on a single microscope varies between 5 and 10% of the absolute numbers, generally less than the variation recorded between samples. Comparing different samples taken at the same depth from the same lithology gives similar results. We have not rigorously assessed variation due to the use of different instruments or different operators, but, although we would expect a greater percentage variation in counting, qualitative tests show that operators agree on relative variations.

Potential sources of error

The results obtained by this method provide strong evidence that it adequately characterizes oil inclusion abundances for inter-sample comparison despite the following potential sources of error.

1. *Missed inclusions.* Inclusions can be hidden by opaque phases or be too small to identify. If the reservoir is reasonably homogeneous, the amount of opaque material should be approximately constant and the number of hidden inclusions a constant proportion of the total. Similarly, unrecognized inclusions will probably form a constant proportion of the total (see discussion on inclusion size distributions in Roedder (1984)).

2. *Variations in sample thickness.* This can vary between 50 and 100 μm in the extreme, giving a factor of two potential error, but variation is normally much less (*c.* 10 μm maximum difference as estimated from interference colours) in a single set, because of the mechanics of sample production. Nevertheless, an error of up to 25% could be generated.

3. *Variation in grain size and shape.* The critical factor influenced by grain size and shape is the grain surface area present within the wafer. This will influence the counting of primary inclusions. For a given field of view, surface area will increase with a decrease in grain sphericity and with a decrease in grain size, and we might expect to see more oil inclusions in finer-grained samples with less spherical grains. this problem can be mitigated by selecting samples of comparable grain size and texture. Every effort was made to ensure this when core sampling, but perfect homogeneity is impossible. There is less of a problem with secondary inclusions within grains because there is little difference in volume sampled, in spite of differences in grain size and shape.

4. *Variations in amount of detrital mineral present.* Quartz is the major detrital fraction present and variation in amount is minor. The error on inclusion counts in quartz due to this factor is likely to be small. The amounts of feldspars and dispersed carbonate cements may show considerable variability between fields of view, and the potential for under- or over estimating numbers is far greater than for quartz. Within the dataset, numbers of inclusions in feldspar do appear to be influenced by such sampling errors.

5. *Porosity variation.* At higher porosities, grains are further apart and fewer grains appear in any field of view. This potentially reduces the numbers of inclusions counted compared with a rock of lower porosity. In fact, the data show the opposite trend: more inclusions are counted in higher porosity samples.

6. *Intra-sample heterogeneity of inclusion trapping.* The small volume sampled may not be representative of the sediment as a whole. At a microscale, the distribution of oil inclusions is highly heterogeneous but 50 fields of view sample several peaks and troughs in inclusion numbers and no gradient was observed in any of the Ula

Formation samples, suggesting that the method is an effective way of smoothing intra-sample variability for these particular sediments. However, variations in microfracture density and the abundance of secondary oil inclusions could be significant in the interpretation of other sample sets. Another reservoir, field or tectonic setting might show completely different microfracture and secondary oil inclusion densities to the Ula Formation. Variations in microfracture density between samples should be monitored using a combination of fluorescence, transmitted light and cathodoluminescence petrography.

7. *Preferred orientation of inclusion arrays.* This was identified as a potential problem for the observation of secondary inclusions in healed microfractures. However, providing that the wafer volume contains a representative sample of the microfracture populations in the rock, then the total area of microfracture planes preserved for observation should be independent of sample orientation. Planes intersecting the wafer at high angle will individually have a smaller area but there will be more of them, while planes intersecting the wafer at a small angle will be fewer but expose a larger area. Problems might occur if the microfractures were sufficiently widely spaced that they did not intersect the wafer, which would be exacerbated if the angle of intersection was low, but this should reflect microfracture density rather than orientation (see point 6).

Migration of hydrocarbons in the Tampen Spur area, Norwegian North Sea: a reservoir geochemical evaluation

I. HORSTAD[1], S. R. LARTER[2] & N. MILLS[1]

[1] *Saga Petroleum as, Kjørbovn. 16, PO Box 490, N-1301 Sandvika, Norway*

[2] *Newcastle Research Group in Fossil Fuels & Environmental Geochemistry (NRG), Drummond Building, University of Newcastle, Newcastle upon Tyne NE1 7RU, UK*

Abstract: This paper describes the application of reservoir geochemical protocols to a study of oil-field and regional petroleum compositional variations in the major fields in the Tampen Spur area. The Tampen Spur high is located in the middle of several oil-mature to gas-mature source-rock kitchens and most of the structures might receive petroleum from several different directions. Based on the definition of several different petroleum populations within the Tampen Spur area, we define five major petroleum charging systems. Correlating the petroleum population variations with geological information allows us to derive likely migration/fill scenarios for many of the oil-fields and requires us to invoke a new Upper Jurassic carrier system in parts of the area connecting Brent Group reservoirs across major faults. Based on this hypothesis, two exploration wells were drilled on structures to test this carrier and petroleum was found in one of the structures. An understanding of the migration pathways and petroleum drainage efficiency of the basins is of great economic interest, because most of the smaller sub-structures are marginally economic. Thus, it is important to provide an effective risk analysis and ranking of the satellite prospects prior to drilling. This study has demonstrated that filling and migration in a geological province are controlled by a series of complex processes and that the filling history of the structures within the Tampen Spur is more complicated than suggested earlier.

Based on a detailed oil–oil correlation by conventional geochemical techniques (gas chromatography–flame ionization detection (GC–FID), gas chromatography–mass spectrometry (GC–MS) and stable carbon isotope analyses), we propose a detailed migration model for the northern part of the Tampen Spur area based on both geological and geochemical information. Thus, this paper represents a summary of a geochemical study of the Tampen Spur area and does not attempt to address all aspects of the geochemistry and geology of the East Shetland Basin and Tampen Spur in general.

Exploration history

It is far beyond the scope of this paper to give a detailed description of the geological history of the Tampen Spur area and the reader is referred to other papers that have discussed this subject in more detail. (Kirk 1980; Eynon 1981; Graue *et al.* 1987; Roberts *et al.* 1987; Karlsson 1986; Nystuen & Fält 1995; Erichsen *et al.* 1987; Olaussen *et al.* 1993).

The Tampen Spur area is located in the northern part of the Norwegian North Sea, between the north-south trending Viking Graben and the East Shetland Basin, and represents a continuation of the East Shetland Platform (Fig. 1). Several giant oilfields have been discovered in the area, of which the Statfjord Field (Fig. 2), discovered in 1974 on the border between the Norwegian and the UK sector, is the largest. The Statfjord Field is one of the largest oilfields in the North Sea, with a STOOIP of more than 1200×10^6 Sm^3. The main reservoirs in the Statfjord Field are the Middle Jurassic Brent Group, the Early Jurassic Cook Formation and the Late Triassic to Early Jurassic Statfjord Formation (Fig. 3) (Kirk 1980; Eynon 1981; Graue *et al.* 1987; Roberts *et al.* 1987).

The Gullfaks Field (Fig. 2) was discovered in 1978 and in addition to the Brent Group, the Cook Formation and the Statfjord Formation, the Triassic Lunde Formation was found to be oil-bearing in this structure, demonstrating that oil migration and accumulation have occurred locally via at least four different stratigraphic levels in this area (Erichsen *et al.* 1987; Olaussen *et al.* 1993; Horstad *et al.* 1990). The current estimates of the STOOIP in the Gullfaks Field are about 500×10^6 Sm^3 and, as for the Statfjord Field, about 80% of the reserves are present in the Brent Group. The Snorre Field (Fig. 2) is located on the northern rim of the Tampen Spur and was discovered in 1979. In this area of the Tampen Spur, the Late Jurassic to Early Cretaceous uplift resulted in removal of the Cook Formation and the Brent Group, and in the northern part of the field the erosion removed the Statfjord

From Cubitt, J. M. & England, W. A. (eds), 1995, *The Geochemistry of Reservoirs*, Geological Society Special Publication No. 86, pp. 159–183.

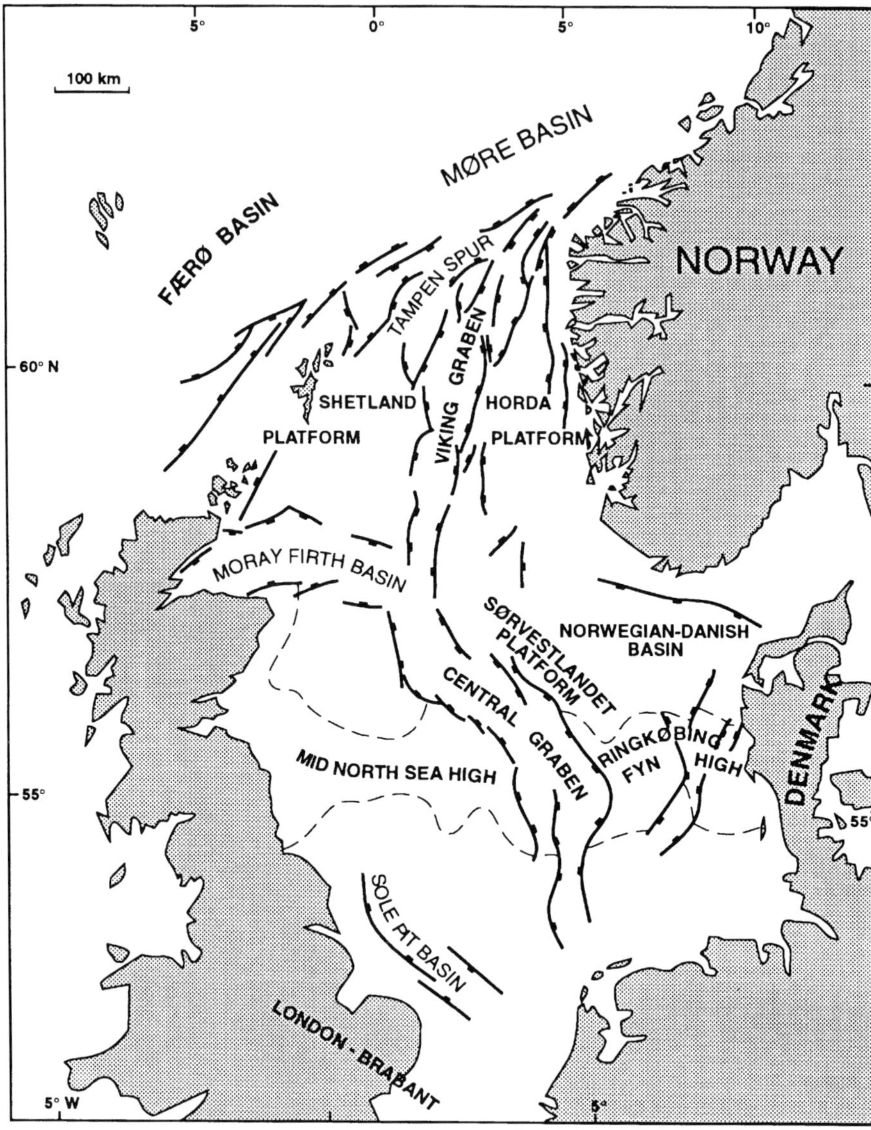

Fig. 1. Geographic location of the Tampen Spur in the North Sea.

Formation and cuts down into the Triassic Lunde Formation. The main reservoir unit in the Snorre Field is the Lunde Formation which contains about 75% of the total STOOIP of 490×10^6 Sm^3 (Hollander 1987), the rest being accumulated within the Statfjord Formation in the southern part of the field.

In addition to these three major accumulations upon the Tampen Spur high, several smaller structures have been discovered in between the larger fields. Statfjord Øst was discovered in 1976, Statfjord Nord in 1977, Tordis in 1987, Vigdis from 1986 to 1990 and Gullfaks Vest (34/10-34) in 1991 (Fig. 2). Along the border towards the Viking Graben, several oil and gas discoveries have been made in down-thrown terraces at several different structural levels. The Visund Field was discovered in 1986 and consists of several different segments with large differences in the petroleum composition and fluid contacts. Petroleum has been found in the Lunde Formation (condensate), the Statfjord Formation (oil) and the Brent Group (gas with oil

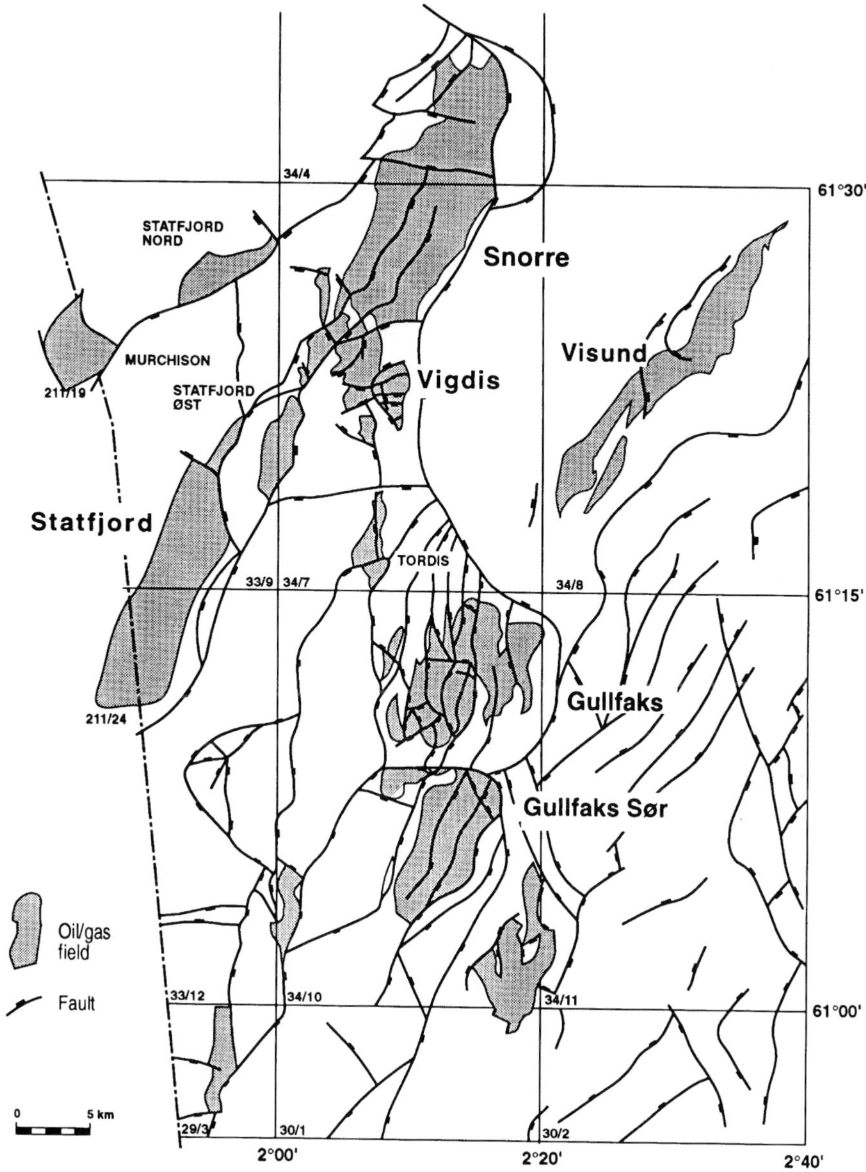

Fig. 2. Detailed map of the Tampen Spur area.

leg) and this is a good example of how complicated the migration is in the northern part of the North Sea. The Gullfaks Sør Field was discovered from 1978 to 1989 and is as complex as the Visund Field with respect to petroleum composition and fluid contacts. In this structure, petroleum has been found in the Lunde Formation, the Statfjord Formation, the Cook Formation and the Brent Group. The deepest producing discovery was made in 1985 on the 34/10 Gamma structure, where a relatively dry gas was produced from the Brent Group at 4090 m. Thus, the Tampen Spur contains a series of oil, condensate and gas discoveries with gas/oil ratios (GOR) ranging from 50 Sm^3/Sm^3 to 11 500 Sm^3/Sm^3, reservoir burial depths ranging from shallower than 2 km to deeper than 4 km, and variations in petroleum type from light, *n*-alkane-rich typical North Sea crude oil to biodegraded oils.

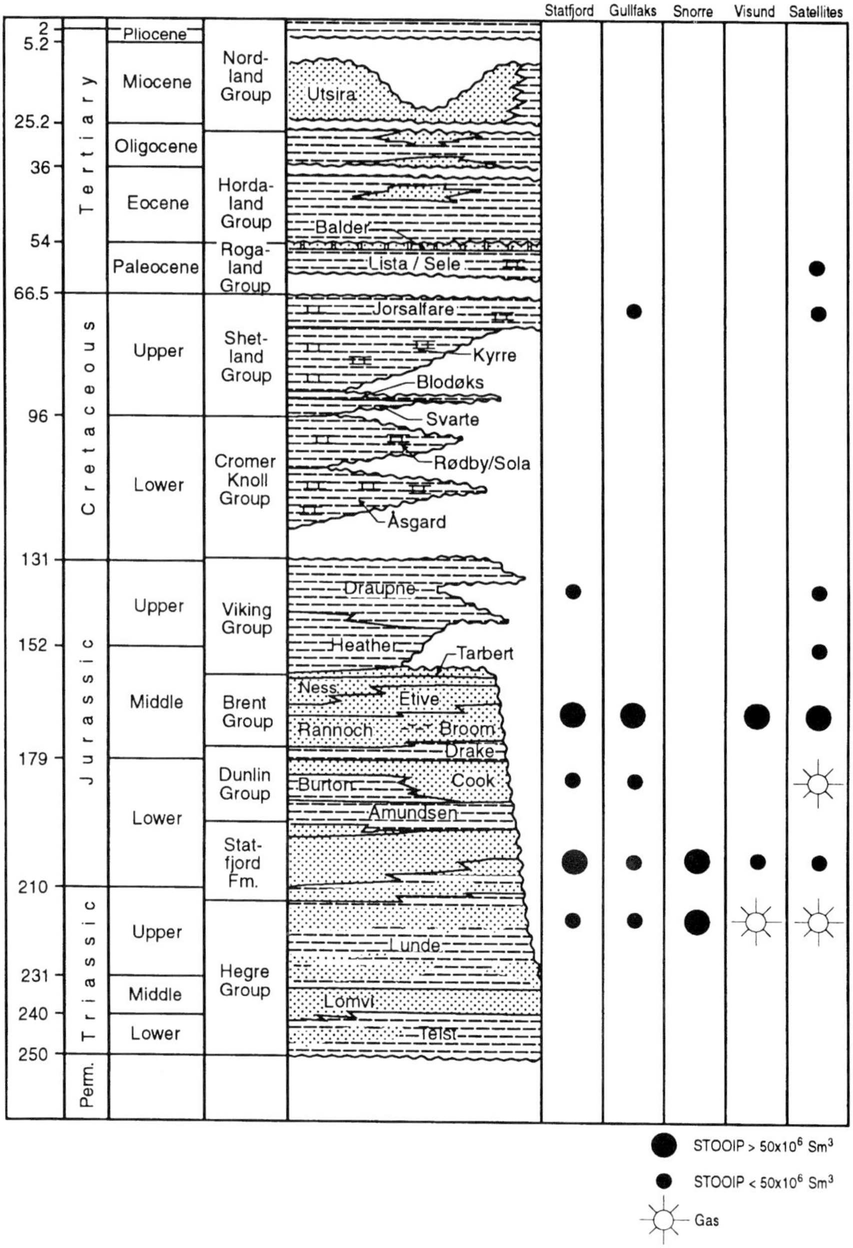

Fig. 3. General stratigraphy and reservoir levels in the Tampen Spur area.

Several of the oilfields on the Tampen Spur lie adjacent to multiple source kitchens and could have been filled from more than one direction.

The Tampen Spur represents a structural high which is surrounded by several source kitchens where the Upper Jurassic source rocks are oil-mature to gas-mature (Fig. 4) and capable of expelling petroleum at several different stages from the Late Cretaceous to present. Virtually all the structures in the area are located in such a position that they may have received petroleum from more than one direction and it is of vital importance to the further exploration of the area to understand the petroleum migration and emplacement systems in

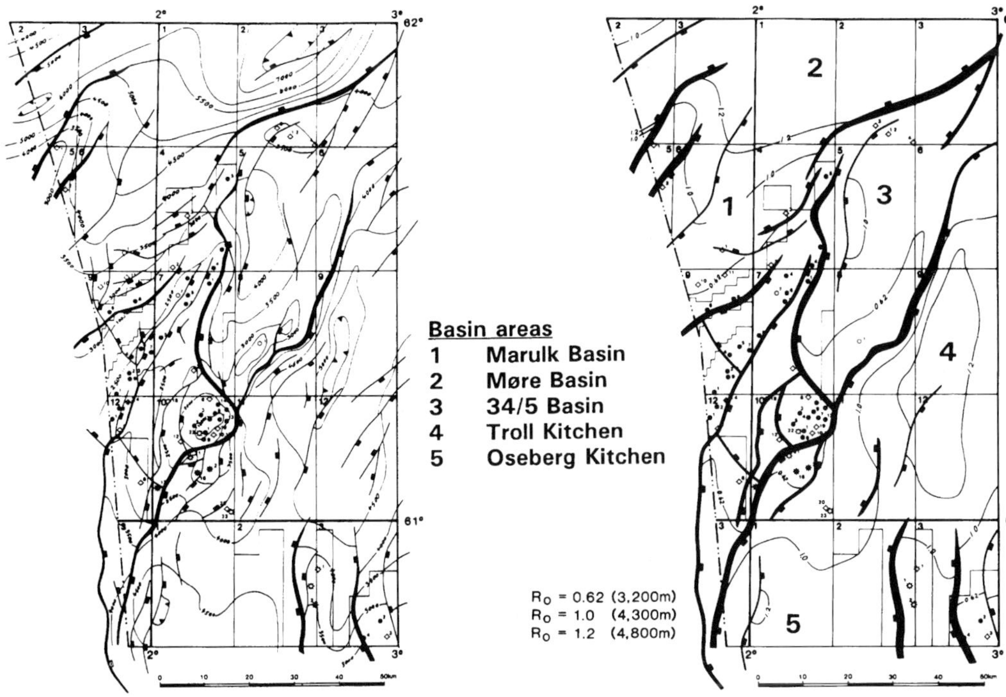

Fig. 4. Base Cretaceous depth map and major source-rock kitchens in the northern North Sea (modified from Karlsson 1986).

the area. The Marulk Basin is located to the northwest of the Snorre Field, the Møre Basin to the north, the 34/5 Basin to the northeast and the Viking Graben to the east (Fig. 4) and all have expelled petroleum towards the Tampen Spur high. In addition, a major spill route from the Oseberg Kitchens to the south, through the North Alwyn Field, Brent Field and Statfjord Field has been proposed (Thomas *et al.* 1985; Karlsson 1986).

Geological development of the Tampen Spur area

The Lunde Formation represents the upper part of the Hegre Group and has been divided into three units (Fig. 3). The main reservoir is present in the Upper Lunde Formation. The Upper Lunde Formation consists of alternating sandstones, oxidized claystones and palaeosols, deposited on an arid to semi-arid, wide alluvial floodplain. The net/gross sand ratio varies significantly through the formation and is controlled by climatic factors, sea-level changes, regional basin subsidence, rate of sediment influx, local faulting and local lake-level changes (Nystuen & Fält 1995). The lower part of the Lunde Formation represents a braided stream facies which gradually changes to a more meandering river system. The transition from the Lunde Formation to the Statfjord Formation represents a gradual change towards a more humid climate and a river system dominated by braided streams. Towards the upper part of the Statfjord Formation the marine influence decreases and the overlying Dunlin Group (Amundsen Formation, Burton Formation, Cook Formation and Drake Formation) represents fully marine conditions. The Dunlin Group is overlain by the Brent Group delta system, which built out in a northward direction in the Middle Jurassic. The Brent Group was transgressed by the Viking Group marine shales (Heather Formation and Draupne Formation). These were deposited in a tectonically active regime with major rifting in several pulses during the Upper Jurassic (Dahl & Solli 1993). Footwall uplift and exposure resulted in erosion of the Triassic to Middle Jurassic section and down-flank deposition of the Upper Jurassic sandstones interbedded with shales and claystones in structurally low positions (Solli in press). These sandstone intervals now represent prospective stratigraphic traps and also one of the keys to the understanding of the petroleum migration in the area. The Cretaceous and Tertiary sections represent marine depositional conditions and is has earlier been suggested that this time period was dominated by passive subsidence. However, new data suggest that producible reservoirs are present at these levels and that the geological evolution of the Tampen Spur in the Tertiary might have been much more complex than previously realized. A good illustration of this complexity is the apparent Tertiary biological degradation of the oil within the Brent Group in the Gullfaks Field, which suggests that oxygen-rich meteoric water passed through the structure after oil emplacement (Horstad *et al.* 1990, 1992).

Most of the discovered fields in the Tampen Spur area represent rotated fault blocks which were cut by NNE–SSW and NE–SW faults during the two rift phases in the Upper Jurassic (Early Bathonian–Late Kimmeridgian and Late Kimmeridgian–Ryazanian) (Dahl & Solli 1993). During this time, the whole area was uplifted and the Triassic to Jurassic sedimentary section was eroded. Generally, the depth of erosion increases to the north and east and we consider that in the northern part of the Snorre Field more than 1000 m of sediments were eroded (Dahl & Solli 1993). Most of the sediments were transported and redeposited downslope on the footwall block and built Upper Jurassic to Cretaceous sedimentary wedges (Solli in press).

The petroleum-bearing structures in the area are sealed either by Lower Jurassic Dunlin Group shales deposited above the Statfjord Formation, or by Upper Jurassic and Cretaceous shales and marlstones which drape the whole area. The quality of the cap rocks varies considerably across the Tampen Spur and this has a major impact on the migration of petroleum. Cap-rock leakage is common in the area; for example, the seismic data quality on the eastern part of the Gullfaks Field is seriously affected by large amounts of gas escaping into the post-Jurassic section, which results in high seismic wave absorption and poor signal penetration (Saeland & Simpson 1981). The gas above the Gullfaks Field appears to be a mixture of thermogenic gas from the Jurassic and Triassic reservoirs and shallow biogenic gas produced within the post-Jurassic section (H. Irwin personal communication). Between the Snorre Field and the Gullfaks Field, gas and heavier petroleum have been encountered above a Jurassic structure which has only residual petroleum in the Statfjord Formation (well 34/7-2). Gas leakage has also been reported to occur above the Snorre Field (Leith *et al.* 1993) and the extent of the vertical leakage and dysmigration from the structures is controlled by the nature of the cap rock (entry capillary pressure), the intensity of faulting and the height of the petroleum column (Leith *et al.* 1993). Leak-off data in the area suggest that in structures such as Gullfaks, seal failure by fracturing has occurred. Thermogenic gaseous hydrocarbons are typically observed leaking 400 to 700 m into Cretaceous cap rock above the reservoirs in the Tampen Spur area. The low petroleum potential of the Cretaceous section in most instances clearly reveals the leakage of the hydrocarbons from the hydrocarbon column as the sole source of cuttings gas anomalies.

Recently, several discoveries have been made within the Upper Jurassic and Tertiary sections in the area and it is of great importance to understand the three-dimensional oil and gas migration network as exploration continues in the area. Previously, migration was thought to occur only within the major reservoir sections and it was difficult to explain how the petroleum could migrate from one structure to another. New wells in the area have demonstrated that the Upper Jurassic to Cretaceous sandstones play an important role in the understanding of the petroleum migration and fill–spill relationships in the area, with abundant oil staining commonly found in Upper Jurassic sands.

The main oil source rock in the Tampen Spur area is the Upper Jurassic Draupne Formation (Kimmeridge Clay equivalent) which demonstrates an excellent source rock–oil correlation with the reservoired oils in the area. The underlying Heather Formation locally represents a classical Type II kerogen-rich source rock with high oil potential, but usually possesses a much lower kerogen quality than the Draupne Formation. Head-space gas analysis of a deep well in the footwall (downthrown) block east of the Snorre Field shows that the shales in the Draupne Formation have total gas yields in excess of 25 000 μl gas/kg rock, while 8000 μl gas/kg rock are recorded in the Heather Formation (Fig. 5). Thus, even at the same location within the basin, the extent of petroleum generation from the Heather Formation (buried several hundred metres deeper) seems to be exceeded by that in the Draupne Formation. We conclude from the extensive oil–source rock correlation studies and from quantitative considerations that most of the oil accumulated in the Tampen Spur area has been generated from the Draupne Formation. Significant contributions from the Heather Formation or coal sequences remain to be documented in the area. Most of the oils in the Tampen Spur high are undersaturated with respect to gas, but in the structures located on the down-faulted terraces towards the Viking Graben, gases and condensates become more dominant (Visund, 34/10 Gamma). In the deepest structures (e.g. 34/10 Gamma), dry gas with only traces of heavier components is trapped and the Heather Formation

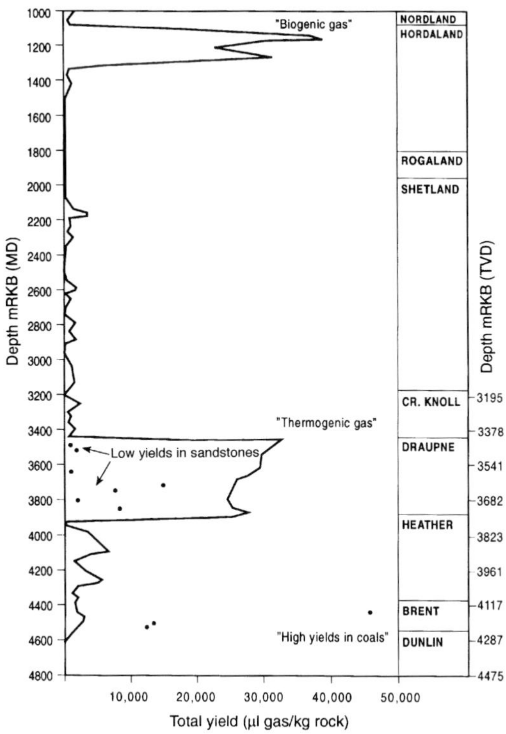

Fig. 5. Total gas as a function of depth from head-space analysis of cuttings from well 34/7-15S.

might represent an important source rock with respect to gas here. However, the generation and expulsion of gas from the Viking Group source rocks remains poorly understood and should be studied further.

The Tampen Spur high is located in the middle of several oil-mature source-rock kitchens (see numbered areas in Fig. 4) and might receive petroleum from several different directions. Although all oils appear to have been sourced from the same source-rock horizon, small differences in facies and maturity-dependent biological-marker parameters are to be expected among the different oils. It is not well understood whether biological marker maturity-related parameters of expelled oils are a function of the source-rock temperature at the time of oil expulsion or at the time of oil generation. However, it is clear that the potential petroleum expulsion mechanisms may become important in controlling minor differences in biological-marker signature in expelled oils. It has been demonstrated that the thickness and distribution of sand bodies within source-rock sequences have a large effect on the petroleum potential gradient and therefore the direction, efficiency and timing of petroleum migration (England *et al.* 1987; Mackenzie *et al.* 1987; Leythaeuser *et al.* 1988). Thus, the distribution of Upper Jurassic sands and siltstones within the source rocks in the area might well result in differences in drainage efficiency, drainage history, expulsion, timing and minor differences in the maturity of the expelled petroleum between different source-rock basins even at the same temperature. Despite concern over the detailed understanding of the mechanisms controlling the biological-marker signature in source rocks, empirical evidence gives us confidence in the use of this approach in oil-source rock correlation.

Several models have been proposed to describe the migration of petroleum into the giant oilfields within the East Shetland Basin in the northern North Sea. The most accepted models were published by Thomas *et al.* (1985) and Karlsson (1986), both suggesting a major spill along the North Alwyn–Brent–Statfjord–(Snorre)–Gullfaks trend. These models were based on Gussow's model (Gussow 1954), with a gradual change in the distribution of gas and oil along the migration route and a shallowing of the oil–water contacts along the same path. Along the fill–spill route from the North Alwyn Field to the Gullfaks Field, only the North Alwyn and the Brent Fields contain a separate gas cap and it is not obvious that Gussow's model can be applied directly to reservoirs with undersaturated oils with a separate gas cap, such as is the case in the Norwegian part of the North Sea. Clearly, the pressure–volume–temperature and phase relationships in this area are much more complex than the simple system described by the Gussow model, and this model should therefore be applied with great caution.

The combination of sample availability, reservoir interval distribution and the presence of multiple source-rock kitchens make the Tampen Spur area very suitable for a reservoir geochemical study. An understanding of the migration pathways and petroleum drainage efficiency of the basins is of great economic interest, because most of the smaller sub-structures are marginally economic. Thus, it is important to provide an effective risk analysis and ranking of the satellite prospects prior to drilling.

Methods

During the last five years, interest in reservoir geochemistry has shown a steady increase. At present, reservoir geochemical studies range from large-scale intra-reservoir petroleum compositional heterogeneities to sub-metre scale variations within a single reservoir zone or even down to the fluid inclusion scale (Larter & Aplin 1995). Compositional variations examined vary from gross parameters such as GOR and petroleum asphaltene content (Wilhelms & Larter, 1995) to variations in individual compounds such as biological-marker hydrocarbons, sulphur, nitrogen and oxygen species (Larter *et al.* 1990; Horstad *et al.* 1990; Li *et al.* 1992).

Sampling

Both tested oil samples (DST, RFT and FIT) and solvent extracts from conventional reservoir cores have been used to construct the oil population maps of the area. This is necessary because the availability of produced oils is limited and because it is essential to include dry wells which only have oil shows and therefore have not been tested. Although significant differences in the relative distribution of petroleum compound classes (saturated hydrocarbons, aromatic hydrocarbons, NSO compounds and asphaltenes) (Horstad *et al.* 1990) and the distribution of nitrogen compounds such as arbazoles (Stoddart *et al.* 1995) have been reported to occur between core extracts and oils, our data from clastic reservoirs suggest that this does not have any influence on the distribution of aliphatic hydrocarbon biological markers and stable carbon isotopes in mixed core extract and tested oil sample sets (Fig. 6). However, the possibility of such effects should not be ignored, and duplicate sampling and analysis of both cores and tested oils is preferred whenever possible. Because the construction of a petroleum population map in a large area typically requires 50 to 200 GC–MS runs, it is difficult to avoid the use of data from different sources. The quality of the input data is important and because the geochemical differences that are detected between the different structures tend to be small in homogeneous source provinces, such as the Tampen Spur area, direct comparison of data from different consultants should be avoided. Our experience is that in some cases significant, small, but indeed real, compositional differences have been masked because the samples were analysed by different laboratories. In other studies, samples that belong to the same oil population may be grouped as two populations due to lack of reproducibility of geochemical data between different laboratories. Our oil population map of the Tampen Spur area is based on more than 300 GC–MS analyses from four different consultants. To allow synthesis of all these data, all sub-studies include several duplicate or reanalysed samples from previous studies, which allows a proper calibration of the different datasets. Thus, the construction of these maps is a continuous process and updating is required whenever new wells are drilled. When a new discovery is made in an area where an oil population map is available, we routinely reanalyse 5–15 relevant samples from the area as a single batch together with the oil from the new discovery. This allows a proper characterization and classification of the new oil and updates the oil population map of the area.

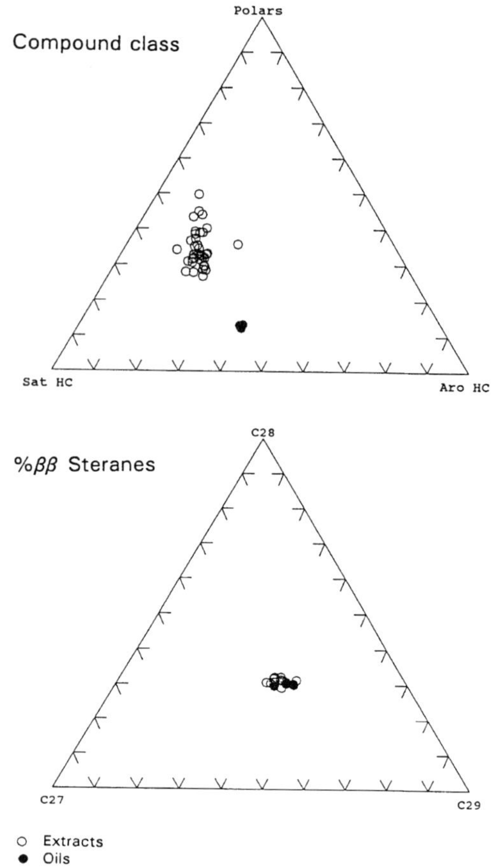

Fig. 6. Example of fractionation of compound classes between reservoir core extracts and tested oil samples from the Brent Group reservoir in the Gullfaks Field well 34/10-1. Note that biological markers (i.e. distribution of C_{27}–C_{29} ββ-steranes) do not show any signs of fractionation.

Analytical techniques

The general reservoir geochemical protocol followed is a modification of that developed by the Petroleum Geochemistry Group at the University of Oslo, Norway (Larter *et al.* 1990; Horstad *et al.* 1990; Karlsen & Larter 1991). All conventional cores have been screened by Iatroscan (thin layer chromatography–flame ionization detection, TLC–FID) (Karlsen & Larter 1991) prior to selection of samples for detailed GC–FID, GC–MS and stable carbon isotope analyses of medium-pressure liquid chromatography (MPLC) separated fractions. Typically, screening includes five to ten samples from each reservoir interval in each well, and two to five of these are subsequently submitted for detailed analyses of the saturated hydrocarbon and aromatic hydrocarbon fractions. Some of the data have been acquired as part of full-field reservoir geochemical studies and such studies typically include 50 to 100 samples from each well. As mentioned previously, GC–MS analysis performed at different times or at different laboratories must be calibrated by reanalysing samples from each data subset. The temperature programmes applied for GC–MS analysis of the hydrocarbon fractions in the different laboratories have varied slightly and we have experienced that even with the same GC column and the same temperature programme, variable GC resolution of the biological markers can be documented, both between different laboratories and as a function of time in the same laboratory. This has a large impact on the quantification of individual compounds, especially for steranes (*m/z* 217 and 218) and monoaromatic steroids (*m/z* 253), where coelution is a serious problem for nominal mass GC–MS. The selected GC programme used here is a compromise between peak resolution and peak shape/definition, optimized for the compounds that are of interest. Currently, we use a DB-5 equivalent column for GC analysis of the aromatic hydrocarbons. The GC programme consists of four different gradients to allow maximum resolution of the naphthalenes and phenanthrenes: start temperature 80°C held for 1 min, 1°C min^{-1} to 103°C followed by 3°C min^{-1} to 140°C, then 1°C min^{-1} to 180°C and 10°C min^{-1} to 300°C. The GC programme for saturated hydrocarbons is 80°C for 3 min, then 4°C min^{-1} to 300°C with 12 min hold time at 300°C, on a non-polar column (DB-1 equivalent column). GC starting temperature for oil samples has been set at 40°C. The GC programme applied in the GC–MS analysis of biological-marker hydrocarbons allows maximum resolution of steranes and hopanes and has a single temperature gradient on a non-polar column (DB-1): starting at 180°C for 1 min, then 1.7°C min^{-1} to 310°C with 10 min hold time at 300°C.

In general, it is not possible to define a universal set of biological-marker ratios which can be used to differentiate all oils into various populations or to define how large differences among oils should be before they are classified into different populations. We have applied different statistical methods such as the Student *t*-test and principal component analysis (Davis 1986) to decide whether the differences perceived are statistically valid or not. Such analyses require a large and homogeneous dataset composed of a single subset of samples run by the same contractor in the same sample batch. If data from two different laboratories are analysed in the same principal components analysis, the main discrimination will usually be between the two (laboratory defined) data subsets and not between the multiple oil populations. The choice of biomarker hydrocarbons that should be used in any situation is based on experience within the area, and a large dataset should be applied in order to choose the most useful ratios. Within the Tampen Spur area, the 25-norhopanes are present only in trace amounts and cannot be applied to discriminate oils, while on the east flank of the Viking Graben (quadrants 30, 31 and 35) these compounds are extremely useful for discriminating between the different oil populations. The differences between the oil populations are typically expressed by variations in several biological-marker ratios, but are best illustrated by a few of the conventional ratios. Within the Tampen Spur area the 17α(H),21β(H)-28-30-bisnorhopane is one of the most useful compounds for discriminating between oils. The origin and occurrence of bisnorhopane has been debated since it was first reported in the late 1970s (Seifert & Moldowan 1978; Grantham *et al.* 1980; Moldowan *et al.* 1984), and while most of the other hopanes are ubiquitous in most source rocks and oils, the presence of bisnorhopane is variable. However, it is usually present in significant concentrations in the Upper Jurassic source rock in the North Sea, but seems to be more common in the Draupne Formation than in the Heather Formation. The relative amount of bisnorhopane in an oil seems to be more dependent upon the facies of the source rock from which the oil was expelled than its maturity at the time of expulsion. When the maturity range of the analysed oils is large, the relative concentration of 28,30-bisnorhopane might reflect a significant maturity component. The diahopanes, diasterenes and stable carbon isotope signatures have also proven to be useful correlation parameters in this area.

Field filling

Based on the theoretical considerations made by England *et al.* (1987), a reservoir will fill in a sequential manner. The petroleum will first enter the most permeable zones with the lowest pore-entry pressures and subsequently fill the reservoir in a set of advancing petroleum fronts (Fig. 7). Because the filling takes place over a certain period of time (e.g. 5 to 30 Ma in the North Sea), the source rock will continue to subside and an

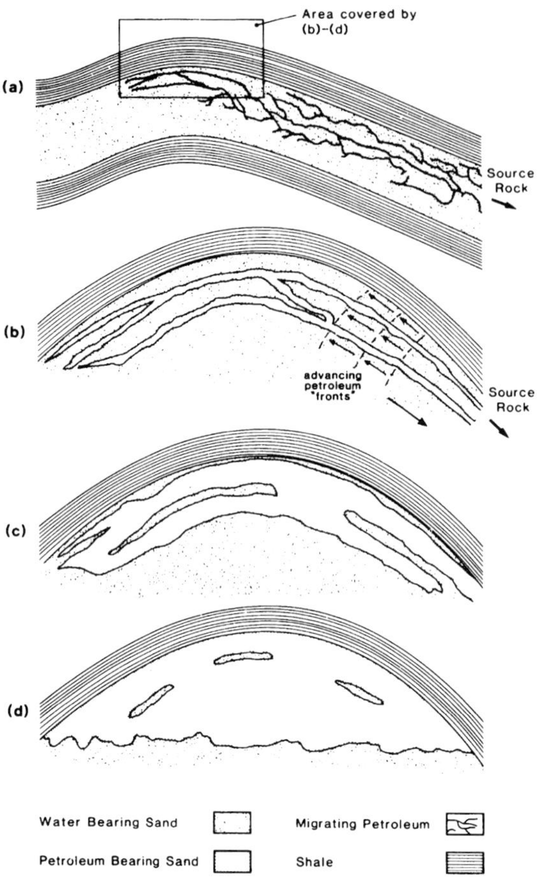

Fig. 7. Schematic filling sequence of a reservoir (from England *et al.* 1987).

increasingly mature petroleum will enter the trap. Thus. if a reservoir is filled from two different directions and the two oils do not subsequently mix within the reservoir, it should be possible to map the distribution of the two oils within the reservoir (Fig. 8). In addition, if the oil within the trap has been expelled over a significant range of maturities, it might be possible to detect gradients of maturity within each population which might give information on the filling directions into the field. Our data suggest that such differences exist in nature and that the geochemical analysis can be used to map heterogeneities in oil compositions in high quality well-connected clastic reservoirs (Horstad *et al.* 1990). This can be used to increase the understanding of field filling and regional migration routes in the area. Recent experience in more complex, less well-connected clastic reservoirs, both in this area and elsewhere in the North Sea, suggests that field-filling processes can be much more locally controlled and complex to unravel by reservoir geochemistry. Thus, fieldwide gradients in composition are sometimes not present for good geological reasons.

Several mechanisms will act to reduce the previous heterogeneities in the petroleum columns, but these processes (principally diffusion and density-driven overturning) are not efficient enough to homogenize the petroleum over the kilometre scale lateral distances of a large oil-field. There is no single technique that can be used to resolve the petroleum migration history in a province and a multidisciplinary approach is the only healthy approach to such a challenge. In order to attempt this, it is necessary to make a composite of structural maps, reservoir sand distribution maps, structurally possible migration paths and the distribution of oil populations in the area to evaluate possible migration and field-filling scenarios. Similarly, both inorganic and organic geochemical methods may be relevant, depending on the field (Larter & Aplin 1995). Based on combined evaluation and analysis of all the available information, the most likely migration pathways might then be proposed in the area. It is

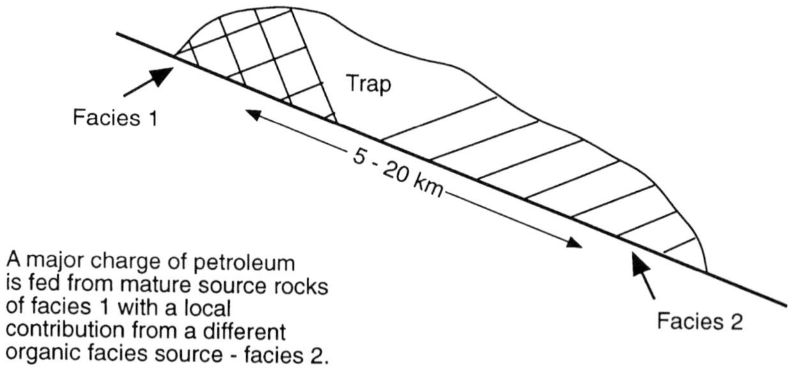

Fig. 8. Potential heterogeneties on a hypothetical reservoir caused by filling from two different sources.

important to test the evolving migration model continuously by analysis of new oil samples and incorporation of new geological information in the area.

Within the northern North Sea, most of the oil reservoirs are located at the crestal part of rotated fault blocks and therefore most of the wells are drilled in a structurally high position with only a few wells being aimed at down-flank prospects. Thus the available samples from the basins are heavily biased towards shallow areas and no conventional samples are available from higher pressure and temperature regimes within the deeper part of the basins. Although migration of oil and gas from the basin towards shallower structures at lower temperatures and pressures may result in phase changes (England *et al.* 1987; England & Mackenzie 1989; Larter & Mills 1991) the bulk of the samples studied here represent fluids accumulated as single-phase liquid petroleum, thus the geochemical data can be interpreted conventionally.

Oil populations

Based on detailed geochemical mapping of data from reservoir core extracts and tested oil samples from most of the exploration wells within the Tampen Spur area, it has been possible to differentiate the oils into several populations. Previously, it was suggested that the Gullfaks Field filled from two different source basins (Horstad *et al.* 1990). The Cook Formation and Statfjord Formation filled from the north and/or east while the Brent Group filled form the west. The proposed filling directions into the Gullfaks Field were based on tentative maturity gradients within each of the two major oil populations in the field. The oil within the Gullfaks Field was subsequently biodegraded and the extent of biodegradation varies across the field, from apparently undegraded oil within the Statfjord Formation in the eastern part to the most severely degraded oil within the Brent Group in the westernmost part of the field (Horstad *et al.* 1990). The extent of biological degradation was considered too mild to affect the large number of biological-marker parameters used. Later studies, which included more samples from neighbouring fields, have confirmed this model for the filling of the Gullfaks Field.

In this study, 86 different tested oil samples from 39 wells (Table 1) and *c.* 220 reservoir core extracts from 48 wells (Table 2) in the Tampen Spur area were analysed in four batches by GC–FID, GC–MS and stable carbon isotope analysis of saturated and aromatic hydrocarbon fractions. The selection of core samples for detailed geochemical characterization was based on Iatroscan (TLC–FID) screening of more than 1000 samples. The oils from the Tampen Spur that have been included in this study have been differentiated into five major oil populations (Table 3) which migrated into the area from

Table 1. *Number of tested oil samples analysed from each well*

Well	33/9	34/4	34/7	34/8	34/10
–1		1	3	1	3
–2					
–3			3	5 + 2	3
–4		3	1	4 + 2	2
–5		2	1		1
–6		1	3		
–7		2	2		3
–8			3		1
–9			1		4
–10			4		
–11					3
–12			2		
–13	2		1		2
–14	1		2		1
–15					
–16			2		
–17A			1		
–18			2		
–19			2		
–20					
–21/21A			2		

Table 2. *Number of reservoir core extracts analysed by GC–FID and GC–MS from each well*

Well	33/9	33/12	34/4	34/7	34/8	34/10
–1					7	11
–2		4	2	3		
–3		3	3	17	7	3
–4	2		10	3	10	2
–5			7	2	9	1
–6	2		9	4		
–7	2	3		10		1
–8	2			6		1
–9	3					1
–10	2			9		
–11						
–12	2			5		1
–13				6		
–14	4			1		
–15	1			4		
–16				3		
–17A				3		
–18				8		7
–19				3		
–20				1		
–21/21A				10		

Table 3. *Selected biological marker ratios and stable carbon isotope ratios from tested oil samples and reservoir core extracts from the different oil populations in the Tampen Spur area*

Group	Oil population	$\delta^{13}C$ Sat HC*	Bisnor/nor hopane* (*m/z* 191)	%$\beta\beta$ in C_{29} steranes† (*m/z* 217)	%20S in $C_{29}\alpha\alpha$ steranes† (*m/z* 217)	C_{30} dia hopane/ C_{30} dia hopane +$C_{29}\beta\alpha$† (*m/z* 191)	$C_{27}\beta\alpha$ dia 20S/($C_{27}\beta\alpha$dia 20S + $C_{27}\alpha\alpha\alpha$20T† (*m/z* 217)
I	Statfjord Nord	–28.8 to –29.3	0.44–0.50	0.54–0.57	0.47–0.53	0.64–0.75	0.63–0.71
II	Snorre Lunde Fm.	–30.3 to –31.1	0.67–0.70	0.55–0.57	0.54–0.59	0.61–0.64	0.63–0.66 Lunde Fm.
	Stfj. Fm.			0.54–0.55	0.55–0.56	0.52–0.56	0.55–0.60 Stfj. Fm.
III	Statfjord Øst	–28.4 to –29.8	0.54–0.61	0.52–0.57	0.46–0.56	0.47–0.60	0.55–0.60
IV	Statfjord (Statfjord Fm.)	–29.8 to –30.2	0.51–0.53	0.51–0.53	0.43–0.45	0.48–0.51	0.41–0.51
V	Visund	–29.2 to –29.7	0.26–0.37	0.56–0.67	0.56–0.64	0.73–0.86	0.73–0.79

* Mainly controlled by source rock facies within the study area
† Mainly controlled by maturity within the study area

different directions and at different stratigraphic intervals (Fig.9). The mapping of oil populations in the area has demonstrated that the Tampen Spur has a more complex filling history than proposed earlier (Thomas *et al.* 1985; Karlsson 1986) and that it is possible to evaluate the filling of such a complex area by a combined geological and reservoir geochemical approach. Previous methods suggest possible migration routes or pathways, but cannot be applied to decide from which direction the oil actually migrated into the structures (Nybakken 1991).

Based on the samples included in this study, five oil populations have been described in the Tampen Spur area, differentiated on the basis of stable carbon isotope analysis and molecular marker data (Table 3). With the exception of the Visund Field accumulation, which is more mature (based on

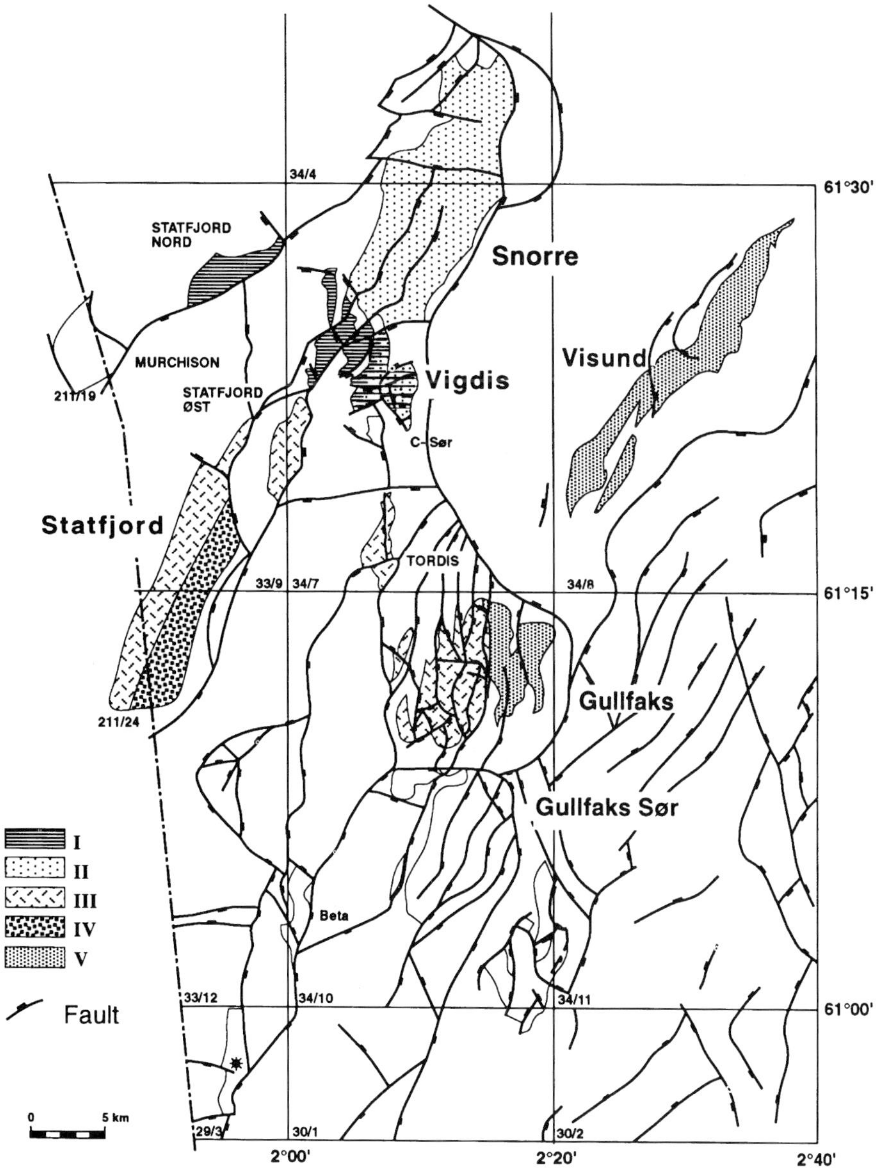

Fig. 9. Petroleum populations identified in the Tampen Spur area.

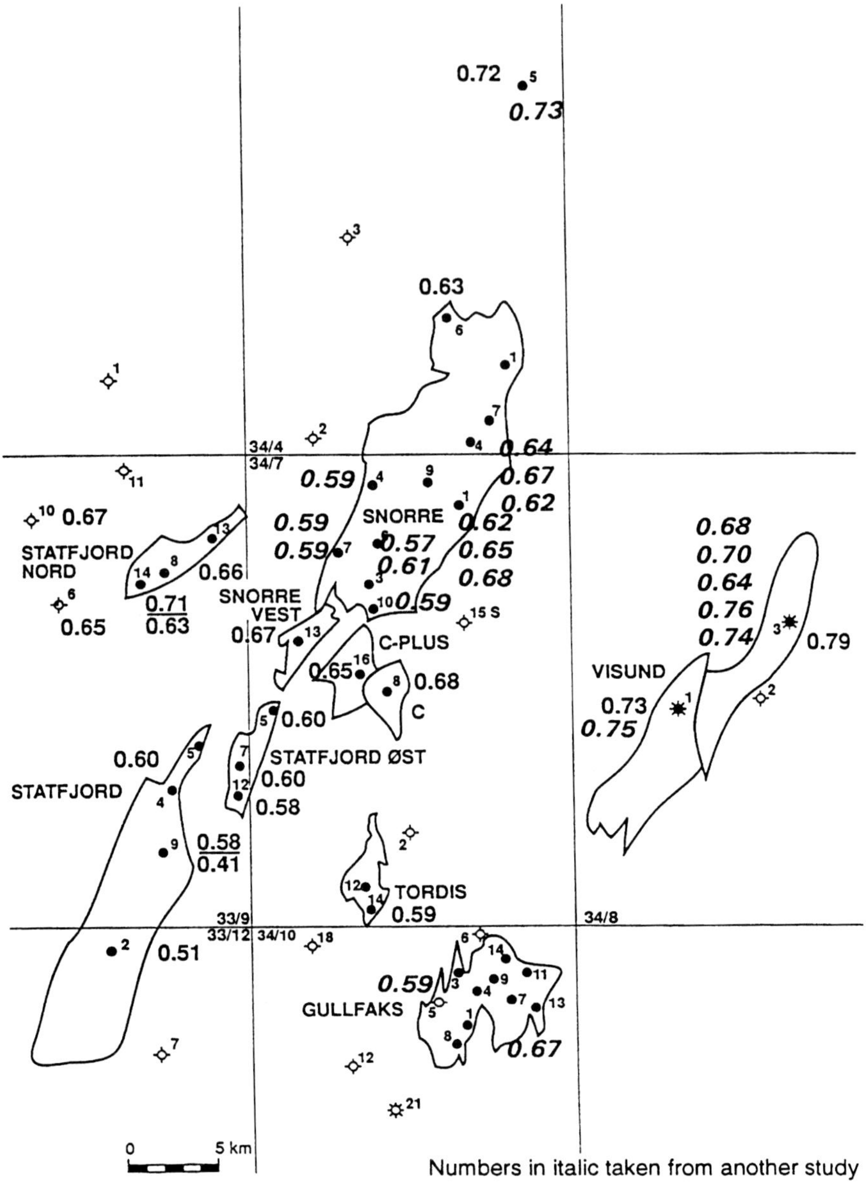

Fig. 10. Variations in the $C_{27}13\beta(H)$, $17\alpha(H)$-diasterane (20S)/[$C_{27}13\beta(H)$, $17\alpha(H)$-diasterane (20S) + $C_{27}14\alpha(H)$, $17\alpha(H)$-sterane (20R)] ratio in the study area. Note that data have been selected from two different studies.

C_{30}diahopane, $C_{27}\beta\alpha$dia20S sterane (Fig. 10) and $C_{29}\beta\beta$ steranes), the other four groups have similar ranges of maturities but significantly different bisnorhopane/norhopane ratios (Fig. 11) and stable carbon isotope signatures. Although we cannot disprove that the oils represent a maturity sequence from a single source rock, the relative amount of bisnorhopane and stable carbon isotope ratios suggest that the oils have migrated into the area from several different source-rock kitchens. This is consistent with the structural configuration in the area.

(1) *The Statfjord Nord oil population* is present within the Upper Jurassic sandstones and the Brent Group in the Statfjord Nord Field and in the Brent Group in Vigdis Vest (formerly Snorre Vest, well

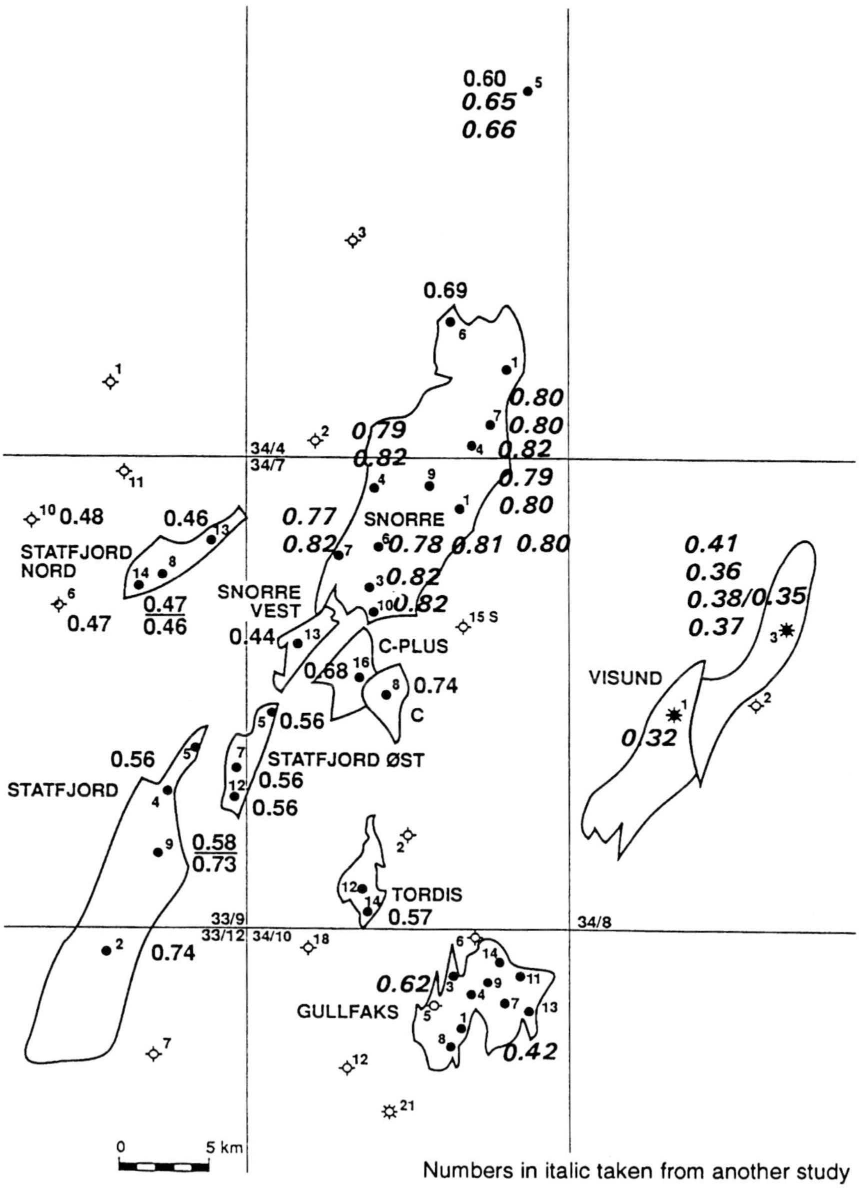

Fig. 11. Variations in the bisnor/norhopane ration in the study area. Note that the data have been selected from two different studies.

34/7-13) and part of Vigdis Middle (former C + II structure, well 34/7-19). This oil is characterized, and can be differentiated from other oils in the area, by a relatively low 28,30 bisnor/norhopane ratio and a heavy carbon isotope signature ($\delta^{13}C$ from –28.5 to –29.5 for the saturated hydrocarbon fraction) (Fig. 12).

(2) *The Snorre Field oil population* is present within the Statfjord Formation and Lunde Formation in the Snorre Field and possibly within the Zeta Structure to the north of the Snorre Field. This oil has a light carbon isotope signature ($\delta^{13}C$ from –30.3 to –31.0 for the saturated hydrocarbon fraction) and a high 28,30-bisnor/norhopane ratio

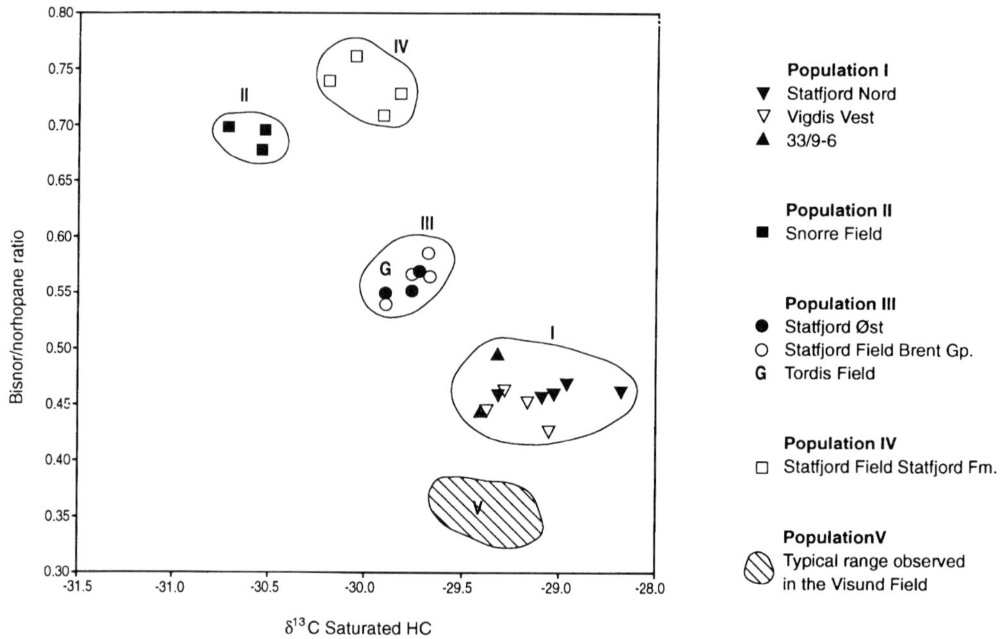

Fig. 12. Stable carbon isotope signature of the saturated hydrocarbon fraction and bisnor/norhopane ratio of selected samples from the Tampen Spur area.

(Fig. 12). The stable carbon isotope signature is usually a reproducible analysis even over a long data acquisition period, and our database consistently shows that the Snorre oils are relatively isotopically light while the Statfjord Nord oils are isotopically heavy (Fig. 13). It is interesting to note that these two oil populations, which are located geographically close to each other, are geochemically so different. The maturity of these two oil populations is the same (Table 3) and the compositional differences are possibly controlled by variations in source facies within the Draupne Formation source rock on the rim of the Møre Basin and the Marulk Basin. Alternatively, the oils within the Statfjord Nord accumulation might have migrated into the Tampen Spur area from the East Shetland Basin to the west, but we do not have any data to prove or disprove this hypothesis. The maturity within the Snorre family oils increases from south to north within the field (Fig. 10) and this suggests that the Lunde Formation reservoir in the northern part of the field drains a deeper and more mature part of the source-rock kitchen than the Statfjord Formation reservoir in the southern part of the field. The oil–water contact varies across the Snorre Field and is deepest in the southwest, suggesting a filling of the structure from the southwest. However, the oil in the Lunde Formation in the northern part is the most mature with a higher GOR, and most of the biological markers show a more mature distribution (Figs. 14 & 15; Table 3). It has been suggested that the increasing GOR in the northern part of the field is a result of a late gas migration into the field from the deep basins to the north. The Snorre Field is undersaturated with respect to gas (GOR varies from 63 to 215 Sm^3/Sm^3) and there is a good relationship between the maturity of the oil and the maturity/wetness of the associated gas, which suggests that the gas and oil have been generated at the same time (Fig. 16). It is therefore suggested that the Snorre Field has been filled through at least two fill points, one in the southwest charging the Statfjord Formation and one in the northwest charging the Lunde Formation. This model is able to explain the variations in the GOR, biological-marker parameters and the different oil–water contacts across the field. The similarity in the biological markers reflecting source facies and the stable carbon isotope signature of the oil in the Statfjord Formation and the Lunde Formation in the Snorre Field, suggests that there are minor variations in the source-rock facies in the drainage area of the Snorre Field.

The Snorre Field is located at the southern rim of the Møre Basin where the Draupne Formation is buried deeper than 5000 m (Fig. 4) and large gas volumes would be expected to drain southwards

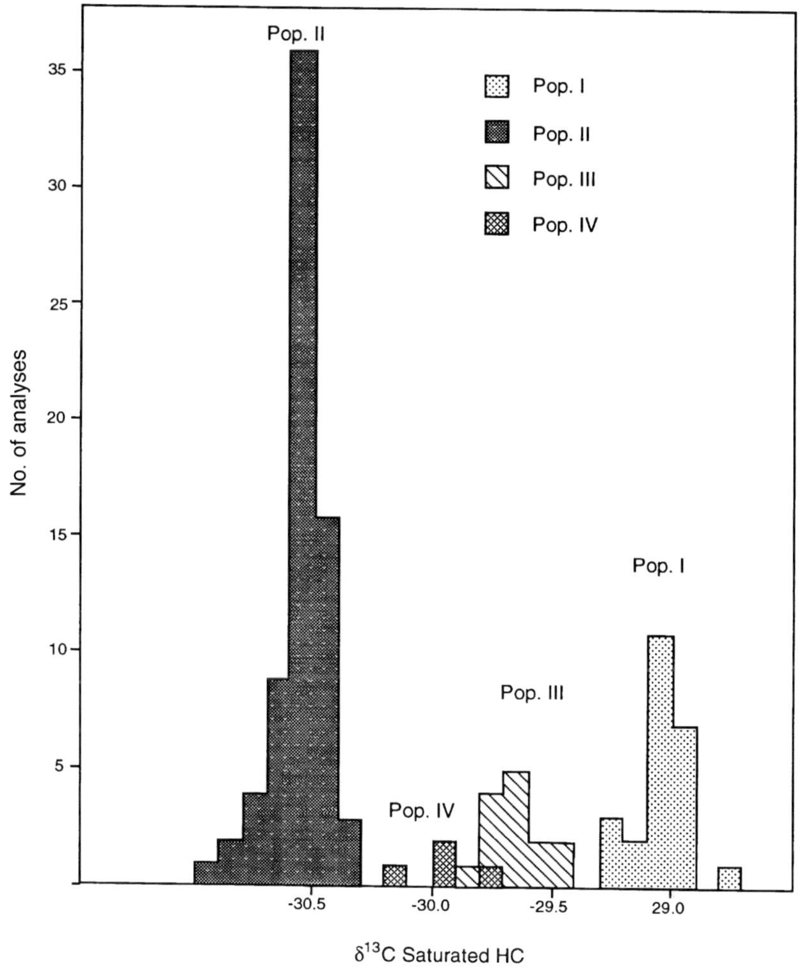

Fig. 13. Stable carbon isotope signature of the saturated hydrocarbon fraction of oils and extracts from four of the petroleum populations in the area.

into the northern part of the Snorre Field. The lack of gas in the Snorre Field might be explained by (i) non-deposition of oil and gas sources in the central part of the Møre Basin, (ii) erosion of the possible source rock, (iii) lack of migration pathways southwards from the Møre Basin into the Snorre Field, or (iv) a preferential drainage of gas in a different direction, controlled by the sand distribution and not by the structural configuration of the area. It is suggested that the absence of source rocks in the central part of the Møre Basin is the most probable cause of this lack of gas. The light carbon isotope signature of the oil within the Snorre Field might be due to the presence of only the upper part of the Draupne Formation in the source basin that is in communication with the major reservoirs in the Snorre Field. This section of the Upper Jurassic is isotopically significantly lighter than the lower part of the Draupne Formation and the Heather Formation in this area (Fig. 17). It would appear from the maturity levels in the Møre Basin and the GOR range in the Snorre Field that the bulk of the Møre Basin is not in communication with the Tampen Spur in the Snorre area. This demonstrates how the mapping of oil populations might be applied to evaluate the presence or absence of active, communicating source rocks in undrilled areas.

(3) *The Statfjord Øst oil population* is present in the Brent Group in the Statfjord Field, Statfjord Øst structure, Tordis Field and Gullfaks Field (Fig. 9). This is almost the same fill–spill route that was

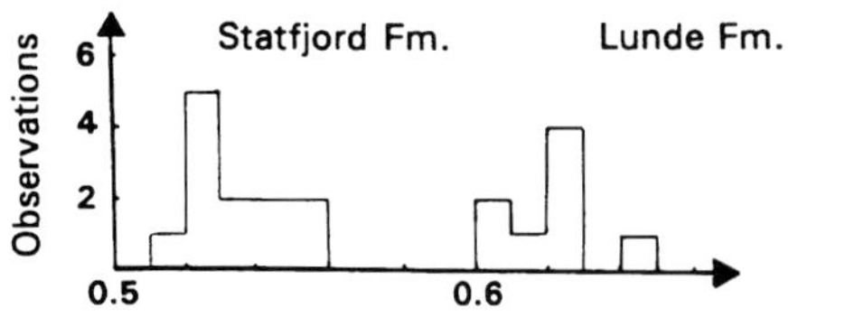

Fig. 14. Grouping of Snorre Field oils into two different subgroups based on maturity differences (i.e. C_{30}-diahopane/(C_{30}-diahopane + C_{29}-moretane).

proposed previously for the whole Tampen Spur (Thomas *et al.* 1985; Karlsson 1986), but this migration route was not confirmed by detailed reservoir geochemical information at that stage. This charging conduit is also in good agreement with the general migration model proposed for the Brent Group in the Gullfaks Field which was based solely on information from the Gullfaks Field (Horstad *et al.* 1990), demonstrating the ability to predict petroleum migration in remote areas. However, it is stressed that our experience elsewhere in the North Sea suggests that some fields have local near-field fill points which are not clearly related to regional drainage directions. The Statfjord Øst oil population has also been tested within a Paleocene sand in one of the sub-structures in the area. This oil population is characterized by an intermediate carbon isotope signature ($\delta^{13}C$ from –29.5 to –30.0 for the saturated hydrocarbon fraction) and an intermediate 28,30-bisnor/norhopane ratio (Fig. 12). Thus, a mixing of the Statfjord Nord oil population with the Snorre Field oil population results in a mixture similar to the Statfjord Øst oil population. Because these three oil populations migrate towards the same apex in the central part of block 34/7, it becomes difficult to elucidate the migration paths in this area. At this stage, it is believed that the three segments in the Vigdis Field have been filled from two different directions and that these two oil populations have mixed in the central part of the field. The Vigdis Vest and the western part of Vigdis Middle (34/7-19) have a Statfjord Nord type of oil, while Vigdis Øst (former C structure, well 34/7-8) has a geochemical signature similar to the Snorre Field. The transition zone has been filled from both directions. Such a migration and filling model of the Vigdis Field is supported by artificial oil mixing (equal volumes) studies performed with oils from the Snorre Field, Statfjord Øst and Vigdis Vest (Fig. 18).

(4) *The oil in the Statfjord Formation in the Statfjord Field* is different to any of the other major

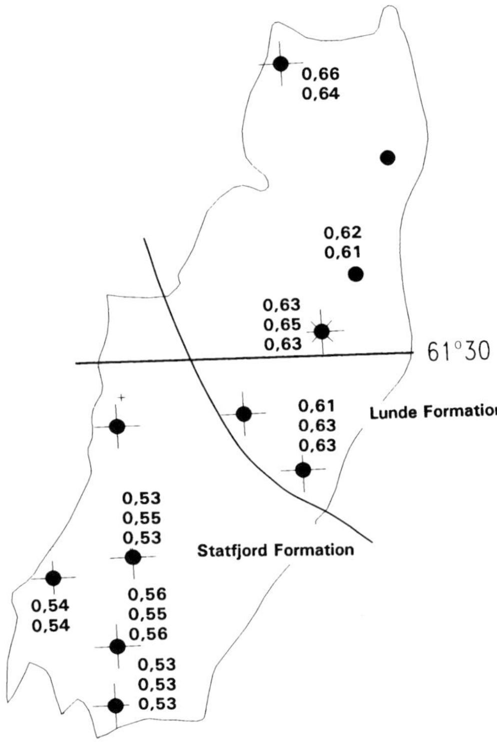

Fig. 15. Variations in the C_{30}-diahopane/C_{30}-diahopane + C_{29}-moretane) ratio across the Snorre Field, showing the higher maturity of the oil within the Lunde Formation in the northern part of the field as compared to the Statfjord Formation in the southern part.

oil populations on the Tampen Spur, but has a similar maturity. This oil has a high 28,30-bisnor/norhopane ratio and an intermediate carbon isotope signature ($\delta^{13}C$ from –29.7 to –30.2 for the saturated hydrocarbon fraction). It is stressed that we have analysed only a few wells from the Statfjord Field and the filling and migration history of the Statfjord Field might be more complex than suggested by our data.

It is interesting to note that the oils within the Statfjord Formation and the Brent Group in the Statfjord Field are so different (groups III and IV in Table 3). This suggests that the Statfjord Field has been filled from two different sources, as was demonstrated earlier in the Gullfaks Field (Horstad *et all* 1990). Maturity differences between the oils in the Statfjord Formation and the Lunde Formation in the Snorre Field are also evident (Fig. 14). The different geochemical signatures of petroleum entrapped at different stratigraphic levels might suggest that migration and filling of the structures in the Tampen Spur area and the East Shetland

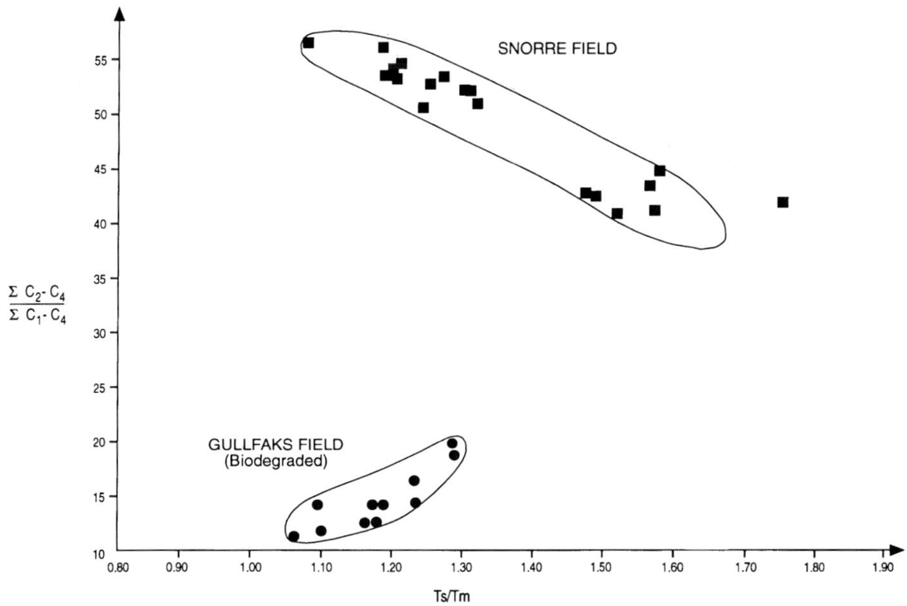

Fig. 16. Gas maturity (wetness) and oil maturity (T_s/T_m) correlation in the Snorre Field, suggesting that the gas is associated with oil.

Basin occur through several different stratigraphic levels. It is possible that the Brent Group, the Statfjord Formation and the Lunde Formation drain different geographical and/or vertical parts of the same source-rock kitchen, resulting in different petroleum charges within each stratigraphic level. This also implies that migration from one formation to another might be restricted even within areas where faulting occurs. Thus, in addition to geochemical mapping, detailed structural maps of

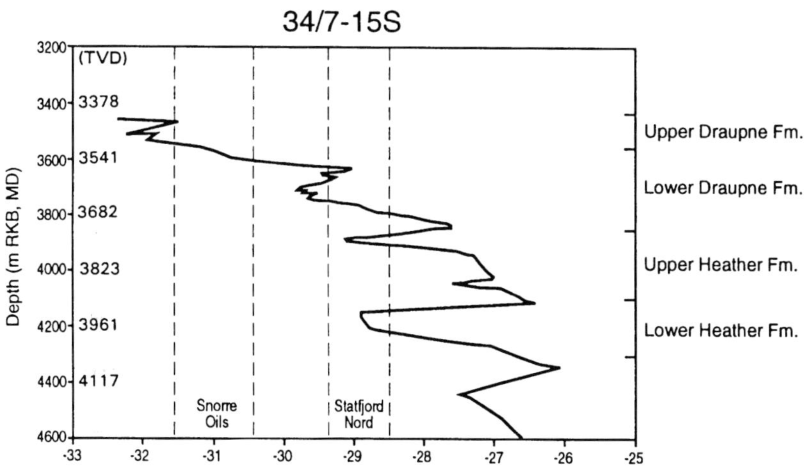

Fig. 17. Stable carbon isotope signature of the saturated hydrocarbon fraction as a function of depth in the Viking Group shales in well 34/7-15S, and the range observed in the Snorre and Statfjord Nord petroleum populations.

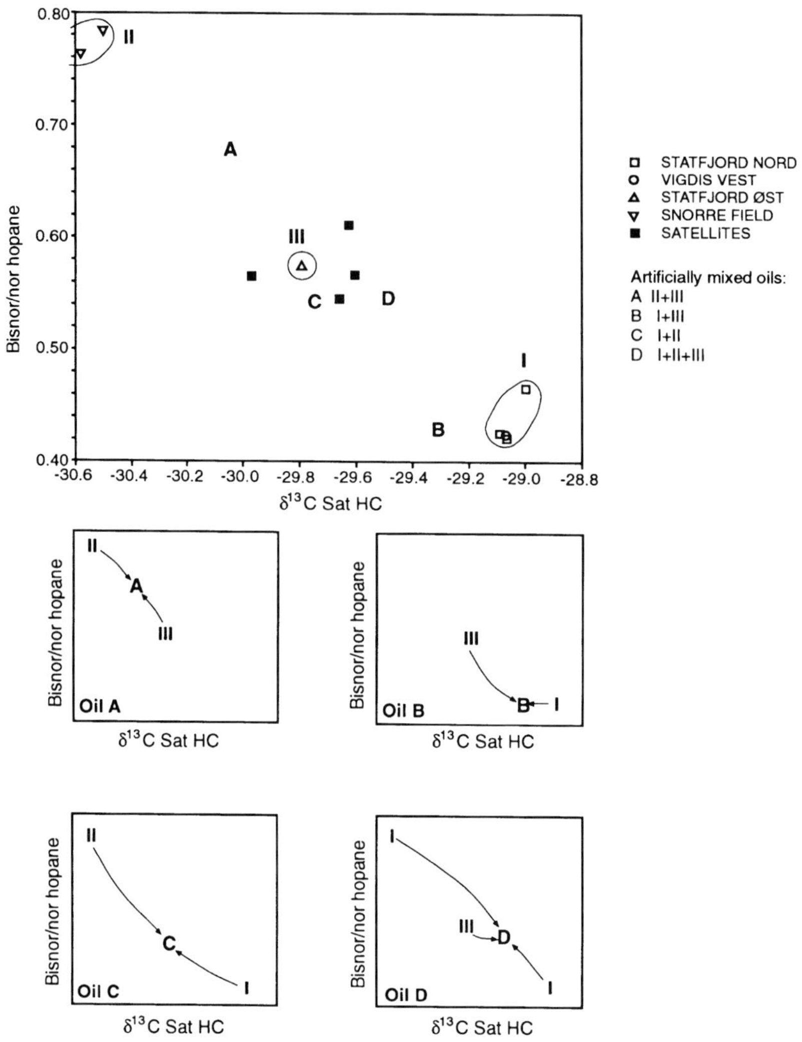

Fig. 18. Stable carbon isotope signature of the saturated hydrocarbon fraction and bisnor/norhopane ratio of selected samples from the Tampen Spur area and artificially mixed oil samples from three major petroleum populations (mixed in equal volumes).

possible migration routes should be constructed at several depths to elucidate migration routes in the area.

(5) *The Visund oil population* is located within the Brent Group in the Visund Field, which represents a down-thrown fault block or terrace towards the Viking Graben, and in the Cook Formation and the Statfjord Formation in the Gullfaks Field. This oil has a higher maturity than any of the other oils on the Tampen Spur (Table 3) and has been expelled from a high maturity source rock, probably near the Viking Graben axis. This is in accordance with the local presence of gas caps in the Visund Field and a relatively high GOR as compared to the structures on the Tampen Spur high, which host mainly highly undersaturated oils. The oil leg in the Visund Field probably spills to the southwest up the large fault towards the eastern part of the Gullfaks Field. Data from new wells suggest that the Visund Field might have been filled from several directions and that petroleum emplacement is complex. However, a spill route towards Gullfaks is in accordance with the proposed filling of the Cook Formation and the Statfjord Formation in the Gullfaks Field (Horstad *et al.* 1990), again demonstrating that oil population mapping can be used to

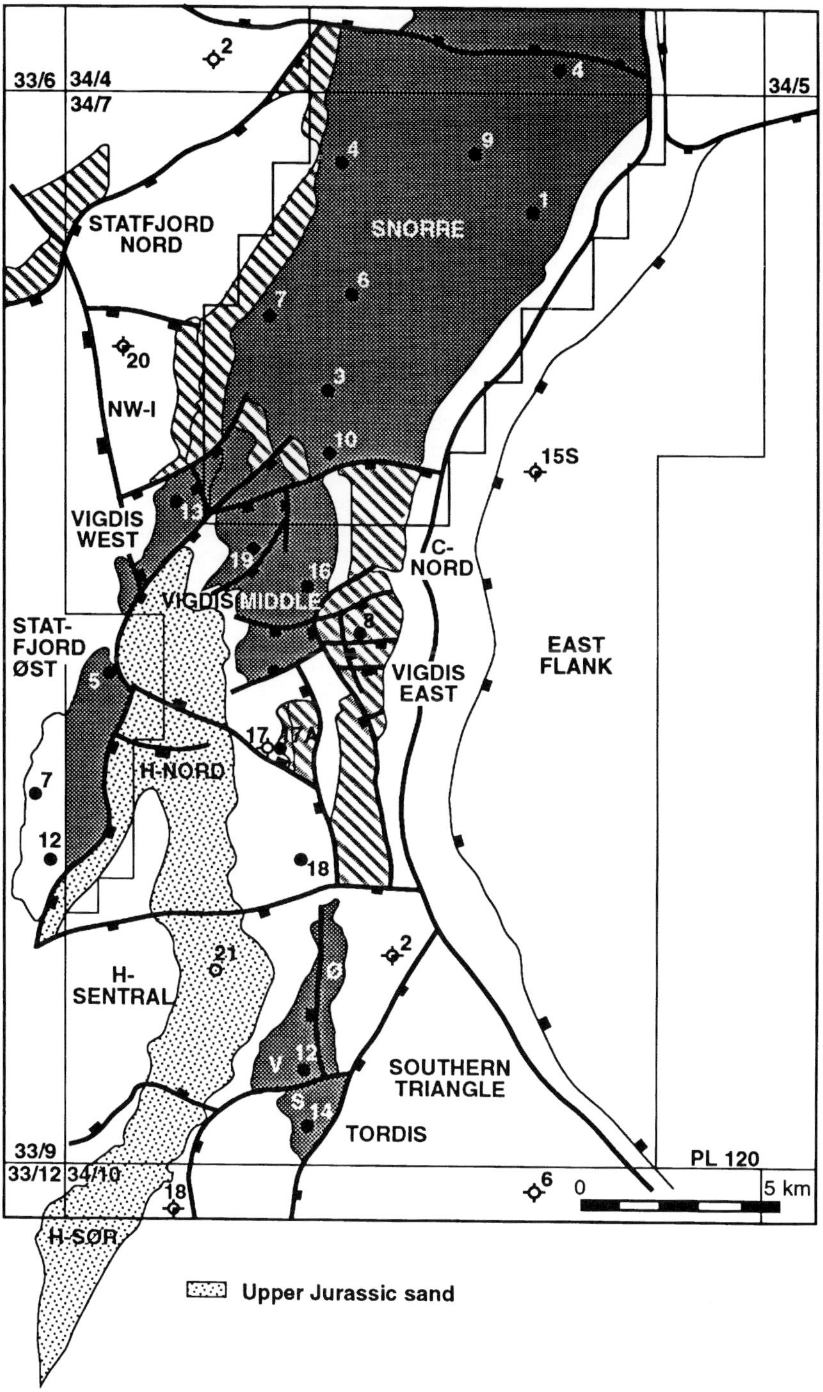

Fig. 19. Location of the two exploration wells (34/7-20 and 21) and the proposed extent of the Upper Jurassic discovery (after Solli in press).

predict oil migration and petroleum type in remote areas.

Ranking of undrilled prospects and model testing

Several prospects have been mapped between the large discoveries within the Tampen Spur, and the ranking sequence of drilling is dependent on a combination of several factors. Two wells were drilled within block 34/7 (licence 089) in 1992 and the petroleum population map that was constructed of the area was applied in the prospect ranking. Both wells were aimed at prospects that were considered to have a favourable position for migration in light of the model discussed previously.

The first well was spudded in July 1992 on an Upper Jurassic pinch-out trap in the northwestern corner of block 34/7 (well 34/7-20 in Fig 19). The petroleum population mapping suggested that the oil proven by three wells in the Upper Jurassic sandstone and the Brent Group in the Statfjord Nord structure and in the Brent Group in the Vigdis Vest segment were similar and could be grouped as a major oil population in the area. This suggested that there is, or has been, an active migration path between the two structures. based on seismic mapping, it was difficult to map and document a potential spill point from Statfjord Nord to Vigdis Vest at Brent Group level, and the Vigdis Vest accumulation was believed to have been sourced by spill from Statfjord Øst to the south rather than from Statfjord Nord to the north. The migration path required spill through a 15 m saddle point between the Statfjord Øst and Vigdis Vest and this was considered to be within the uncertainties of seismic mapping and depth conversion in the area. However, the conclusions from the geochemical mapping were convincing and a new oil-correlation study, including ten oils from the same area, was conducted by another laboratory. The results confirmed the similarities between the Statfjord Nord and Vigdis Vest oils and their dissimilarities with the other oil populations mapped in the area. This resulted in a seismic remapping of the northwestern area to see if it was possible to document a possible spill route between the two structures.

A combination of a sedimentological model and seismic mapping suggested that Upper Jurassic sands might possibly be present between the Statfjord Nord and Vigdis Vest structures and that these could have acted as active migration channels in the area. Because sandstones of Upper Jurassic age were missing on the structural highs, the possibilities of a stratigraphic trap seemed high. Unfortunately, only a few metres of Upper Jurassic sandstone were present at the well location. The Upper Jurassic sandstone was water-bearing , but had good oil shows in the upper part (EOM ranging from 5.4 to 16.4 mg/g rock) and the compositions of these shows were similar to the Statfjord Nord and Vigdis Vest oils. The most likely explanation for the lack of filling of this structure was a lack of proper seal to the stratigraphic trap. The penetrated sand most likely acted as an active petroleum carrier feeding the Vigdis Vest segment. The well was classified as dry with oil shows.

The second well was spudded in October 1992 on a stratigraphic play within the Upper Jurassic section in the western part of block 34/7 (well 34/7-21 in Fig. 19). A relatively thick Upper Jurassic wedge had been mapped in a down-flank position on the rotated footwall block and the sedimentological model suggested that it was likely that this wedge consisted of amalgamated sandstones, siltstones and shales (Solli in press). A horizontal seismic anomaly across the Upper Jurassic wedge represented a possible direct petroleum indicator. This seismic anomaly was later identified as an interference signal between effects due to structural dip and lithology changes. The structure was located between the Statfjord Øst structure and the Tordis field, and was considered to be located in a favourable position for petroleum migration. The petroleum mapping suggested that oil migrated through this area towards the Gullfaks Field in block 34/10 to the south of block 34/7. The well penetrated several sand intervals within the Upper Jurassic section and tested oil. The geochemical composition of the produced oil was similar to the geochemical signature of the oils from the Statfjord Øst structure and the Tordis Field. The prognosis of both sand distribution and petroleum population were in good agreement with the actual results from the well, which was classified as an oil discovery. This again demonstrated the ability to forecast petroleum migration and its utility as an exploration tool within mature petroleum provinces. In particular, confidence in the geochemical model tipped the balance in the assessment of risk with the prospect identifying a new target type in this area.

Conclusions

Based on a combination of conventional geological and geophysical data and a petroleum population map of the oils on the Tampen Spur, a refined migration model was constructed. The migration models proposed earlier were based largely on geological structure maps constructed from seismic and basin modelling. Neither of these methods can distinguish between a physically possible migration route and an active migration route through which the oil actually migrated. Because we have shown that in this area only minor portions of the potential

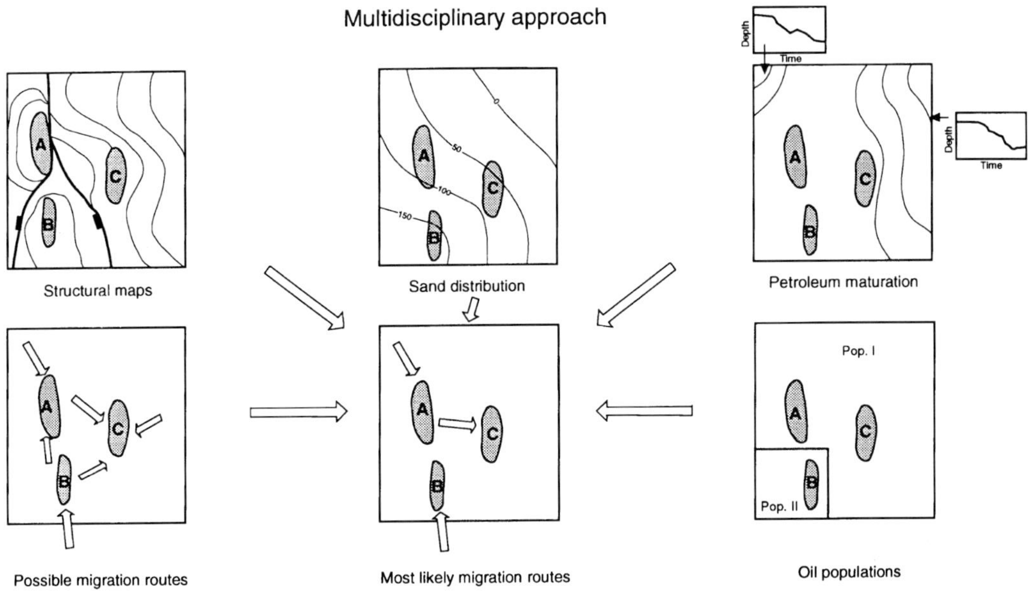

Fig. 20. Input parameters and maps used to evaluate a typical migration history.

carrier system typically carry petroleum (*c* <3% of the volume of the carrier, Larter & Horstad 1992), the distinction between possible and likely carrier routes becomes essential. Obviously, detailed reservoir geochemical information now represents an improvement and a necessary part of any future petroleum migration study.

The oils in the vicinity of block 34/7 have been differentiated into five major petroleum populations and it has been suggested that several of the structures have been filled from more than one direction. Usually, it is possible to distinguish between oils accumulated within different stratigraphic intervals in the same field. This study has demonstrated that filling and migration in a geological province are controlled by a series of complex processes and that the filling history of such structures within the Tampen Spur is more complicated than suggested earlier.

The migration model has been tested and verified by two exploration wells drilled in 1992. Some ability to forecast local-scale petroleum migration has been demonstrated. One of the wells was dry, but contained good oil shows from the same oil population as predicted. A lack of seal resulted in no accumulation and the well may have penetrated an active petroleum migration route. The second well was an oil discovery, again with the same oil type as predicted by the petroleum maps. It is stressed that both accumulations could potentially have received petroleum from several other directions, the oil populations mapping representing an important exploration tool when ranking the prospects in the area. We have also developed a series of procedures that should be followed when a migration model is constructed. It is vital to take all geological and geochemical data into account and apply fully integrated teamwork where both geologists and geochemists participate (Fig. 20). The geochemical dataset has to be of high and uniform quality and if different data subsets are used, several calibration samples should be rerun in every sample set. Likewise, if the model is updated by data from a new well, 5–15 relevant samples should be rerun in the same batch. The paper shows that early predictions of regional petroleum migration conduits can be obtained from analyses of lateral gradients in oilfields (Horstad *et al.* 1990), these models being updated and tested by new data. Reservoir geochemical gradients do appear to have predictive capabilities on a basinal scale. However, we emphasize that in other studies we have performed, complex sedimentology and tectonics may rupture simple interpretations and 'gradients' may not be at all obvious.

The authors wish to thank all the participants on licence PL 089 (Statoil, Deminex, Elf Aquitaine, Esso, Norsk Hydro and Idemitsu Oil Exploration) for agreeing to publish the paper and for their support of the geochemical work performed in the licence. The licence group in Saga Petroleum a.s. are thanked; their contributions, ideas and fruitful discussions have helped a lot. K. Bjørlykke at the University of Oslo is acknowledged for his support to the project and for helpful discussions. Special thanks are due to R. Knarud, H. C. Rønnevik, A. Schwartzbard, T. Solli and N. Dahl. L. Ravdal is acknowledged for drafting assistance, and the laboratory section in Saga Petroleum as is acknowledged for assistance. the paper benefited from reviews by T. Dodd (BP) and an anonymous geologist/geochemist.

References

DAHL, N. & SOLLI, T. 1993. The structural evolution of the Snorre Field and surrounding areas. *In*: PARKER, J. R. (ed.) *The Petroleum Geology of NW Europe: Proceedings of the 4th Conference.* Geological Society, London, 1156–1166.

DAVIS, J. C. 1986. *Statistics and Data Analysis in Geology.* John Wiley & Sons, New York.

ENGLAND, W. A. & MACKENZIE, A. S. 1989. Geochemistry and petroleum reservoirs. *Geologische Rundschau,* **78,** 214–237.

——, ——, MANN, D. M. & QUIGLEY, T. M. 1987. The movement and entrapment of petroleum fluids in the subsurface. *Journal of the Geological Society, London,* **144,** 327–347.

ERICHSEN, T., HELLE, M. & ROGNEBAKKE, A. 1987. Gullfaks. *In*: SPENCER, A. M. ET AL. (eds) *Geology of Norwegian Oil and Gas Fields.* Graham & Trotman, London, 273–286.

EYNON, G. 1981. Basin development and sedimentation in the Middle Jurassic of the northern North Sea. *In*: ILLING, L. V. & HOBSON, G. D. (eds) *Petroleum Geology of the Continental Shelf of North Western Europe.* Heyden & Sons, London, 196–204.

GRANTHAM, P. J., POSTHUMA, J. & DE GROOT, K. 1980. Variations and significance of the C27, and C28 triterpane content of a North Sea core and various North Sea crude oils. *In*: DOUGLAS, A. G. & MAXWELL, J. R. (eds) *Advances in Organic Geochemistry 1979.* Pergamon Press, Oxford, 29–38.

GRAUE, E., HELLAND-HANSEN, W., JOHNSON, J., LØMO, L., NØTTVEDT, A., RØNNING, K., RYSETH, A. & STEEL, R. 1987. Advance and retreat of Brent Delta System, Norwegian North Sea. *In*: BROOKS, J. & GLENNIE, K. (eds) *Petroleum Geology of North West Europe.* Graham & Trotman, London. 915–938.

GUSSOW, W. C. 1954. Differential entrapment of oil and gas – a fundamental principle. *American Association of Petroleum Geologists,* Bulletin, **38,** 816–853.

HOLLANDER, N. B. 1987. Snorre. *In*: SPENCER, A. M. ET AL. (eds) *Geology of Norwegian Oil and Gas Fields.* Graham & Trotman, London, 307–318.

HORSTAD, L., LARTER, S. R., DYPVIK, H., AAGAARD, P., BJØRNVIK, A. M., JOHANSEN, P. E. & ERIKSEN, S. E. 1990. Degradation and maturity controls on oil field petroleum heterogeneity in the Gullfaks Field, Norwegian North Sea. *Organic Geochemistry,* **16,** 497–510.

—, — & MILLS, N. 1992. A quantitative model of biological petroleum degradation within the Brent Group reservoir in the Gullfaks Field, Norwegian North Sea. *Organic Geochemistry,* **19** (1–3), 107–117.

KARLSEN, D. A. & LARTER, S. R. 1991. Analysis of petroleum fractions by TLC–FID: applications to petroleum reservoir description. *Organic Geochemistry,* **17** (5), 603–617.

KARLSSON, W. 1986. The Snorre, Statfjord and Gullfaks oilfields and the habitat of hydrocarbons on the Tampen Spur, offshore Norway. *In*: SPENCER, A. M. ET AL. (eds) *Habitat of Hydrocarbons on the Norwegian Continental Shelf.* Norwegian Petroleum Society, Graham & Trotman, London, 77–85.

KIRK, R. H. 1980. Statfjord Field: A North Sea giant. *In*: HALBOUTY, M. T. (ed) *Giant Oil and Gas Fields of the Decade: 1968–1978.* American Association of Petroleum Geologists, Memoir, **30,** 95–116.

LARTER, S. R. & APLIN, A. C. 1995. Reservoir geochemistry: methods, applications and opportunities. *This volume.*

—— & HORSTAD, I. 1992. Migration of hydrocarbons into Brent Group reservoirs: some observations from the Gullfaks Field, Tampen Spur area North Sea. *In*: MORTON, A. C. ET AL. (eds) *Geology of the Brent Group.* Geological Society, London, Special Publication, **61,** 441–452.

—— & MILLS, N. 1991. Phase controlled molecular fractionations in migrating petroleum charges. *In*: ENGLAND, W. A. & FLEET, A. J. (eds) *Petroleum Migration* Geological Society, London, Special Publication, **59,** 137–147.

——, BJØRLYKKE, K. O., KARLSEN, D. A., NEDKVITNE, T., EGLINTON, T., JOHANSEN, P. E. & LEYTHAEUSER, D. 1990. Determination of petroleum accumulation histories: Examples from the Ula field, Central Graben, Norwegian N. Sea. *In*: BULLER, A. (ed.) *N. Sea Oil and Gas Fields II.* Graham & Trotman, London, 319–330.

LEITH, T. L., KAARSTAD, I., CONNAN, J., PIERRON, J. & CAILLET, G. 1993. Recognition of caprock leakage in the Snorre Field, Norwegian North Sea. *Marine and Petroleum Geology,* **10,** 29–41.

LEYTHAEUSER, D., SCHAEFFER, R. G. & RADKE, M. 1988. Geochemical effects of primary migration of petroleum in Kimmeridge source rocks from Brae field area, North Sea. I: Gross composition of C_{15+} soluble organic matter and molecular composition of C_{15+}-saturated hydrocarbons. *Geochimica et Cosmochimica Acta,* **52,** 701–713.

LI, M., LARTER, S. R., STODDART, D. & BJORØY, M. 1992. Liquid chromatographic separation schemes for pyrrole and pyridine nitrogen aromatic heterocycle fractions from crude oils suitable for rapid characterisation of geochemical samples. *Analytical Chemistry,* **64,** 1337–1344.

MACKENZIE, A. S., PRICE, I. & LEYTHAEUSER, D. 1987. The expulsion of petroleum from Kimmeridge clay source rocks in the area of the Brae Oilfield, UK continental shelf. *In*: BROOKS, J. & GLENNIE, K. (eds) *Petroleum Geology of North West Europe.* Graham & Trotman, London, 865–877.

MOLDOWAN, J. M., ARNOLD, E. & CLARDY, J. 1984. Structure proof and significance of steroisomeric 28,30-bisnorhopane in petroleum and petroleum source rocks. *Geochimica et Cosmochimica Acta*, **48,** 1651–1661.

NYBAKKEN, S. 1991. Sealing fault traps – an exploration concept in a mature petroleum province: Tampen Spur, northern North Sea. *First Break*, **9** (5), 209–222.

NYSTUEN, J. P. & F & FÄLT, L. M. 1995. Upper Triassic – Lower Jurassic reservoir rocks in the Tampen Spur area, Norwegian North Sea. *In*: HANSLEIN, S. ET AL. (eds) *Petroleum Exploration and Exploitation in Norway*. Norwegian Petroleum Society, Elsevier, Amsterdam, in press.

OLAUSSEN, S., BECK, L., FÄLT, L. M., GRAUE, E., JACOBSEN, K. G., MALM, O. A. & SOUTH, D. 1993. Gullfaks Field – Norway East Shetland Basin, Northern North Sea. *In*: FOSTER, N. H. & BEAUMONT, E. A. (eds) *Structural Traps VI* American Association of Petroleum Geologists, 55–83.

ROBERTS, J. D., MATHIESON, A. S. & HAMPSON, J. M. 1987. Statfjord. *In*: SPENCER ET AL. (eds) *Geology of Norwegian Oil and Gas Fields*. Graham & Trotman, London, 319–340.

SÆLAND, G. T. & SIMPSON, G. S. 1981. *Interpretation of 3-D Data in Delineation of a Subunconformity Trap in Block 34/10, Norwegian North Sea*. American Association of Petroleum Geologists, Memoir, **32,** 217–235.

SEIFERT, W. K. & MOLDOWAN, J. M. 1978. Application of steranes, terpanes and monoaromatics to the maturation, migration and source of crude oils. *Geochimica et Cosmochimica Acta*, **42,** 77–95.

STODDART, D. P., HALL, P. B., LARTER, S. R., BRASHER, J., LI, M. & BJORØY, M. 1995. The reservoir geochemistry of the Eldfisk Field, Norwegian North Sea. *This Volume.*

THOMAS, B. M., MØLLER-PEDERSEN, P., WHITTAKER, M. F. & SHAW, N. D. 1985. Organic facies and hydrocarbon distribution in the Norwegian North Sea. *In*. THOMAS, B. M. ET AL. (eds) *Petroleum Geochemistry in Exploration of the Norwegian Shelf*. Graham & Trotman, London, 3–26.

WILHELMS, A. & LARTER, S. R. 1995. Overview of the geochemistry of some tar mats from the North Sea and USA: implications for tar-mat origin. *This volume.*

Modelling density-driven mixing rates in petroleum reservoirs on geological time-scales, with application to the detection of barriers in the Forties Field (UKCS)

W. A. ENGLAND[1,2], A. H. MUGGERIDGE[1,3], P. J. CLIFFORD[1,4] & Z. TANG[1]

[1] *BP Exploration Operating Company Ltd., BPX Technology Provision, Chertsey Road, Sunbury-on-Thames, Middlesex, TW16 7LN, UK*

[2] *Present address: BP–Statoil Alliance, N-7004 Trondheim, Norway*

[3] *Present Address: SSI, Egham, Surrey, TW20 8RY, UK*

[4] *Present Address: BP Exploration, Anchorage, Alaska, USA*

Abstract: This paper discusses how to deduce the strength of intra-field barriers in reservoirs which have significant present-day differences in reservoir fluid density. The method combines a knowledge of the geological time available since filling, with estimates of the rate of density-driven mixing. If the observed density differences are inconsistent with a barrier-free reservoir model, we show how limits can be set on the strength of the barrier needed to account for any observed lack of mixing. A case history from the Forties Field is presented, illustrating these principles. It is shown that a transmissibility barrier of at least 1% must be included to account for the present-day lack in density equilibration between Forties and SE Forties. The results are validated against pressure decline data since production start-up. Two methods of estimating density-driven mixing rates are given. One is based on an analytical formula applicable to a rectangular geometry, and the other uses an unmodified black oil simulator. One significant result is the weak dependence of mixing rates on vertical permeability over a wide range of values, for many typical reservoir configurations. This implies that in many geological circumstances, subsurface density differences (coexisting at the same depth) can be used to infer the presence of intra-reservoir barriers.

Variations in hydrocarbon density and composition between wells are observed in many oil reservoirs under appraisal. In this paper, we discuss how to exploit these differences to locate and quantify the nature of internal flow barriers which can hinder the production of oil and gas. For example, Fig. 1 shows the observed subsurface reservoir fluid densities of samples taken from the Forties Field, (England 1990). Plotting the data against depth reveals two distinct sets of points: those from the main field and those from South East Forties. At any given depth, there is a difference in density between the fluids in South East Forties, and the main part of the field. This is mechanically unstable with respect to density overturning.

This paper discusses the general problem of how the rate of this process may be estimated on a geological time-scale and used to set limits on the type of barriers present in the field (assuming that the time available since filling is known). This information is valuable to engineers planning the positions of production and water injection wells, as any unforeseen barriers may seriously reduce the value of a well.

Variations in fluid properties result from the way in which the reservoir originally filled and the time-scale of mixing since that time, as proposed by England *et al.* (1987). The variations in composition are often inherited from one-sided migration into a structure, as in the case of Forties, where the most mature source-rock kitchen is off-set to one side of the reservoir. In the presence of barriers to complete mixing, the most mature petroleum is not able to mix completely with the fluids which arrived earlier, leading to a higher gas/oil ratio (GOR), more mature fluids being found on the side of the reservoir nearest to the fill point. In Forties, this GOR difference is responsible for the density difference between the two parts of the field (Fig. 1).

Mixing rates

Reservoir fluids tend to mix via molecular diffusion and density-driven convection towards a state of mechanical and compositional equilibrium. In this state, assuming that thermal convection is absent, fluid composition and density are constant across the field at any given depth (thermal convection in oil reservoirs with normal vertical thermal

From CUBITT, J. M. & ENGLAND, W. A. (eds), 1995, *The Geochemistry of Reservoirs*, Geological Society Special Publication No. 86, pp. 185–201.

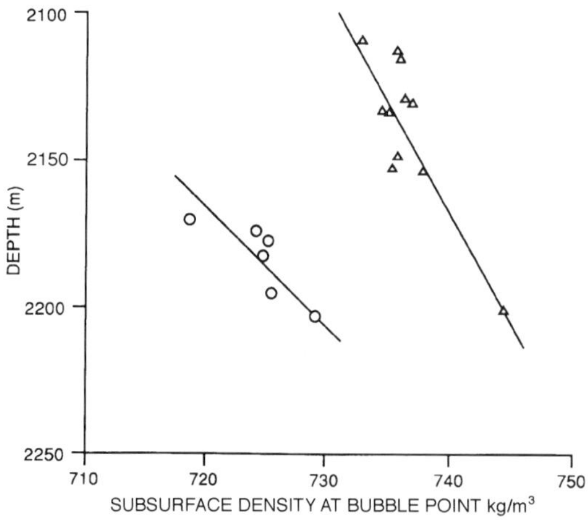

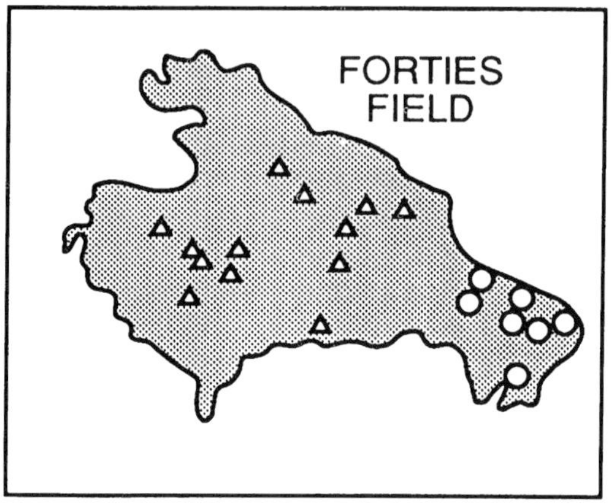

Fig. 1. Subsurface reservoir fluid densities of samples taken from the Forties Field (England & Mackenzie 1989). Samples from SE Forties and the main part of the field have different subsurface densities owing to the way in which the field filled.

gradients is unlikely as the critical Rayleigh number is not usually exceeded (England *et al.* 1987). However there is no critical Rayleigh number for reservoirs with lateral temperature gradients and thermal convection may then become important (Jacqmin 1990).

In the presence of barriers to density-driven mixing, fluid homogenization is often observed to be incomplete. Molecular diffusion is not generally effective at causing fieldwide mixing (England *et al.* 1987) and is mainly ignored in the work described in this paper, which concentrates on calculating the rate at which bulk convection can eliminate density differences inherited from reservoir filling. It should be remembered that molecular diffusion is an important auxiliary mixing process, particularly over the shorter distances associated with the vertical dimension of the reservoir, i.e. within individual wells.

The time-scale of reservoir mixing is a function of the fluid and reservoir properties, such as bed dimensions, permeability and viscosity. In particular, the rate of mixing due to density-driven convection is altered by large-scale variations in permeability across the reservoir. For example, the differences in oil densities observed between the Forties and South East Forties suggests that there must be some sort of mechanical barrier to flow between the two parts of the reservoir, although the precise nature of the barrier is hard to define without using modelling methods.

Estimating fieldwide connectivity

The objective is to be able to estimate fieldwide connectivity (including the 'strength' of barriers such as faults etc.) from a knowledge of the timing of reservoir filling and the variations in fluid properties across it. England & Mackenzie (1989) suggested simple analytical methods for estimating the time-scales of mixing as a result of thermal convection, molecular diffusion and density-driven convection. However, these were purely order-of-magnitude estimates for simplified reservoir descriptions. Numerical simulation is required in order to validate these calculations and to model more complicated situations. For example, suppose that a reservoir under appraisal is initially supposed to be barrier-free, but has significant fluid density differences. Modelling of mixing rates may show that most density differences would be eliminated on a time-scale of 1 Ma. However, if local geology shows that most of the oil has been in place for 10 Ma, the barrier-free reservoir engineering model needs to be modified in some way, usually by the inclusion of an intra-field barrier.

However, reservoir mixing occurs on a time-scale of hundreds of thousands of years whilst conventional reservoir simulation programs were originally designed to model oil production over a period of tens of years. Indeed, some of the physical effects important in reservoir mixing, such as thermal gradients and molecular diffusion, cannot be represented in a conventional black oil reservoir simulation model. Jacqmin (1990) developed a program specifically to investigate thermal convection and its interaction with gravitational segregation, which applies to fields with significant lateral thermal gradients. In contrast, we have modelled convection due to inherited density differences and have ignored gravitational segregation.

The aim of this work is to determine whether conventional reservoir simulation programs can be used to model density-driven mixing on geological time-scales; in addition, an improved analytical solution to the problem of density-driven convection is presented. This was used to benchmark the results of numerical simulation. Simulation was used to determine how the time-scale predicted by the analytical solution was altered by additional factors such as molecular diffusion and the presence of shales. Finally, the methods were applied to the Forties Field case study.

Analytical solutions to density-driven mixing rates in idealized (rectangular) geometries.

The analytical solution presented here was originally derived by Gardner *et al.* (1962) to model segregated flow in a miscible displacement along a thin tilted bed. The analysis is equally appropriate to the problem of density-driven mixing on a geological time-scale. Considering a vertical cross-section through a hypothetical rectangular reservoir, imagine that the reservoir has filled with oil from left to right and that filling was faster than density-driven mixing, leading to a density difference between the two parts of the field. This results in the left-hand half of the reservoir being filled with less dense oil and the right-hand half with denser oil (as less dense oils tend to be produced as the source rock matures and GOR and API density increase). The analysis examines the movement of the interface between the two oils as a function of time. Of course, natural filling processes inevitably lead to a more complex distribution of fluids, but this calculation establishes an order of magnitude for density-driven mixing that can be rapidly calculated (i.e. without resorting to numerical simulation).

Analytical solution (Gardner et al. 1962)

Figure 2 shows the process of density-driven convection in the hypothetical rectangular reservoir. Starting with the denser fluid in the right-hand part, as time progresses the interface remains linear and gradually moves by tilting towards its equilibrium, horizontal position. This movement is described by the parameter $(2x/h)$, which increases from zero as the interface moves towards the horizontal. Figure 2 defines $(2x/h)$, and

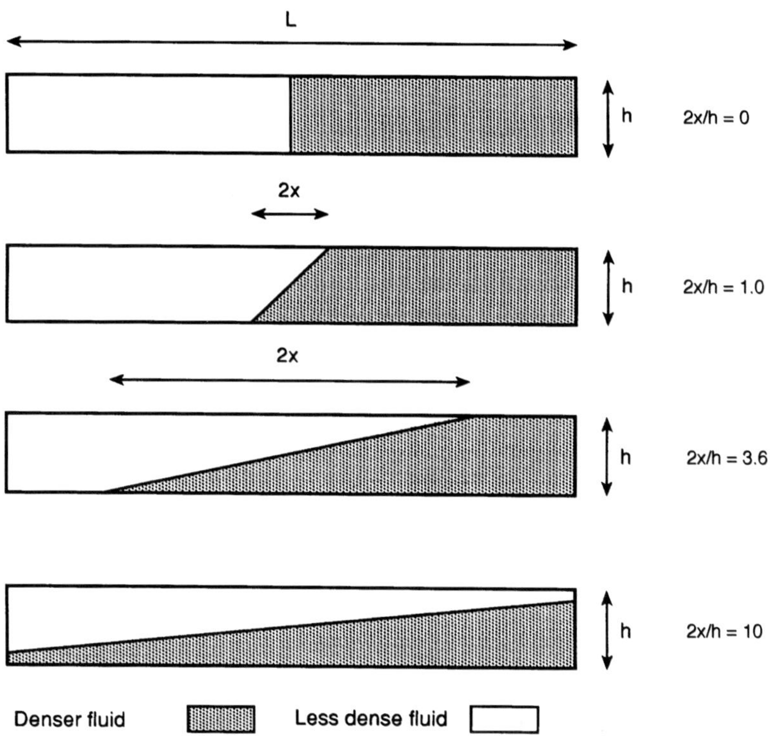

Fig. 2. Terms used to define the movement of an interface between fluids of different densities, after Gardner *et al.* (1962).

shows how it increases for several stages of the process.

Gardner *et al.* (1962) presented analytical solutions for this situation for 'short times', 'long times' and a formula which blended the two and which was validated against bead-pack experiments. The blended formula is convenient to use in practice, because there is no need to determine whether one is in the 'short time' or 'long time' regime:

$$\left(\frac{2x}{h}\right)^2 = \left(\frac{16}{3}\right) F^2 \frac{K_H}{K_V} \frac{(t/t_0)^2}{(1 + t/t_0)} \quad (1)$$

where

$$t_0 = \frac{4}{3} \frac{\phi h \mu F}{K_V \, g \Delta \rho}$$

F is a complex function of the viscosities (μ) of the two different fluids involved. However, in the rest of this report we assume that both the fluids have the same viscosity and F is therefore close to 1.0. K_H and K_V are the horizontal and vertical permeabilities of the reservoir rock and t_0 is a 'characteristic time' which determines the rate at which mixing takes place. ϕ is porosity, g is the acceleration due to gravity and Δρ is the density difference between the two fluids involved.

Equation 1 was evaluated using reasonable values of the parameters involved, similar to those used by England & Mackenzie (1989), applicable to a black oil reservoir. the results in Fig. 3 show how the predicted degree flow of movement of the interface, measured by $(2x/h)$, varies with time. The input parameters shown in Table 1 represent the 'base cause'. The effect of variations in vertical permeability was calculated for a number of different values of K_V, between 100 mD and 0.01 mD, to explore the effect of variations in overall reservoir permeability. The effect of variations in K_V was proposed to be an important control on density-driven mixing times (England & Mackenzie 1989).

In order to monitor the degree of mixing, the average fluid densities of imaginary wells at the extreme left- and right-hand sides of the reservoir were considered to indicate the degree of mixing. A simple calculation showed that a value of the mixing parameter $(2x/h)$ equal to 100 corresponds to a reduction of the original (left–right) density difference to 20% of its initial value. In other words, a value of $(2x/h)$ equal to 100 represents near-complete mixing.

Figure 3 shows that for the base case and values of K_V between 100 mD and 1.0 mD, mixing times are essentially identical, *c.* 0.2 Ma (for $(2x/h)$ = 100). Even for values of K_V as small as 0.01 mD, the mixing time-scale is still less than 1 Ma. in other words, reducing the vertical permeability from 100 mD to 1.0 mD has no significant effect on the time-scale for density-driven mixing. This apparently counter-intuitive result reflects the negligible component of vertical flow in the long time regime.

These results are more simply represented by the 'long time' formula given by Gardner *et al.* (1962):

$$\left(\frac{2x}{h}\right)^2 = \frac{4K_H \, g\Delta\rho F t}{\phi \mu h} \quad (2)$$

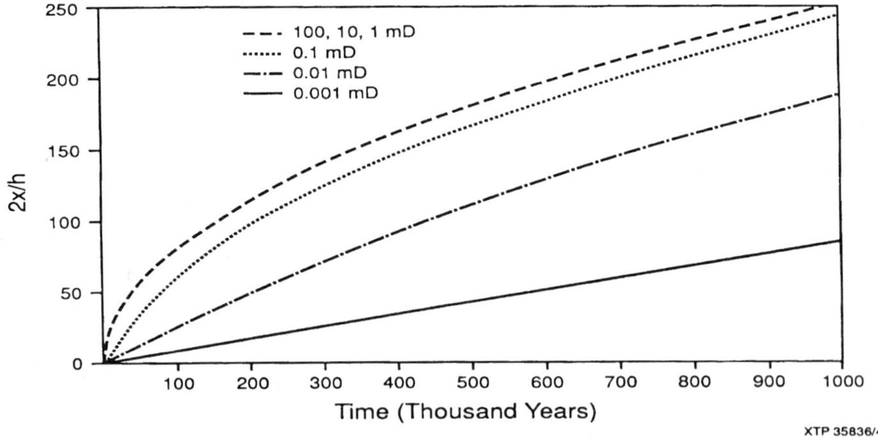

Fig. 3. Effect of changes in vertical permeability (Equation 1). Horizontal permeability fixed at 100 mD.

Table 1. *Parameters used to describe base case rectangular reservoir model*

Parameter	Value	Units	SI Equivalent
Model length (l)	2000	m	
Model height (H)	100	m	
Fluid viscosity (μ)	1	cp	(1×10^{-3} Pa s)
Rock porosity (ϕ)	0 2		
Vertical permeability (K_V)	100	mD	(10^{-13} m^2)
Horizontal permeability (K_H)	100	mD	(10^{-13} m^2)
Fluid density difference ($\Delta\rho$)	10	kg m^{-3}	
Viscosity function (F)	1.0		
Acceleration due to gravity (g)	9.81	m^2 s^{-2}	
Oil saturation (S)	1.0		

which is valid when $(4x^2K_V/h^2K_H)$ is much greater than 1 (note the slight change in conditions compared to Gardner *et al.* (1962)). If the mixing parameter $(2x/h)$ equals 100 and K_H has a value of 100 mD, this implies that Equation 2 is valid when K_V is much greater than 0.01 mD, which can be verified from Fig. 3. For example, Equation 2 predicts that 164 000 years are needed to attain a value of $(2x/h)$ equal to 100, which agrees with the value calculated from Equation 1 (see Fig. 3). In the following section, Equation 2 will be compared with the order-of-magnitude estimate for density-driven mixing which was given in England & Mackenzie (1989).

Comparison with estimates made by England & Mackenzie (1989)

England & Mackenzie (1989) gave the following equations to make order-of-magnitude estimates of the rate at which density-driven convection occurs in reservoirs. The equation was based on Darcy's law and an estimate of the pressure difference driving density-driven mixing. Using the definitions given previously and listed in Table 1:

$$\tau_{DENS} = \frac{\phi L^2 \mu}{S < K > g \Delta\rho H} \qquad (3)$$

where

$$\frac{(L + H)}{<K>} = \frac{L}{K_H} + \frac{H}{K_V}.$$

τ_{DENS} is the order-of-magnitude estimate of the density-driven mixing process as defined in England & Mackenzie (1989), S is the oil saturation and $<K>$ represents a harmonic average of a hypothetical mixing pathway involving vertical flow over the height, H, of the reservoir shown in Fig. 2, and the horizontal flow along its length, L.

Table 2 compares the mixing times calculated using the analytical solution of Gardner *et al.* (1962)(Equation 1), with the order-of-magnitude estimates (Equation 3). In the case of Gardner *et al.*'s (1962) formula, Equation 1 was used with $(2x/h)$ set to 100 to represent 'good mixing'. The data are shown as a log–log plot in Fig. 4. Considering the case where horizontal and vertical

Table 2. *Comparison of density-driven mixing time-scales*

		Estimated time-scale (thousands of years)	
K_V (mD)	K_H (mD)	Gardner *et al.* (1962)*	England & Mackenzie (1989)
100	100	160	26
10	100	160	37
1	100	160	149
0.1	100	220	1270
0.01	100	450	12 500

* Mixing criterion: $2x/h = 100$

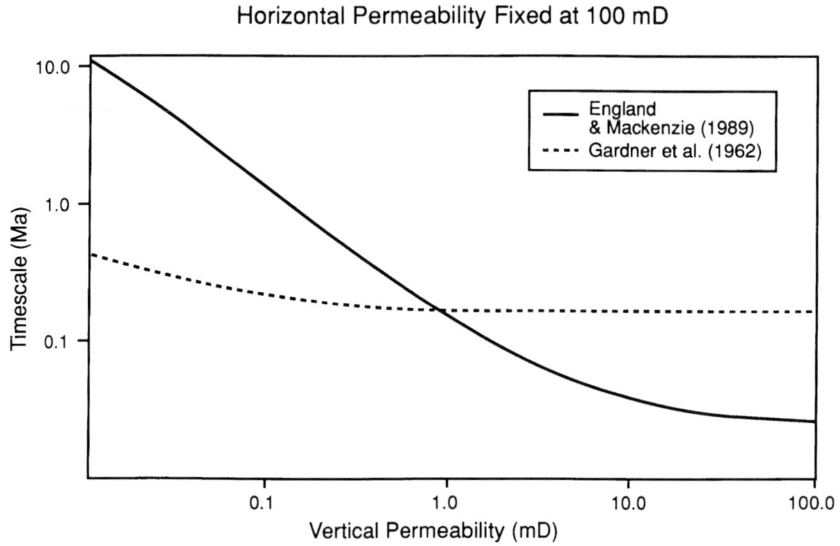

Fig. 4. Comparison of density-driven mixing time-scales predicted by Gardner *et al.* (1962) and England & Mackenzie (1989).

permeabilities are equal, the order-of-magnitude estimate of England & Mackenzie (1989) is reasonably close to the analytical solution. Choosing a different $(2x/h)$ criterion for what actually constitutes a 'well mixed reservoir' would shift the times calculated using Gardner *et al.*'s (1962) equation by a factor of two or three.

It is clear, however, that the order-of-magnitude estimates of England & Mackenzie (1989) consistently overestimate the effect of low vertical permeabilities because of a simplified derivation which did not account for the different cross-sections of the vertical and horizontal flowpaths. The difference between the absolute values of the two equations at 100 mD is not surprising given the simplifications made in the derivation of Equation 3, which was intended as an order-of-magnitude estimate only.

In the case of a reservoir with $L = 2000$ m and $H = 100$ m, an interface position of $(2x/h) = 100$ corresponds to $(2x/h) = 5L/H$. Substituting this value of $(2x/h)$ into the 'long time' formula (Equation 2) and rearranging gives:

$$t = \frac{25}{4}\frac{\phi L^2 \mu}{K_H g \Delta\rho H}. \qquad (4)$$

This has a similar form to the order-of-magnitude estimate of England & Mackenzie (1989) (Equation 3). The main difference is the lack of dependence on K_V, at least in the 'long time' regime where K_V is much greater than 0.01 mD (which is the range for which Table 2 was calculated).

In conclusion, Gardner *et al.*'s analytical equation for the movement of the interface between two immiscible fluids of different densities has been investigated for a hypothetical rectangular reservoir. The calculated rates for the cases shown in Table 1 and Fig. 3 were fast on a geological time-scale. A significant result was the relatively weak effect that vertical permeability has in determining the rate of movement of the interface.

Previous work, (England & Mackenzie 1989) using order-of-magnitude estimates had correctly estimated the time-scales for nearly isotropic reservoirs, but overestimated the effect of low vertical permeabilities in slowing the movement of the interface between the two fluids.

As will be discussed later, this result has a great effect on the correct interpretation of fluid density data, such as that from Forties and SE Forties (Fig. 1). Using the order-of-magnitude approach of England & Mackenzie (1989), it would have been difficult to say whether the lack of mixing of the fluids in the two regions of Forties was due to (a) a sealing barrier or reservoir pinch out, or (b) a general worsening of reservoir quality, and consequent large decrease in vertical permeability due to a higher density of stochastic shales and decreasing sand-body heterogeneity. However, as discussed later, because the effect of changes in vertical permeability is weaker then previously estimated, the chance of distinguishing (a) and (b) is greatly improved.

Numerical solutions in a rectangular geometry

Method of setting up problem for a reservoir simulator

This model represents a simple rectangular, vertical cross-section through a reservoir. The reservoir has filled from the left and this is represented by two oils of different density filling different halves of the reservoir, with a vertical interface between them. The heavier oil fills the right- hand part of the reservoir. For simplicity, the oils are assumed to be incompressible and there is no mobile water within the reservoir. The ECLIPSE reservoir simulation package (Intera Information Technologies Ltd, Highlands Farm, Henley on Thames, Oxon RG9 4PS, UK) was used for this study, but there is no reason to suppose that other programs would not work equally well.

The 'API tracking' option has to be used in order to be able to simulate the flow and mixing of different types of oils. The principal feature of this option is that it uses an Implicit Pressure Explicit Saturation (IMPES) method of solution. Most runs with ECLIPSE use the fully implicit method of solution. As a result, particular care was taken to ensure that the maximum possible time-step did not make the solution unstable. At the same time, we had to ensure that the time-step was not so small that the simulation time was excessive. This meant that the maximum time-step had to be scaled up or down when the grid was refined or the absolute permeability was changed.

Results for rectangular grid

Using a Cray computer, the simulation described above took less than 500 seconds of CPU time. The base case was rerun with different vertical permeabilities, the idea being to validate the ECLIPSE black oil simulations against the analytical solution of Gardner *et al.* (1962). The results are summarized in Table 3, which also includes the results of high resolution MISTRESS simulations described in the next section.

Agreement is seen to be satisfactory, considering the limitations of finite grid and time-steps discussed previously. The black oil simulations gave values of the interface position parameter ($2x/h$) which were about 70% of that predicted by the analytical solution given in Equation 1. The effect of reducing vertical permeability was correctly handled, as seen by comparing runs with K_V of 100 mD and 1.5 mD which gave essentially identical results.

High resolution numerical solution

Having confirmed that the time-scales predicted by analysis and by a conventional black oil simulation program agreed, it was important to investigate how other physical effects, such as diffusion and rock property heterogeneity, might alter the time-scales for density overturning. This required a simulation program that was capable of modelling physical diffusion. In addition, a relatively fine grid model was necessary to ensure that numerical diffusion was insignificant when compared with physical diffusion. Numerical diffusion is the process by which inadequately fine gridding leads to the spreading out of (physically) sharp interfaces.

The high resolution program MISTRESS, described by Christie (1989), was used for these simulations. This uses an IMPES finite difference formulation incorporating Flux Corrected Transport to ensure that the program is both fast and accurate. It was originally designed to model the development of viscous fingers in miscible displacements on the scale of a grid block in a conventional simulation program and hence does not have some of the functionality of a conventional black oil reservoir simulator. This means that the code is faster and finer grids can be used to model the details of the fluid flow.

Table 3. *Comparison of numerical simulations of density-driven mixing with the analytical solution of Gardner* et al. *(1962)*

$K_V = K_H = 100$ mD (isotropic reservoir)

Elapsed time (years)	MISTRESS ($2x/h$)	ECLIPSE ($2x/h$)	Analytical ($2x/h$)
2500	10.6	–	12
5000	14.9	12	17
7500	17.9	–	21
10 000	20	17.7	25

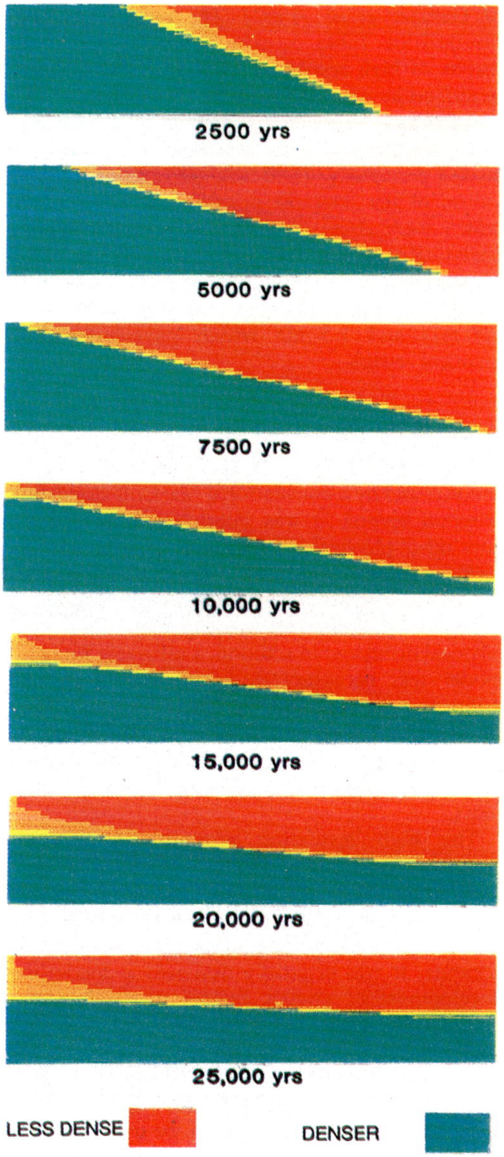

Fig. 5. Base case simulation using high resolution MISTRESS numerical simulator.

As a basic check, simulations using MISTRESS were performed using the base case input parameters shown in Table 1. The graphical results are shown (Fig. 5) and the positions of the interface, $(2x/h)$, listed in Table 3. As expected, agreement with the analytical solution (Equation 1) is slightly better than for the simpler black oil simulator runs. This is probably due to the more advanced time-step selection algorithm used by MISTRESS and the finer grid. There was also a lessening in the blurring of the interface between the two fluids in the high resolution runs.

Inclusion of stochastic shales

A stochastic shale distribution with an effective vertical permeability of 0.15 times the horizontal permeability was introduced. The shale distribution was constrained so that at least one grid block (on a 60 x 30 grid) separated each shale. The shales were represented in the simulation program by transmissibility barriers. The fluid and rock properties were otherwise unchanged from the base case.

Figure 6 shows the distribution of the two oils in the shaley reservoir after 20 000 years. There was little difference when compared with the base case. The interface shape is slightly perturbed but the rate of overturning has not been significantly changed. The analytical solution of Gardner *et al.* (1962) predicts that the movement of the interface will not be altered (with respect to the isotropic base case) for this ratio of vertical to horizontal permeabilities. This suggests that, at least for this example, the effective vertical permeability correctly represents the impact of shale distribution on this type of flow.

Inclusion of molecular diffusion

A typical value for physical diffusion was taken to be $1.7 \times 10^{-10}\,m^2\,s^{-1}$. Figure 7 shows the oil distribution after 20 000 years when the effects of molecular diffusion are included in the simulation. A comparison with the distribution for the same time (Fig. 5) without diffusion shows that the originally sharp interface has been spread out so that it covers approximately one-third of the reservoir thickness.

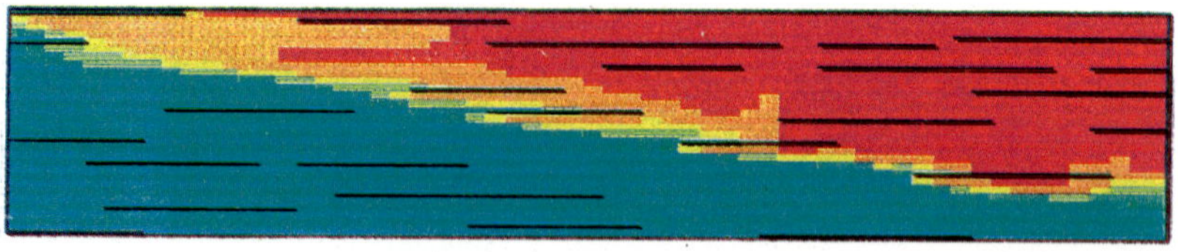

Fig. 6. MISTRESS simulation in a reservoir with random shales, after 20 000 years elapsed time. See Fig. 5 for key.

Fig. 7. MISTRESS simulation with molecular diffusion included,, after 20 000 years elapsed time. See Fig. 5 for key.

The rate of overturning, which is driven by the density differences between the two fluids, has been slowed because of the mixing between the oils and the resulting reduction in the density gradient. However, the analytical solution still gives a reasonable estimate of the time-scale for gravitational overturning.

Graduated oil properties

A real reservoir is unlikely to start off with two oils of different properties separated by a sharp interface because source rocks mature gradually over a long period of time so reservoir filling is a continuous process. Similarly, it can be argued that there will be some mixing even while filling continues. A more realistic representation of the initial conditions allows the oil density to vary linearly with distance in the direction of filling. Figure 8 shows the density distribution resulting from these initial conditions after 20 000 years. The fluid and rock properties are unchanged from the base case.

The smaller density gradient has reduced the rate of overturning so that after 20 000 years there is still a significant difference in densities across the length of the reservoir. In the base case, the oil distribution was virtually constant across the reservoir after this period of time. This suggests that while the analytical solution of Gardner *et al.* (1962) can give reasonable order-of-magnitude estimates, numerical simulation is required to investigate the effects of initial conditions. A good representation of the initial conditions in turn requires a reasonable migration modelling capability. Alternatively, present-day density distributions can be used to provide a first estimate.

These results show that well written and carefully set up black oil simulators can successfully model fluid mixing on a geological time-scale. The analytical solution of Gardner *et al.* (1962) gives improved results to that of England & Mackenzie (1989) especially in anisotropic reservoirs.

Application to the Forties Field

Setting up the model on the simulator

The ECLIPSE black oil simulator was used for this part of the study because the full functionality of a commercial program was required to input complex engineering data from the Forties Field.

The different densities between the fluids in Forties and SE Forties tend to cause fluid overturning, i.e. less dense fluid in the SE rises to the top of the denser fluid in the main reservoir and the denser fluid will move into the base of the lighter oil in SE Forties (Fig. 1). The principal fluid movement, therefore, will be between the main part and the SE of the reservoir. Consequently, the x-axis of our simulation was aligned to the principal flow direction (Fig. 9). Because the study focuses on the relative movement of the two fluids between the two parts of the field, a reservoir grid was chosen which covers most of SE Forties and just the adjacent part of the main field.

The x-direction grid was chosen with a gridblock size which varies across the model, with a finer resolution close to the boundary between SE Forties and the main part of the field, (Fig. 9). This ensured good resolution in the region of interest,

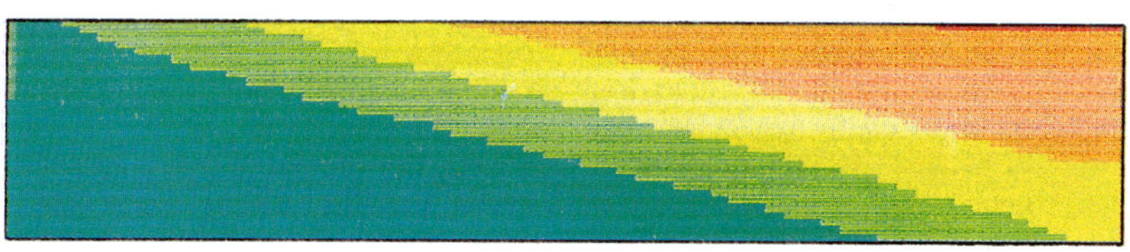

Fig. 8. MISTRESS simulation with an initial graduated density distribution, after 20 000 years elapsed time. See Fig. 5 for key.

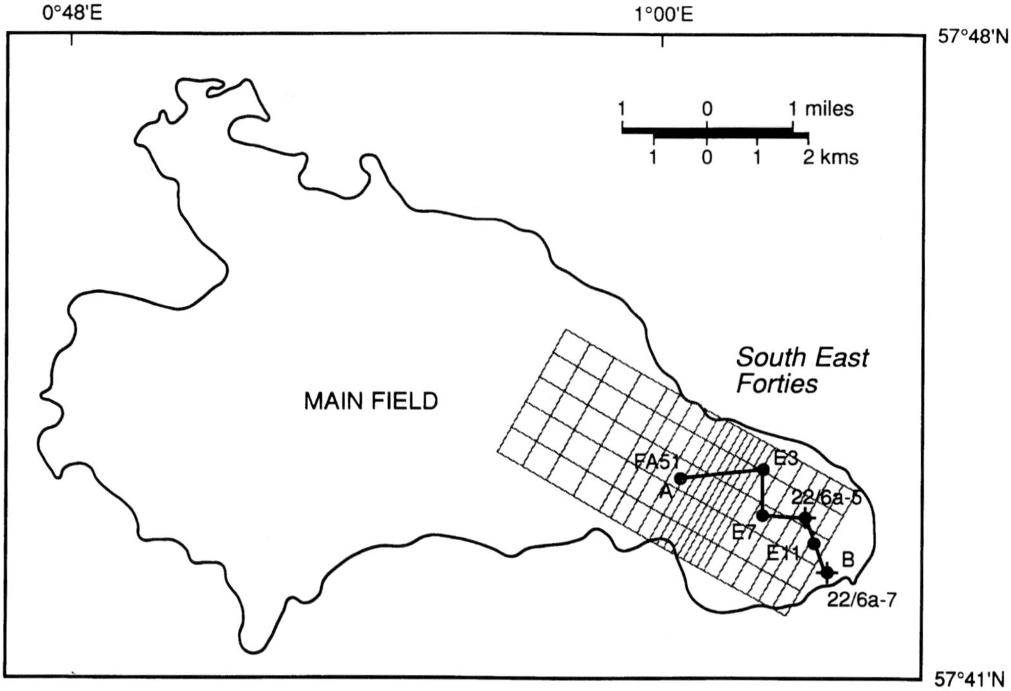

Fig. 9. Plan view of grid used to model density-driven mixing in the Forties Field.

while also minimizing computing time. Because the major flow is in the *x–z* plane, only coarse grid blocks in the *y*-direction were needed to cover the major part of SE Forties. Five grid blocks of 1992 ft were used to account for the level of permeability variation involved.

Vertical subdivision

The vertical grid-block direction fits the seven formation layers of the reservoir used to model production from Forties. At the beginning of this study, seven grids in the vertical direction were used to correspond with the formation layers. However, early runs showed that this gave unsatisfactory results due to inadequate vertical resolution. We felt that this problem was caused by the nature of natural convective mixing, which may demand an accurate definition of the boundary between regions of counterflow. We therefore tried increasing the number of vertical grid blocks, i.e. each of the formation layers was equally subdivided into two or three to form 14- and 21-layer models.

The 14- and 21-layer models give similar results, and both were much better than the seven-layer model. The computing time for the 21-layer model is considerably higher than that of the 14-layer model (the 21-layer model is three times slower than the 14-layer model), therefore a 14-layer model was selected based on the consideration of resolution and computing time.

In summary, a three-dimensional (3D) model with a total of $22 \times 5 \times 14$ grid blocks was chosen for the study. Parameters such as permeability, net-to-gross ratios, porosities and formation thickness were assigned to each grid block by overlaying the grid on the contour maps for these variables, provided from the current Forties reservoir engineering model.

The API tracking option was used to model the mixing of different types of oil, which have different surface densities and pressure-volume-temperature (PVT) properties, as described for the rectangular reservoir geometry in the previous section. This option allows reservoir fluids to keep their identities, i.e. physical properties, during the process of moving through the reservoir (this method replaces the conventional method of ‘PVT regions’ for oil, which assigns physical values to fluid according to the cell they are in). Apart from solving the governing equations for flow in porous media, the API tracking option will solve an extra mass conservation equation at the end of each time-step, updating the oil surface density in each grid block to model the mixing of the different oil types.

The different types of oil are assumed to be completely miscible with each other. This option can trace the movement of reservoir fluids and makes the present work possible.

Because this study models the mixing over a geological time-scale, i.e. thousands to millions of years, great care was taken to ensure minimum computing time and good convergence by choosing appropriate parameters for time-step selection.

Results

This section presents the estimated rates of density-driven mixing for the Forties Field. Most of the work was performed using the ECLIPSE black oil simulator and the grid described in the previous section. Simulation was first carried out for a simple hypothetical case without any barriers. Then we investigated various ways of explaining the

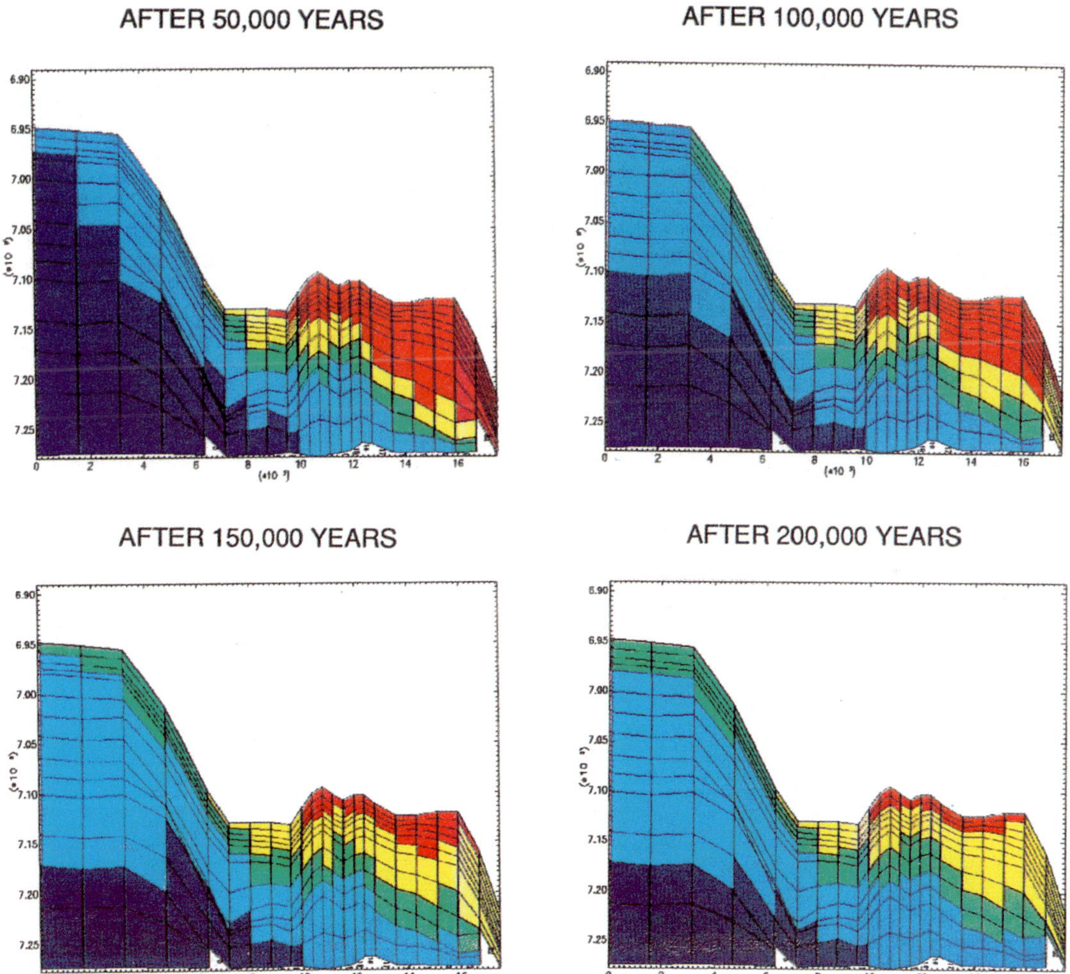

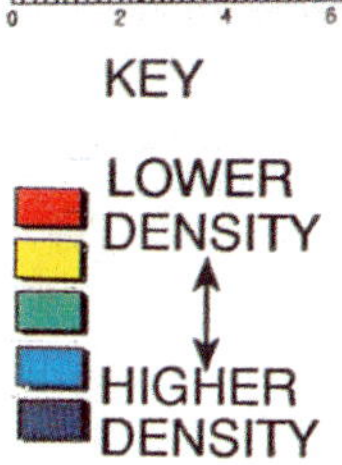

Fig. 10. Simulation results for a hypothetical barrier-free Forties reservoir. The *y*-axis represents depth; the *x*-axis is distance through the grid shown in Fig. 9, near to and approximately parallel to A–B.

presence of a lateral difference in density flow between the two parts of the Forties reservoir.

Reservoir without shale barriers (base case)

In this case, the left side of the reservoir (from grid block 1 to 12 in the x-direction) is initially filled with the denser oil of the main Forties Field, and the remainder is filled with the less dense oil of the SE Forties. There is no transmissibility barrier and the oil–water contact is taken to be the same across the whole reservoir (2217 mss). For simplicity, the initial interface between the two types of oils was assumed to be sharp and vertical. This represents the base case, i.e. one based on the Forties full-field engineering model which at the time had no transmissibility barrier between the two parts of the field.

Figure 10 shows the simulation results after 50 000, 100 000, 150 000 and 200 000 years, which is relatively short compared with the period of time since oil began to fill the Forties Field (believed to be around 10 Ma). The dark blue colour on the left represents the denser oil in the main reservoir and the red colour on the right represents the less dense oil. As can be seen, the mixing process occurs continuously over this time-scale, with more and more lighter oil entering the top of the main reservoir and the denser oil sliding to the base of the SE Forties. The overturning time is geologically fast: after 0.2 Ma most of the light oil, except near the top of the SE field, has moved into the main reservoir. By 0.4 Ma (not illustrated), the two oils in the two parts of the reservoir are almost fully mixed; this implies an average density difference between Forties and SE Forties reduced from 18 kg/m^{-3} to 3 kg/m^{-3}. This timescale is of the same order of magnitude as the analytical prediction of Gardner *et al.* (1962) of the 0.1 Ma, discussed previously for a rectangular reservoir.

The simulation results from this barrier-free base case show that substantial density-driven mixing should have occurred within 0.4 Ma. The presence of the current-day measured density differences demonstrates that a model without barriers is incompatible with observation. In the following sections, various models are tested to explain the observed differences in fluid density in the Forties Field.

Reservoir with shale barrier

From the structural cross-section map for the Forties/SE Forties boundary (Fig. 11), it is noticeable that between well E3 (in the main field) and E7 (in SE Forties) there is a continuous shale formation, from the top of the K layer in the main sand to the bottom of the FG layer in SE Forties. This shale appears to block the passage of fluid, at least in the same cross-sections, and poses serious resistance to fluid flow. Between these two wells, there are also some disconnected shales (apart from the continuous shale mentioned previously). If they become connected they could also restrict fluid flow. No faults are seen on the 3D seismic cross-section of this area. The presence of continuous and random shales, therefore, could hinder the mixing of the two types of oil.

10% transmissibility barrier. Assuming the shale formation does not completely prevent fluid flow, a case was run in which the effective transmissibility between the E3 and E7 well region was set to be 10% of the value without shale. In other words, a

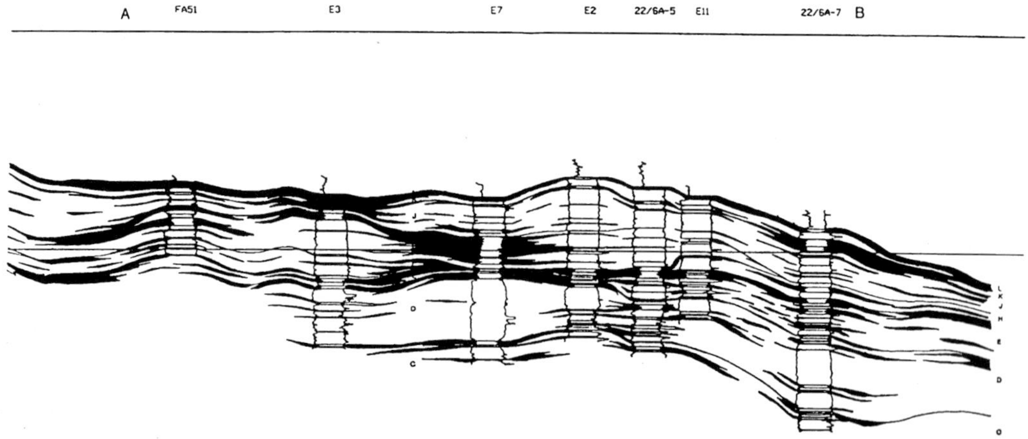

Fig. 11. Structural cross-section near the Forties/SE Forties boundary, A–B in Fig. 9.

transmissibility barrier of 0.1 was included in the model. This reduces the all flows across the interface to 0.1 of what they would be in the absence of transmissibility barrier, with the same pressure difference between the centres of the grid blocks on the other side of the barrier. Modelling results showed that the mixing speed is not significantly reduced by this 10% barrier. The mixing of the two fluids is still fast relative to the time available since oil emplacement. Thus, a 10% transmissibility barrier is insufficient to account for the current-day differences in subsurface densities.

1% transmissibility barrier. A further case was run in which the effective transmissibility between well E3 and E7 was reduced to 1% of the value without shale. With this region of lower transmissibility, simulation results indicated severely restricted flow. As can be seen from Fig. 12, the overturning of the two fluids has been significantly slowed down. The fluid density distribution at times of 0.1, 0.2, 0.3 and 0.4 Ma changes slowly with time. The degree of mixing at the end of 0.4 Ma is more or less the same as that of the case without any barrier after 50 000 years. Even by the end of 2 Ma (Fig. 13d) the two fluids have still not fully mixed. The average density difference between Forties and SE Forties has been reduced only from 18 kg/m^{-3} to 7 kg/m^{-3}. Note that the initial density difference would in fact have been higher than that assumed for modelling purposes, which we took from the present-day density difference. However, as the rate

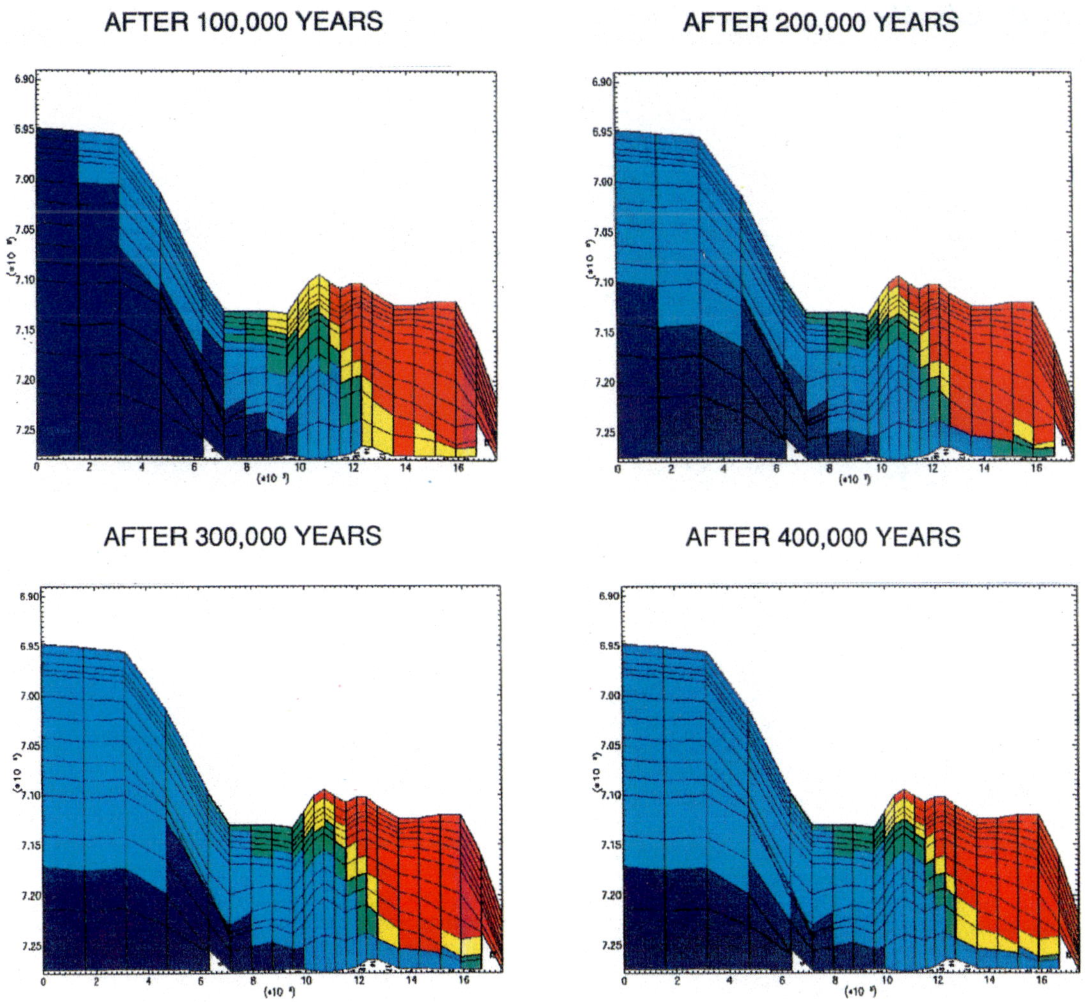

Fig. 12. Simulation results for the Forties reservoir with a 1% transmissibility barrier assumed between SE Forties and the main part of the field (first 400 000 years). See Fig. 10 for key.

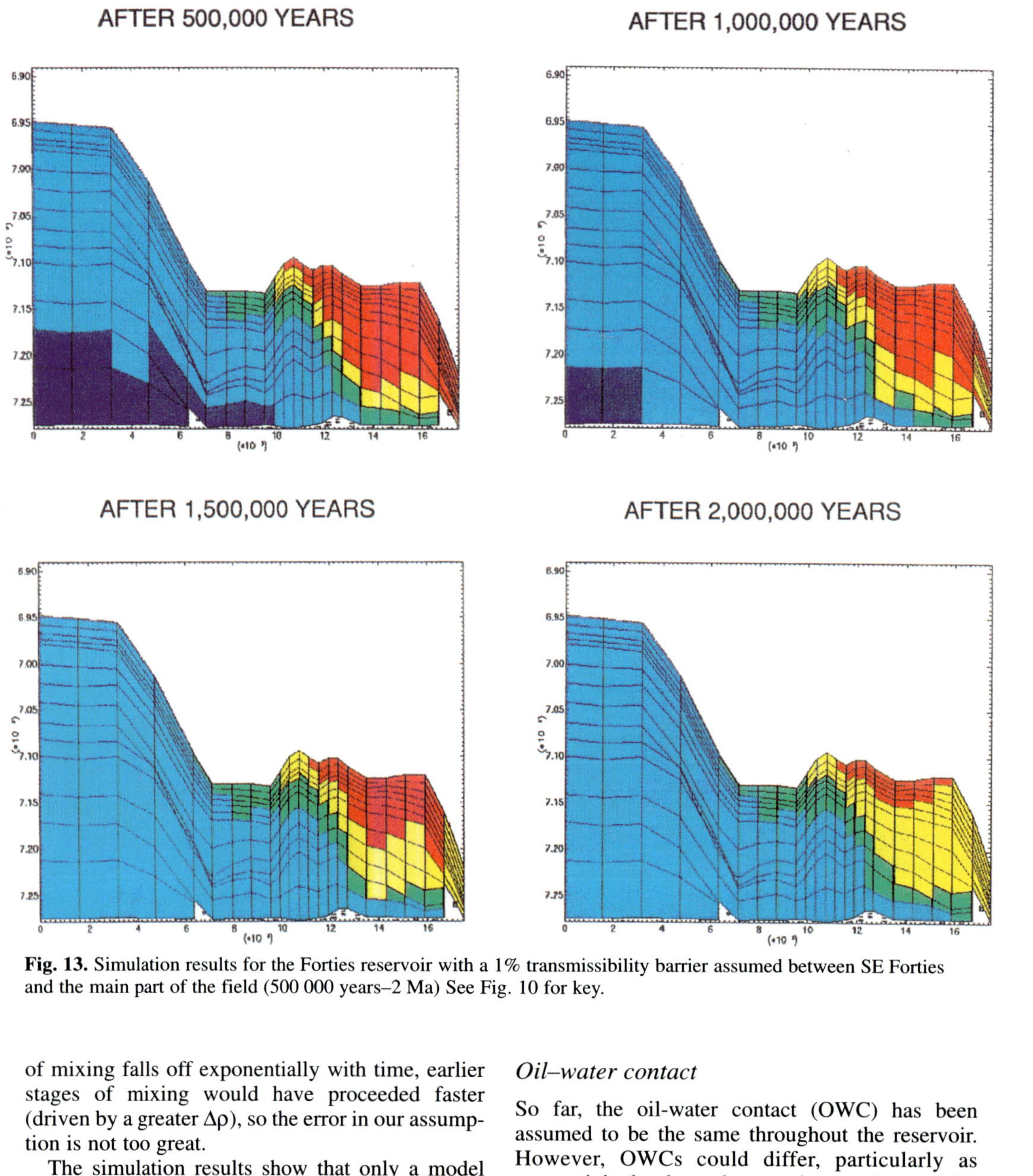

Fig. 13. Simulation results for the Forties reservoir with a 1% transmissibility barrier assumed between SE Forties and the main part of the field (500 000 years–2 Ma) See Fig. 10 for key.

of mixing falls off exponentially with time, earlier stages of mixing would have proceeded faster (driven by a greater $\Delta\rho$), so the error in our assumption is not too great.

The simulation results show that only a model with at least a 1% transmissibility barrier between the main and SE part of the reservoir can possibly result in substantial fluid density difference persisting for 2 Ma. The other reservoir models predict fully mixed reservoir fluids within 1 Ma, which is inconsistent with the geological time available since filling. In this way, certain reservoir models can be eliminated from further consideration and the range of possible models can be narrowed down by setting limits on intra-field barrier strengths.

Oil–water contact

So far, the oil-water contact (OWC) has been assumed to be the same throughout the reservoir. However, OWCs could differ, particularly as connectivity has been shown to be poor between SE Forties and the main field. The presence of barriers often raises a question: if there is a strong barrier, why should the OWC be the same across all of the Forties reservoir, as is observed to be the case? In this section we investigate the rates of OWC level equilibration and the effect of the process on the mixing of different types of fluids.

We have already seen that a transmissibility barrier of 1% or less is sufficient to be consistent with the present-day density distribution. This transmissibility barrier is only applied to the part of the reservoir above the OWC level, as no substantial shale exists below the OWC. In the test, we start with an initial condition in which the OWC level in Forties is 127 ft higher than that in SE Forties. Figure 14 shows the results of the simulation. It can be seen from Fig. 14a, only 100 years after the start of the calculation, that the water levels have not yet changed significantly, except in the regions close to the boundary between the two parts of the Forties. As time passes, the OWC levels get closer. As can be seen in Fig. 14b (after 1000 years), there is no sharp discontinuity in OWC level across the reservoir. By the end of 10 000 years, the region in the main reservoir close to the boundary has a lower OWC than before, while in SE Forties close to the boundary the water level goes up. However, in the region further from the boundary, the water level remains unchanged. Gradually and finally, the OWC across the reservoir reaches equilibrium (Fig. 14c & d). The whole process for reaching OWC equilibrium only takes 80 000 years. Compared with the time for near-complete mixing of two types of oil, the time-scale required for OWC equilibrium is short, even though a strong transmissibility barrier exists.

It is evident from the simulation previously described that the water levels reach equilibrium

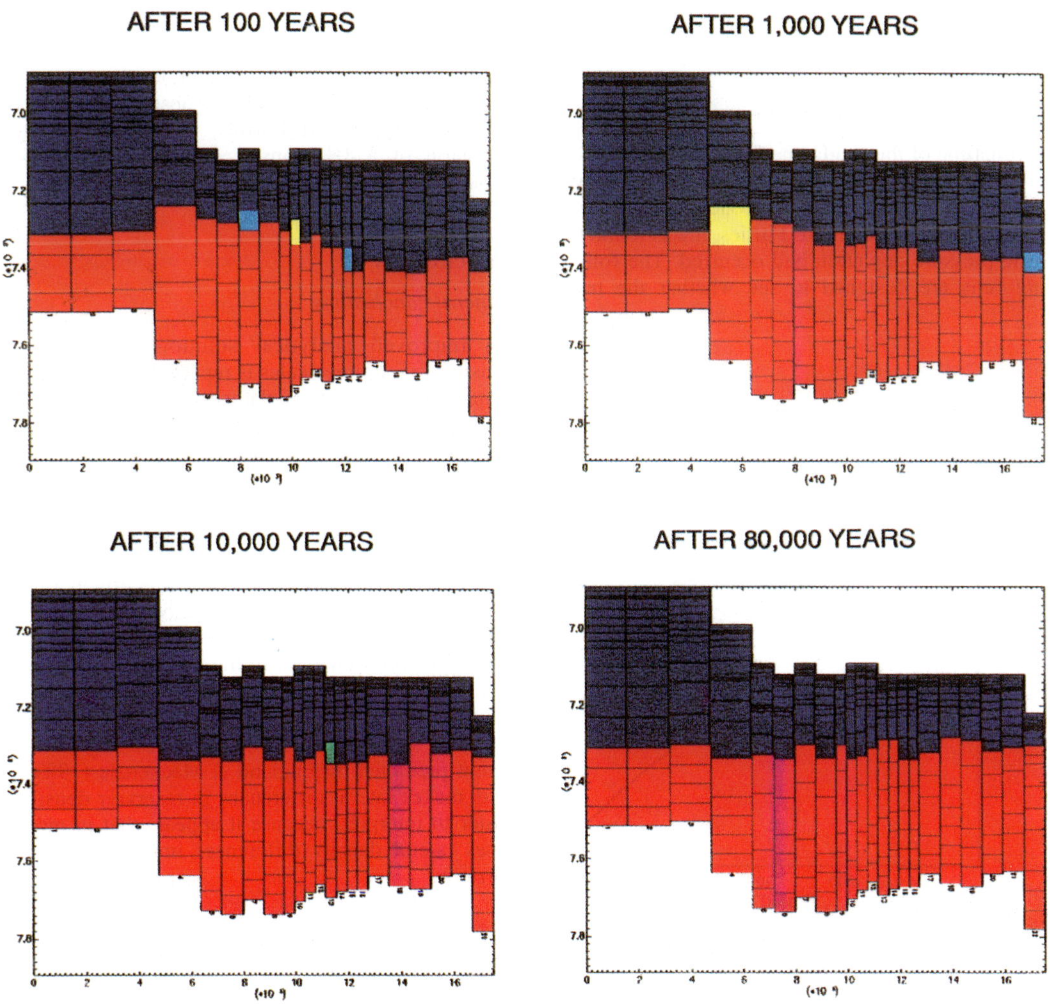

Fig 14. Simulation results for the Forties reservoir oil–water contact equilibration.

at a high speed. This can be explained from the analytical solution of fluid mixing in a rectangular reservoir. As can be seen from Equation 2, the higher density difference($\Delta\rho$) results in faster mixing. The initial density difference between the two types of oil is only 18 kg m^{-3} whereas the difference between the oil and the water is around 300 g m^{-3}. Moreover, the lack of shale presence below the OWC level inevitably speeds up mixing. These two factors result in the fast equilibration of OWC levels.

In some reservoirs, it may be possible to exploit the observation of petroleum density differences coexisting with an equilibrated OWC to set upper and lower limits on transmissibility barriers that are consistent with geologically plausible filling histories. Similarly, small differences in pressure may also be utilized in a similar way to reservoir fluid differences, because both are unstable to fluid flow over geological time-scales.

Comparison with production data

At the time of this study, the Forties Field had been producing oil for 16 years. At first, only the main field produced oil, but when production wells were subsequently drilled in SE Forties, it was found that little pressure depletion had occurred. The following runs are designed to see whether the transmissibility barrier deduced from the density-driven mixing rate is compatible with the pressure data.

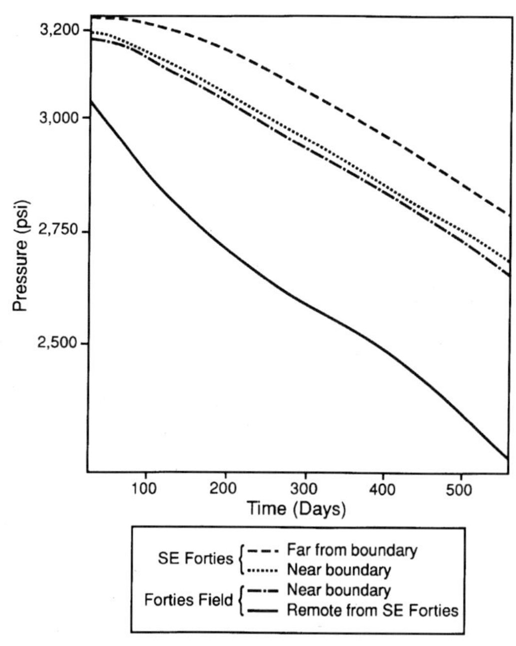

Fig. 15. Pressure change for the barrier -free model.

Figure 15 shows the predicted pressure response with time after the main field began to produce oil at a net rate of 10 000 bbl/day continuously, assuming that there is no transmissibility barrier between SE Forties and the rest of the field. A single production well is located at the left boundary of the model. As can be seen, with no barrier assumed, the pressure in both parts of the reservoir drops at almost the same rate. However, with a 1% transmissibility barrier incorporated in the model, the pressure response in the two parts of the reservoir is significantly different (Fig. 16). After nearly two years of production, the pressure drop in the main field is 1000 psi. However, in SE Forties (close to the boundary with the main field), the predicted depletion is only 300 psi and further away from the boundary the pressure is hardly altered.

This confirms that a model for Forties without barriers is inconsistent with the observed pressure decline data for SE Forties. Furthermore, the inclusion of a 1% transmissibility barrier gives good agreement with the observed pressure data.

Conclusions

General conclusions

An improved analytical solution to estimate density-driven time-scales is reported. The use of a numerical reservoir simulator to estimate density-driven time-scales has been validated against the analytical solution for an idealized rectangular reservoir.

The effects of stochastic shales, molecular diffusion and the initial density distribution on the time-scales of mixing have been studied, using the high resolution MISTRESS code. The effects of a stochastic shale distribution could be satisfactorily reproduced using an effective vertical permeability. Although slight changes in the mixing time-scales were seen, they were not significant.

Low (grid-block scale) vertical permeability does not slow down density-driven mixing to the extent predicted by England & Mackenzie (1989). This makes lateral differences in reservoir density easier to interpret, because it appears that in most cases lateral density differences imply intra-field barriers.

Conclusions from Forties case study

This study has used present-day subsurface fluid density variations to demonstrate that a significant transmissibility barrier exists between the SE Forties Field and the main Forties Field. This

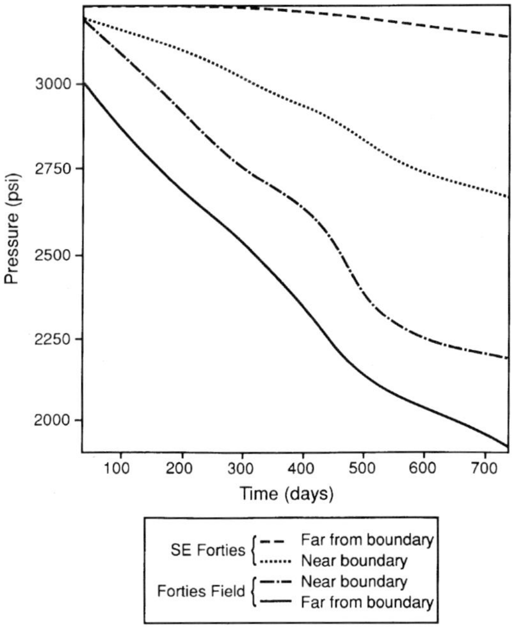

Fig. 16. Pressure change for a model with a 1% transmissibility barrier.

barrier must be effective in resisting the flow across the boundary between the two parts in the field. A transmissibility barrier of 1% or smaller must be applied to the Forties/SE Forties boundary to account for present day observations.

The oil-water contact levels are found to reach equilibrium quickly on the geological time-scale, even with the presence of a strong transmissibility barrier (1%) between the two parts of the reservoir. The time-scale for the water level to reach equilibrium is in the order of 100 thousand years in contrast to more than 2Ma needed for mixing of oils. The rate of pressure decline in SE Forties during the first years of production from Forties predicted by the reservoir model with a 1% barrier also appears to be consistent with the measurements made in the field.

The study used the Forties Field as a test case, but the method could be used for fields at an appraisal stage, where production data are not available. In many simple cases, the analytical solution (Equation 2) will give satisfactory results without the need to use numerical simulators.

The following procedure is suggested. If different parts of a reservoir have different subsurface densities, there is a strong possibility that a barrier is present. Further work should then be performed to estimate (1) the time-scale of density-driven mixing (either using a black oil simulator or Equation 2), and (2) the geological time available for mixing. A comparison of the two times should allow one to see whether barriers are needed to explain current-day reservoir fluid density data. If the time-scale for complete mixing is much shorter than the geological time available for mixing, simulations should then be performed to set limits on the degree of connectivity within the reservoir.

This study has demonstrated for the first time a method of using present-day reservoir fluid density differences to deduce the presence of intra-field flow barriers. The simulation method can provide a good estimate of reservoir connectivity at an early stage of reservoir appraisal. The data required are modest: reservoir fluid density differences, possible reservoir dimensions and permeability maps. This method of using numerical simulation is a significant advance of technology which can be readily adopted by practising reservoir engineers.

We acknowledge the generous assistance of Graeme Bennett and the members of the Forties Field group in allowing access to their reservoir engineering data, and thank BP Exploration and the Forties partners for permission to publish.

References

CHRISTIE, M. A. 1989. High resolution simulation of unstable flows in porous media. *Society of Petroleum Engineers, Reservoir Engineering* ,**287,** 297–303.

ENGLAND, W. A. 1990. The organic geochemistry of petroleum reservoirs. *Advances in Organic Chemistry,* **16,** 415–425.

—— & MACKENZIE, A. S. 1989. Some aspects of the geochemistry of petroleum fluids. *Geologische Rundschau,* **78**(1), 292–303.

——, ——, MANN, D. M. & QUIGLEY, T. M. 1987. The movement of entrapment of petroleum in the subsurface. *Geological Society, London, Journal,* **144,** 327–347.

GARDNER, G. H. F., DOWNIE, J. & KENDALL, H. A. 1962. Gravity segregation of miscible fluids in linear models. *Society of Petroleum Engineers, Journal,* **95,** 95–104.

JACQMIN, D. 1990. Interaction of natural convection and gravity segregation in oil/gas reservoirs. *Society of Petroleum Engineers, Reservoir Engineering,* **288,** 233.

Petroleum geochemistry of the Haltenbanken, Norwegian continental shelf

D. A. KARLSEN[1], B. NYLAND [2], B. FLOOD [2], S. E. OHM[2,5], T. BREKKE[3], S. OLSEN[4] & K. BACKER-OWE[1]

[1] *Department of Geology, University of Oslo, PO Box 1047 Blindern, N-0316 Oslo, Norway*

[2] *Norwegian Petroleum Directorate, PO Box 600, N-4001 Stavanger, Norway*

[3] *Brekke Chemometrical, Leirstad 27, 6800 Førde, Norway*

[4] *Rogaland Research - RF, PO Box 2503, Ullandhaug, N-4004 Stavanger, Norway*

[5] *Present address: Elf Aquitaine Norge a/s, PO Box 168, N-4001 Stavanger, Norway*

Abstract: Thirty-three oil/condensate samples, representing most of the known petroleum accumulations in the Haltenbanken region, were analysed by a comprehensive geochemical scheme to determine possible genetic relationships. The organic geochemistry of the samples is comparatively invariate if one excludes the characteristics attributed to maturity differences and phase-induced fractionation. On this basis, in a gross sense, the C_{10+} fractions of most of the petroleum samples are suggested to be derived from the Spekk Formation. However, the ratio of pristane to phytane and the stable carbon isotope composition of the petroleums proved useful to differentiate this picture. The database was subjectively divided into three main genetic groups which show dependence on the geographical location of the fields.

Group 1, comprising the samples from Njord, Smørbukk Sør well 6506/12-5, Tyrihans N, Tyrihans S, Trestakk and the 6506/12-3 DST 6 sample above the Smørbukk Sør main reservoir, is clearly differentiated from Group 2, comprising samples from Midgard, 6407/2-2 DST 1 and from Alve 6507/3-1 DST 3. Group 3, comprising samples from Draugen, Heidrun, Mikkel plus the Smørbukk Sør petroleum 6506/12-3 DST 1, forms an intermediate population between Groups 1 and 2. In general, Group 2 is confined to the easterly, more proximal part of the basin, while Group 1 fields have drainage areas in distal, more deeply buried regions of the basin. Available data on the geochemistry of the two major source rocks in the region, the Spekk Formation and the deeper Åre Formation, lead us to conclude that the C_{15+} fractions of the Group 1 petroleums are derived predominantly from a distal marine anoxic facies of the Spekk Formation. Group 2 forms the other end member, supposedly sourced mainly from more terrestrially influenced proximal marine (partly dysaerobic?) isotopically heavier shales of the Spekk Formation. Group 3 contains characteristics of both source-rock facies. We see no evidence in biomarker distributions of the current database for significant contributions to the heavy ends of the reservoired petroleums from Åre coals. However, we cannot, based on our limited source-rock database, exclude contributions in some localities from Åre Formation shales. Furthermore, we have no evidence to exclude contributions to the light hydrocarbon range from the Åre Formation, e.g. condensate/gas contributions. We feel that the Spekk Formation varies significantly in geochemistry, both laterally and vertically, i.e. according to Walters law, and that this formation is less homogeneous with respect to geochemistry and the depositional and early diagenetic environment than commonly described. The metal composition of the petroleums generally substantiates these interpretations on laterally varying dysaerobic/anoxic conditions of the Spekk Formation, even if it proved difficult in cases to compare metal distributions in petroleums with drastically different gas/oil ratios.

Because most fields, despite overall similarities, contain petroleum with specific singular minor characteristics, we suggest that, with the exception of the Draugen and Midgard Fields, lateral migration has been predominantly short- to medium-range with respect to intra-field distances. A core extract from well 6609/11-1 (Helgeland basin) deserves special attention, as our data suggest that this core at 2561 m contains a zone of terrestrially derived oil, possibly lacushine, most likely sourced from a Triassic or Palaeozoic formation.

The purpose of this paper is to integrate a consistent database of geochemical data on 33 petroleums from the Haltenbanken region (Fig. 1) with published material on petroleum distributions and hydrocarbon generation in the region. This was conducted in order to shed light on possible genetic relations between condensates and oils from the Haltenbanken. Furthermore, it was hoped that this work would reveal some of the controls of petroleum-type occurrences within the region.

From CUBITT, J. M. & ENGLAND, W. A. (eds), 1995, *The Geochemistry of Reservoirs*, Geological Society Special Publication No. 86, pp. 203–256.

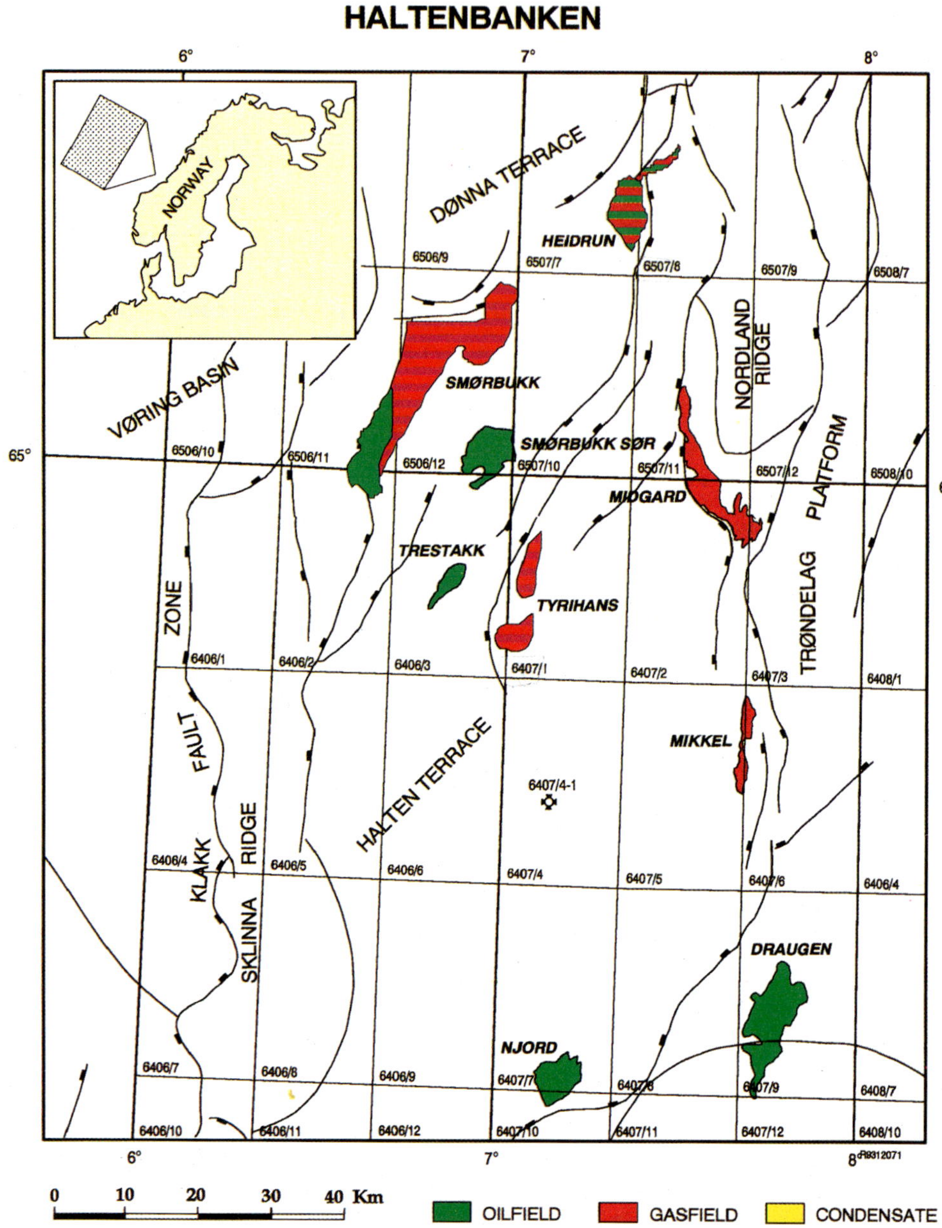

Fig. 1. Map of the greater Haltenbanken area, showing the main petroleum accumulations analysed in this work, as well as some of the main structural elements.

Since the first block was awarded in the Halten Terrace in 1980 and Saga made the first discovery (6507/11-1, gas/condensate) in 1981, the region has proved to be a prolific hydrocarbon basin, with present cumulative proven resources of approximately 336×10^6 Sm^3 oil and about 316×10^9 Sm^3 gas, for a total of about 0.61×10^9 tonne oil equivalents.

Structurally, the central part of the Haltenbanken, the Halten Terrace (Fig. 1) is separated from the Trøndelag Platform to the east by the Kristiansund–Bodø Fault Zone. To the west, the

West Haltenbank High separates the Halten Terrace from the deeper Møre–Vøring Basin. The Haltenbanken comprises a highly faulted oblique-slip tectonically induced basin (Gabrielsen *et al.* 1984; Bøen *et al.* 1984; Bugge *et al.* 1984; Bukovics & Ziegler 1985). Differential subsidence of the Halten Terrace versus the Trøndelag Platform occurred in Early and Late Cretaceous times, while the entire Trænabanken–Haltenbanken region underwent rapid subsidence in Late Pliocene–Pleistocene times (Jackson & Hastings 1986). The reader is referred to Hollander (1984) and Gowers & Lunde (1984) for discussion of the geology of the Haltenbanken and Trænabanken regions, and to Bjørlykke *et al.* (1986) and Ehrenberg (1990) for a treatment of the diagenesis and reservoir quality of the Haltenbanken sandstones.

Reservoir play

Most petroleum accumulations in the Haltenbanken region have been found to be structurally related to horsts or rotated fault blocks, e.g. Midgard, Tyrihans (Nord), Heidrun, Smørbukk and Njord, which are genetically linked to extensional faulting during the Kimmerian phase. In some cases, such as Smørbukk Sør, Tyrihans South, 6407/4-1 and 6507/7-1, the reservoirs comprise dome-shaped, possibly salt-induced, structures. Structures like these are thought to have formed over Triassic salt, during the Late Jurassic through Neocomian time, (Jackson & Hastings 1986). The Draugen Field (Provan 1992), being a gentle anticlinal structure on the eastern platform (Trøndelag Platform), is exceptional in this region.

The main reservoir units of this region are the sands of the Jurassic Fangst Group (Fig. 2), e.g. the Smørbukk and the Heidrun Fields. However, occasionally, the main reservoirs are constrained to the Båt Group, e.g. Njord. Again, Draugen, with its Upper Jurassic Rogn Formation sandstones as the main reservoir, represents an exception.

Source rocks: hydrocarbon generation

The Upper Jurassic Spekk Formation (Fig. 2) attains possibly as much as 400 m of accumulated thickness in the (to date) undrilled structural depressions (Figs 3, 4 & 5). It is generally rated immature in the eastern platform areas (the Trøndelag Platform) and overmature to the west, i.e. towards the Vøring Basin (Fig. 3). It is believed to have reached peak oil generation during the Late Tertiary in the Halten Terrace (Campbell & Ormaasen 1987). Thus, the Spekk Formation is rated mature in the regions where most of the petroleum accumulations in the Haltenbanken are found (Fig. 3). It is rated immature in the closest vicinity of the Draugen and Midgard Fields. However, both of these petroleum accumulations have mature Spekk Formation in their drainage areas, as has the Heidrun Field (Figs 3, 4 & 5).

Peak oil generation was estimated to have occurred over large areas in the Haltenbanken area at *c.* 3400 m of burial (Leadholm *et al.* 1985). Initial oil expulsion is rated by Cohen & Dunn (1987) to have occurred at *c.* 120°C, equivalent in their dataset to *c.* 3100 m of burial, while Heum *et al.* (1986) have 0.7% vitrinite reflectance equivalence (VRE) as the threshold for primary migration from this formation. In their dataset, 0.7% corresponds to *c.* 3900 m of burial, while this depth is considered by Cohen & Dunn (1987) to be associated with initial oil to gas cracking ($T \approx 150°C$). The latter authors estimate initial refractory gas generation to start at a depth of *c.* 4600 m. This is similar to Heum *et al.* (1986), who put the deepest limit of oil generation at 1.0% VRE, which is more or less equivalent to 4700 m in their dataset.

Our database on biomarkers from the Spekk Formation contains core extracts from only one core. Yet these data suggest that significant oil expulsion has occurred from the Spekk Formation at a depth shallower than 3900 m. This is inferred, as the biomarker distribution in the core extracts from this sequence is clearly more mature than the most mature condensate in the current database.

The kerogen of the most prolific Hot Shale Kimmeridge equivalent part of the Spekk Formation is a highly favourable source for oil, with a gas/oil generation index (GOGI = C_1–C_5/C_6–C_{36+}) of about 0.19 ± 0.05 ($n = 8$) (Cohen & Dunn 1987). The kerogen of the Spekk Formation is characterized by a $\delta^{13}C$ value of about -28.5 ± 1.6 ($n = 10$, Cohen & Dunn 1987), thus closely matching the values (-28.2 ± 2.7, $n = 3$) found by Elvsborg *et al.* (1985). The pristane/phytane ratio of extracts from this formation is in the range of 0.86–1.60 and the percentage sterane distribution centres around *c.* $C_{27} = 34$, $C_{28} = 33$, $C_{29} = 33$ (Cohen & Dunn 1987).

The deeper Melke Formation is generally considered to have an insignificant oil potential compared to both the Spekk Formation and the deeper Åre Formation (Heum *et al.* 1986; Cohen & Dunn 1987). In particular, Heum *et al.* (1986) rated the Melke Formation as potentially releasing no more than about 1.5 kg of petroleum/tonne rock (mainly gas C_{5-} and some condensate C_6–C_{14}), whereas approximately 30–35 kg of petroleum/tonne rock (gas ≈ 15%, condensate ≈ 45%, oil, i.e. C_{15+}, ≈ 40%) was estimated to have been released from the Spekk Formation during progressive burial to *c.* 4% vitrinite reflectance.

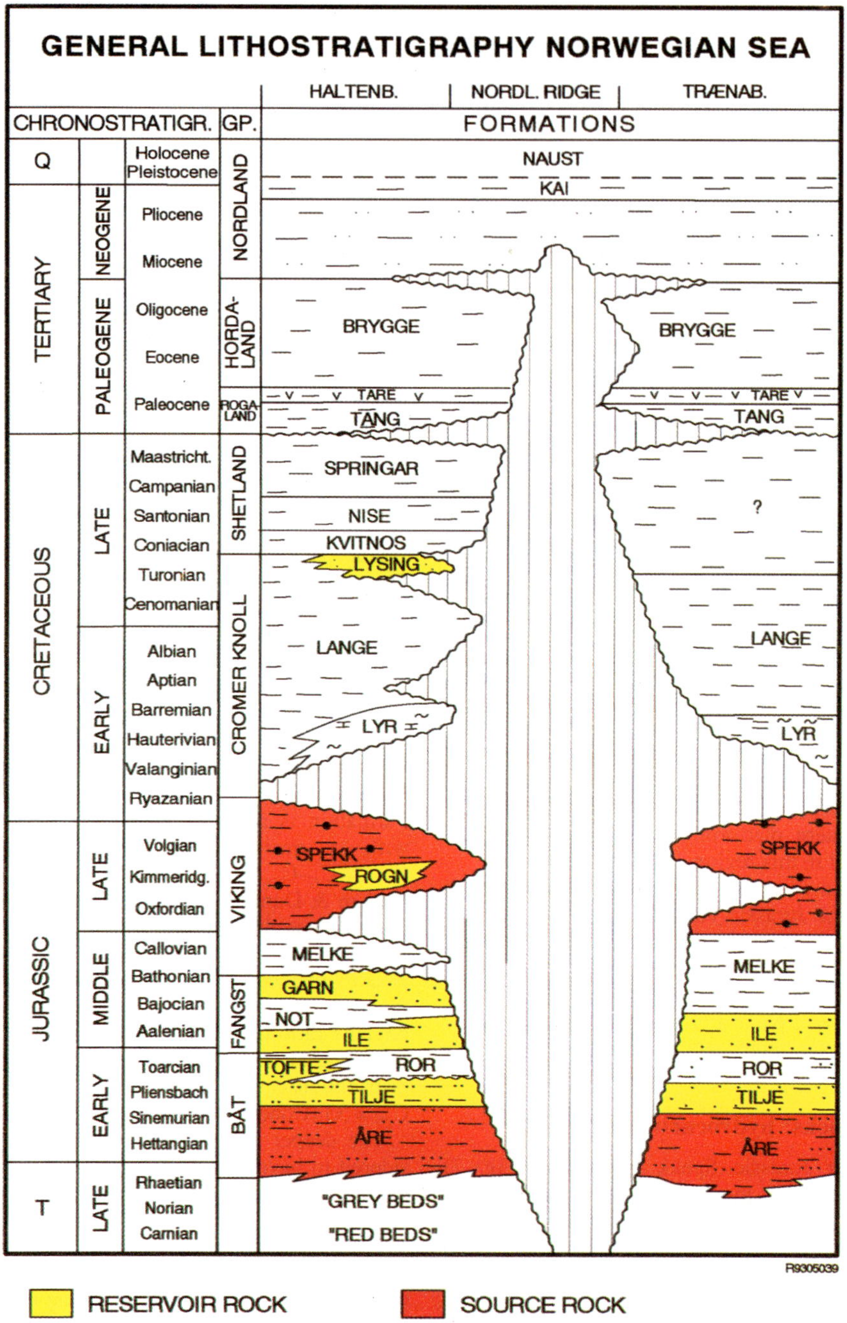

Fig. 2. Generalized stratigraphy and nomenclature of the Haltenbanken region. The two traditionally proposed source rocks and the main reservoir formations are indicated.

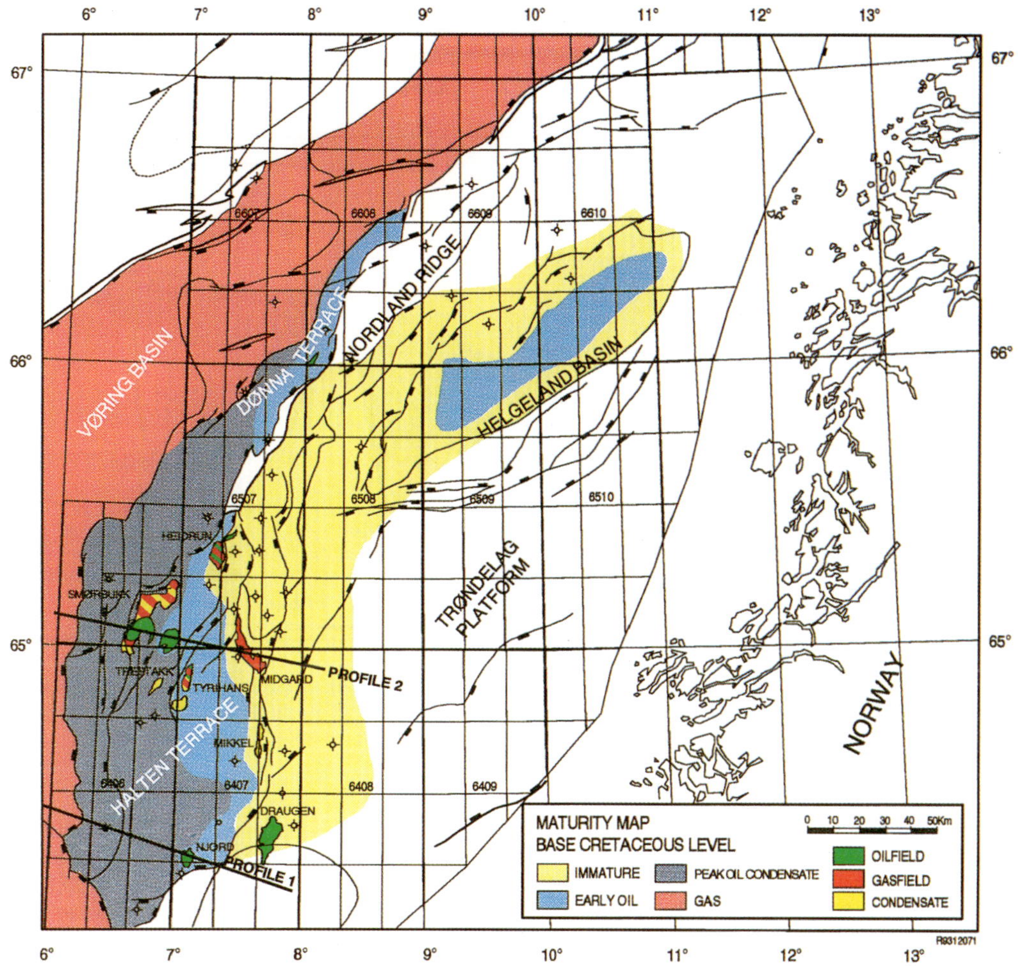

Fig. 3. Regional field location and maturity map at base Cretaceous, i.e. top Spekk Formation.

The more extensive generation which is necessary to saturate the source rock sufficiently for primary migration to occur in lean source rocks, lacking a three-dimensional continuous kerogen network (such as Melke Formation), compared to the situation in a better quality source rock (e.g. Spekk Formation), is important with respect to the timing of expulsion and also to the gas/oil ratio (GOR) of the product (Durand *et al.* 1987; Durand 1988; Stainforth & Reinders 1990). The Melke Formation is possibly a good example of a lean source rock, as explusion mainly in the form of gas is considered to occur only after attaining a maturity of *c.* 1.7% VRE (Heum *et al.* 1986). The quantity of possible C_{15+} hydrocarbons generated before the floor of the 'oil window' is reached, is in this case too low to saturate the source rock sufficiently for expulsion to occur. A cut-off value of about 20% saturation has been used for such comparably lean source rocks (Durand 1988) and, in this case, it is implied that this level was only reached at a maturity of *c.* 1.7% VRE, i.e. corresponding to temperature conditions under which oil would have been fully transformed to condensate/gas.

The coals and shales in the deeper Åre Formation are considered to have expelled about 135 kg of petroleum/tonne rock (gas ≈ 30%, condensate ≈ 50%, oil ≈ 20%) during burial to *c.* 4% vitrinite reflectance (Heum *et al.* 1986). The deeper shales and coals in the Lower Jurassic (the Åre Formation) are indicated to have a GOGI in the

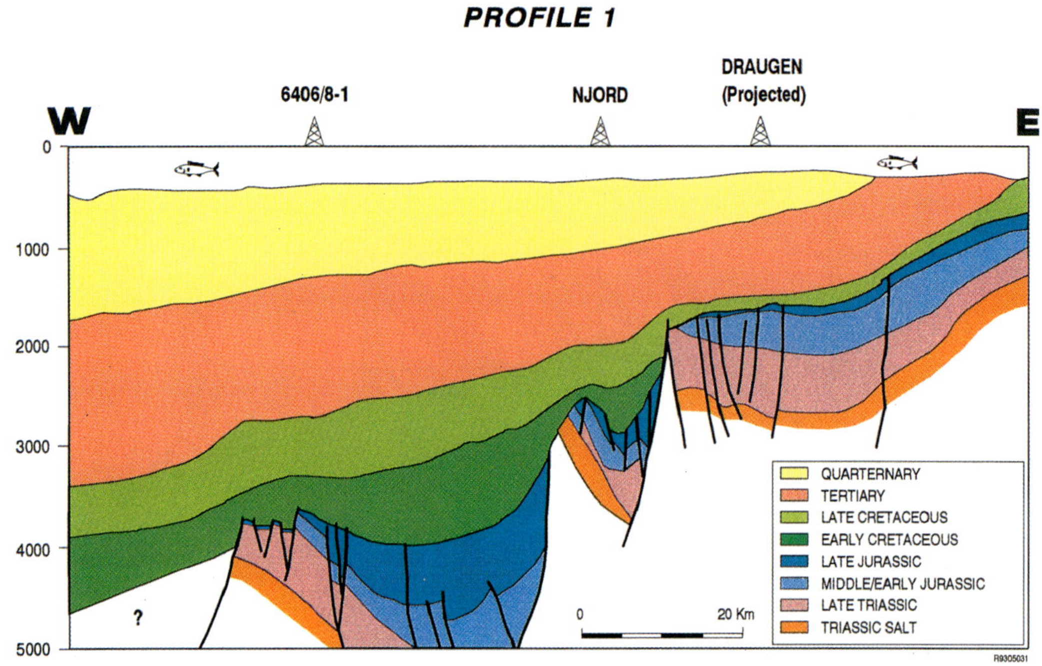

Fig. 4. Profile 1 indicated in Fig. 3. The simplified seismic section shows some of the major tectonic elements and the progressive burial of the late Jurassic Spekk Formation towards the west, from the Trøndelag Platform, through Draugen and Njord towards the southern part of the Vøring Basin. Depth in metres.

range of about 0.33 ± 0.06 (n = 5), i.e. giving rise to a mainly 'mixed oil and gas' (Cohen & Dunn 1987). This formation is likely to be mature in some of the eastern proximal basinal regions where burial of the Spekk Formation is too shallow (Whitley 1992). The kerogen of this unit, though obviously of heterogeneous composition as this formation includes rooted coals, drifted coals, mudstones and shales, was found by Cohen & Dunn (1987) to have a measured $\delta^{13}C$ value of about -26.1 ± 1.1 (n = 11). This is comparable to the data published by Elvsborg *et al.* (1985) (-25.8 ± 0.7 n = 4). The pristane/phytane ratios of extracts from this formation are in the range of 1.60 to 6.60 and the percentage sterane distribution is mostly *c.*C_{27} = 15, C_{28} = 20, C_{29} = 65 (Cohen & Dunn, 1987).

Thus, the oil potential of the Åre Formation is

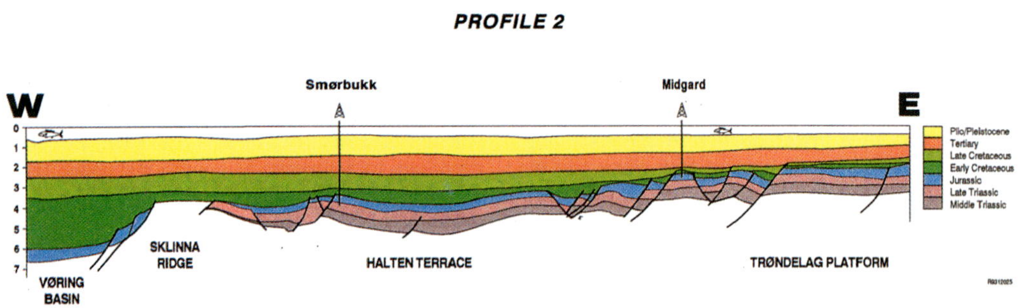

Fig. 5. Profile 2 indicated in Fig. 3. The simplified seismic section shows some of the major tectonic elements and the progressive burial of the late Jurassic Spekk formation towards the west, from the margin of the Trøndelag Platform, through the Midgard Horst, over the Smørbukk Sør accumulation and into the Vøring Basin. Depth in kilometres.

considered by some workers to be at least locally important (Heum *et al.* 1986; Thompson *et al.* 1985). Mo *et al.* (1989) estimated the Åre Formation to be quantitatively more important in terms of oil generated in the drainage area of the Midgard Field (generating about 6.5 times as much oil as the Spekk Formation). Mo *et al.* (1989) assumed, however, that much of this oil was not expelled fast enough to escape cracking to gas (condensate) at greater depths in response to fast burial during the last 3.5 Ma. Maturity modelling (Mo *et al.* 1989) suggests that peak oil generation from the Åre Formation in the deepest portions of the drainage area occurred during Late Cretaceous. The generation can be assumed to continue today in the progressively subsiding basin flanks towards the Midgard Horst (laterally migrating oil kitchen).

Likewise, Forbes *et al.* (1991) ascribed most of the oil generation in the drainage area of the Smørbukk Sør accumulation to the Åre Formation (Åre is about six times as important volumetrically with respect to expelled C_{15+} hydrocarbons compared to the Spekk plus the Melke Formations). They further suggest that expulsion from the Spekk and Melke Formations in this region became significant as late as 5 Ma BP, while in the Åre Formation significant expulsion (quantitatively comparable to the Spekk and Melke formations) occurred probably as early as 40 to 50 Ma BP.

Espitalie *et al.* (1991) rate the petroleum potential (10^6 t HC km^{-2}) of the coals in the Åre Formation as four times higher than that of the Viking Group (Spekk plus Melke Formations). This is due to the high productivity of the Åre Formation and also to its greater cumulative thickness. Espitalie *et al.* (1991) suggest that the Åre Formation has expelled 2.5 times more hydrocarbons than have the Spekk and Melke Formations. The suggested breakdown between coals and shales in Åre, in terms of expelled hydrocarbons, is proposed to be 17 to 1, while the ratio between expulsion for the Spekk and Melke Formations is 2.4 (Espitalie *et al.* 1991).

Over much of the Haltenbanken area the Åre Formation is on the order of 0.2–0.5% R_c higher in maturity than the overlying Spekk Formation (Leadholm *et al.* 1985; Heum *et al.* 1986; Whitley 1992). Obviously, great variations exist due to differential burial depth of the units (Fig. 4), e.g. the difference between the tops of these two formations varies from *c.* 500 m in the Midgard Horst to approximately 1000 m in the area 25 km farther south (Mo *et al.* 1989). Accordingly, generation is assumed to have been initiated earlier in the deeper Åre Formation than in the shallower Spekk Formation, i.e. during Mid- to Late Tertiary times (Campbell & Ormaasen 1987). Heum *et al.* (1986) suggested that expulsion in the Spekk Formation is initiated at a vitrinite reflectance value corresponding to 0.7% R_c, while these authors suggested that in the deeper coals and shales, expulsion occurs between 0.5 and 0.85% R_c and between 0.7 and 0.85% R_c, respectively.

The differences are related to the generaly more narrow activation energy distribution of the kerogens in the Spekk Formation (typical type II distribution compared to the broader distribution in the type II/III kerogens of the Åre Formation (Forbes *et al.* 1991)) and the hydrocarbon retaining qualities of coals (Hvoslef *et al.* 1988). Still, it is clear that it may be incorrect to ascribe one activation energy distribution to the Spekk Formation (Eirik Vik, pers. comm.). This is because the organic facies of the Spekk Formation can be expected to vary significantly both laterally and vertically, in a fashion perhaps similar to that of the Draupne Formation in the Viking Graben, which is known to exhibit significant vertical heterogeneities in kerogen composition (Huc *et al.* 1985) and in kerogen pyrolysate composition (Bailey *et al.* 1990). Thus, in the Viking Graben case, lateral organic-facies heterogeneities of the Kimmeridge Clay or the Draupne Formation are expected, in accordance with Walters law, and do indeed occur in other regions as reported by Bailey *et al.* (1990), Grantham *et al.* (1980) and Scotchman (1991).

A similar situation is likely to occur in the Haltenbanken region and existing geochemical databases of the Spekk Formation are furthermore likely to be biased towards the organic facies found at the drilled structural highs. Thus, the Spekk Formation occurring at palaeo- and present basinal highs is not likely to be the best analogue for the more prolific organic facies expected to be occurring in the basinal depressions (Figs 4 & 5).

Available data tentatively suggest that there are, in a gross sense, substantial similarities in the organic geochemistry and the kinetics of hydrocarbon generation between the Spekk/Melke Formations in the Haltenbanken area and the Draupne/Heather Formations from the Viking Graben (Cohen & Dunn 1987; Espitalie *et al.* 1991; Forbes *et al.* 1991). In particular the $\delta^{13}C$ values of kerogen in the Spekk Formation match closely the data published by Huc *et al.* (1985) for the Draupne in the Viking Graben. These authors found a co-occurring decrease in $\delta^{13}C$ and δ^2H from -30‰ to -25‰ and from -125‰ to -100‰, respectively, in kerogens isolated from a *c.* 100 m section through the Draupne and top Heather Formations (Viking Graben). This, and supporting evidence, was interpreted to reflect a more proximal terrestrially derived organic-matter input with increasing depth through this section (Scotchman 1991). Similarly, Bailey *et al.* (1990) described vertical variations in

$\delta^{13}C$ of kerogens and pyrolysates from the top of the Hot Shale (mostly -30‰ to -29‰) into the Heather, Brent, Dunlin and Statfjord Formations (mostly -27‰ to -24‰).

Lateral migration of petroleum kitchen areas in a proximal direction, up a sloping basin flank in response to subsidence, may lead to situations where the source rock at any time can potentially be expelling petroleum of a range of maturities (Cornford *et al.* 1986). The progressively more mature petroleum signature of the expelled oil is superimposed on *in situ* laterally varying facies signatures of the source rock. Thus, one could envisage that some of the structures in the Haltenbanken area could contain not only a mixture of petroleum from potentially different source rocks, but also petroleum reflecting potentially laterally varying source signatures from both the Spekk and the Åre Formations.

According to Heum *et al.* (1986), all petroleums in the Haltenbanken area, except the undersaturated Draugen oil, represent systems in phase equilibrium, i.e. oils generally saturated with gas and vice versa (oils at their bubble-point and gases at their dewpoint). Heum *et al.* (1986) interpret this to reflect that there is currently a balance between gas and oil generation and migration in the basin and that this is reflected in the petroleum charges of the fields, as these have drainage areas extending into more mature regions of the basin.

Analytical

Oil and condensate samples were analysed by Iatroscan thin layer chromatography–field ionization detection (TLC–FID) (Karlsen & Larter 1991) to determine their gross composition and evaporative loss. The amount of saturated and aromatic hydrocarbons and the amount of polars (resins plus asphaltenes) in the C_{15+} fraction was quantified and normalized to 100%. Whole oil/condensate samples were analysed on a 30 m DB-1 column (i.d. = 0.25 mm) in a Varian 3500 GC–FID (40°C initial temperature, gradient of 4°C min^{-1} to a final temperature of 310°C), using split injection.

Squalane and anthracene were added as internal standards to the oil and condensate samples before MPLC, according to Radke et al. (1980). Care was taken not to let the samples run dry on the Rotavapor during sample work-up. The quantitated data presented in the chromatograms (Figs 20 to 36) should be used only for comparison between samples in this database. Precision of analysis, including addition of internal standard, MPLC, work-up and analysis, generally gives results within a 15% level for oil and condensates. However, analysis of high GOR samples, i.e. gas condensates, which can be strongly affected by losses during sample handling, can give less accurate quantitative data. No correction is made for the slight differences in peak area response between the standards and the quantified compounds.

The resulting saturated and aromatic fractions were analysed by gas chromatography–field ionization detection (GC–FID) for quantification of individual hydrocarbon components. Baseline setting was controlled and manually corrected on all reported compounds before quantification. The saturated fraction was analysed on a 30 m DB-1 column (i.d. = 0.25 mm) in a Varian 3500 GC-FID, while the aromatic hydrocarbon fractions were analysed using a Varian 3700 GC-FID equipped with a 60 m DB-5 column. Temperature programming was 80°C (for whole oils, 40°C) with 1 min hold time, then at 1°C min^{-1} to 103°C, followed by a 3°C min^{-1} gradient to 140°C, a 1°C min^{-1} gradient to 180°C and finally 10°C min^{-1} to 310°C. The analysis was performed with hydrogen carrier gas and split injection.

A HP 5890 GC gas chromatograph (helium carrier gas) with a 60 m DB-1 column (i.d. = 0.25 mm) interfaced with a Finnigan MAT Incos 50 quadrupole mass spectrometer (electron impact mode, 70 eV) was used for biomarker work (*m/z* 191, 218, 217, 231, 253). The gas chromatograph was programmed from 180°C to 310°C at 2°C min^{-1}. Split injection was used. Before analysis *n*-alkanes were removed from the samples using silicalite (West *et al.* 1990). Aromatic fractions were analysed for *m/z* = 231 and 253 on a Fisons MD 800 GC-MS using a 60 m DB-SMS column (i.d. = 0.25 mm, programmed 80–180°C with 10°C min^{-1} and 180–310°C with 1.7°C min^{-1}). Selected samples (28 out of 33) were also analysed on a high resolution VG TS 250 mass specrometer interfaced with a Hewlett Packard 5890 GC, with a 50 m SE54 column (i.d. = 0.22 mm). The following ions were monitored: *m/z* 177.1642, 191.1800, 205.1956, 259.2473, 217.1956, 218.2034, 231.1174, 253,1956. The instrument was operated in electron impact mode (70 eV). Temperature programming of the gas chromatograph was from 45°C to 150°C at 35°C min^{-1} followed by a gradient of 2°C min^{-1} to 310°C. Splitless injection was used.

The isotopic composition ($\delta^{13}C$ and $\delta^{2}H$) of oils and condensates was analysed on the dead crude/condensates directly (approximately the C_{10+} fraction) utilizing combustion in sealed ampullas and measurement of carbon dioxide and hydrogen (reduction of water by zinc) in a Finigan MAT 251 mass spectrometer. All samples were analysed two to three times and the reported data are the best estimates. Precision was ± 0.1 for the $\delta^{13}C$ measurements and ± 4 for the $\delta^{2}H$ analysis.

Twenty-eight selected samples were analysed for $\delta^{13}C$ composition of individual saturated hydrocarbons (GC–IRMS). A VG Isochrome II system with a Hewlett Packard 5890 GC containing a 50 m OV-1 column (i.e. = 0.22 mm) was used (Bjorøy *et al.* 1991). Temperature programming was from 50°C to 300°C at 4°C min^{-1}. Splitless injections were used. The pristane, phytane and the normal alkanes *n*-C_{14} to C_{21} were analysed. The precision of the $\delta^{13}C$ measurements of individual hydrocarbons is estimated in the ± 0.2 range for these samples.

Results and discussion

Gas/oil ratios

Table 1 shows the database used in this study. Figure 6 illustrates the range of gas/oil ratios (GOR) of the samples used in this study, superimposed on a generalized phase diagram from

Table 1. *Database used in this study*

Sample no.	Sample code	Well	Test	Field	Formation	Depth interval for DST (m RKB)	Density		API	Subsurface GOR (m^3/m^3)	Evaporation Loss†
							Gas*	Oil condensate ($g\ cm^{-3}$)			
1	TR1	6406/3-2	DST2	Trestakk	Garn	3937–3995	0.81	0.83	39.0	200	40
2	T2	6407/1-2	DST1	Tyrihans Sør	Garn	3659–3669	0.735	0.792	47.2	869	55
3	T1	6407/1-3	DST1	Tyrihans Nord	Garn	3698–3703	0.665	0.872	30.8	148	30
4	I1	6407/6-3	DST1	Mikkel	Garn	2479–2489	0.770	0.819	41.3	2120	60
5	I2	6407/6-3	DST2	Mikkel	Not	2546–2555	0.755	0.752	56.7	2500	85
6	I3	6407/6-3	DST3	Mikkel	Ile	2570–2577	0.720	0.741	59.5	2500	85
7	N1	6407/7-1	DST3	Njord	Tilje 1	2946–2973	0.704	0.834	38.2	207	30
8	N2	6407/7-1	DST5	Njord	Ile	2758–2781	0.786	0.834	38.2	218	30
9	N3	6407/7-2	DST1	Njord	Tilje + Åre	2870–2879	0.744	0.815	42.1	180	30
10	N4	6407/7-3	DST2	Njord	Tilje 1	2990–3014	0.740	0.810	43.2	190	30
11	N5	6407/7-3	DST3	Njord	Tilje 2	2852–2868	0.731	0.803	44.7	389	45
12	N2	6407/7-4	DST2A	Njord	Tilje 3	2990–3008	0.700	0.831	38.8	181	30
13	N7	6407/7-4	DST2B	Njord	Tilje 3	2990–3008	0.735	0.832	38.6	170	30
14	D1	6407/9-2	DST2	Draugen	Garn	1651–1657	0.96	0.825	40.0	38	40
15	D2	6407/9-5	DST1	Draugen	Rogn	1654–1661	0.785	0.825	40.0	18	40
16	H5	6507/3-1	DST3	Alve	Garn	3611–3636	0.71	0.80	45.4	3432	55
17	H6	6507/7-2	DST2	Heidrun	Tilje	2417–2439	0.617	0.922	22.0	48	20
18	H7	6507/7-2	DST6	Heidrun	Tofte	2232–2245	0.650	0.700	70.6	2636	95
19	H1	6507/7-4	DST1	Heidrun	Garn	2494–2499	0.688	0.970	14.4	84	30
20	H1	6507/7-4	DST2	Heidrun	Garn	2470–2476	0.685	0.879	29.5	84	30
21	H3	6507/7-4	DST3	Heidrun	Garn	2449–2465	0.705	0.970	14.4	88	30
22	H4	6507/7-5	DSTA2	Heidrun	Garn	2355–2375	0.680	0.868	31.5	89	30
23	X2	6407/4-1	DST2		Garn	3889–3919	0.790	0.800	45.4	1600	60
24	M4	6407/2-2	DST1	Midgard	Garn	2476–2486	0.690	0.770	52.3	8352	85
25	M1	6507/11-3	DST1	Midgard	Ile	2514–2520	0.720	0.849	35.2	172.5	40
26	M2	6507/11-3	DST2	Midgard	Ile	2496–2509	0.690	0.809	43.4	4330	85
27	M3	6507/11-3	DST3	Midgard	Garn	2413–2420	0.675	0.809	43.4	43.30	85
28	S1	6506/12-3	DST1	Smørbukk Sør	Tilje	4222–4241	0.850	0.815	42.1	460	50
29	S2	6506/12-3	DST6	Smørbukk Sør	Lysing	3162–3173	0.795	0.815	42.1	197	45
30	B1	6506/12-5	DST2	Smørbukk Sør	Garn	4004–4010	0.895	0.827	39.6	280	30
31	B2	6506/12-5	DST2A	Smørbukk Sør	Garn	4004–4010	0.795	0.835	38.0	280	30
32	B3	6506/12-5	DST2B	Smørbukk Sør	Garn	4004–4010	0.795	0.827	39.6	280	30
33	B4	6406/12-5	DST4	Smørbukk Sør	Finnvær Gp.	3174–3178	0.718	0.800	45.4	150	40

* Relative to air

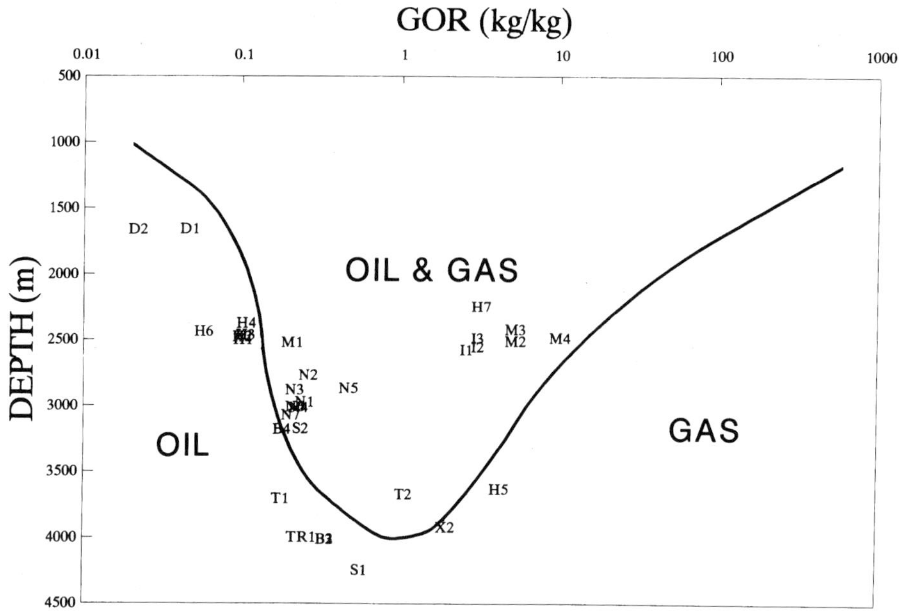

Fig. 6. Gas/oil ratios (GOR) of petroleum samples analysed in this work superimposed on phase diagram from England & Mackenzie (1989). The line separating gas from oil fields was originally drawn on the basis of fields in the Viking Graben, Norwegian Continental Shelf. The figure is tentatively used to portray that while most oils are presently close to saturation or in communication with gas, the Draugen samples are grossly undersaturated and some fields are apparently supersaturated, i.e. N5 from Njord, Mikkel samples (I), Tyrihans Sør (T2). This may reflect that these fields are in a different pressure regime compared to most of the other fields.

England & Mackenzie (1989). Figure 7 shows the GOR frequency distribution of the petroleums analysed.

While it is clear that in the subsurface the petroleums are either fluids, i.e. oil, or gases, a division of the fluids such as shown in Fig. 6 is useful for geochemical investigations on petroleums and, as will become evident later, genetic correlations in particular. Many oils and gases (condensates), except the Draugen, plot in the vicinity of the saturation line for hydrostatic conditions for oils from the Viking Graben, used to establish the empirical diagram (Fig. 6). This suggests that most oils are close to their bubble-point and that most gases are close to their dew-point, assuming that the reservoirs are at hydrostatic pressure. The figure also indicates that the composition of some of these fluids, as well as the regional temperature and pressure gradients, are comparable, in a gross sense, to the conditions for the samples discussed by England & Mackenzie (1989).

The M1 oil-leg sample from Midgard plots close to the oil region in Fig. 6, whereas the condensate samples M2, M3 and M4 plot close to the gas region. The gas/condensate Midgard Field with its 11.5 m oil leg in the Beta segment represents a system where the gas/condensate is saturated and in equilibrium with the oil leg (Mo *et al.* 1989). Petroleums such as N5 from Njord, Mikkel samples (I) and Tyrihans Sør (T2) plot in the supersaturated region. This may reflect the fact that these fields have developed overpressure or that they are currently in direct communication with overpressured drainage areas (Fig. 8). In this map, medium overpressure is indeed indicated for 6407/4-1 (X2), 6507/3-1 and Njord, while the insert in Fig. 8, taken from Ehrenberg *et al.* (1992), further suggests that the deep accumulations Smørbukk Sør (S1, B1, B2 and B3) and Trestakk (TR1), occur in the transitional to overpressured regions.

One field deserves special attention. The GOR of the Draugen fluids (Table 1, and Ellenor & Mozetic 1986) is lower than the values generally expected to be expelled from a type II kerogen even at incipient expulsion (England & Mackenzie 1989). Indeed, our results suggest that the Draugen oils are not less mature than oils from Mikkel (6407/6-3) and Tyrihans Nord (6407/1-3), both of which have significantly higher GOR values (Table 1). Heum *et al.* (1986) suggested that Draugen drained oil

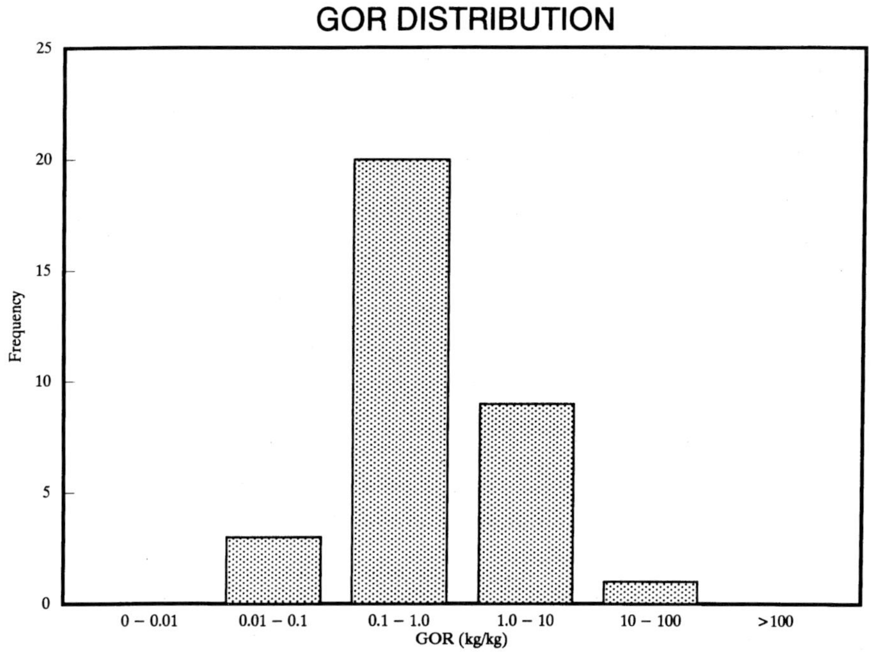

Fig. 7. Frequency distributions of analysed petroleums with GOR classifications according to England & Mackenzie (1989).

from the Spekk shales to the west, whereas the dip of the Åre strata possibly focused gas/condensates westwards, away from the Draugen. If the Draugen Field were filled at shallower depth without receiving later petroleum, an undersaturated situation, as at the present, could result due to progressive burial, i.e. the GOR would not change. Thus, the grossly undersaturated Draugen oil suggests that this accumulation is not in contact with the deeper parts of the original drainage area today and/or that the Draugen might represent the oil leg of a partitioned petroleum phase, and that the seal of this hydrostatic field (Provan 1992) can retain higher-boiling hydrocarbons more effectively than gas.

Continued movement in the fault zones to the west of the Draugen, occurring during or after oil migration, may have disrupted later migration to the field by juxtapositioning unfavourable carriers. It is also possible that diffusive loss of methane has ocurred from this rather shallow accumulation, as methane can leave even shale cap rocks in a geologically short period by diffusion-related mechanisms (Krooss 1985, 1988). Geochemical data, discussed later, on two Draugen oils (6407/9-2 and 6407/9-5) do not reveal any maturity or significant facies differences between these oils, suggesting that oil arrived at this laterally extensive accumulation as a comparatively time-constrained migration event. The fact that the oils in these two wells appear to be slightly biodegraded (see later discussions) might suggest that there is no significant recent/current influx of oil to this structure.

However, earlier gas influx to the accumulation may have occurred over a wider time interval and may be responsible for the different GOR values, which suggest filling from the west/north (Karlsen & Larter 1989), and for the varied isotope values of methane $\delta^{13}C$ which in the field range from about –51‰ (suggesting mixture with early reservoired biogenic gas) in the south, to a more thermogenic signature (-46‰) in the northern part of the field (S. R. Larter & W. A. England pers. comm.).

When comparing petroleum samples of such varied GOR as in the current database, due attention must be paid to potential phase-induced fractionation or partitioning of petroleum compounds ocurring either during migration or in the reservoir (Silverman 1965; England *et al.* 1987; England & Mackenzie 1989). Such phenomena are known to mask facies and maturity relationships if these are based on empirical utilization of molecular and isotopical parameters (Thompson 1987, 1988; Thompson *et al.* 1990; Galimov 1973).

In the Midgard well (6507/11-3) gas/condensate is found in close proximity to the oil leg and the two

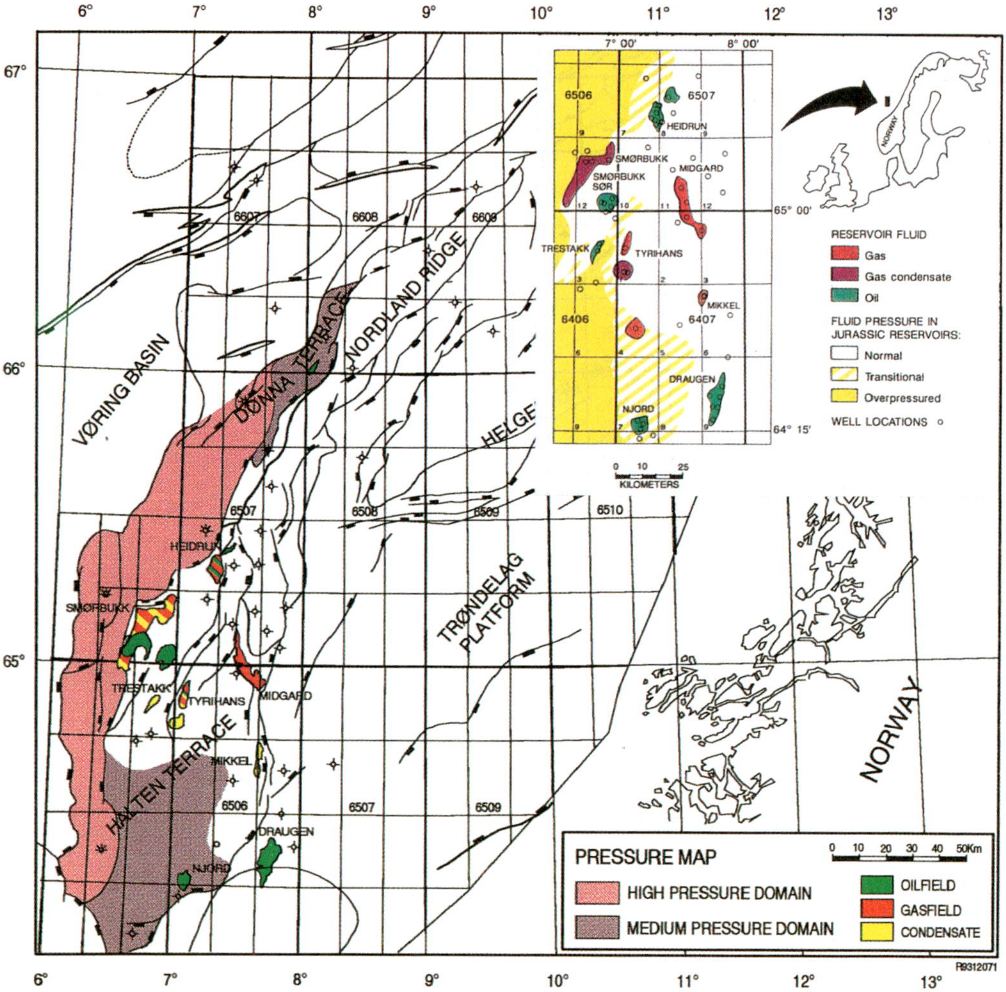

Fig. 8. Regions of normal, medium and overpressure. Also shown, in the insert (with the courtesy of AAPG), are similar data from Ehrenberg *et al.* (1992).

systems are evidently in equilibrium (Mo *et al.* 1989). Data discussed later clearly indicate that these fluids are affected by partitioning not only on a molecular level, but also in a gross compositional sense.

In two-phase systems, hydrocarbons with high vapour pressures (or, more accurately, fugacities) are found to be relatively enriched in the gas phase, while hydrocarbons of lower vapour pressures are found to be relatively enriched in the oil leg. The degree of fractionation, or partitioning, of the compounds between the two phases depends on the pressure, temperature and composition of the fluids. Thus, one would expect in this case that the relative amount of C_{20} and C_{21} triaromatic steroids relative to extended triaromatic steroids is higher in the gas cap than in the oil leg in the Midgard Field. This is indeed what is observed (see later). Similar relationships are observed for the ratios of tricyclic terpanes to pentacyclic triterpanes, trisnorneohopane (Ts) to trisnorhopane (Tm), C_{27} to C_{29} steranes, the lighter compounds 2-methylnaphthalene to 1-methylnaphthalene, phenanthrene to methylphenanthrenes and pristane to phytane (see later discussions).

More subtle evidence for intra-reservoir partitioning (less dramatic shift of the ratios) is observed in the 6407/7-3 DST 3 from Njord (Tilje

2), compared to the 6407/7-3 DST 2 (Tilje 1). The former fluid, which has a higher GOR ratio than the other fluids from Njord, has significantly higher *n*-alkane contents, higher total concentration of saturated hydrocarbons, a higher ratio of saturated to aromatic hydrocarbons, higher contents of mono- plus diaromatic compared to poly-aromatic hydrocarbons and lesser amounts of polar compounds (Table 2). Thus, this more gas-enriched fluid is also slightly enriched in compounds of higher vapour pressures, compared to the other Njord fluids. It is possible that the Tilje 2 part of the reservoir in this compartment, where the 6407/7-3 DST 3 fluid currently occurs, has been better shielded from gas stripping which may have occurred in the other Njord compartments and the reservoir unit in 6407/7-3 (DST 2, Tilje 1). The gas chimney above the Njord Field may be related to this observation. Similar evidence for partitioning is observed in Mikkel (6407/6-3), when comparing DST 1 with the higher GOR sample DST 3.

Thus, gas escaping from structures by dis-migration, or degassing from a structure in a pulse-like fashion due to too high a build-up of the petroleum head, may be another cause of partitioning of reservoired petroleum. In the current database, the gas chimney above Njord might be an example of this, while other well known examples of such phenomena in the Viking Graben include the Hild Field (Skålnes 1993) and the Gamma structure in the Gullfaks area.

Deep fields, such as 6407/4-1, may actually be

Table 2. *Gross compositional data of Haltenbanken petroleums and evaporative loss of C_{15-} compounds during Iatroscan analysis*

Sample no.	SAT%	ARO.1%*	ARO.2%†	POL%‡	SAT/ARO	ARO.1/ARO.2	ARO/POL	Evaporation Loss%§
1	62.8	20.3	8.6	8.3	2.2	2.4	3.5	40
2	70.2	21.3	2.2	6.3	3.0	9.7	3.7	55
3	58.1	22.7	11.4	7.8	1.7	2.0	4.4	30
4	57.1	22.9	13.4	6.6	1.6	1.7	5.5	60
5	70.9	25.5	2.4	1.2	2.5	10.6	23.3	85
6	71.3	26.7	1.0	1.0	2.6	26.7	27.7	85
7	74.6	15.6	6.3	3.5	3.4	2.5	6.3	30
8	74.8	16.6	5.2	3.4	3.4	3.2	6.4	30
9	74.6	15.4	6.6	3.4	3.4	2.3	6.5	30
10	75.2	14.0	5.7	5.1	3.8	2.5	3.9	30
11	81.2	13.6	3.6	1.6	4.7	3.8	10.8	45
12	75.5	15.6	6.3	2.6	3.5	2.5	8.4	30
13	75.0	16.8	5.8	2.4	3.3	2.9	9.4	30
14	68.2	21.3	6.6	3.9	2.4	3.2	7.2	40
15	67.4	21.8	6.9	3.9	2.4	3.2	7.4	40
16	78.0	20.8	0.5	0.7	3.7	41.6	30.4	55
17	44.3	25.0	12.9	17.8	1.2	1.9	2.13	20
18	63.4	29.3	≈0	7.3	2.2	≈	4.0	95
19	52.3	26.5	14.2	7.0	1.3	1.9	5.8	30
20	51.8	26.3	15.1	6.8	1.3	1.7	6.1	30
21	50.6	27.3	15.4	6.7	1.2	1.8	6.4	30
22	62.4	22.6	10.0	5.0	1.9	2.3	6.5	30
23	82.0	16.3	0.5	1.2	4.9	32.6	14.0	60
24	71.4	25.6	≈0	3.0	2.8	≈	8.5	85
25	74.3	20.8	3.6	1.3	3.1	5.8	18.8	40
26	79.0	17.7	0.1	3.2	4.5	177.0	5.6	85
27	78.1	18.6	0.1	3.2	4.2	186.0	5.8	85
28	75.2	16.2	4.1	4.5	3.7	4.0	4.5	50
29	79.5	15.3	2.5	2.7	4.5	6.1	6.6	45
30	70.8	18.7	7.6	2.9	2.7	2.5	9.1	30
31	67.7	19.0	7.4	5.9	2.6	2.6	4.5	30
32	69.4	18.4	6.3	5.9	2.8	2.9	4.2	30
33	83.1	12.1	2.2	2.6	5.8	5.5	5.5	40

* Percentage mono- and diaromatics in C_{15+}
† Percentage polyaromatics in C_{15+}
‡ Percentage asphaltenes and resins in C_{15+}
§ Percentage loss of C_{15-} during the analysis

less influenced by partitioning due to the shorter vertical migration distance involved in filling this deep field, unless intra-reservoir degassing has occurred, i.e. leaking reservoirs. However, the typical, close-to-supercritical conditions which prevail for some of the deepest accumulations might actually lead to a situation where small changes in pressure can result in considerable compositional modifications.

Fractionation during migration, where a gas containing an oil phase (i.e. a condensate) leaves the source rock to migrate in a decreasing temperature and pressure regime, will necessarily eventually result in at least one gas and one oil system. In a quantitative sense, the gas originally in solution in a supercritical petroleum phase (GOR = 0.35 kg/kg) expelled from a Spekk-type source at 4 to 5 km may expand to fill a volume equivalent to six times the initial system, at a depth of about 1 km (England & Mackenzie 1989). This may result in a situation where most shallow reservoirs might, from this simple volumetric relation, end up being filled with gas. Deeper structures might remain filled with oil, as the pressures and temperatures here allow more gas to be dissolved in the oil and vice versa. In this context, it may seem surprising that no free gas is found in the Draugen Field. A likely speculation is that free gas in this shallow structure has been lost by dismigration. If this is the case, there should be residual oil saturation below the oil column in the Draugen Field.

Between these two cases are those fields where co-migrating oil in the gas-exolving petroleum is accommodated as an oil leg after establishing a new pressure–temperature equilibrium, phase separation and mechanical equilibrium in accordance with not only the new temperature and pressure regime, but also possibly with the composition of earlier reservoired fluids. In this case, the gas/condensate dissolved in the arriving oil would of course represent a cogenetic system. This situation might be relevant to the Midgard structure (Elvsborg *et al.* 1985; Mo *et al.* 1989) and possibly also in Heidrun between 6507/7-2 DST 2 and the shallower, high-GOR DST 6 (Tofte Formation), while the Mikkel accumulation might represent a less pronounced manifestation of the same phenomenon.

Gross composition of petroleums (C_{15+})

Figure 9 and Table 2 show the gross composition of the analysed petroleums. Note in Fig. 10 the correlation between GOR and the evaporative loss of the C_{15+} fraction of the bottled fluids occurring during Iatroscan analysis (Karlsen & Larter 1991). There is likewise some correlation between evaporative loss and API (Fig. 11), which also tentatively indicates the maturity of the samples (cf. Longman & Palmer 1987). This suggests that rapid analysis of oils on Iatroscan TLC–FID can be used to get first-order estimates of GOR, API gravity and maturity of non-biodegraded petroleums.

The maturity assessment is, however, based on the assumption that more mature petroleums are more evaporative, i.e. of lower density. Thus, Fig. 11 indicates tentatively that there is a maturity trend from the biodegraded Heidrun oil (H6) as the least mature, through Tyrihans Nord (T1), some of the Heidrun oils (H4 and H2), some of the Smørbukk Sør petroleums (B1 and B2), the Njord petroleums (N), the Draugen oils (D), Tyrihans Sør (T2), 6407/4-1 (X), to the Mikkel condensates (I2 and I3) and the gas condensate 6507/7-2 DST 6, which are indicated to be of highest maturity. Yet, the biomarker distributions of the Mikkel condensates (I1 and I2) clearly reflect that these petroleums are in fact of lowest maturity in this dataset (see later). The lowest API sample, Heidrun 6507/7-2 DST 2, is clearly severely biodegraded with respect to the *n*-alkanes, but without the appearance of 10-desmethyl hopanes in *m/z* 177. The significant unresolved 'hump' of this Heidrun sample is observed in Fig. 27. Similar humps are seen but less pronounced in the other investigated Heidrun oils. The rising baseline, which indicates the mass of compounds in the unresolved complex mixture, is believed to reflect simply the reduced amount of *n*-alkanes with respect to the more complex saturated hydrocarbons inherited from the source-rock system and is simply more easily observed in biodegraded oils than in non-biodegraded analogues (cf. Killops & Al-Juboori 1990).

Thus, Heidrun sample 6507/7-2 DST 2 is by far the most biodegraded of the investigated Heidrun samples, all of which (excluding 6507/7-2 DST 6 from the Tofte Formation), about 200 m shallower, show evidence for biodegradation.

Except for sample 6507/7-2 DST 2 from Heidrun, with its high polar contents of 17.8% (more than twice as high as for the other Heidrun samples), the main difference between the samples, observable from Fig. 9, is a gradual increase in the amount of saturated hydrocarbons. Thus, there is apparently no major compositional step between the samples. Accordingly, the data could be interpreted to suggest that the gross composition of the samples, with the possible exception of 6507/7-2 DST 2, is strongly influenced by phase separation occurring during migration, as such processes are known to lead to significant gross compositional modifications (Krooss *et al.* 1991; Thompson 1987, 1988; Thompson *et al.* 1990; Silverman 1965; England & Mackenzie 1989).

It is, in our opinion, necessary to exercise caution if gross compositional data of petroleums with such

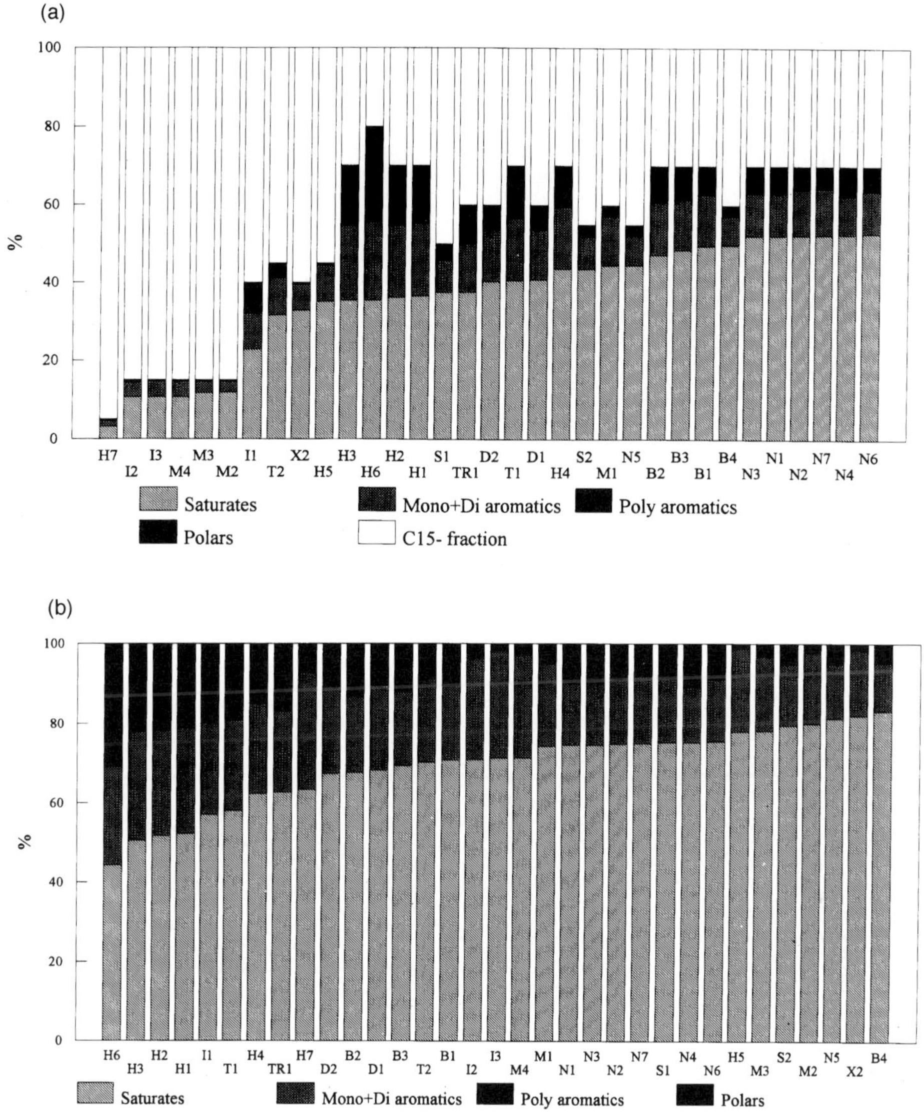

Fig. 9. **(a)** Gross compositional data (saturated hydrocarbons, mono- plus diaromatic hydrocarbons, polyaromatic hydrocarbons and polars) of the samples analysed. Also shown is the evaporative loss of the C_{15-} fraction. For sample identification see Table 1; HUS is the in-house standard. **(b)** As for (a) but showing for clarity only the C_{15+} fraction.

contrasting GOR are to be plotted in traditional ternary diagrams (saturated HCs, aromatic HCs and polar compounds) used for genetic classification, because evaporative fractionation and related processes will greatly influence the gross composition of the petroleums.

Note that in Fig. 9a, where the evaporative loss during analysis is shown, the amount of polyaromatic compounds is more or less similar among most of the Heidrun samples. This indicates that the polar content of 6507/7-2 DST 2 is probably related to intra-reservoir modifications, as such a polar-enriched oil would be expected to have comparatively higher polyaromatic contents (with respect to the other Heidrun samples) if the polars were to migrate with the oil into the reservoir. This is due to the mutual solubility relationship between polar compounds and polyaromatic hydrocarbons.

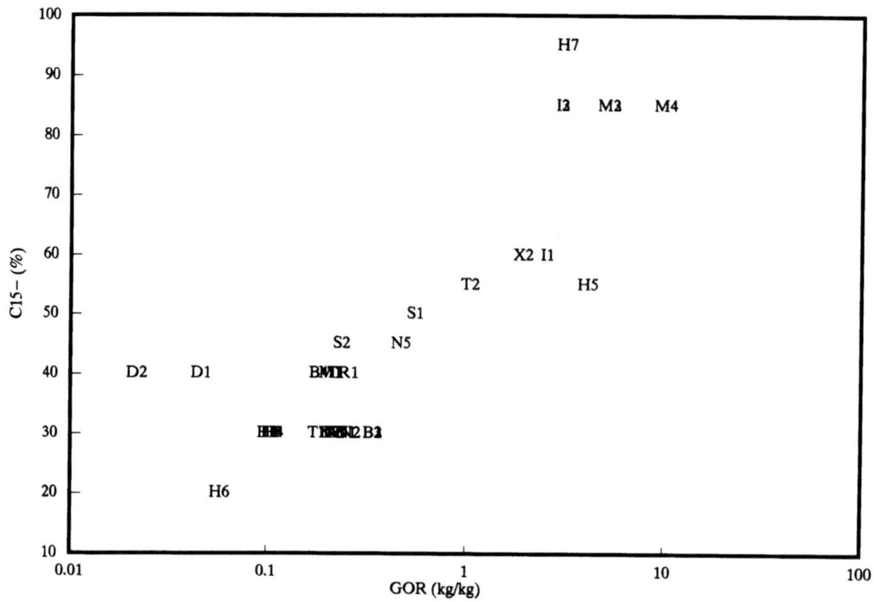

Fig. 10. Relationship between GOR and evaporative loss of the C_{15-} fraction of fluids analysed on Iatroscan TLC–FID. For sample identification see Table 1.

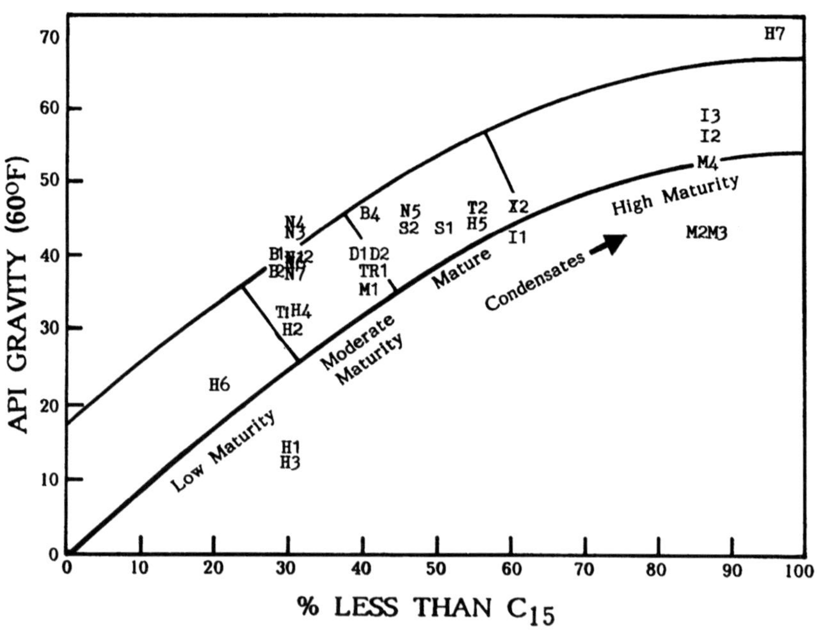

Fig. 11. Cross-plot of API gravity (calculated from density data) and evaporative loss of the C_{15-} fraction of the Haltenbanken fluids during Iatroscan analysis, plotted in a figure based on data from Longman & Palmer (1987). The plot differentiates the condensate/gas systems from oils and indicates to some extent the maturities of the individual petroleum systems. However, most molecular parameters suggest that the Mikkel (I) samples are the least mature in the dataset despite their high GOR and API.

In some of the gas/condensate/light oil samples such as 6507/7-2 DST 6, 6407/2-2 DST 1 Midgard, 6507/11-3 DST 3 Midgard and 6507/11-3 DST 2 Midgard, the polyaromatic content is close to zero and in most cases only limited amounts of polar compounds are found. This results in an abnormally high ratio of aromatic hydrocarbons to polars, and also, due to the mutual solubility relationship between polars and polyaromatic hydrocarbons, a high ratio of mono- plus diaromatic hydrocarbons to polyaromatic hydrocarbons in these samples (Table 2).

In 6407/6-3 DST 2 and 6407/6-3 DST 3, both from Mikkel, and 6407/4-1 DST 2, which are also high-GOR samples, the ratio of monoaromatic hydrocarbons to polyaromatic hydrocarbons is lower. This indicates to us, based also on geochemical data to be discussed later, that these samples are not evaporative condensates to the same extent as those discussed previously. In fact, judged by their biomarker distributions, 6407/4-1 DST 2 has the highest maturity of the samples in the database (Fig. 44), while the two Mikkel samples (DST 2 and DST 3) are of lowest maturity. Thus, the GOR of the petroleums can be a poor indicator of maturity, while the contents of polyaromatic hydrocarbons may be helpful in differentiating between low/moderate maturity evaporative condensates and high maturity condensates.

Figure 12 displays the gross compositional difference in the database. In short, the relative amount of polyaromatic hydrocarbons over mono- plus diaromatic hydrocarbons and the polar contents apparently helps to differentiate normal oils from mixed petroleum systems such as 6507/7-2 DST 6 (H7). Furthermore, the biodegraded high polar Heidrun oil 6507/7-2 DST 2 (H6), separates from the evaporative condensates 6507/11-3 DST 2 and 3 from Midgard and the mature condensate 6407/4-1 DST 2.

The amount of polyaromatic compounds (Fig. 13) correlates with the carbon preference index (CPI) of the petroleums measured on the *n*-alkanes in the range from C_{24} to C_{32}. The CPI ranges from 0.81 to 1.23 and is lowest in the Tyrihans Nord and the Heidrun samples. Sample 6407/1-3 DST 1 (Tyrihans Nord) deserves special attention with respect to CPI as this oil has an odd-to-even predominance in the C_{14}–C_{17} range, while a clear even-to-odd predominance is observed in the

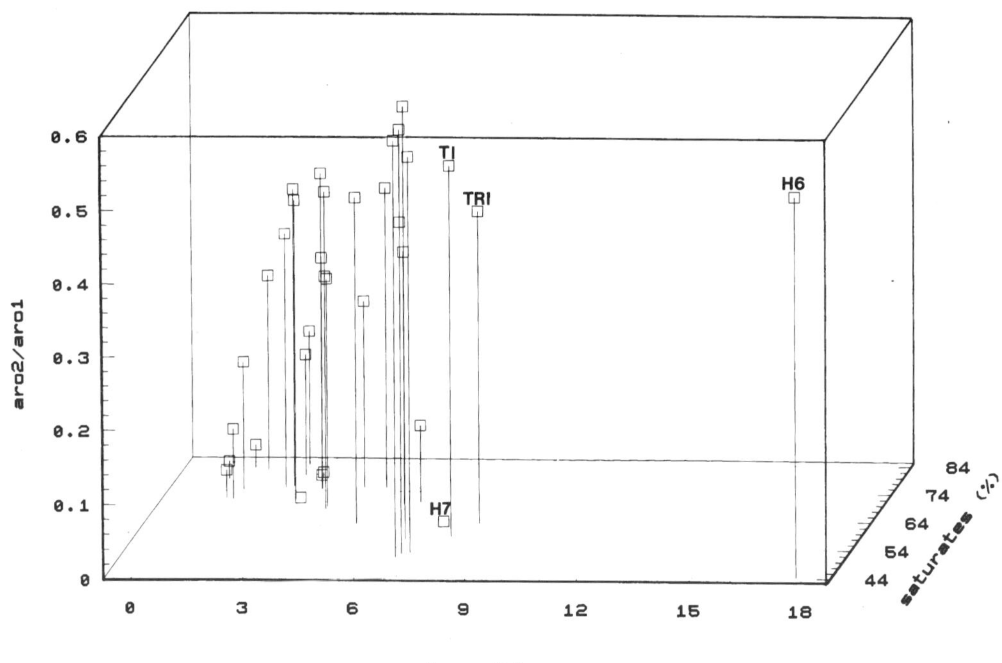

Fig. 12. Gross composition of the petroleums plotted on a box diagram, suggesting that the ratio between polyaromatic hydrocarbons and mono- and diaromatic hydrocarbons (aro2/aro1) and the contents of polar compounds is helpful for distinguishing condensates (high percentage saturated hydrocarbons and low aro2/aro1) from oils.

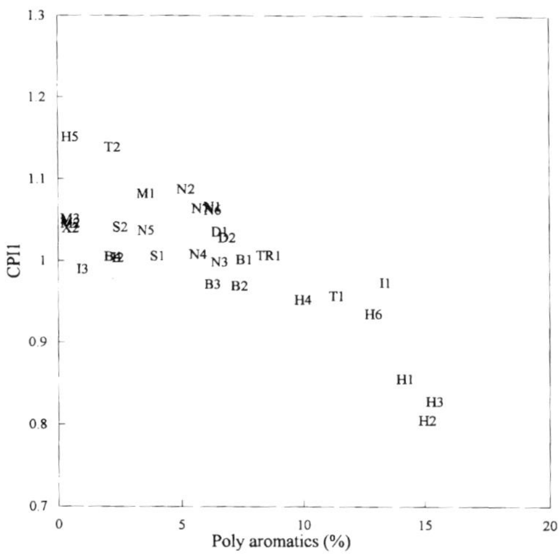

Fig. 13. Cross-plot of the carbon preference index (CPI for n-C_{24} to n-C_{32}) which is controlled mainly by organic facies and maturity, versus the content of polyaromatic hydrocarbons in the petroleums. The data show that the amount of polyaromatic hydrocarbons in the petroleums is high in the Heidrun samples H1, H2 and H3 and that odd predominance is slightly more prevailing than even predominance in the database.

C_{18}–C_{27} range, as shown in Fig. 22, implying mixed organic source-rock facies for this petroleum. A similar even-to-odd predominance is observed in core extracts from the reservoir units in 6506/12-1 from Smørbukk, a condensate field not included in this study. A possible carbonate-influenced source rock may be responsible for the even-to-odd predominance (cf. Shimoyama & Johns 1971, 1972) which is rather seldom observed on the Norwegian continental shelf.

The cause of the even predominance in the Heidrun oils might likewise be source influenced. These CPI data indicate that the Heidrun samples (6507/7-4 DST 1, 2 and 3) and the Midgard samples (6407/2-2, and 6507/11-3), Alve (6507/3-1) plus the Tyrihans Sør sample (6407/1-2 DST 1) have the lowest maturity, followed by the Heidrun sample (6507/7-5 DST 2A), Tyrihans Nord (6407/1-3 DST 1), the Mikkel samples (6407/6-3) and the Smørbukk Sør, Njord and the Draugen samples.

Note that the Mikkel samples of varying GOR do not differ markedly in CPI, though the percentage contents of polyaromatic hydrocarbons vary significantly. The same trend is observed for the Midgard samples (6507/11-3), which vary enormously in GOR. This indicates that the condensates (6507/11-3 DST 2 and 3) in Midgard are of similar maturity to the oil leg (6507/11-3 DST 1).

Isotope composition of whole oils

The isotope compositions, $\delta^{13}C$ and δ^2H, of the C_{10+} fraction of the oils and condensates are shown in Table 3. Taking into account the effects of maturity and phase separation on the carbon isotope composition of petroleums (see later discussions), the database can, on the basis of the carbon isotope signatures, subjectively and conveniently be separated into two apparent end member groups and one intermediate group. **Group 1**: ($\delta^{13} < -29$‰) Njord, Smørbukk Sør 6507/11-5, Tyrihans N, Tyrihans S, Trestakk, 6407/4-1 and 6506/12-3 DST 6. **Group 2**: ($\delta^{13} \approx -27$‰) Midgard and Alve. **Group 3**: ($\delta^{13} \approx -28$‰) Heidrun, Smørbukk 6506/12-3 DST 1, Draugen and Mikkel.

Sample 6407/4-1 DST 2, which, based on maturity parameters is clearly the most mature in the dataset, is placed in Group 1, because we feel

Table 3. *Stable carbon and hydrogen isotope data of the petroleum samples*

Sample	$\delta^{13}C$ (‰)	$\delta^{2}H$ (‰)
6406/3-2 DST 2 Trestakk	–29.5	–124
6407/1-2 DST 1 Tyrihans S.	–30.0	–123
6407/1-3 DST 1 Tyrihans N.	–29.5	–123
6407/6-3 DST 1 Mikkel	–28.4	–122
6407/6-3 DST 2 Mikkel	–28.8	–125
6407/6-3 DST 3 Mikkel	–28.6	–122
6407/7-1 DST 3 Njord	–29.6	–128
6407/7-1 DST 5 Njord	–29.0	–131
6407/7-2 DST 1 Njord	–29.5	–134
6407/7-3 DST 2 Njord	–29.7	–135
6407/7-3 DST 3 Njord	–29.4	–127
6407/7-4 DST 2A Njord	–29.3	–130
6407/7-4 DST 2B Njord	–29.4	–133
6407/9-2 Draugen	–28.8	–134
6407/9-5 Draugen	–28.8	–134
6507/3-1 DST 3 Alve	–27.5	–124
6507/7-2 DST 2 Heidrun	–29.0	–122
6507/7-2 DST 6 Heidrun	–28.2	–123
6507/7-4 DST 1 Heidrun	–28.9	–134
6507/7-4 DST 2 Heidrun	–28.6	–130
6507/7-4 DST 3 Heidrun	–28.5	–125
6507/7-5 DST 2A Heidrun	–28.8	–130
6407/4-1 DST 2	–28.2	–128
6407/2-2 DST 1 Midgard	–27.6	–131
6507/11-3 DST 1 Midgard	–27.5	–130
6507/11-3 DST 2 Midgard	–27.5	–134
6507/11-3 DST 3 Midgard	–27.9	–131
6506/12-3 DST 1 Smørbukk Sør	–28.9	–123
6506/12-3 DST 6 Smørbukk Sør	–29.4	–123
6506/12-5 DST 2 Smørbukk Sør	–29.2	–124
6506/12-5 DST 2A Smørbukk Sør	–29.1	–123
6506/12-5 DST 2B Smørbukk Sør	–29.5	–127
6506/12-5 DST 4 Smørbukk Sør	–29.5	–133

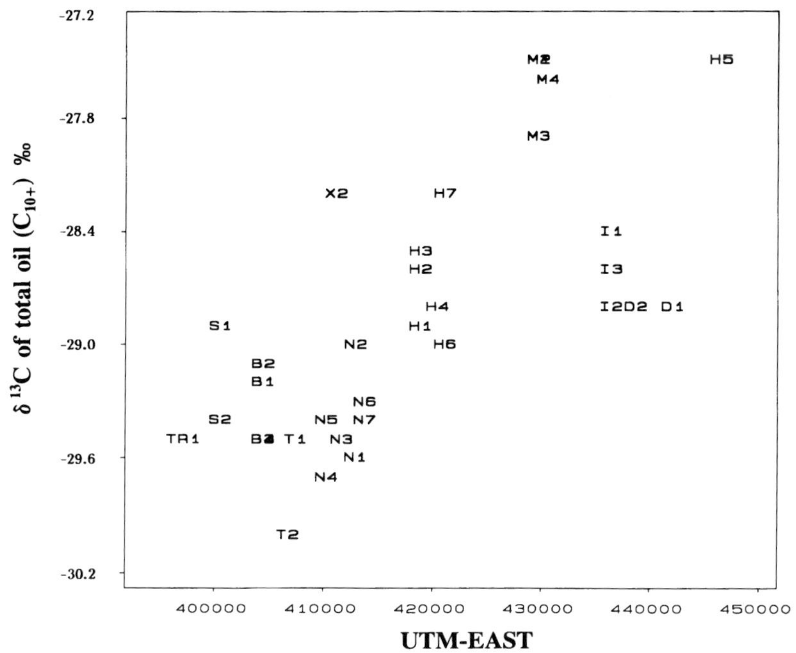

Fig. 14. Stable carbon isotope $\delta^{13}C$ values of the dataset (C_{10+}) plotted versus the UTM-east parameters of the sample localities. The data show that the $\delta^{13}C$ values of the petroleums are strongly influenced by the east–west coordinate of the field locations and we feel that this influence is predominantly a reflection of the organic facies of the source-rock systems which sourced the individual petroleums. This possibly reflects the fact that the early diagenetic environment of the source rock(s) changes from a more anoxic, purely marine facies in the distal portion of the basin, to a more proximal, more terrestrialy influenced (partly dysaerobic?) facies towards the east.

that the isotope signature of this condensate is shifted about 1‰ towards heavier values due to the maturity influence.

Figure 14 shows the $\delta^{13}C$ composition of the petroleums plotted as a function of the UTM-east map reference. We interpret this correlation to suggest that in a regional sense, most fields in this area received petroleum rather locally. The pattern could tentatively, in a genetic sense, reflect that the isotopically lighter hot shale facies of the Spekk Formation (Table 4) is not well developed (non-deposition or eroded) or not buried sufficiently deep towards the east, so that petroleum expelled in this region is carrying a signature more influenced by the stratigraphically deeper and older, isotopically heavier and more terrestrially and oxic-influenced Spekk Formation/Melke Formation shale facies. As mentioned earlier, we postulate such a facies shift to occur on the basis of analogies with the Kimmeridge Clay/Draupne Formation/Mandal Formation in the Viking Graben/Central Graben (Bailey *et al.* 1990), and what is known from drilled sections of the Spekk Formation in the Haltenbanken region, e.g. the facies shift in the Spekk Formation observed by Whitley (1992) as a shift in maceral composition from sapropelic shale in 6507/7-1 (northwest of Heidrun) to vitrinite-rich shale/siltstone on the Heidrun Field.

Thus, the indicated trend reflects a change in the carbon isotope signature of the Spekk Formation

Table 4. *Compilation of stable isotope data of the Spekk and Åre Formation source rocks*

Sample	$\delta^{13}C$ (‰)	$\delta^{2}H$ (‰)
6407/7-2 2839.5 m Coal Njord*	–24.5	–127
6407/7-2 2889.0 m Coal Njord*	–27.4	–128
6407/7-4 3086.5 m Coal Njord*	–25.4	–131
6407/7-4 3136.9 m Coal Njord*	–24.1	–120
Åre Fm. n = 4†	–25.8 ± 0.7	
Åre Fm. Kerogens n = 11‡	–26.1 ± 1.1	
Spekk Fm. Kerogens n = 3†	–28.2 ± 2.7	
Spekk Fm. Kerogens n = 10‡	–28.5 ± 1.6	

* Non-extracted coals, centre portion
† Cohen & Dunn (1987)
‡ Elvsborg *et al.* (1985)

from isotopically heavier terrestrial input in the proximal eastern part of the basin, to isotopically lighter organic matter in the distal west. Taking Walters law into consideration, a change from lighter carbon isotope values in the top of the Spekk Formation to heavier in the lower parts of the Spekk Formation and possibly also the Melke Formation, could be postulated. Contributions to the heavier end of the C_{10+} fractions of the reservoired petroleums from the Åre Formation coals are rated as minor based on the biomarker signatures discussed later. However, we have no data to exclude contributions from the coals to the gasoline/gas range of the reservoired petroleums, nor to exclude contributions from intra-Åre shales, which potentially can be locally qualitatively and quantitatively important.

Sample 6407/4-1 DST 2, which has all the characteristics of a mature condensate (Figs 30 & 44), is believed to be isotopically slightly heavier than its western location should suggest, mainly because it is the most mature petroleum in the dataset, as evidenced by cracking of extended hopanes.

In the case of the Smørbukk Sør (6506/12-3) petroleum DST 1 and the nearly 1100 m shallower DST 6 from the Cretaceous Lysing Formation (S1 and S2), the difference in their isotope signature is entirely explainable within the context of enrichment of the heavy isotope in the higher GOR petroleum (S1 GOR = 460 m^3/m^3, S2 GOR = 197 m^3/m^3) according to the general principles outlined by Galimov (1973, 1985).

It may thus seem surprising that the Midgard (6507/11-3, M1, M2 and M3) petroleums do not differ significantly in their $\delta^{13}C$ value, despite the great differences in GOR between the gas cap and the oil leg (Table 2). The molecular distributions discussed later support our postulations that oil was present earlier in the gas zones in this field and that the oil was later displaced by gas. However, fluid-inclusion studies of the vertical distribution of fluorescent petroleum inclusions in this field, according to Karlsen *et al.* (1993) and Nedkvitne *et al.* (1993), should be undertaken before drawing any conclusion on this matter.

Sample 6507/3-1 (H5), which is located far to the north of the other Haltenbanken samples (Fig. 3), may be isotopically heavy mainly due to a source effect, i.e. Alve is supposedly being sourced mainly from a terrestrially influenced facies variety of the Spekk Formation.

Figure 15 shows the carbon isotope composition of the petroleums as a function of the depth to the reservoirs. There is apparently a trend indicating that samples from reservoirs between approximately 1600 m and 3700 m contain oil or condensable fluids (C_{10+}) which become isotopically lighter with increasing depth, while reservoirs from levels deeper than 3700 m have a more weakly developed opposite trend (i.e. isotopically heavier oil or condensable C_{10+} fluids) with increasing depth.

The data are tentatively interpreted to reflect that the deeper reservoirs, down to *c.* 3700 m, have significant volumes of mature, isotopically lighter Spekk Formation within their drainage areas, while the shallower eastern reservoirs have received oil from a more terrestrially influenced Spekk Formation source-rock facies, as outlined above.

The Midgard samples (6507/11-3 and 6407/2-2) are interpreted to depart slightly from the left trend-line in Fig. 15, suggesting that the petroleum in this field has a smaller contribution from an anoxic purely marine Spekk Formation than any of the other investigated samples except Alve. According to Mo *et al.* (1989), only the Åre Formation has reached sufficient maturity and expelled sufficient quantities of C_{10+} hydrocarbons within the drainage area of the Midgard Field. Still, our data on biomarker distributions from Åre Formation coals do not support the interpretation that the HCs come from coals. Instead, we postulate that the Spekk Formation is the main oil generator even within the drainage area of the Midgard horst and that migration here is (or was) more long-ranging and earlier than commonly believed. Alternatively, the deeper parts of the Melke Formation or the shales in the Åre Formation may have contributed to the reservoired oil. Still, we have no evidence to exclude the possibility that the Åre Formation coals may have contributed to the reservoired gas and the light hydrocarbon (e.g. the C_{10-}) fractions of the petroleums.

The remaining samples in the lower part of the right trend-line shown in Fig. 15 are the most westerly placed accumulations in this dataset. The isotope composition of the heavy petroleum fraction (C_{10+}) of these deep fields is interpreted to reflect contributions to the reservoired petroleum from a less terrestrially influenced source rock, i.e. from a distal open-marine-type anoxic shale.

Note that the Draugen oils (D1 and D2) may have had their source area at a similar depth to the source areas of the petroleum in the Mikkel Field (6407/6-3, I1, I2 and I3 at *c.* 2500 m), which is the closest accumulation.

When comparing Fig. 14, which shows $\delta^{13}C$ of the petroleums versus UTM-east, which we interpret to reflect mainly increasing oil expulsion of the hot shale facies of the Spekk Formation towards the west, and Fig. 15, which shows $\delta^{13}C$ plotted versus depth to the accumulations, we feel inclined to suggest that most petroleums have experienced moderate lateral migration. The main exceptions to this are the Draugen petroleums (D1 and D2) and, to a lesser extent, the Mikkel (I1,

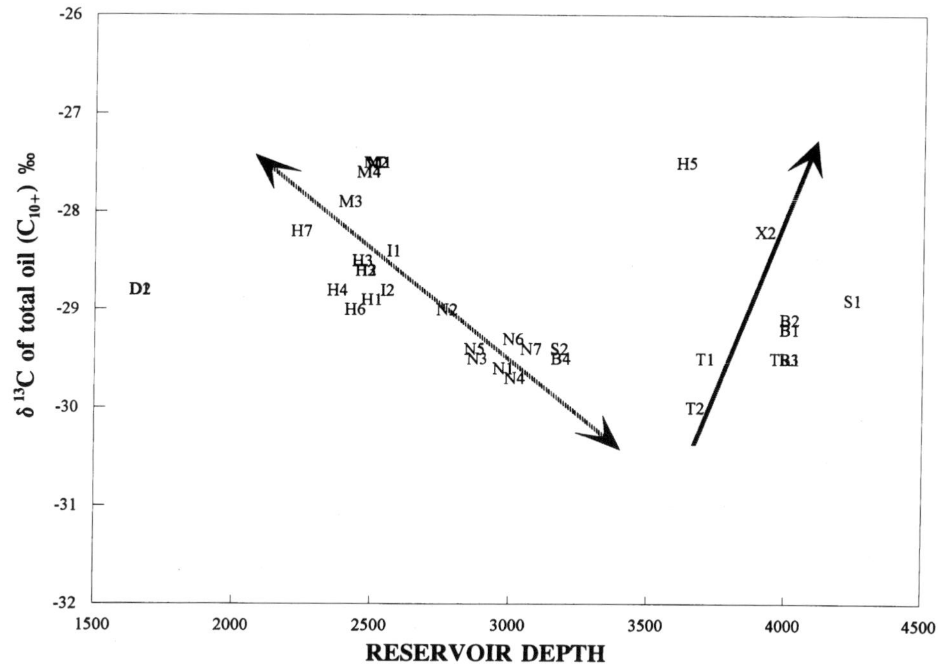

Fig. 15. Stable carbon isotope $\delta^{13}C$ values of the dataset (C_{10+}) plotted versus depth of the accumulations. The figure tentatively indicates, in combination with Fig. 14, that most fields apparently have laterally limited migrational distances (except the Draugen). Furthermore, the isotope compositions of the petroleums are explainable in the context of dominantly distal marine anoxic-derived Spekk Formation C_{10+} contributions to the reservoired petroleums in the deeper western part of the region. In the eastern fields, a more terrestrially influenced facies of the same formation, or less likely of the Åre Formation, is assumed to have contributed significantly to the C_{10+} fractions as measured on the reservoired petroleums.

I2 and I3) and Midgard (M1–M4) petroleums. This interpretation is supported by the modelled maturity map (Fig. 3). We postulate further that none of the petroleums seem to have experienced drastically more vertical migration than the other samples. The only possible exceptions are the 6407/3-1 sample (H5) and the Midgard (6507/11-3) samples (M1, M2, M3 and M4).

The gross compositional make-up of the samples, in terms of percentage saturated hydrocarbons, mono- plus diaromatic hydrocarbons, polyaromatic hydrocarbons and polars, varies considerably (Table 2). Owing to the thermodynamically ordered carbon isotope distribution of the source organic matter (Galimov 1985), even genetically related petroleums, i.e. a phase-separated condensate and black oil leg, can be expected to differ isotopically in cases of pronounced compositional fractionation.

Thus, genetically, associated condensates and oils have been found to exhibit a 2–3‰ enrichment in the heavy isotope in the condensate (Galimov 1973). The reason for this is the enrichment of the condensate in more transformed hydrocarbons (gasoline-range hydrocarbons) compared to the associated oil (kinetic isotope effect). The gasoline-range hydrocarbons in the condensates are themselves the source of isotopically lighter gaseous molecules and hence they are depleted in the light isotope (Galimov 1985). Also, effects such as enrichment in the condensate in the heavier isostructural saturated hydrocarbons relative to *n*-alkanes (intramolecular isotope effect; Galimov 1973, 1985) will make the condensate heavier than the associated oil leg.

Owing to the generally close relationship between the $\delta^{13}C$ value of the constituents of the polar compounds (asphaltenes plus resins) in the petroleum (Galimov 1973), variations in the relative amounts of these compounds are not expected to complicate our attempts at genetic correlations between oils nor between the oils and source-rock kerogens.

Asphaltenes and resins of petroleums that have not undergone extensive thermal cracking, which is the case for the fluids in this database (except perhaps 6407/4-1), can in this context be regarded as soluble kerogen (Tissot 1981; Bandurski 1982; Behar & Pelet 1984; Solli & Leplat 1986). It follows that their isotopic composition is highly relevant for any oil–oil, oil–source rock correlation purposes, because the isotope composition of these compounds reflects more closely the isotope composition of the source-rock kerogen. Often, the isotopic composition of the polars (asphaltenes and resins) is relatively similar to the oil as a whole (Galimov 1973). The low concentration of polars and polyaromatic compounds in condensates (Table 2, Fig. 9) could thus be expected to manifest itself in giving the condensates isotopically lighter carbon isotope values. As discussed previously, however, an inverse relationship is often observed (Galimov 1973) and this is ascribed to partitioning or the kinetic isotope effect of mature condensates. It follows that the isotope composition of the condensates is likely to be influenced by these two effects. Under weak partitioning conditions, the condensate is expected to be isotopically similar or lighter compared to the parent oil, while in the case of pronounced partitioning, it will become heavier, as discussed previously. It is clear to us that a database such as the present one cannot be interpreted either purely statistically nor in a geological context without bearing these effects in mind.

It follows that the condensate and the black oils from Heidrun are possibly more closely related than their total carbon isotopic signatures suggest (condensate being heavier; see Table 3). In fact, as discussed later, these samples show clear indications of being genetically related (same maturity by CPI, Fig. 13) and no indications of different source facies are found in their biomarker distribution.

We suggest that the predominant cause of the carbon isotope variation is the source effect. A depletion of heavy carbon isotopes from *c.* two to four per mil in bitumens relative to kerogens for post-Carboniferous sediments (Galimov 1973, 1980) may be used as a first approximation for correlating oils and kerogens.

Thus, bearing in mind the difference described earlier in $\delta^{13}C$ between Spekk and Åre Formation kerogens, Group 1 petroleums could be interpreted to contain C_{10+} compounds mainly sourced from the Spekk Formation. Group 3, being slightly heavier, could accordingly contain significant contributions both from the Spekk and the Åre Formations, while Group 2 could be interpreted to contain C_{10+} compounds predominantly derived from the Åre Formation.

However, the biomarker data on extracts from Åre Formation coals do not support this interpretation. We propose instead that the isotopically heavier signature towards the east is reflecting a facies shift in the Spekk Formation from isotopically lighter, prolific marine sapropelic hot shales in the west to a more vitrinitic, proximal dysaerobic, terrestrially influenced facies in the east. It is, furthermore, likely that occasional non-deposition/erosion or development of a more condensed sequence in the Spekk Formation occurs locally towards the east (Figs 4 & 5), and that oil-range hydrocarbon contributions from the Spekk Formation here are more biased towards the isotope signatures expected for the deeper and older Melke Formation.

There is no correlation between δ^2H and $\delta^{13}C$ of the oils in the database. The stable hydrogen isotope (δ^2H) composition of the petroleums becomes, however, slightly lighter towards the southeast, suggesting that in the Haltenbanken area there may be a facies component in this direction. It is known that different types of kerogens and, in particular, different groups of coal macerals vary significantly with respect to their δ^2H and $\delta^{13}C$ signatures (Schwarzkopf & Schoell 1985) and, furthermore, that bitumens from terrestrial systems often tend to be lighter with respect to hydrogen than the marine biota (Redding *et al.* 1980; Yeh & Epstein 1981). Based on the limited data on the hydrogen isotope signature of the Åre Formation coals (Table 4), one can only speculate about the reasons for the light δ^2H values in the Åre coals. These values are in the range of the sapropelic coals and about -20‰ lighter than typical vitrinite and inertinite analysed by Schwarzkopf & Schoell (1985). The maturity of these coals was determined by vitrinite reflectance measurements ($R_o = 0.59$ $n = 7$, 125 readings). The results might suggest that a significant proportion of the macerals in Åre coals, which may generate liquid hydrocarbons during marginally mature conditions, are comparatively low in deuterium and represent liptinitic-derived material (Thompson *et al.* 1985; Hvoslef *et al.* 1988). Thus, the *n*-alkanes released would be expected to have comparatively light δ^2H values.

Evidently, it is necessary to investigate the δ^2H signatures of the Spekk, Melke and Åre Formations in more detail and to expand the lateral database on the isotope signatures of Spekk, Melke and Åre Formation kerogens.

Carbon isotope composition of individual n*-alkanes*

Twenty-seven oils selected from the database and a coal extract from Njord were analysed for carbon isotope composition of individual *n*-alkanes in the saturated fraction, in order to investigate in detail the genetic relations within the C_{14}–C_{21} hydrocarbon range, which is outside the hydrocarbon range where biomarkers occur. Furthermore, it was hoped that this analysis could shed some light on the issue of contributions from the Åre Formation coals to the condensate-range hydrocarbons, e.g. also to *n*-alkanes in the C_{14}–C_{21} hydrocarbon range.

The data are presented in Table 5 and in Figs 16 to 18. It is observed that in many petroleums, the normal alkanes are progressively isotopically lighter with increasing carbon number. This is to be expected, as these *n*-alkanes are transformed from longer parent molecules by cracking and are themselves a source for isotopically lighter gaseous compounds produced by kinetically controlled cracking in the source rock (Galimov 1985). In our opinion, the samples 6507/11-3 DST 2 and DST 3 (M2 and M3, GOR = 4330 m^3/m^3) and 6507/11-3 DST 1 (M1, GOR = 172.5 m^3/m^3) from Midgard are cogenetic, despite their different GORs, because of the similar carbon isotope distribution over the alkane range investigated (Fig. 17). Thus, the $\delta^{13}C$ distribution of the individual *n*-alkanes, and the comparatively isotopically lighter isoprenoids, clearly shows that the gas condensates are genetically closely related to the oil leg in Midgard. Therefore, the idea that the isotope composition of the light ends analysed reflects a mixing of oil with an isotopically heavier condensate in these samples is not favoured. Yet, the 6407/2-2 DST 1 (M4, GOR = 8352 m^3/m^3) sample from Midgard (Fig. 16) appears to be of different origin compared to the other Midgard samples.

A similar trend and 'pattern' to those observed for the Midgard samples M1, M2 and M3 is observed in the condensate sample 6407/4-1 (Fig. 17), which is interpreted to be the sample of

Table 5. *Stable carbon isotope data on individual* n-*alkanes, pristane and phytane*

Sample no.	Component									
	n-C_{14}	*n*-C_{15}	*n*-C_{16}	*n*-C_{17}	Pristane	*n*-C_{18}	Phytane	*n*-C_{19}	*n*-C_{20}	*n*-C_{21}
1	−30.6	−30.7	−30.7	−31.4	−30.8	−31.2	−30.7	−30.7	−30.9	−30.9
2	−30.5	−30.3	−30.3	−30.9	−30.6	−30.8	−30.6	−30.5	−30.5	−30.5
3	−29.5	−29.8	−29.7	−30.3	−30.6	−30.4	−30.5	−30.4	−30.4	−30.5
4	−30.4	−29.4	−29.8	−30.1	−29.9	−29.9	−29.9	−29.7	−29.1	−29.2
5	−29.8	−29.8	−30.0	−30.4	−30.4	−30.5	−30.8	−30.4	−30.3	−30.9
6	−28.7	−29.4	−29.3	−29.6	−30.4	−29.7	−30.8	−29.7	−29.7	−30.3
8	−29.0	−28.8	−29.0	−29.5	−29.9	−29.3	−29.1	−29.3	−28.9	−28.8
9	−30.2	−30.2	−30.3	−31.1	−30.7	−31.1	−30.2	−30.5	−30.0	−30.7
11	−29.9	−30.2	−30.3	−30.6	−31.0	−31.0	−30.7	−30.4	−30.9	−31.3
14	−28.7	−28.7	−29.0	−29.4	−28.9	−28.8	−28.4	−28.6	−28.4	−28.8
15	−29.3	−29.4	−29.3	−29.7	−30.0	−30.1	−29.7	−29.7	−30.4	−30.3
16	−27.4	−27.5	−27.7	−28.2	−28.4	−28.2	−28.1	−28.1	−28.0	−28.0
17	−28.2	−28.3	−28.6	−28.8	−29.2	−29.0	−29.0	−28.7	−29.0	−29.0
18	−28.9	−28.1	−28.1	−29.3	−28.8	−29.1	−28.8	−28.4	−28.4	−28.6
19	−27.7	−28.1	−28.0	−28.4	−29.3	−28.8	−28.2	−28.2	−28.2	−28.5
20	−28.9	−29.0	−29.0	−29.1	−29.6	−29.3	−29.8	−29.2	−29.6	−29.5
21	−28.9	−29.1	−29.8	−29.8	−30.5	−30.2	−30.9	−30.6	−30.7	−31.3
22	−29.1	−29.2	−29.2	−29.5	−30.0	−29.7	−30.1	−29.6	−29.6	−29.6
23	−28.1	−28.3	−28.4	−29.0	−29.8	−28.6	−29.6	−29.1	−29.4	−29.5
25	−27.9	−28.1	−28.3	−28.5	−29.5	−28.5	−29.4	−28.3	−28.2	−28.0
26	−28.2	−28.3	−28.4	−28.7	−29.5	−28.7	−29.2	−28.9	−29.0	−29.2
27	−27.8	−28.2	−28.2	−28.1	−29.4	−28.2	−29.5	−28.2	−28.3	−27.9
28	−29.0	−29.1	−29.0	−29.2	−30.1	−29.5	−30.0	−29.6	−29.8	−29.8
29	−29.7	−29.8	−30.0	−30.2	−30.5	−29.9	−30.0	−29.8	−29.8	−29.8
30	−29.5	−29.6	−29.5	−29.7	−30.5	−29.8	−30.5	−29.6	−29.5	−29.5
32	−29.5	−29.6	−29.5	−29.8	−30.6	−29.8	−30.6	−29.8	−30.1	−30.2
33	−28.9	−29.2	−29.4	−29.8	−30.2	−29.5	−30.0	−29.8	−29.9	−30.0
6407/7-2 2839.5 m Coal ext. R_m = 0.55%	−29.2	−29.3	−29.4	−29.8	−30.0	−30.1	−29.0	−30.1	−29.7	−30.2

* $\delta^{13}C$ ratios in ‰ relative to the PDB standard

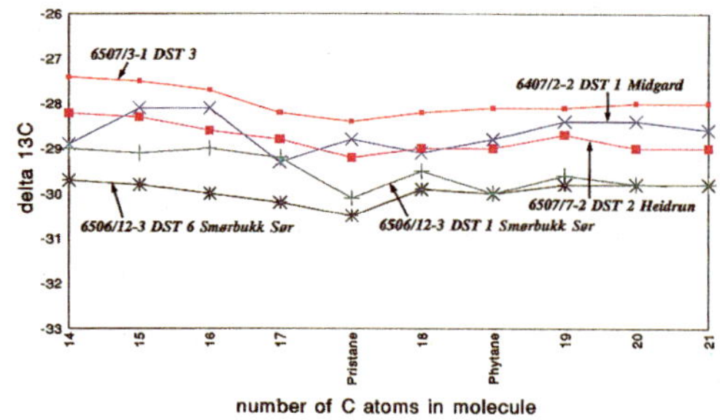

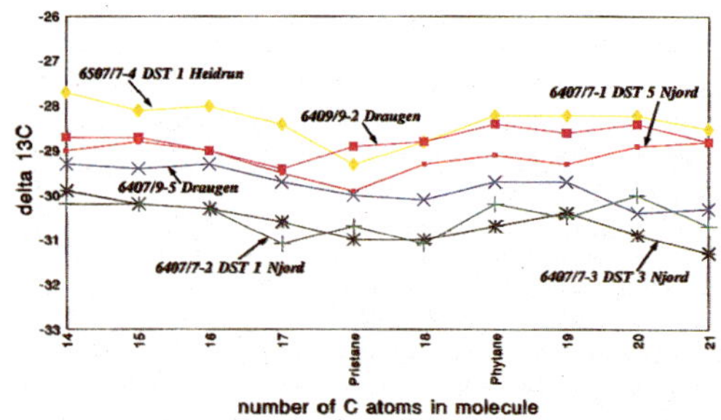

Figs 16–18. Stable carbon isotope $\delta^{13}C$ values of individual *n*-alkanes and the isoprenoids pristane and phytane. Three main groups are distinguished.

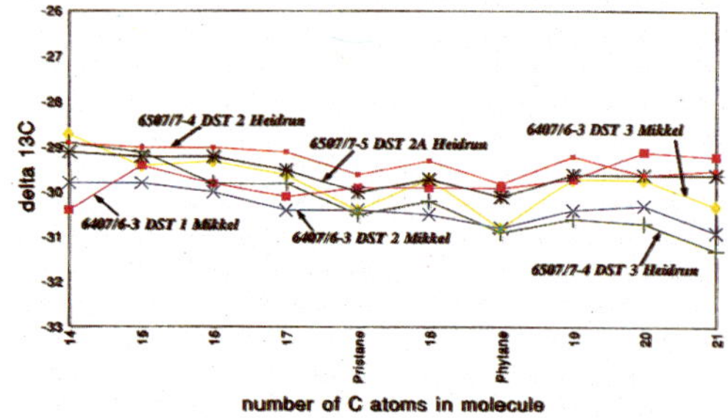

Fig. 18

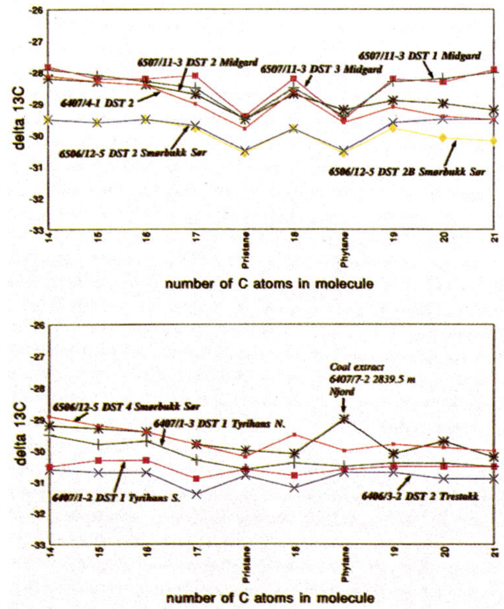

Fig. 17

highest maturity in this database. Note that this sample does not display a ‘pattern’ significantly different from the Midgard and the Smørbukk Sør petroleums, despite the fact that the latter is, overall, isotopically lighter. Note also that there are, in some cases, discrepancies in the pattern of closely associated samples in the heavier end (e.g. DST 2 and DST 2B samples from Smørbukk Sør). This is partly related to a lower precision in the measurements of low concentration *n*-alkanes. Thus, small discrepancies in patterns in the heavier end should not be overinterpreted.

Therefore, also belonging to the ‘Midgard family’ are the 6506/12-5 samples (B1 to B4) from Smørbukk Sør, and also the mature condensate 6407/4-1 DST 2 (X2, GOR = 1600 m^3/m^3) which plots together with the Midgard condensates (M2 and M3). Slightly related to the Midgard oil leg family are the 6506/12-3 DST 2 and 6 (S1 and S2) samples from Smørbukk Sør. However, these samples can also, in a gross sense, be associated to the Njord (N), Draugen (D), Trestakk (TR1), Tyrihans Sør (T2), Heidrun (H, NB but not H5), Tyrihans (Nord) and Mikkel petroleums (I). Within this group the Tyrihans Sør and the Trestakk petroleums are extremely similar. The minor differences existing between them with respect to stable isotope composition of total oils and molecular distributions are more likely to reflect differences in the palaeo-petroleum composition of the fields than differences related to the later infill of petroleum from the Spekk Formation. Thus, our data tentatively indicate that these two accumulations are part of the same migration system, and that there is, based on these isotope data, no evidence to suggest migration of C_{10+} hydrocarbons from Tyrihans Sør/Tyrihans Nord into Midgard, contrary to the suggestions of Heum *et al.* (1986) who based their conclusions on pressure–volume–temperature (PVT) modelling.

Note that the proposed biodegradation does not

seem to have modified the isotope signatures of the lightest measured *n*-alkanes from Draugen, (Figs 18 & 25) the Smørbukk Sør sample (6506/12-5 DST 2B, Fig. 17) which has a flat *n*-alkane profile $< n\text{-}C_{15}$, nor for the extremely biodegraded Heidrun sample (6507/7-2 DST 2, H6, Figs 16 & 27).

The 6407/7-1 DST 5 sample from Njord differs from the other Njord samples and is in some respects similar to the Draugen sample 6407/9-2. This may indicate that the Njord structure represented by 6407/7-1 DST 5 has a different filling history to other Njord substructures. This interpretation is strongly supported by the molecular composition of this Njord sample relative to other Njord samples.

The difference in the isotope composition of the two Draugen samples in their heavier end is surprising in view of their similar biomarker and aromatic and saturated hydrocarbon chromatograms. A larger database from the laterally extensive Draugen structure should be investigated before any conclusions are drawn. However, it is not unlikely that the southern sub-structure penetrated by the 6407/9-5 well where the Garn Formation, unlike in the northern wells (6407/9-1, 9-2, 9-3), is non-existent (Provan 1992), has seen a different migration history compared to the northern part of the field. The Draugen Field was interpreted by Karlsen & Larter (1989) to have been filled from west-northwest. Thus, the discrepancy may reflect the fact that the 6407/9-5 sample is genetically related to the 6407/7-2 DST 1 and 6407/7-3 DST 3 samples from Njord, whereas the 6407/7-1 DST 5 from Njord is more closely related to oils in the northern Draugen wells (represented by 6407/9-2) produced by migration from similar source rocks.

Detailed examination also reveals that the normal alkane profile of 6507/7-4 DST 2 (H1) from Heidrun is isotopically slightly heavier than the rest of the Heidrun samples (Fig. 18). The condensate 6507/3-1 DST 3 from Alve (Fig. 16) is isotopically heavier and so significantly different from the rest of the sample set with respect to $\delta^{13}C$ *n*-alkane distribution that a significant facies difference is inferred.

Note that the $\delta^{13}C$ *n*-alkane profile of the coal extract 6407/7-2 2889.0 m ($R_m = 0.55\%$) falls in the range of some of the isotopically lighter oils. This tentatively indicates that this coal produces *n*-alkanes with a $\delta^{13}C$ signature similar to that of some of the oils in the database. Thus, apparently these data suggest that *n*-alkane contributions to reservoired fluids from coals may be difficult to trace and that the carbon isotope signatures of the *n*-alkanes from such coals are not isotopically heavy compared to those of the reservoired petroleums.

Conclusions of genetic linkage in the sample set based on stable isotopes

Based on the $\delta^{13}C$ values of individual *n*-alkanes, the samples were subjectively, for convenience differentiated into three groups, with only the Alve sample (6507/3-1 DST 3) being uniquely different to the rest. **Group A**: Njord, Draugen, Trestakk, Tyrihans Sør, Tyrihans Nord. Group A consists of most of Groups 1 + 3 from whole oil isotope classification. **Group B**: Midgard, Smørbukk Sør, 6407/4-1. **Group C**: Heidrun and Mikkel, which have characteristics intermediate between groups A and B. Note that the samples from Smørbukk Sør and 6407/4-1 are separated from Group 1, defined on the basis of the $\delta^{13}C$ of the total extract, and are now placed with Midgard.

It is clear from the previous discussion that grouping of the samples differs somewhat depending on whether one uses the $\delta^{13}C$ *n*-alkane distribution or the $\delta^{13}C$ of the whole oils for correlation. This is to be expected in mixed systems, as the gross composition of the petroleums varies as discussed.

It is likely that the polar compounds (resins plus asphaltenes) and the aromatic hydrocarbons retain original facies signature better than the transformed *n*-alkanes. But, on the other hand, it is clear that the heavier compounds, i.e. the resins and the asphaltenes, are also those expected to be most apt to carry over an isotope signature of earlier dismigrated reservoired oil, to a later arriving oil or condensate. Accordingly, these compounds, which have low mobility during dismigration, can easily be mixed with later arriving petroleum.

The results suggest that there is no convincing evidence for an isotopically heavy condensate contribution to the petroleum in the deep 6407/4-1, Smørbukk Sør and Trestakk, which are the three deepest petroleum accumulations in the database (> 3800 m).

It is also necessary to analyse in more detail a varied sample set from the Spekk and the Åre Formations, before deciding on the relative contributions of the Spekk versus Åre Formations to the petroleums in the Haltenbanken region. In particular, it may be necessary to analyse the carbon isotope composition of the aromatic fractions and possibly also the carbon isotope signatures of extracts from coals prepared by different degrees of crushing, i.e. similar to extraction of bitumen from 'closed' versus 'open' pores, as generally discussed by Beletskaya (1978), Sajgo *et al.* (1983) and Price & Clayton (1992). This could shed light on fractionation during primary migration (Leythaeuser *et al.* 1987), which may be relevant if complex cyclo-alkane ring systems, such as the biomarkers, are retained more effectively in the coal submicropores than slender *n*-alkanes with their smaller kinetic diameters.

Molecular distributions in the petroleums

The data discussed previously focused on parameters which characterize major proportions of the petroleum such as GOR, gross composition of the liquid phase and the carbon and hydrogen isotope compositions of the petroleums. This section focuses on the detailed distribution of molecular parameters in the database, including biomarkers, and on concentrations of individual compounds in the petroleums.

Figure 19 displays the concentration of individually quantified saturated and aromatic hydrocarbons in the oils as a principal component plot. Sample coding is shown in Table 1. The two axes at 45 degrees to the two principal components are interpreted as enrichment of some of the petroleums in saturated hydrocarbons versus aromatic hydrocarbons, and light hydrocarbons versus heavy hydrocarbons, respectively. The increase in saturated compounds may reflect phase separation which will produce, on one hand, a more paraffinic gas/condensate, corresponding to the high-GOR samples (e.g. M2, M3 and M4 from Midgard), and, on the other, a residual oil leg enriched in polars and aromatic hydrocarbons (e.g. M1 from Midgard). This axis may also represent a source effect as petroleum samples from the western locations generally have higher *n*-alkane concentrations, e.g. Mikkel (I), Njord (N) and the 6506/12-3 DST 6 (S2) above the main reservoir Smørbukk Sør.

Figures 20 to 36 display representative chromatograms of the saturated and the aromatic petroleum fractions from each field in the database. Also shown are the *m/z* 191 and the *m/z* 231 ion chromatograms. These represent tricyclic terpanes, tetracyclic terpanes, pentacyclic triterpanes and triaromatic steroid distributions in the oils.

Selected maturity and facies parameters are shown in Tables 6 and 7 and are displayed in a principal component plot (Fig. 44). It is beyond the scope of this paper to discuss comprehensively all molecular characteristics of each petroleum. However, we will point to some observations made on the samples from this database which span a wide variety of petroleums, from gas/condensates to dead black oils. The gas/condensate 6507/7-2 DST 6 (H7) has a biomarker distribution which is believed to represent a mixture of earlier reservoired petroleum with a gas/condensate (Fig. 28). Only one obvious high maturity condensate occurs in the database, namely 6407/4-1, with evidence for thermal destruction of extended hopanes (Fig. 30). Most other condensates, such as M2, M3 and M4 from Midgard 6507/11-3 have biomarker distributions characterized by moderate maturity.

Biodegradation has severely affected the Heidrun oil in 6507/7-2 (Fig. 27). Whitley (1992) suggests that the petroleum column in this central well was biodegraded when the field was *c*. 1000 m shallower than today, and that this biodegraded fluid was mixed with later arriving oil. The extensively biodegraded oil in 6507/7-2 and the significant 'hump' in the chromatograms of the saturated fractions from the Heidrun flank wells 6507/7-4 (Fig. 29) and 6507/7-5, might reflect the fact that the reservoir compartment penetrated by the 6507/7-2 well, upon being buried sufficiently deeply to pass the phase-out floor for biodegradation, did not receive large volumes of fresh non-biodegraded oil from the potential source basin to the southwest. Non-biodegraded oil was probably more effectively mixed with biodegraded oil down-flank (wells 6507/7-4 and 6507/7-5). These occurrences of biodegraded versus less biodegraded oil might also reflect the lack of intra-reservoir communication between these compartments, e.g. the central part of the structures penetrated by the 6507/7-2 well was shielded or represents a dead-end with respect to later arriving oil. Incipient biodegradation has in addition affected the Draugen petroleums from 6407/9-2 (Fig. 25) and 6407/9-5. This conclusion is based on the depletion of *n*-alkanes in the C_{15-} range occurring in both samples.

Generally, the database displays lower pristane/phytane ratios in the west (Trestakk (Fig. 20), Smørbukk Sør (Figs 34 & 35) and Tyrihans Sør (Fig. 21)) and higher ratios in the east (Midgard (Figs 31 to 33) and Draugen (Fig. 25)) despite the fact that most petroleums in the west are of higher maturity. This trend is likely to reflect a change in organic facies.

Note the even predominance in the C_{19} to C_{26} range in the Tyrihans Nord petroleum (Fig. 22, marked by an asterisk) yet an odd predominance in the C_{17-} range. Even predominance is known to occur in oils sourced from carbonate-influenced facies (Shimoyama & Johns 1971, 1972). An even predominance is also observed in reservoir core extracts from the Smørbukk Field. This signature may reflect a facies shift in the Spekk Formation from generally high phytane content, both relative to pristane and *n*-alkanes and absolutely to lower values in the east and north.

Maturity of the petroleums

The presence of well defined and relatively high concentrations of monoaromatic steroid hydrocarbons in the samples from Draugen (6407/9-2

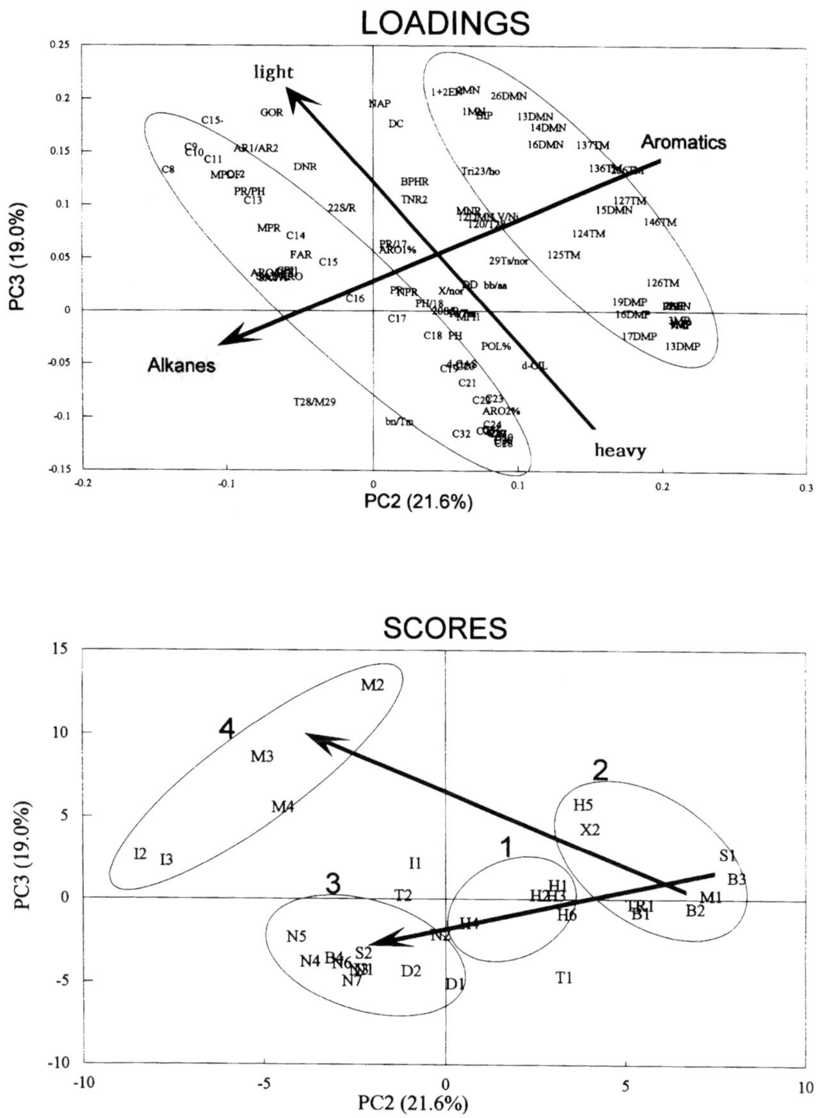

Fig. 19. Principal component plot of quantified saturated (C14 denoting an *n*-alkane with 14 carbon atoms) and aromatic hydrocarbons (MN = methylnaphthalenes, DMN = dimethylnaphthalenes, TMN = trimethylnaphthalenes, PHEN = phenanthrene, MP = methylphenanthrenes and DMP = dimethylphenanthrenes; the prescript refers to the different isomers). Also shown are various other biomarker and isotope parameters. The plot displays the paraffinic character of most of the Njord samples (N) except sample N2, the higher GOR Mikkel samples (I2,I3), the Smørbukk Sør (B4) and the S2 sample above the Smørbukk Sør accumulation. the composition of the samples reflects both the organic facies and enrichment of saturated compounds in higher GOR samples. Note in particular, enrichment of light boiling saturated compounds (e.g. C_{14}) and light boiling aromatics, e.g. MN relative to phenanthrenes and higher boiling *n*-alkanes in the condensates from Midgard (M2, M3) relative to those in the oil leg from Midgard (M1). H6 is a biodegraded Heidrun sample lacking most of the lighter *n*-alkanes, Fig. 27.

and 9-5), Mikkel (6407/6-3), Tyrihans Nord (6407/1-3), Tyrihans Sør (6407/1-2), Heidrun (6507/7-4, -5 and -2 DST 2) and Njord (e.g. 6407/7-2) signifies that these petroleums are significantly less mature than the rest of the sample set (Figs 37 & 38).

The possibility of mistaking other compounds for the monoaromatic steroid hydrocarbons in the

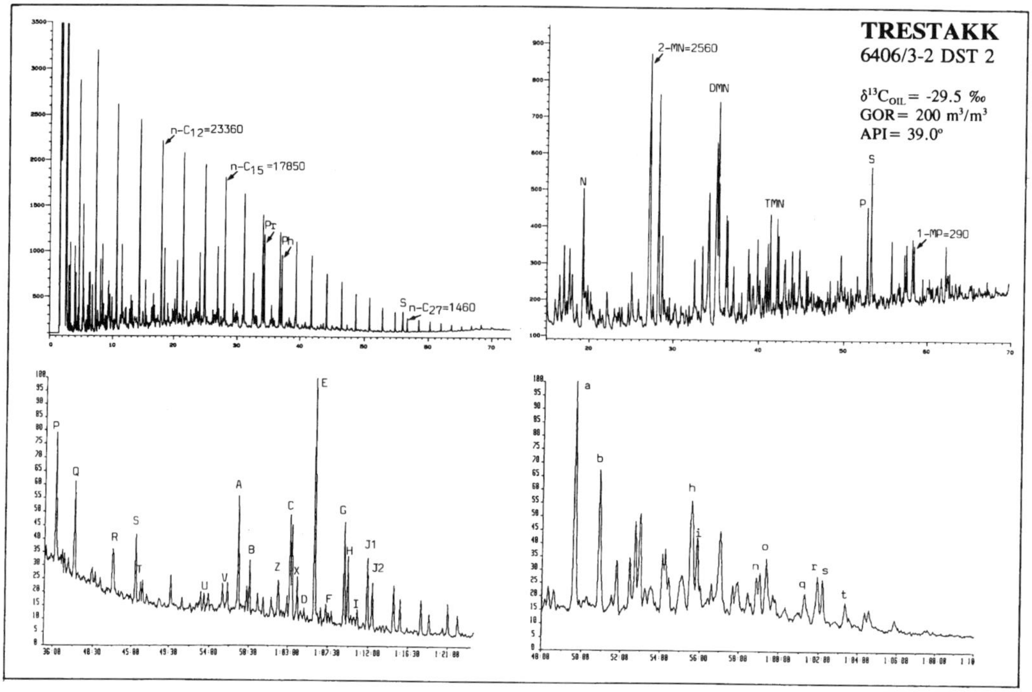

Figs 20 to 36. GC–FID and GC–MS traces of selected samples from the database. Shown are GC–FID traces of saturated and aromatic fractions from oils; S denotes internal standards (squalane and anthracene) added to the samples before MPLC separation. The GC–MS ion-chromatograms show the distribution of 'tricyclic terpanes, tetracyclic terpanes, pentacyclic triterpanes and steranes (see Tables 8 & 9 for peak identification). Note, in all samples, the occurrence of bisnorhopane, compound Z. Also note the correlation between the ratios of Ts/Tm = A/B, hopane X/normoretane = X/D and the ratio of the C_{29}Ts/C_{29} hopane, i.e. compound C and its nearly co-eluting compound (between C and X). We propose that there is no major facies difference between the petroleums in this region and that the Spekk Formation is the most likely source rock for all petroleums.

mature samples, e.g. 6407/4-1 DST 2 (Fig. 37), necessitates caution in calculating the ratio between monoaromatic and triaromatic steroids (Table 7). In particular, Abbott & Maxwell (1988) demonstrated by laboratory experiment that the aromatization rate constant for non-rearranged monoaromatic steroids is faster than for the rearranged isomers at any one particular temperature. Furthermore, C_{29} monoaromatic steroids co-elute with their non-rearranged counterparts if both series are present in geological material. Therefore, when measuring the degree of aromatization (amount of triaromatic steroid / (triaromatic steroid + monoaromatic steroid)), one can easily obtain too low a value if a more resistant rearranged monoaromatic steroid is misidentified as a non-rearranged monoaromatic steroid.

Abbott & Maxwell (1988) found no evidence for transformation of non-rearranged to rearranged non-aromatic steroids. Indeed, Beach *et al.* (1989) suggest, based on laboratory thermal alteration of steroids on bentonite at 200°C, that the increase in the ratio of short-chained (C_{20} and C_{21}) to long-chained triaromatic steroid hydrocarbons (e.g. C_{28}), commonly observed in increasingly mature oils and sediment extracts, reflects preferential (faster) degradation of the long-chain homologue (C_{28}) and not transformation to short-chain homologues.

Figure 3 shows a regional maturity map calculated at base Cretaceous, which demonstrates, in light of the mono- to triaromatic steroid parameter discussed previously, that most fields except the Draugen and Midgard have short to medium lateral migration distances. Furthermore, the geological profiles (Figs 4 & 5) display progressively deeper burial of the Spekk Formation towards the west.

The other aromatic maturity parameters, such as the methylnaphthalene ratio (MNR), the dimethylnaphthalene ratio (DNR), the trimethylnaphthalene ratio (TNR2) and the methylphenanthrene index 1 (MPI1), were found to have a low degree of co-variation or to be relatively invariate in the dataset (Table 6). Furthermore, these parameters do not correlate well with the biomarker maturity

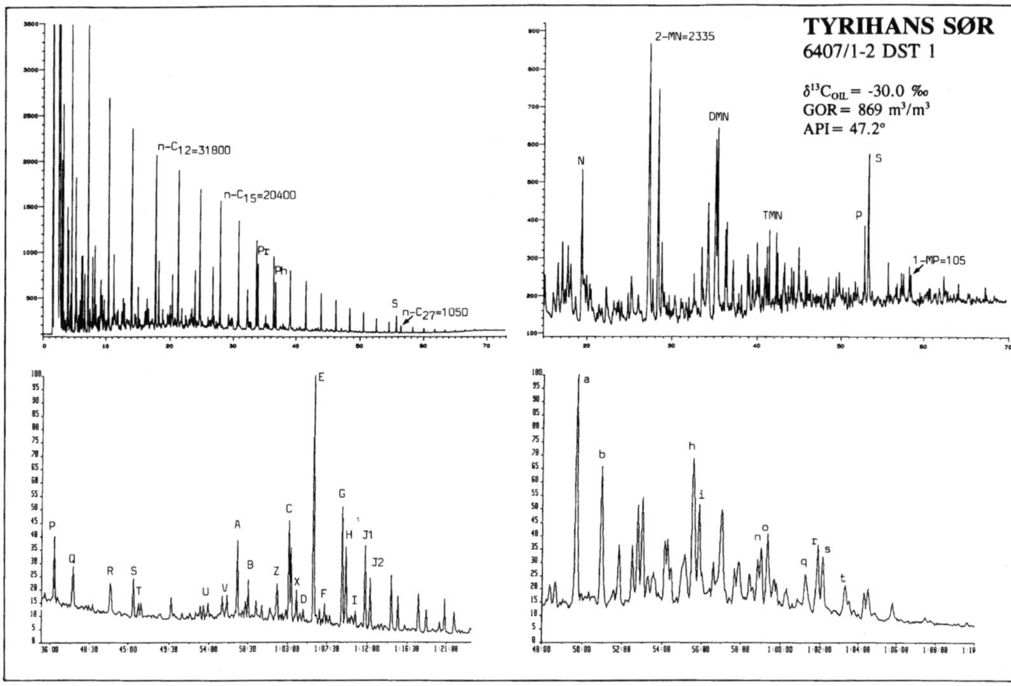
TYRIHANS SØR
6407/1-2 DST 1
δ13COIL = -30.0 ‰
GOR = 869 m3/m3
API = 47.2°
n-C12=31800
n-C15=20400
Pr
Ph
S
n-C27=1050
2-MN=2335
DMN
N
TMN
P
S
1-MP=105
P Q R S T U V A B Z C X D E F G H I J1 J2 S
a b h i n o q r s t

Fig. 21

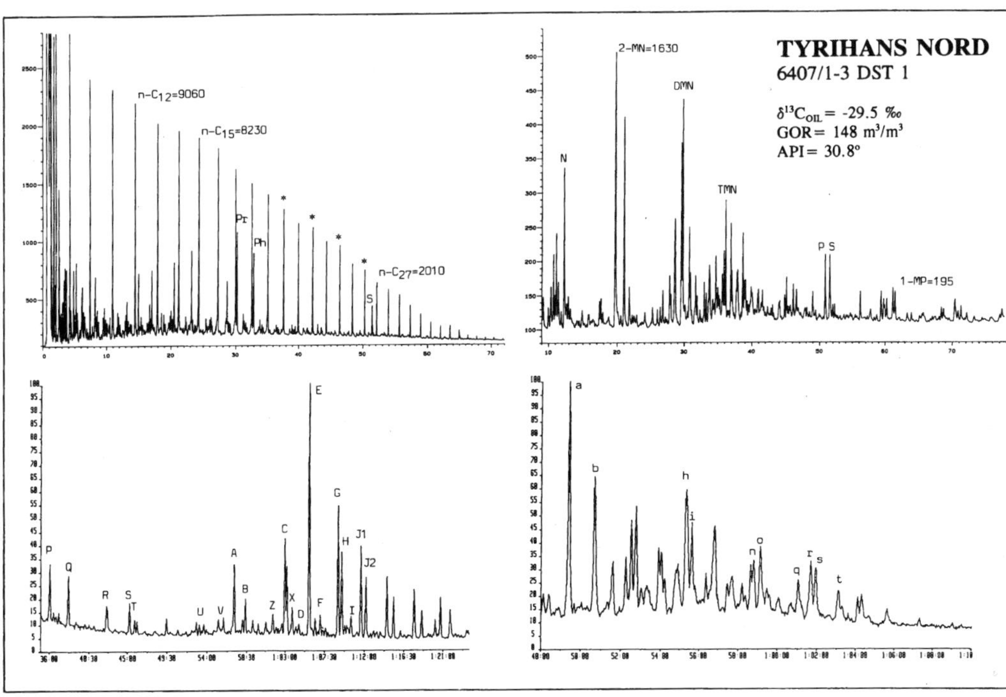
TYRIHANS NORD
6407/1-3 DST 1
δ13COIL = -29.5 ‰
GOR = 148 m3/m3
API = 30.8°
n-C12=9060
n-C15=8230
Pr
Ph
*
n-C27=2010
S
2-MN=1630
DMN
N
TMN
P S
1-MP=195
P Q R S T U V A B Z C X D E F G H I J1 J2
a b h i n o q r s t

Fig. 22

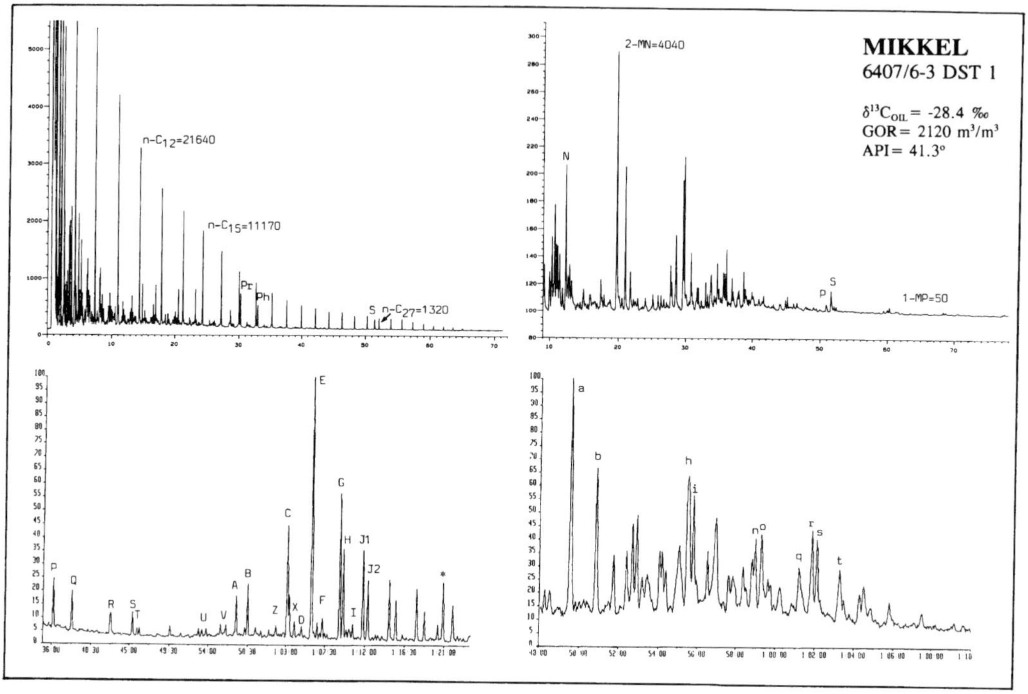
MIKKEL
6407/6-3 DST 1
$\delta^{13}C_{OIL}$ = -28.4 ‰
GOR = 2120 m³/m³
API = 41.3°
n-C12=21640
n-C15=11170
Pr
Ph
S
n-C27=1320
2-MN=4040
N
p
S
1-MP=50
P
Q
R
S
T
U
V
A
B
Z
C
X
D
E
F
G
H
I
J1
J2
*
a
b
h
i
n
o
q
r
s
t

Fig. 23

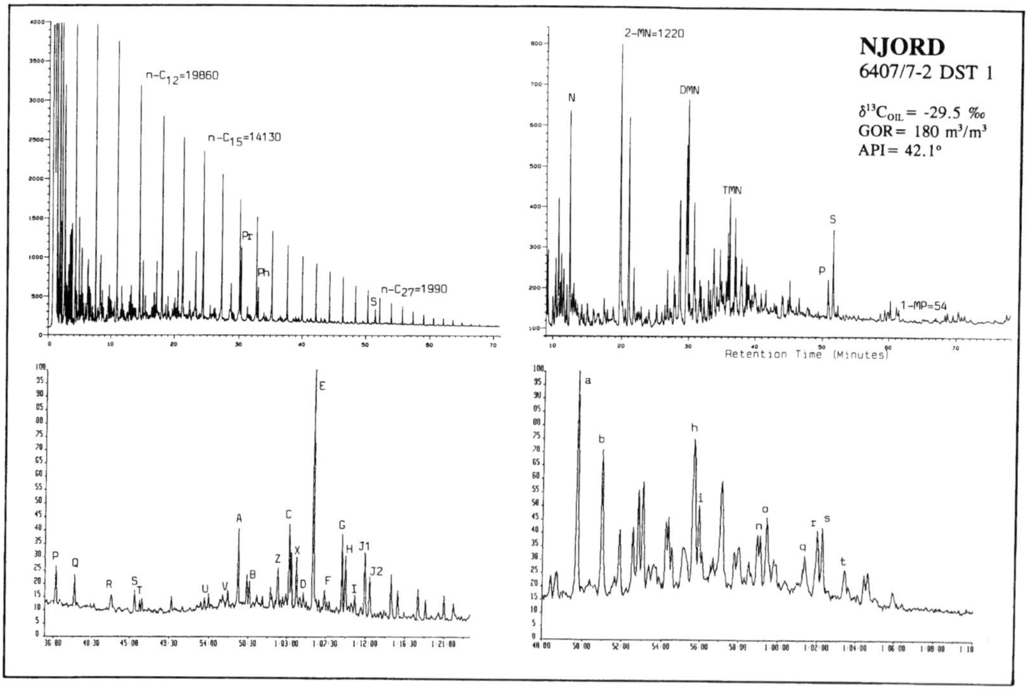
NJORD
6407/7-2 DST 1
$\delta^{13}C_{OIL}$ = -29.5 ‰
GOR = 180 m³/m³
API = 42.1°
n-C12=19860
n-C15=14130
Pr
Ph
S
n-C27=1990
2-MN=1220
N
DMN
TMN
p
S
1-MP=54
Retention Time (Minutes)
P
Q
R
S
T
U
V
A
B
Z
C
X
D
E
F
G
H
I
J1
J2
a
b
h
i
n
o
q
r
s
t

Fig. 24

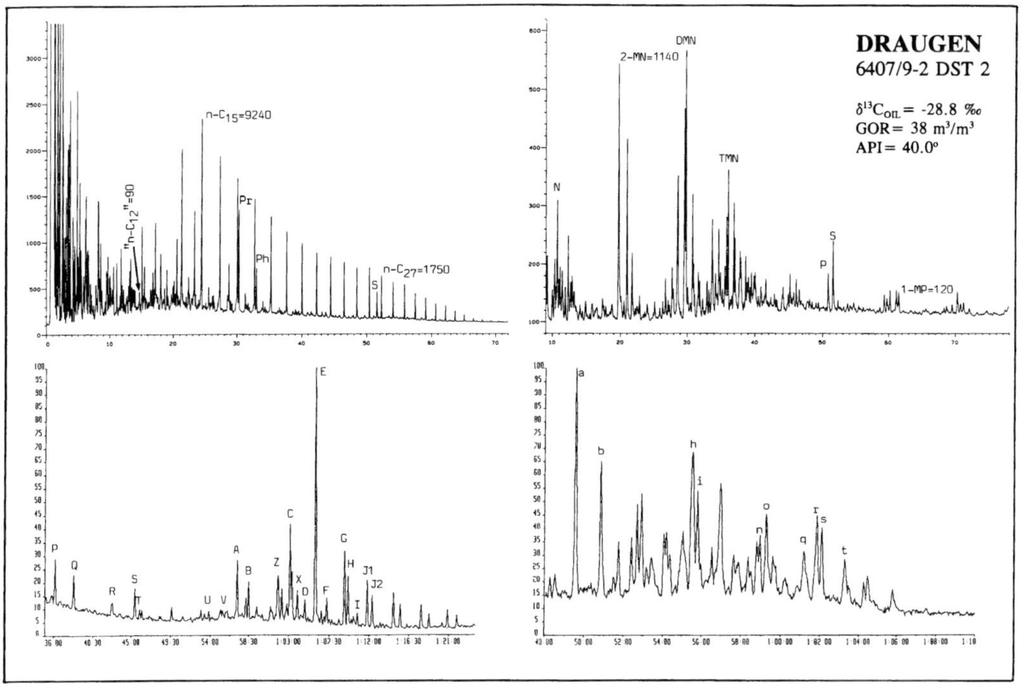
DRAUGEN
6407/9-2 DST 2
δ13COIL = -28.8 ‰
GOR = 38 m3/m3
API = 40.0°
n-C15=9240
"n-C12"=90
Pr
Ph
S
n-C27=1750
2-MN=1140
DMN
TMN
N
p
S
1-MP=120
P Q R S T U V A B Z C X D E F G H I J1 J2
a b h i o r s n q t

Fig. 25

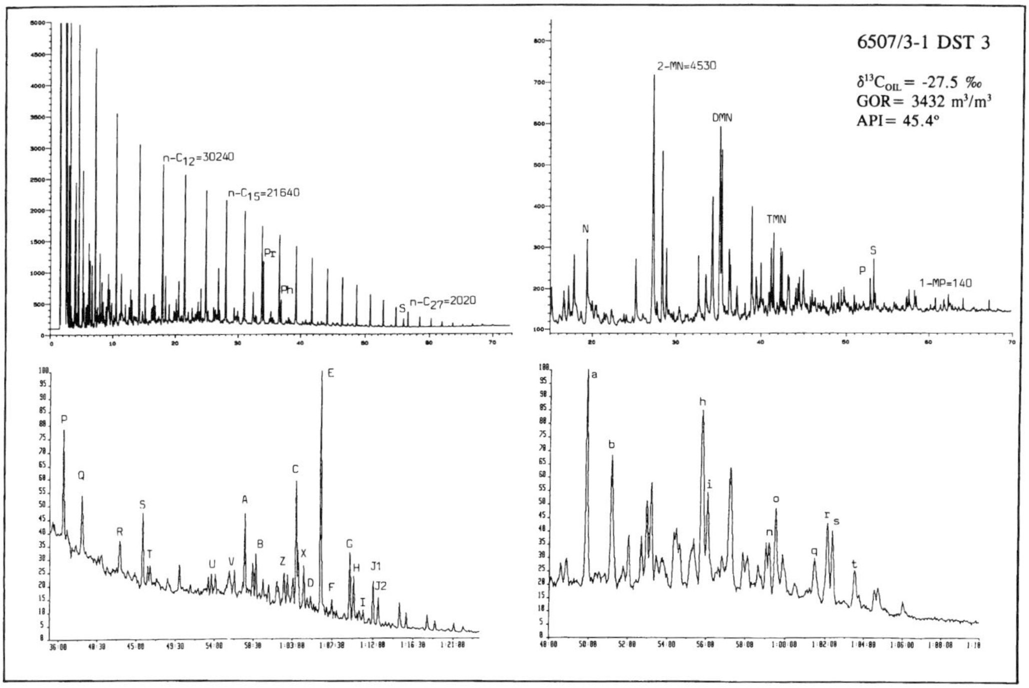
6507/3-1 DST 3
δ13COIL = -27.5 ‰
GOR = 3432 m3/m3
API = 45.4°
n-C12=30240
n-C15=21640
Pr
Ph
S
n-C27=2020
2-MN=4530
DMN
TMN
N
p
S
1-MP=140
P Q R S T U V A B Z C X D E F G H I J1 J2
a b h i o r s n q t

Fig. 26

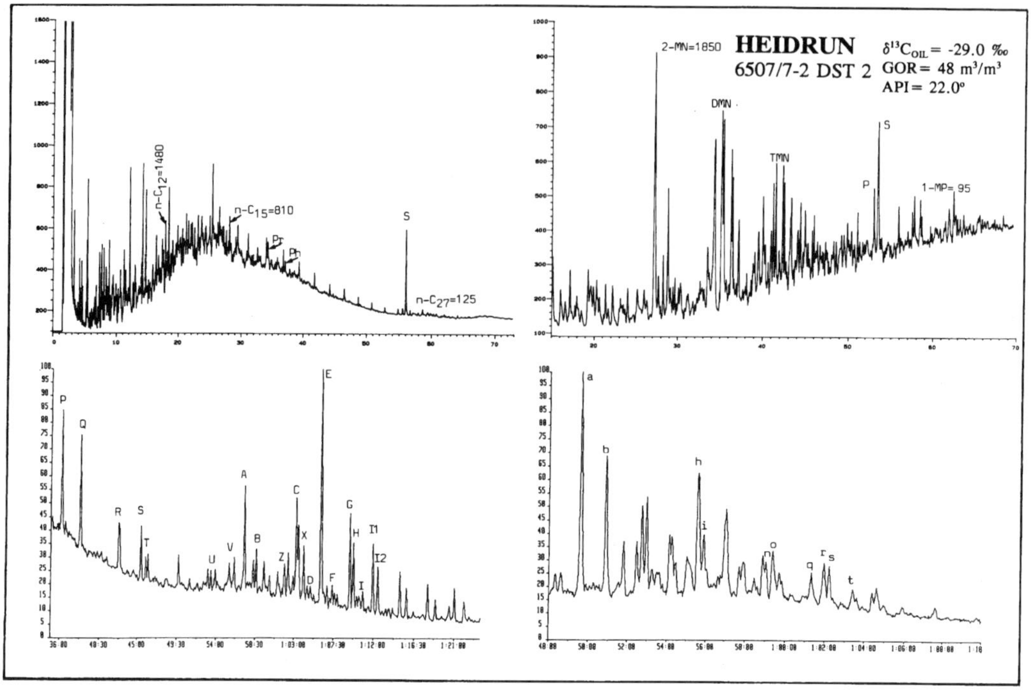

HEIDRUN
6507/7-2 DST 2
δ13COIL = -29.0 ‰
GOR = 48 m3/m3
API = 22.0°
2-MN=1850
DMN
TMN
P
S
1-MP= 95
n-C12=1480
n-C15=810
Pr
Ph
S
n-C27=125
P
Q
R
S
T
U
V
A
B
Z
C
X
D
E
F
G
H
I
I1
I2
a
b
h
i
n
o
q
r
s
t

Fig. 27

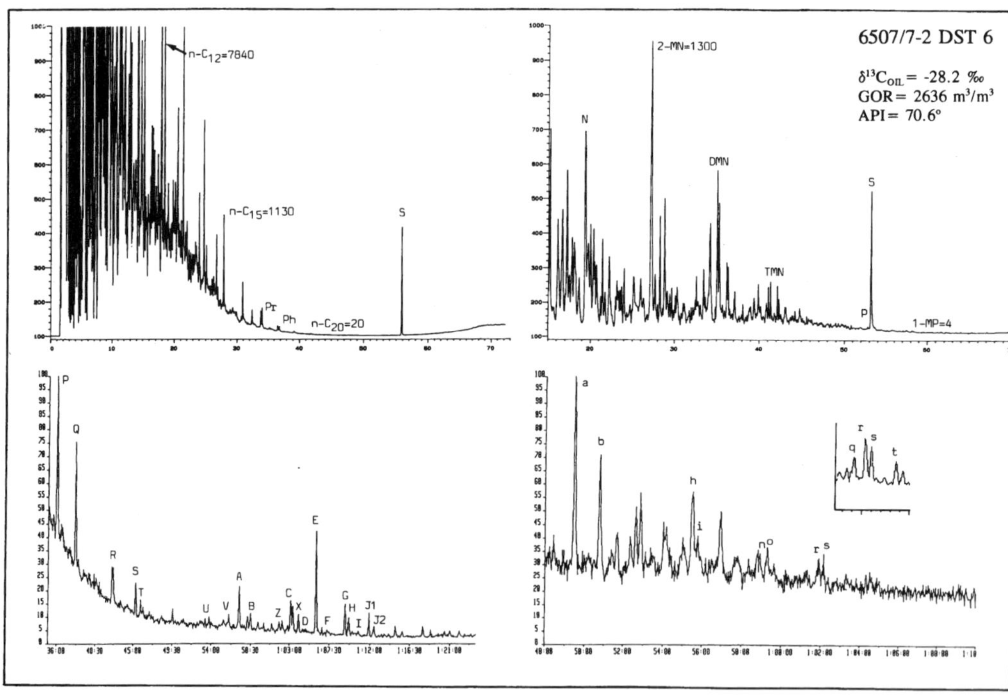

6507/7-2 DST 6
δ13COIL = -28.2 ‰
GOR = 2636 m3/m3
API = 70.6°
2-MN=1300
N
DMN
TMN
S
P
1-MP=4
n-C12=7840
n-C15=1130
S
Pr
Ph
n-C20=20
P
Q
R
S
T
U
V
A
B
Z
C
X
D
E
F
G
H
I
J1
J2
a
b
h
i
n
o
r
s
q
t

Fig. 28

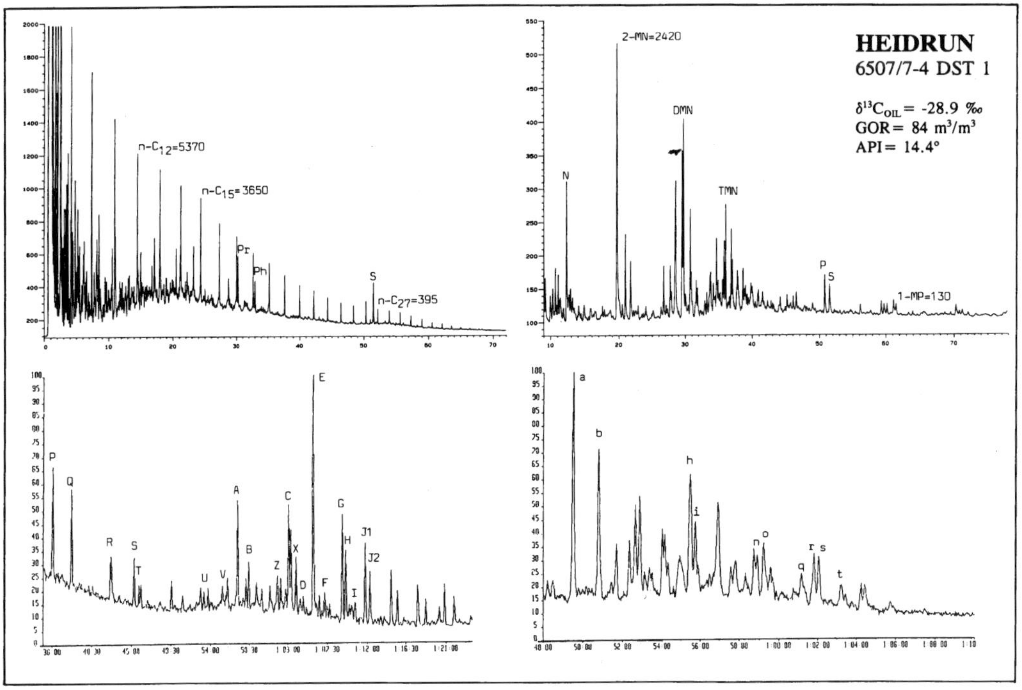
HEIDRUN
6507/7-4 DST 1
$\delta^{13}C_{OIL}$ = -28.9 ‰
GOR = 84 m³/m³
API = 14.4°
n-C12=5370
n-C15=3650
Pr
Ph
S
n-C27=395
2-MN=2420
N
DMN
TMN
P
S
1-MP=130

Fig. 29

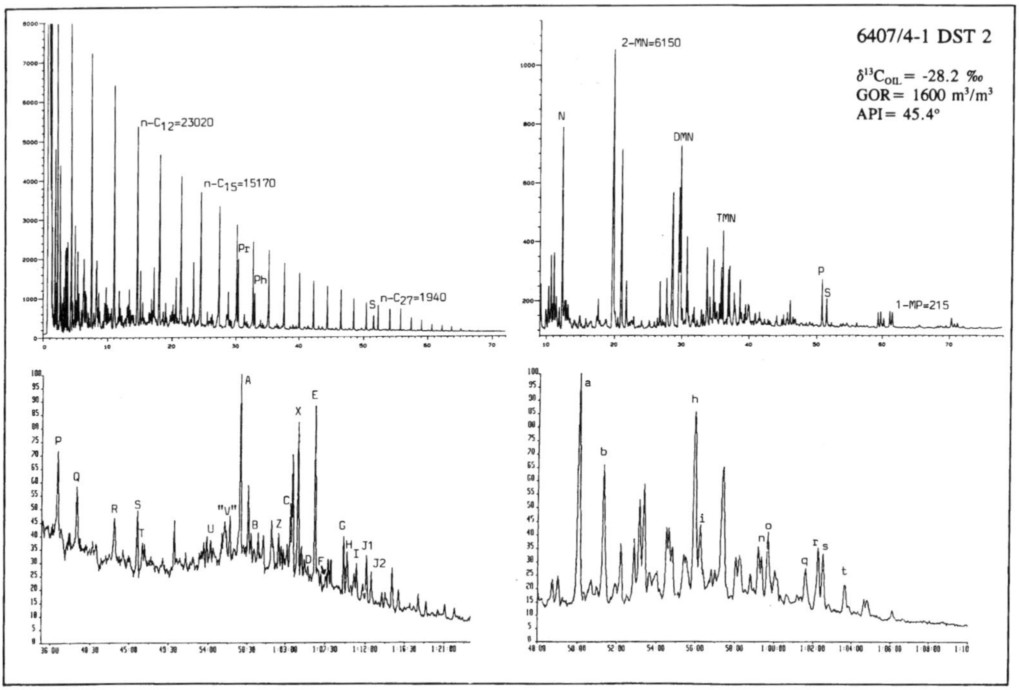
6407/4-1 DST 2
$\delta^{13}C_{OIL}$ = -28.2 ‰
GOR = 1600 m³/m³
API = 45.4°
n-C12=23020
n-C15=15170
Pr
Ph
S
n-C27=1940
2-MN=6150
N
DMN
TMN
P
S
1-MP=215

Fig. 30

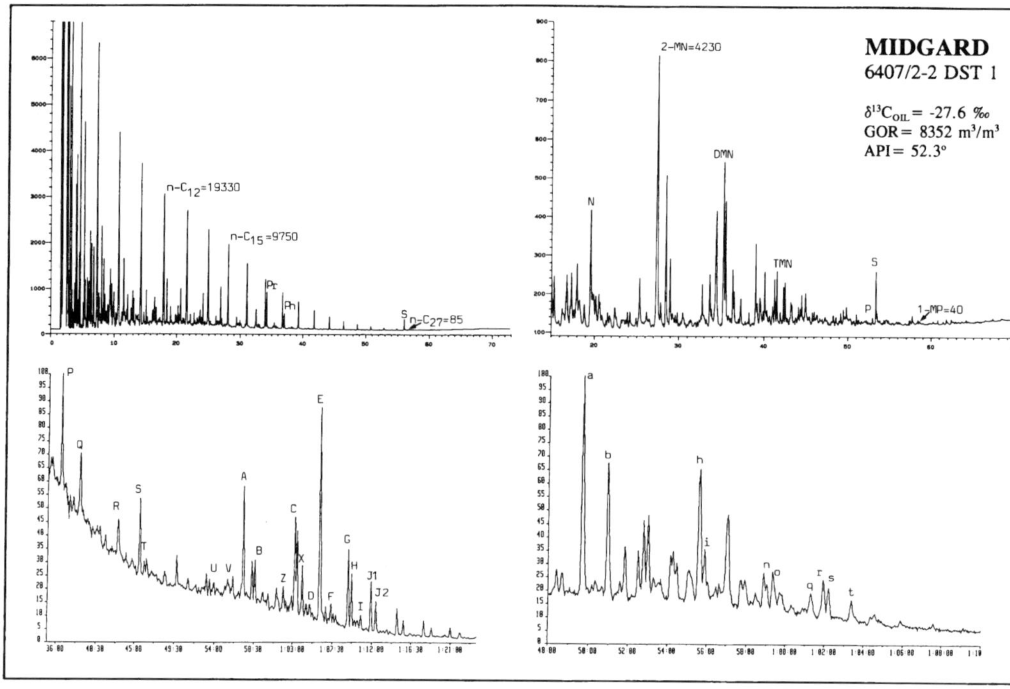
MIDGARD
6407/2-2 DST 1
$\delta^{13}C_{OIL}$ = -27.6 ‰
GOR = 8352 m^3/m^3
API = 52.3°
n-C_{12}=19330
n-C_{15}=9750
Pr
Ph
n-C_{27}=85
2-MN=4230
DMN
TMN
1-MP=40

Fig. 31

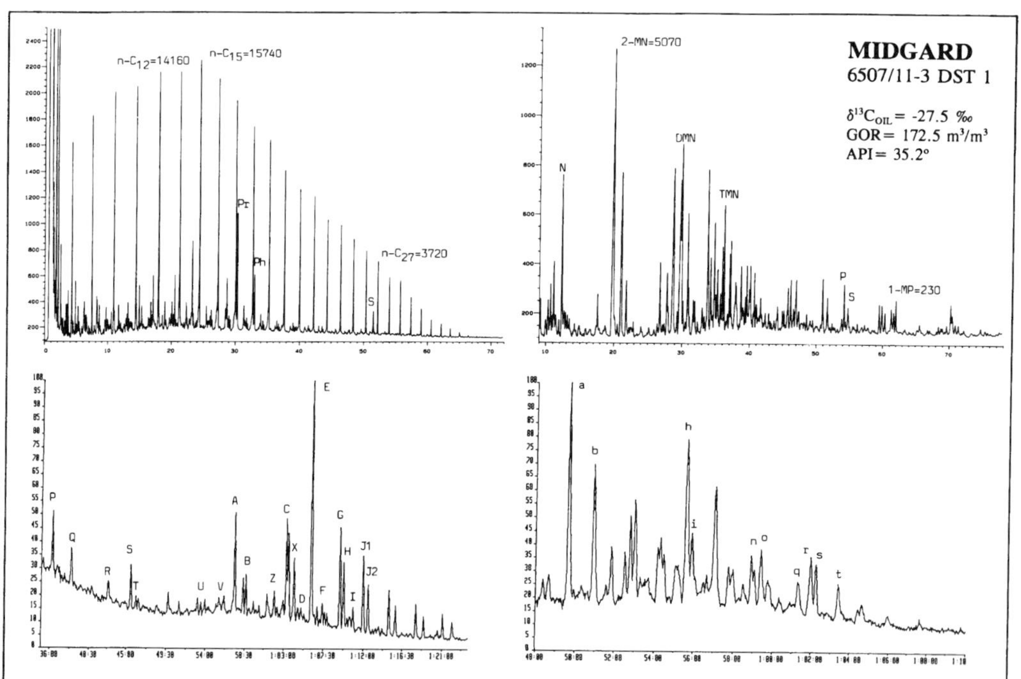
MIDGARD
6507/11-3 DST 1
$\delta^{13}C_{OIL}$ = -27.5 ‰
GOR = 172.5 m^3/m^3
API = 35.2°
n-C_{12}=14160
n-C_{15}=15740
Pr
Ph
n-C_{27}=3720
2-MN=5070
DMN
TMN
1-MP=230

Fig. 32

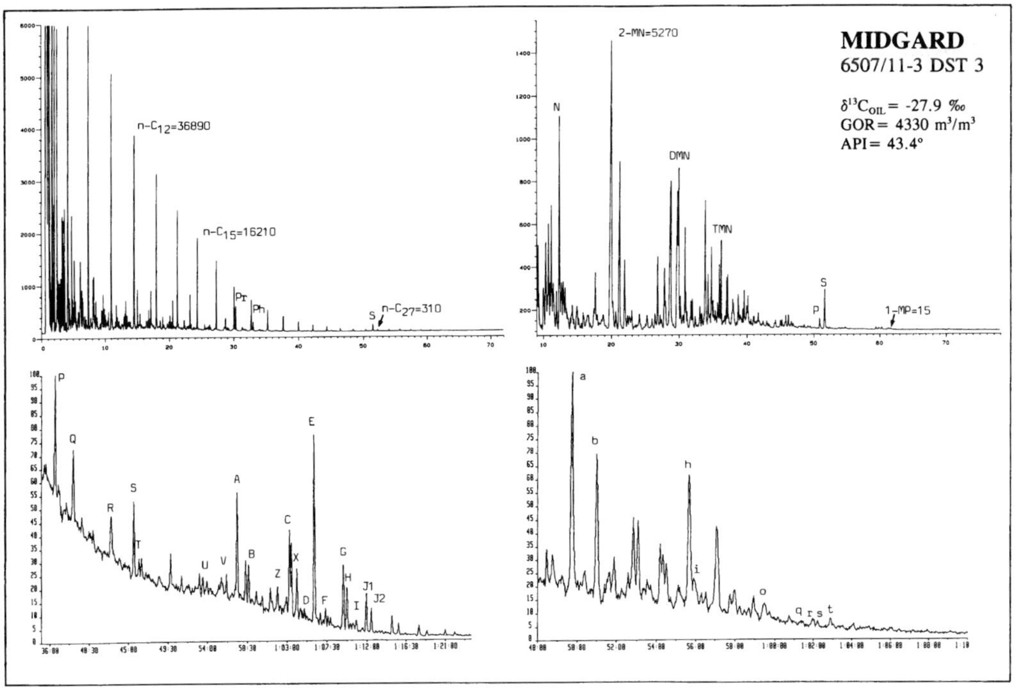
MIDGARD
6507/11-3 DST 3
δ13C OIL = -27.9 ‰
GOR = 4330 m3/m3
API = 43.4°
n-C12=36890
n-C15=16210
Pr
Ph
S
n-C27=310
2-MN=5270
N
DMN
TMN
P
S
1-MP=15

Fig. 33

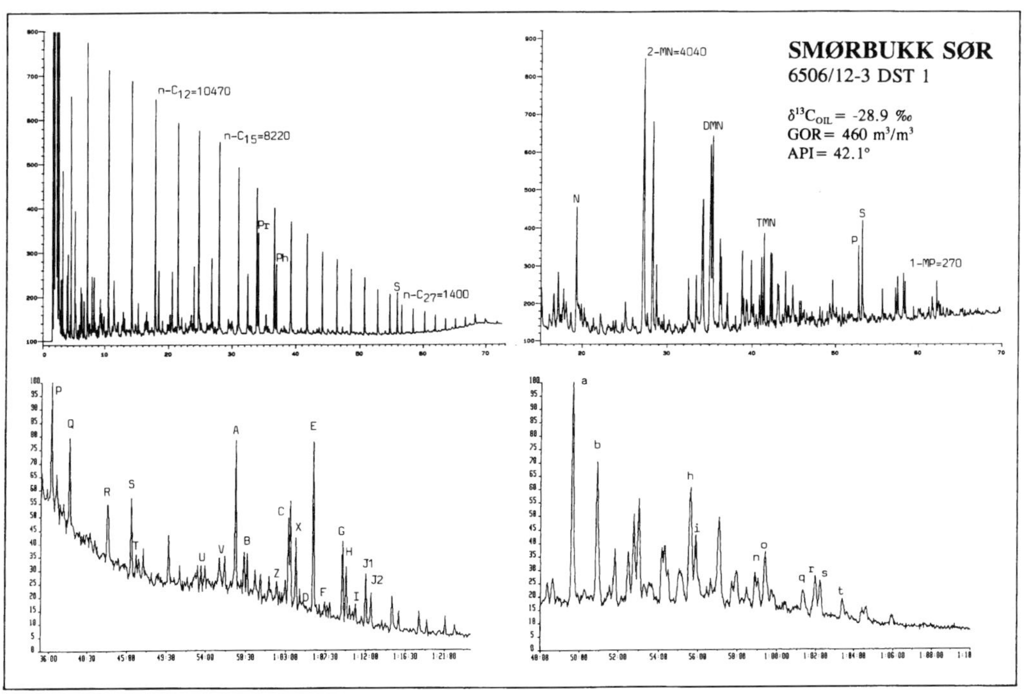
SMØRBUKK SØR
6506/12-3 DST 1
δ13C OIL = -28.9 ‰
GOR = 460 m3/m3
API = 42.1°
n-C12=10470
n-C15=8220
Pr
Ph
S
n-C27=1400
2-MN=4040
N
DMN
TMN
P
S
1-MP=270

Fig. 34

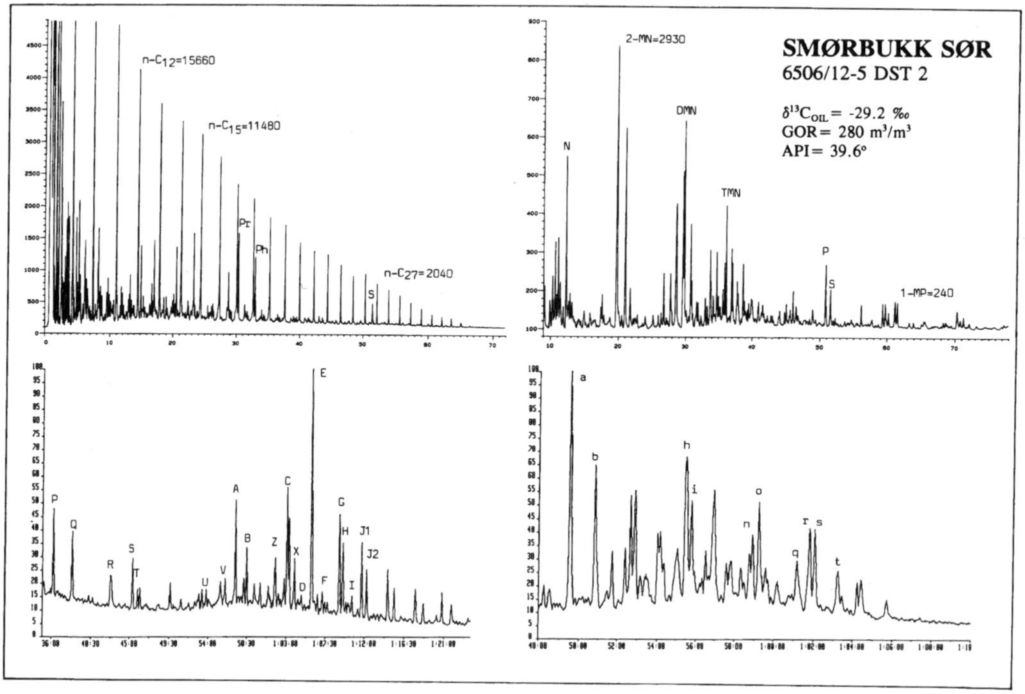
n-C12=15660
n-C15=11480
Pr
Ph
n-C27=2040
S
2-MN=2930
N
DMN
TMN
P
S
1-MP=240
SMØRBUKK SØR
6506/12-5 DST 2
δ13COIL= -29.2 ‰
GOR= 280 m3/m3
API= 39.6°
P
Q
R
S
T
U
V
A
B
Z
C
X
D
E
F
G
H
I
J1
J2
a
b
h
i
n
o
q
r s
t

Fig. 35

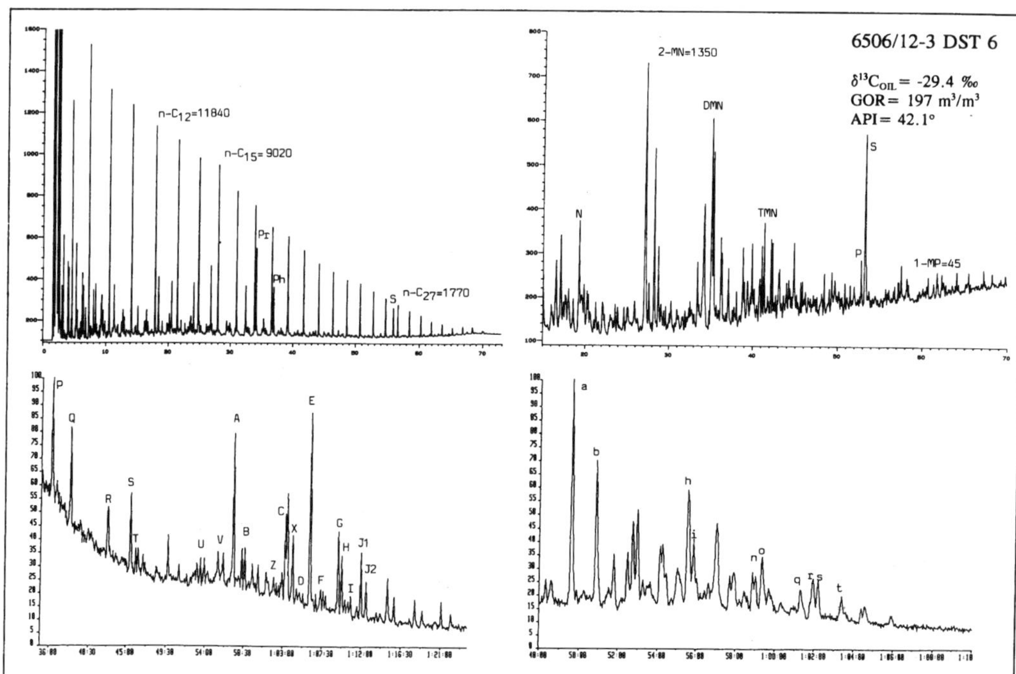
n-C12=11840
n-C15=9020
Pr
Ph
S
n-C27=1770
2-MN=1350
N
DMN
TMN
P
S
1-MP=45
6506/12-3 DST 6
δ13COIL= -29.4 ‰
GOR= 197 m3/m3
API= 42.1°
P
Q
R
S
T
U
V
A
B
Z
C
X
D
E
F
G
H
I
J1
J2
a
b
h
i
n
o
q
r
s
t

Fig. 36

Table 6. *Aromatic maturity parameters of the Haltenbanken database*

Sample code	Sample no.	MNR	DNR	TNR2	MPR	MPI1	MPDF
TR1	1	1.58	2.87	0.75	1.06	0.72	0.45
T2	2	1.60	2.76	0.80	1.41	0.77	0.50
T1	3	1.54	3.99	0.50	1.03	0.75	0.50
I1	4	1.91	6.27	0.59	1.51	0.74	0.57
I2	5	1.55	13.83	0.54	2.21	0.74	0.63
I3	6	1.70	5.33	0.63	1.66	0.75	0.59
N1	7	1.47	4.64	0.61	1.13	0.64	0.49
N2	8	1.71	4.62	0.66	1.02	0.57	0.45
N3	9	1.54	4.08	0.63	1.42	0.63	0.50
N4	10	1.44	3.36	0.64	1.25	0.54	0.49
N5	11	1.60	4.25	0.65	1.14	0.52	0.51
N6	12	1.61	3.89	0.61	1.88	0.58	0.52
N7	13	1.52	3.35	0.57	1.26	0.56	0.50
D1	14	1.43	4.62	0.69	0.77	0.60	0.41
D2	15	1.54	3.93	0.64	0.76	0.65	0.43
H5	16	2.11	4.05	0.80	1.55	0.74	0.50
H6	17	2.96	3.50	0.76	2.07	0.81	0.51
H7	18	5.51	3.57	0.81	1.58	0.52	0.51
H1	19	3.63	5.59	0.70	0.96	0.64	0.49
H2	20	3.20	4.52	0.61	1.05	0.65	0.47
H3	21	3.23	4.26	0.68	1.08	0.64	0.46
H4	22	2.74	12.74	0.66	1.01	0.69	0.46
X2	23	2.60	6.88	0.76	1.07	0.63	0.49
M4	24	2.59	6.00	1.03	0.59	0.56	0.46
M1	25	3.25	7.92	0.76	1.,28	0.88	0.54
M2	26	2.57	13.33	0.73	2.06	0.53	0.63
M3	27	2.39	9.41	0.74	1.94	0.63	0.62
S1	28	1.78	4.01	0.84	1.65	0.85	0.51
S2	29	2.14	3.65	0.84	2.11	0.96	0.54
B1	30	1.66	5.40	0.67	1.14	0.71	0.50
B2	31	1.67	5.31	0.58	1.08	0.68	0.49
B3	32	1.69	5.39	0.60	0.90	0.55	0.45
B4	33	1.99	4.55	0.60	1.43	0.63	0.53

MNR 2-MN/1-MN (Alexander *et al.* 1985)
DNR (2,6-DMN + 2,7-DMN)/1,5-DMN (Alexander *et al.* 1985)
TNR2 (1,3,7-TMN + 2,3,6-TMN)/(1,3,5-TMN + 1,3,6-TMN + 1,4,6-TMN) (Alexander *et al.* 1985)
MPR 2-MP/1-MP (Radke 1987)
MPI1 1,5(2-MP + 3-MP)/(PHEN + 1-MP + 9-MP) (Radke 1987)
MPDF (2-MP + 3-MP)/(2-MP + 3-MP + 1-MP + 9-MP) (Kvalheim *et al.* 1987)
$R_c = 0.40 \cdot \text{MPI1} + 0.60$ (Radke 1987)

parameters as a result of several effects. One of these is that the parameters reflect maturity differences in portions of the petroleums leading to complications in mixed systems, as discussed earlier. This may suggest that most of the petroleums represent integration of less mature black oil with relatively high biomarker concentrations, with condensates with low biomarker contents and relatively higher concentrations of mature aromatic maturity parameters. A second complication is that some of these parameters are not sensitive measures of expulsion maturities for source rocks at relatively low maturity and they are also significantly influenced by the organic facies, e.g. they function best for type II/III and III source rocks (Radke 1987, 1988) therefore the lack of correlation between biomarker parameters and the methylphenanthrenes may not be surprising if the marine Spekk Formation is the source rock.

The maturity calculated from the methylphenanthrenes (Fig. 39) indicates, however, that the bulk of the C_{14-15} hydrocarbons in most of the petroleums was expelled at a maturity of about 0.7 to 0.9 vitrinite reflectance, with a tail towards higher maturities for some of the condensates and the Midgard oil leg. The Midgard system is

Table 7. *Selected facies and maturity parameters plus carbon and hydrogen isotope data*

Sample no	Sample code	Sample (well test field)	Parameter 1	2	3	4	5	6	7	8	9	10	δ13C	δ2H
1	TR1	6406/3-2 DST2 Trestakk	0.76	0.56	0.64	0.00	1.00	0.67	0.40	0.52	0.34	0.48	–29.5	–124
2	T2	6407/1-2 DST1 Tyrihans S.	0.74	0.57	0.65	0.52	0.78	0.67	0.49	0.53	0.20	0.43	–30.0	–123
3	T1	6407/1-3 DST1 Tyrihans N.	0.83	0.56	0.58	0.55	0.53	0.66	0.35	0.55	0.18	0.42	–29.5	–123
4	I1	6407/6-3 DST1 Mikkel	0.71	0.56	0.57	0.37	0.41	0.43	0.18	0.47	0.16	0.27	–28.4	–122
5	I2	6407/6-3 DST2 Mikkel	0.71	0.41	0.52	0.49	0.29	0.43	0.19	0.49	0.24	0.29	–28.8	–125
6	I3	6407/6-3 DST3 Mikkel	0.75	0.50	0.56	0.39	0.53	0.47	0.23	0.49	0.32	0.28	–28.6	–122
7	N1	6407/7-1 DST3 Njord	0.89	0.53	0.57	0.66	0.58	0.77	0.55	0.53	0.16	0.41	–29.6	–128
8	N2	6407/7-1 DST5 Njord	0.89	0.52	0.55	0.59	0.58	0.72	0.50	0.64	0.18	0.39	–29.0	–131
9	N3	6407/7-2 DST1 Njord	0.86	0.58	0.57	0.60	0.65	0.75	0.59	0.55	0.14	0.37	–29.5	–134
10	N4	6407/7-3 DST2 Njord	0.91	0.58	0.57	0.62	0.70	0.75	0.58	0.54	0.20	0.41	–29.7	–135
11	N5	6407/7-3 DST3 Njord	0.91	0.58	0.56	0.54	0.65	0.78	0.58	0.55	0.17	0.39	–29.4	–127
12	N6	6407/7-4 DST2A Njord	0.84	0.52	0.56	0.59	0.61	0.65	0.57	0.54	0.18	0.37	–29.3	–130
13	N7	6407/7-4 DST2B Njord	0.83	0.52	0.56	0.66	0.52	0.67	0.56	0.55	0.17	0.38	–29.4	–133
14	D1	6407/9-2 Draugen	0.70	0.55	0.54	0.52	0.45	0.60	0.56	0.55	0.15	0.34	–28.8	–134
15	D2	6407/9-5 Draugen	0.72	0.56	0.55	0.58	0.43	0.60	0.57	0.57	0.12	0.33	–28.8	–134
16	H5	6507/3-1 DST3 Alve	0.77	0.55	0.65	0.00	1.00	0.66	0.41	0.68	0.30	0.33	–27.5	–124
17	H6	6507/7-2 DST2 Heidrun	0.80	0.60	0.61	0.59	0.72	0.71	0.44	0.52	0.34	0.42	–29.0	–122
18	H7	6507/7-2 DST6 Heidrun	0.87	0.50	0.68	0.59	0.72	0.72	0.40	0.63	0.61	0.35	–28.2	–123
19	H1	6507/7-4 DST1 Heidrun	0.84	0.51	0.54	0.63	0.48	0.70	0.42	0.49	0.32	0.44	–28.9	–134
20	H2	6507/7-4 DST2 Heidrun	0.83	0.48	0.54	0.65	0.48	0.71	0.46	0.51	0.29	0.42	–28.6	–130

21	H3	6507/7-4 DST3 Heidrun	0.84	0.48	0.55	0.64	0.46	0.73	0.46	0.51	0.26	0.44	–28.5	–125
22	H4	6507/7-5 DST2A Heidrun	0.82	0.48	0.54	0.58	0.57	0.71	0.43	0.56	0.35	0.43	–28.8	–130
23	X2	6407/4-1 DST2	0.89	0.60	0.57	0.35	1.00	0.88	0.61	0.59	0.33	0.64	–28.2	–128
24	M4	6407/2-2 DST1 Midgard	0.88	0.52	0.61	0.00	1.00	0.72	0.37	0.79	0.35	0.46	–27.6	–131
25	M1	6507/11-3 DST1 Midgard	0.85	0.47	0.55	0.46	0.64	0.72	0.39	0.58	0.20	0.46	–27.5	–130
26	M2	6507/11-3 DST2 Midgard	0.86	0.57	0.54	0.51	0.75	0.73	0.40	0.64	0.29	0.46	–27.5	–134
27	M3	6507/11-3 DST3 Midgard	0.89	0.57	0.57	0.40	0.86	0.73	0.41	0.65	0.39	0.46	–27.9	–131
28	S1	6506/12-3 DST1 Smørbukk Sør	0.95	0.56	0.62	0.00	1.00	0.78	0.33	0.54	0.42	0.54	–28.9	–123
29	S2	6506/12-3 DST6 Smørbukk Sør	0.90	0.56	0.61	0.00	1.00	0.79	0.31	0.52	0.38	0.55	–29.4	–123
30	B1	6506/12-5 DST2 Smørbukk Sør	0.88	0.63	0.63	0.35	0.85	0.64	0.46	0.51	0.26	0.41	–29.2	–124
31	B2	6506/12-5 DST2A Smørbukk Sør	0.88	0.56	0.60	0.37	0.86	0.64	0.45	0.55	0.37	0.43	–29.1	–123
32	B3	6506/12-5 DST2B Smørbukk Sør	0.89	0.53	0.58	0.47	0.88	0.65	0.46	0.55	0.38	0.42	–29.5	–127
33	B4	6506/12-5 DST4 Smørbukk Sør	0.80	0.52	0.55	0.32	0.81	0.71	0.54	0.63	0.21	0.43	–29.5	–133

* Parameters:
1, Hopane X/(hopane X + 17β(H), 21α(H)-30-norhopane) (Cornford *et al.* 1986) (hopane X = diahopane cf. Moldowan *et al.* 1991)
2, 20S/(20S + 20R) of C29 5α(H), 14α(H), 17α(H) sterane isomers (Mackenzie *et al.*1984)
3, ββ/(ββ + αα) of C29 (20R + 20S) sterane isomers (Mackenzie *et al.* 1980)
4, C28 triaromatic steroids/(C28 triaromatic steroids + C29 monoaromatic steroids) (Mackenzie *et al.* 1985)
5, C20/(C20 + C28) triaromatic steroids (Mackenzie *et al.* 1985)
6, Ts/(Ts + Tm) (Seifert & Moldowan 1981)
7, 17α(H), 21β(H)-28,30-bisnorhopane/(17α(H), 21β(H)-28,30-bisnorhopane + Tm)
8, 17α(H), 21β(H), 22(S)-tetrakishomohopane/(17α(H), 21β(H), 22(S)-tetrakishomohopane + 17α(H), 21β(H), 22(S)-pentakishomohopane)
9, C23 tricyclic terpane/(C23 tricyclic terpane + 17α(H), 21β(H)-hopane)
10, 29Ts/(29Ts + 17α(H), 21β(H)-30-norhopane) (identification of 29Ts cf. Moldowan *et al.* 1991)

Table 8. *Peak identification for tricyclic terpaness, tetracyclic terpanes and pentacyclic triterpanes*

Peak	Identification		
A	18α trisnorneohopane (Ts)	$C_{27}H_{44}$	(III, R = i-C_4H_{11})
B	17α trinorhopane (Tm)	$C_{27}H_{46}$	(II, R = H)
Z	Bisnorhopane§	$C_{28}H_{48}$	(IV)
C	αβ norhopane (normoretane)	$C_{29}H_{50}$	(II, R = C_2H_5)
D	βα norhopane	$C_{29}H_{50}$	(III, R = C_2H_5)
E	αβ hopane	$C_{30}H_{52}$	(II, R = i-C_3H_7)
F	βα hopane (moretane)	$C_{30}H_{52}$	(III, R = i-C_3H_7)
G	22S αβ homohopane	$C_{31}H_{54}$	(II, R = i-C_4H_9)
H	22R αβ homohopane	$C_{31}H_{54}$	(II, R = i-C_4H_9)
I	βα homomoretane	$C_{31}H_{54}$	(III, R = i-C_4H_9)
J	22S αβ bishomohopane	$C_{32}H_{56}$	(II, R = i-C_5H_{11})
	22R αβ bishomohopane	$C_{32}H_{56}$	(II, R = i-C_5H_{11})
K	22S αβ trishomohopane	$C_{33}H_{58}$	(II, R = i-C_6H_{23})
	22R αβ trishomohopane	$C_{33}H_{58}$	(II, R = i-C_6H_{13})
L	22S αβ tetrakishomohopane	$C_{34}H_{60}$	(II, R = i-C_7H_{15})
	22R αβ tetrakishomohopane	$C_{34}H_{60}$	(II, R = i-C_7H_{15})
M	22S αβ pentakishomohopane	$C_{35}H_{62}$	(II, R = i-C_8H_{17})
	22R αβ pentakishomohopane	$C_{35}H_{62}$	(II, R = i-C_8H_{17})
P	Tricyclic terpane	$C_{23}H_{42}$	(V, R = i-C_4H_9)
Q	Tricyclic terpane	$C_{24}H_{44}$	(V, R = i-C_5H_{11})
R	Tricyclic terpane (17R, 17S)	$C_{25}H_{46}$	(V, R = i-C_6H_{13})
S	Tetracyclic terpane	$C_{24}H_{42}$	(VI)
T	Tricyclic terpane (17R, 17S)	$C_{26}H_{48}$	(V, R = i-C_7H_{15})
N	Tricyclic terpane	$C_{21}H_{38}$	(V, R = C_2H_5)
O	Tricyclic terpane	$C_{22}H_{40}$	(V, R = C_3H_7)
Y	25,28,30-trisnorhopane/moretane	$C_{27}H_{46}$	(VII)
X	Unknown triterpane	$C_{30}H_{52}$	

α and β refer to hydrogen atoms at C-17 and C-21, respectively, unless indicated

Table 9. *Peak identification for steranes*

Peak	Identification		
a	20S βα diacholestane	$C_{27}H_{48}$	(I, R = H)
b	20R βα diacholestane	$C_{27}H_{48}$	(I, R = H)
c	20S αβ diacholestane	$C_{27}H_{48}$	(II, R = H)
d	20R αβ diacholestane	$C_{27}H_{48}$	(II, R = H)
e	20S βα 24-methyl-diacholestane	$C_{28}H_{50}$	(I, R = CH_3)
f	20R βα 24-methyl-diacholestane	$C_{28}H_{50}$	(I, R = CH_3)
g	20S αβ 24-methyl-diacholestane	$C_{28}H_{50}$	(II, R = CH_3)
	+ 20S ααα cholestane	$C_{27}H_{48}$	(III, R = H)
h	20S βα 24-ethyl-diacholestane	$C_{29}H_{52}$	(II, R = C_4H_5)
	+ 20R αββ cholestane	$C_{27}H_{48}$	(IV, R = H)
i	20S αββ cholestane	$C_{27}H_{48}$	(IV, R = H)
	+ 20R αβ 24-methyl-diacholestane	$C_{26}H_{50}$	(II, R = CH_3)
j	20R ααα cholestane	$C_{27}H_{48}$	(III, R = H)
k	20R βα 24-ethyl-diacholestane	$C_{29}H_{52}$	(I, R = C_7H_5)
l	20R αβ 24-ethyl-diacholestane	$C_{29}H_{52}$	(II, R = C_7H_5)
m	20S αα 24-methyl-cholestane	$C_{28}H_{50}$	(III, R = CH_3)
n	20R αββ 24-methyl-cholestane	$C_{28}H_{50}$	(IV, R = CH_3)
	+ 20R αβ 24-ethyl-diacholestane	$C_{29}H_{52}$	(II, R = C_4H_5)
o	20S αββ 24-methyl-cholestane	$C_{28}H_{50}$	(IV, R = CH_3)
p	20R ααα 24-methyl-cholestane	$C_{28}H_{50}$	(III, R = CH_3)
q	20S ααα 24-ethyl-cholestane	$C_{29}H_{52}$	(III, R = C_2H_5)
r	20R αββ 24-ethyl-cholestane	$C_{29}H_{52}$	(IV, R = C_2H_5)
s	20S αββ 24-ethyl-cholestane	$C_{29}H_{52}$	(IV, R = C_2H_5)
t	20R ααα 24-ethyl-cholestane	$C_{29}H_{52}$	(III, R = C_2H_5)
u	5α sterane	$C_{21}H_{36}$	(VI, R = C_2H_5)
v	5α sterane	$C_{22}H_{38}$	(VI, R = C_2H_7)

α and β refer to hydrogen atoms at C-5, C-14 and C-17 in regular steranes and at C-13 and C-17 in diasteranes

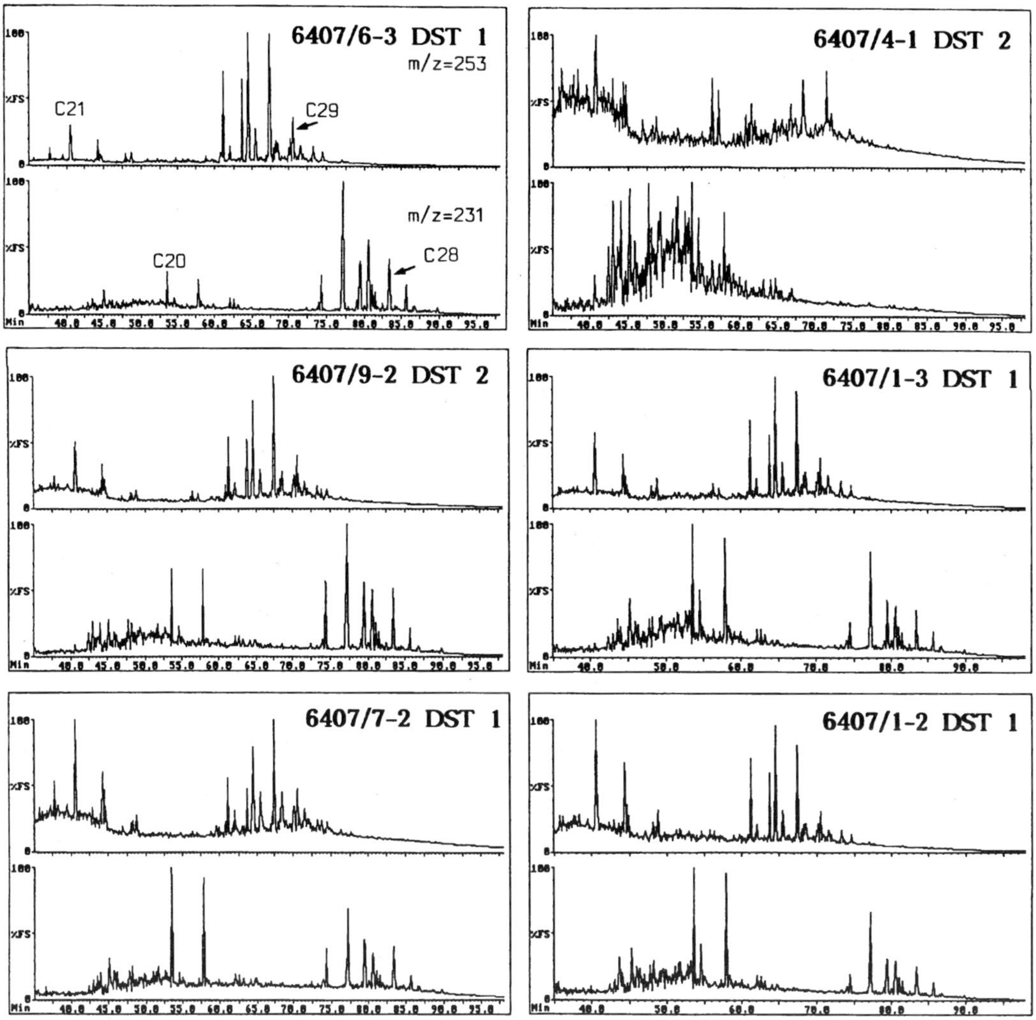

Figs 37 & 38. Well defined monoaromatic steroid hydrocarbons (e.g. the C_{29}) are found in the Draugen, Mikkel, Tyrihans Nord, Tyrihans Sør, Njord and Heidrun petroleums and imply that these petroleums are less mature than the rest of the petroleums in the database, e.g. the Smørbukk Sør, the 6407/4-1 and the Trestakk petroleums. It appears that the ratio of monoaromatic steroids to triaromatic steroids (e.g. parameter 4 in Table 7) and field location coincide well with the regional maturity map of the Spekk Formation (Fig. 3), implying that lateral migration distances are only moderate except for the Draugen and the Midgard accumulations.

suggested to have been severely modified by evaporative fractionation (i.e. enrichment in the condensates of phenanthrene relative to methylphenanthrenes) resulting in an apparently lower maturity for the Midgard gas compared to the Midgard oil leg. If the systems were recombined into a single fluid, it is likely that the maturity of the Midgard fluid, as indicated by the methylphenanthrene index, would only be moderate.

The ratio of hopane X/(hopane X + normoretane) in the petroleums is shown in Fig. 40 and indicates that the Haltenhanken petroleums are more mature than the bulk of North Sea oils. Still, a number of the Haltenbanken petroleums contain good proportions of well defined monoaromatic steroids (Table 7; Figs 37 & 38), suggesting input from comparatively recently matured source rocks, such as the Spekk Formation.

As a result, samples can be arranged in the following sequence in terms of decreasing maturity, if emphasis is placed more on the biomarkers than on the medium-range aromatic

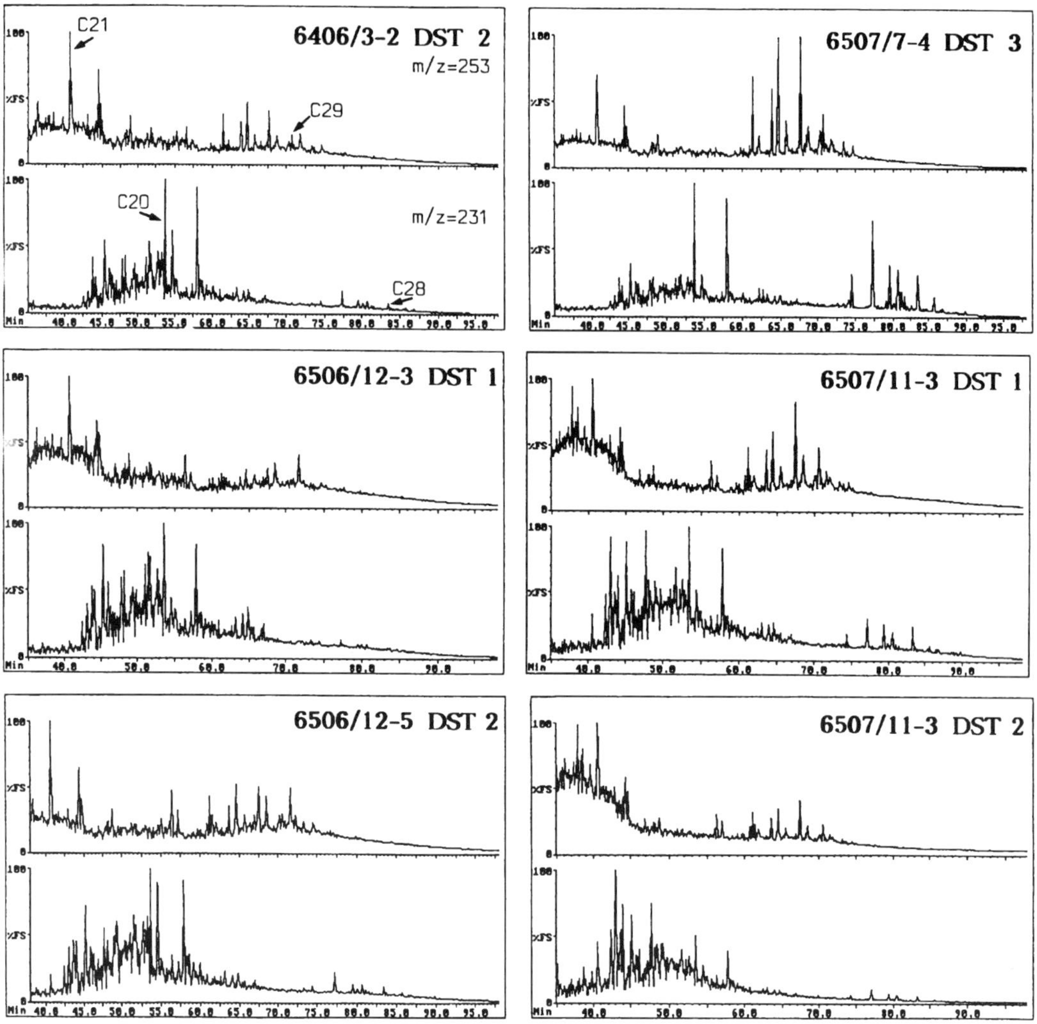

Fig. 38

maturity parameters: 6407/4-1 >> Smørbukk Sør 6506/12-3 > Midgard > Smørbukk Sør 6506/12-5 > Trestakk >> Njord.

Significantly less mature, however, with the indicated trend are: Heidrun ≥ Draugen ≥ Tyrihans Nord (clear even predominance) > Tyrihans Sør, while Mikkel is marginally less mature than this group.

In terms of differences in biomarker signatures, one of the more prominent features is the occurrence of relatively more C_{35} extended hopanes compared to C_{34} extended hopanes in the Mikkel samples (Fig. 23).

In sample 6407/4-1 DST 2 (Fig. 30), there is evidence for thermal destruction of the extended hopanes. Sample 6507/3-1 DST 3 (Fig. 26) is severely modified by partitioning and is clearly of a different facies compared to the rest of the sample set (observe its carbon isotope composition and aromatic GC–FID distribution). In terms of maturity, this sample is tentatively placed together with Njord. The Heidrun sample 6507/7-2 DST 6 (GOR = 2636 m^3/m^3) is clearly a mixture of moderately mature oil and gas condensate (Fig. 28). It contains well defined monoaromatic steroids as do the other Heidrun samples which have GOR values between 48 and 84 m^3/m^3. This interpretation is supported by the approximately zero

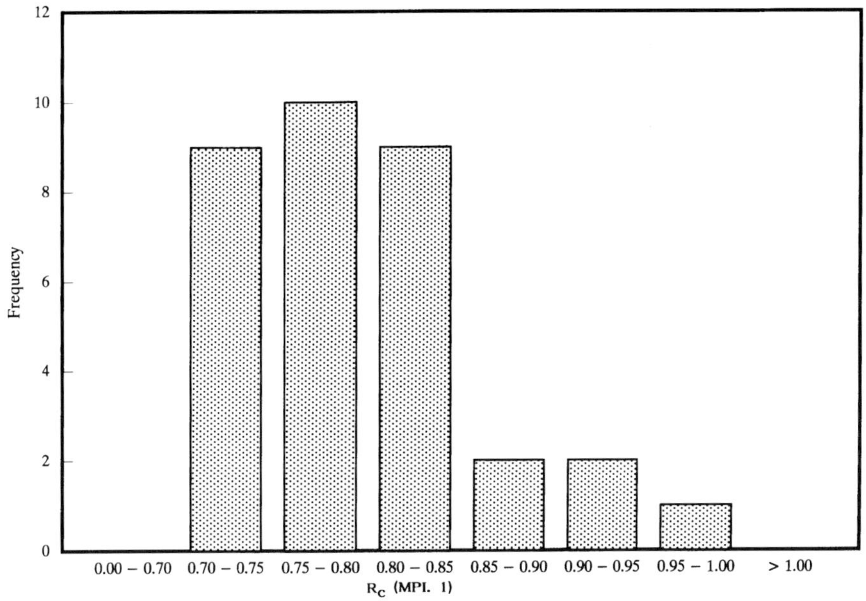

Fig. 39. Frequency distribution of the Haltenbanken petroleums calculated from their phenanthrene and methylphenanthrene distribution (MPI1, Table 6). While this parameter is known not to be a sensitive measure of subtle maturity differences during explusion of petroleum from Spekk Formation type kerogen at low maturity levels, it indicates tentatively that the bulk of the medium range hydrocarbons in petroleums from the Haltenbanken were expelled from their source rock at a maturity of between 0.7 and 0.9% vitrinite reflectance. Evaporative fractionation will, however, affect this parameter strongly as shown for the Midgard samples, where the maturity in the oil leg (M1) equals 0.93%, while in the condensates it ranges from 0.73 to 0.79 due to the relative enrichment in the condensate of phenanthrene relative to the methylphenanthrenes (Figs 31 to 33).

content of polyaromatic hydrocarbons (no defined triaromatic steroids in *m/z* 231, Fig. 12) and surprisingly high (7.3%) content of polars (Fig. 12).

Organic facies parameters in the petroleums

The biomarker signatures of the petroleums in the database are, in a gross sense, relatively uniform, if one disregards the effects which can be attributed to maturity and evaporative fractionation, i.e. the source rock facies of the different petroleums are similar. The occurrence of bisnorhopane is noted in all petroleums, as is the occurrence of tricyclic terpanes. Also, there is generally good correlation between the facies and maturity parameter, Ts/(Ts + Tm), and maturity parameters like hopane X/(hopane X + normoretane) and others (Fig. 44). This would not be expected if there were a dramatic facies shift for the petroleum source rocks in this database.

The conclusion is that no evidence from these data suggests that any of the petroleum samples were sourced from predominantly coaly source rocks. Thus, either the Åre Formation did not contribute to the reservoir fluids, or it did not contribute to the biomarker distributions in the present petroleums. The latter interpretation is favoured as this formation, over large regions, is significantly higher in maturity than the Spekk Formation. Accordingly, the present reservoir petroleums will have biomarker distributions which are influenced more by the relatively higher biomarker concentrations in the moderately mature Spekk Formation rather than the lower concentrations of biomarkers in more mature petroleum derived, at an earlier stage, from the Åre Formation. Such petroleum could potentially have dismigrated from the traps or become diluted by progressively arriving petroleum from the Spekk Formation.

Figure 41 displays stable carbon isotope composition versus pristane/phytane ratio of the oils. The Haltenbanken data show higher pristane/phytane ratios for similar $\delta^{13}C$ values compared to the Viking Graben data of Chung *et al.* (1992). Tentatively, this suggests that petroleums from Haltenbanken have, in their medium range, contributions from a terrestrial source-rock facies, which expelled condensate with a higher pristane/phytane

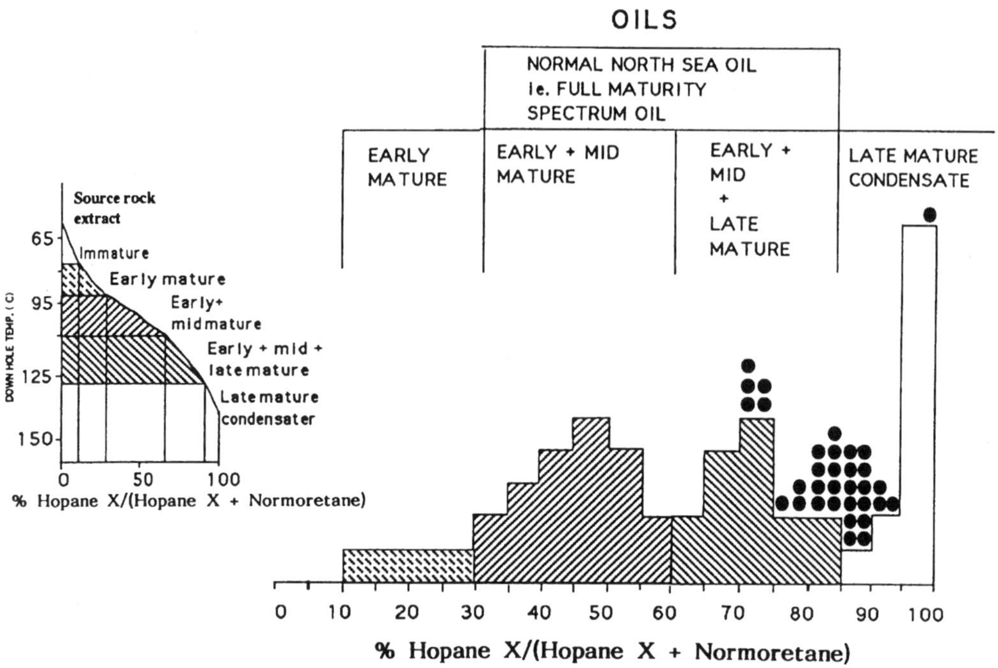

Fig. 40. The abundance of hopane X or diahopane (Moldowan *et al.* 1991) relative to normoretane (Cornford *et al.* 1986) in the Haltenbanken fluid (black dots), superimposed on a figure taken from (Cornford *et al.* 1986). The bulk of the heavy end of the Haltenbanken petroleums was expelled from their source rocks in the late-mid to late maturity stage.

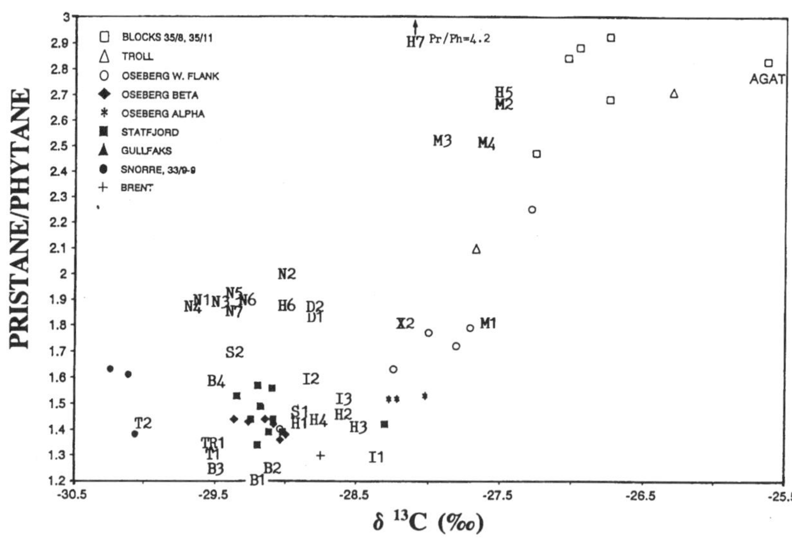

Fig. 41. The stable carbon isotope composition and the pristane/phytane data of the Haltenbanken petroleums plotted in a figure modified after Chung *et al.* (1992). The Njord (N) and Draugen (D) samples are closely associated and possibly form a more oxic facies compared to the Smørbukk Sør, Trestakk, Tyrihans Nord, Tyrihans Sør and Heidrun samples, which plot together with the Mikkel samples and make up a separate group. The isotopically heavier Midgard samples demonstrate the higher pristane/phytane ratio in the condensates (M4, M3, M2) relative to the oil leg (M1). The high GOR samples (H7 and the H5) display the same effect of pristane enrichment relative to phytane due to evaporative fractionation.

ratio. However, the ratio may represent the effects of evaporative fractionation or a facies shift within the Spekk Formation. as discussed previously.

A plot of the pristane/n-C_{17} versus phytane/n-C_{18} (Fig. 42) underlines the fact that the maturity of the C_{17} to C_{18} range in the petroleums is wide and that only Midgard and the 6507/3-1 DST 3 samples contain terrestrial input. Maturity ranges from most mature in Njord (N), Midgard (M), Draugen (D) and Smørbukk Sør (S1, B), to least mature in Heidrun (H), Tyrihans Sør and Mikkel (I). Thus, compared to the maturity ranking discussed previously based mostly on biomarkers, the results here indicate that mixed facies contributions must have affected many of the petroleums, perhaps with the exception of Mikkel and Njord.

Hence, we suggest that several of the Haltenbanken petroleum systems are mixed. The

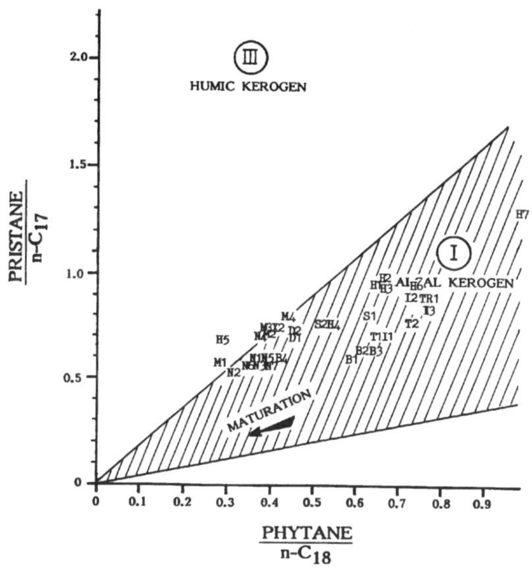

Fig. 42. The pristane/n-C_{17} and phytane/n-C_{18} parameters of the petroleums plotted in a figure taken from Connan & Cassou (1980). The samples again fall mainly into two groups, one indicating a perhaps dysaerobic to anoxic proximal depositional environment (Njord, Draugen, Midgard, partly Heidrun and 6507/3-1) and one more purely marine anoxic distal organic facies (represented by Mikkel, Tyrihans Sør, Tyrihans Nord, Trestakk and partly Heidrun). The tentative relative maturity sequence is also indicated. However, it is worth noting that one of the effects of evaporative fractionation, which is relevant for the highest GOR samples in the database, is the relative enrichment in the gas/condensate of the isoprenoid hydrocarbons pristane and phytane relative to n-C_{17} and n-C_{18} (H7) (Fig. 28). This effect is also observed for the Midgard condensate samples (M4, M3, M2) relative to the Midgard oil leg (M1) (Figs 31 & 33).

biomarkers reflect a differing maturity to pristane/n-C_{17} versus phytane/n-C_{18} data and aromatic maturity parameters (Table 6), and the petroleums have contributions from mixed source facies, e.g. Spekk Formation with vertical and lateral organic facies variation.

The percentage sterane distribution of the petroleums (Fig. 43) shows that the sterane hydrocarbons of the oils are sourced principally from the same organic facies. We take this as strong evidence that the Spekk Formation is the source for the bulk of the oils, especially as the Åre Formation coals have a totally different sterane distribution.

A preliminary study was performed on extracts of coals from the Åre Formation in the 6507/10 block, where the total thickness of coals is known to be more extensive than in the southern part of the Haltenbanken. These coals are buried to *c.* 3500 m. Their extracts are characterized by extremely low contents of tricyclic terpanes to pentacyclic terpanes, lack of bisnorhopane, ten times higher pristane/phytane ratio and a sterane distribution dominated by C_{29} steranes. If this distribution reflects accurately hydrocarbons expelled from the coals approaching 4000 to 4500 m, then the coals did not contribute to the heavy end of the reservoired Haltenbanken fluids. These coal extracts from the northern part of the Haltenbanken are similar to an extract from Åre Formation coal in the Njord Field in the southern part of the Haltenbanken, at a maturity of $R_m = 0.55\%$, which has an n-alkane and isoprenoid distribution similar to a waxy oil from Indonesia discussed by Thompson *et al.* (1985).

While it is clear that a larger database on coals at higher maturities could be investigated, perhaps by different crushing and extraction procedures, it appears that, if the biomarker distribution of this ground coal extract from the Njord Field is representative of the petroleum expelled from Åre coals, then the contribution from the coals to the C_{25+} range biomarkers of Haltenbanken petroleums is minor. This supports Cohen & Dunn (1987) who concluded, based on gochemistry and studies of the gas to oil generation index of the coals and their maceral composition, that the Åre coals are likely to retain their generated hydrocarbons sufficiently long for further burial to occur, so that the expelled products would be condensate/gas with only minor quantities of oil. As indicated in Fig. 17, the light oil end (C_{14} to C_{21} hydrocarbon range) of the coals, if expelled as a condensate, is similar to the rest of the Haltenbanken database.

A preliminary characterization of a few extracts from the Spekk Formation in the 6407 quadrant, from a depth of *c.* 3900 m, displays n-alkane, isoprenoid, methylphenanthrene and dibenzothiophene distributions and facies controlled

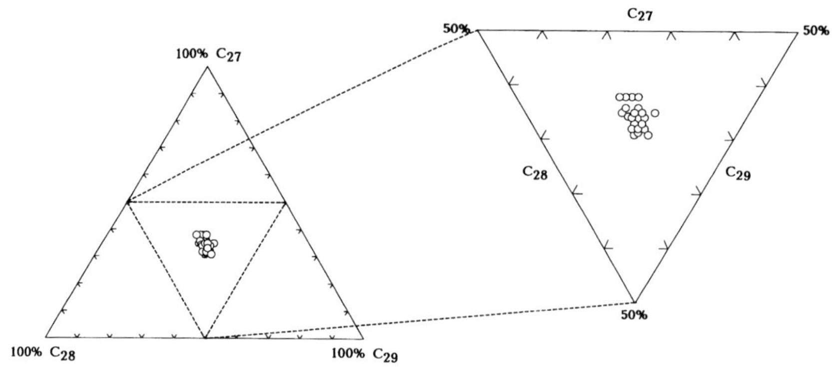

Fig. 43. The percentage sterane distribution measured on *m/z* 218.2034 of the database indicates in a gross sense that a comparatively homogeneous source-rock system contributed to the bulk of the heavy end of the reservoired petroleums. The slight spread in the database reflects the higher thermal resistance to maturity of the C_{27} relative to the C_{29} sterane, and a slight effect ascribed to evaporative fractionation enriching the higher GOR samples in the lower molecular weight steranes. Thus, no evidence is seen for higher terrestrial plant input and the differences in isoprenoid to *n*-alkane ratios are most likely to reflect slight differences in early diagenetic environment. These can be different Eh and pH conditions and a potentially different degree of bacterial input on a predominantly type II marine algal-dominated proto-kerogen.

sterane and hopane biomarker distributions similar to those found in the nearby reservoirs.

We suggest that significant oil expulsion has occurred from these samples of the Spekk Formation at depths shallower than 4000 m, in order for the sterane, tricyclic terpane and the pentacyclic triterpane maturity parameters in the extracts to compare with those typically found in the Haltenbanken petroleums. Dahl & Augustson (1993) estimated significant hydrocarbon generation to have commenced (R_o = 0.5%) at *c.* 3100 m in the Haltenbanken region, a maturity which may have been reached at a depth of *c.* 2500 m in 6407/4-1 and generally in the Haltenbanken area at 2800 to 3000 m, according to Vik & Hermanrud (1993). If oil expulsion starts from the Spekk Formation at 3500 m or, as suggested by Cohen & Dunn (1987), from a little more than 3000 m, the volumetric calculations of fluids expelled from drainage areas into structures could be significantly affected. Thus, it may not be necessary to invoke oil contributions from the Åre Formation to make the mass balance between expelled, lost and reservoired oil.

Furthermore, we feel that the mass balance problems may to some extent be related to difficulties in identifying from the seismic evidence the change from the 'unproductive' Melke Formation to the prolific Spekk Formation (Grigo *et al.* 1993), suggesting that the Spekk Formation can locally be thicker than indicated from the seismic data. In addition, thermal anomalies in the Haltenbanken area, modelled by Grigo *et al.* (1993), may have caused earlier migration to some of the structures in the central and northern parts of this region than commonly believed.

Hermans *et al.* (1992), who modelled secondary migration in the southern and central Haltenbanken region, concluded that the Spekk Formation alone may have sufficient generative capacity to fill most accumulations to the spill point, provided that leakage through cap rocks and along faults is quantitatively unimportant and residual oil saturation in the carrier rock is negligible.

To conclude the discussion on organic maturity and facies parameters of the Haltenbanken fluids, principal components plot Fig. 44 (see Table 7 for explanation of the parameters) is used to display some of the variation in the database. Whilst the Mikkel (I) petroleums and the 6407/4-1 (X2) are opposite extremes in terms of maturity, and maturity parameters are responsible for most of the variation in the database, the bisnorhopane and stable isotopes are important in differentiating facies relationships between petroleums.

Thus, there is apparently evidence in the database to suggest that the Spekk Formation is the main oil contributor in the study area and that, in a gross sense, there is significant facies variation in the Spekk Formation from more terrestrial, potentially dysaerobic facies in the east to more typically marine anoxic conditions in the west. Lateral facies variations in the Spekk Formation were described by Whitley (1992) as a shift in maceral composition from sapropelic shale in 6507/7-1 (northwest of Heidrun) to vitrinite-rich shale/siltstone in the Heidrun Field. Superimposed on this pattern are local variations in source rock as

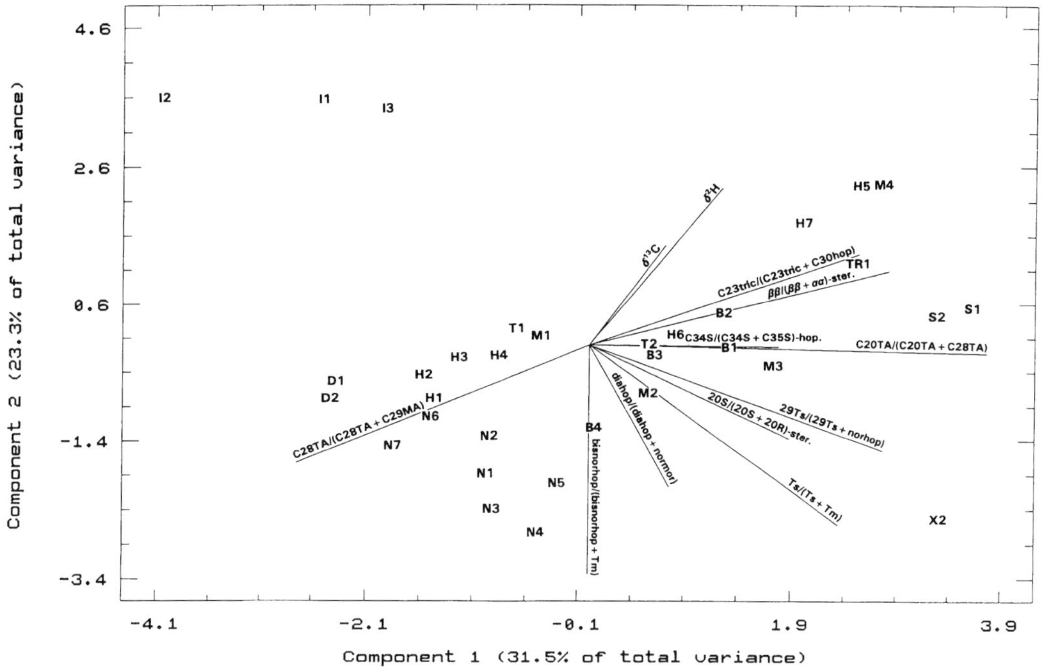

Fig. 44. Principal components plot showing the maturity and facies parameters from Table 7 plus the stable carbon and hydrogen ratio in the database. Note that the Mikkel samples (I) are clearly separated from the rest owing mainly to their low maturity. The Njord samples score on the bisnorhopane facies parameter and the Njord (N), Draugen (D) and the Heidrun samples (H) separate from the Smørbukk Sør, Trestakk, Midgard samples mainly owing to maturity differences. Note that the 6407/4-1 (X2) sample differentiates from the rest mainly owing to its high maturity.

reflected by even predominance in Tyrihans Nord petroleum and core extracts from Smørbukk. Equally likely are significant vertical variations in organic facies of the Spekk Formation, similar to those found in the Draupne Formation, with more isotopically heavier kerogen in the older part of this formation, reflecting a more humic organic matter content and/or more dysaerobic conditions in the diagenetic environment.

Geochemically there is little evidence to suggest that the Åre Formation has expelled oil of quantities which are traceable in the presently reservoired fluids. However, expelled condensate/gas is likely, especially in the north where the accumulated thickness of the Åre Formation is more extensive than in the south. Contributions from this formation to the light and medium range hydrocarbons (gas plus condensate) of the reservoired fluids may occur in other fields than the Midgard though, as yet, little geochemical evidence exists to support this interpretation unequivocally.

Metals in Haltenbanken petroleums

Vanadium/nickel ratios were measured in the oils as an empirical database exists for the use of this ratio in oils and source-rock extracts (Hodgson & Baker 1959; Mello *et al.* 1988; Barwise 1990), and because this parameter can be used as an indicator of Eh and pH of the depositional environment (Lewan 1980, 1984). The samples were analysed by ICP–MS to determine their metal composition according to procedures outlined in Olsen *et al.* (1993). Our preliminary results suggest that the following conclusions can be drawn: $^{51}V/(^{51}V + ^{58}Ni)$ is *c.* 0.1 to 0.35 in Njord, *c.* 0.5 to 0.6 in Draugen and in the Midgard oil leg and *c.* 0.8 to 0.9 in Heidrun, Mikkel, Tyrihans, Smørbukk Sør, Trestakk, Heidrun and Midgard gas/condensate.

It was found that $^{51}V/(^{51}V + ^{58}Ni)$ is negatively correlated with the even predominance of *n*-alkanes and the pristane/phytane ratio, but it is positively correlated with the phytane/n-C_{18} ratio. The correlation coefficient between the $^{51}V/(^{51}V + ^{58}Ni)$ ratio and $\delta^{13}C$ of the petroleums (C_{10+}) is 0.65, indicating that about 42% ($0.65^2 \times 100$) of the variation between the vanadium/nickel ratio and the carbon-isotope facies parameter can be ascribed to a linear relationship between the two variables. This is lower than expected, even if we consider the effects of mixed origin, i.e. the $\delta^{13}C$ parameter reflects the (C_{10+}) petroleum composition, whilst the metal content is expected to be determined by the heavier polar fraction.

The data indicate that most of the western fields have petroleum sourced from a marine anoxic-dominated source rock while the Draugen and the Midgard Fields contain petroleum of a possibly less anoxic marine source-rock facies. These results support the interpretations of the carbon isotope composition of the C_{10+} petroleum fraction.

However, it may be surprising to note that the V/Ni ratio in the Njord petroleums suggests a comparatively less anoxic marine depositional system that the Draugen, Mikkel and Midgard petroleums. Therefore, it is likely that the oils presently reservoired in Njord come from a different source-rock facies compared to the oils in Smørbukk Sør. However, one would postulate, based on their paraffinic character, moderate pristane/phytane ratios and overall biomarker distributions, that they are both likely to have been sourced from a marine source rock such as the Spekk Formation. Consequently, we feel that the Spekk Formation source rock in the Njord drainage area has a more terrestrially influenced kerogen compared to the Spekk Formation facies in the western part of the study area, i.e. the drainage area of the Tyrihans, Trestakk and the Smørbukk Sør Fields.

Preliminary work has shown that metal distributions are correlatable to GOR and thus gross composition. Specifically, Ni, Co, Sr, Ag and Bi correlate positively with the absolute amount of heavy *n*-alkanes in the petroleums as these elements are relatively more abundant in paraffinic petroleums of the western accumulations, such as Njord, Smørbukk Sør and Trestakk Fields.

Preliminary studies performed on two cores representing Spekk and Åre Formations have shown that the Spekk Formation extracts displayed $^{51}V/(^{51}V + ^{58}Ni)$ ratios higher than 0.9, while values for the Åre Formation extracts ranged from 0.5 to 0.99 (Olsen *et al.* 1993). This indicates that the Åre Formation, at least in this location, could have been less anoxic during deposition than earlier believed.

The data might indicate that some of the elements are not restricted to the polar fraction. Typically, V, which tends to occur in relatively high proportions in high-GOR aliphatic systems, such as the 6407/4-1 DST 2 ($^{51}V/(^{51}V + ^{58}Ni) = 0.728$, GOR = 1600 m^3/m^3) and the Midgard gas 6507/11-3 DST 3 ($^{51}V/(^{51}V + ^{58}Ni) = 1.000$, GOR = 4330 m^3/m^3), is substantially lower in the Midgard oil leg (6507/11-3 DST 1; $^{51}V/(^{51}V + ^{58}Ni) = 0.561$, GOR = 173 m^3/m^3). In the high-GOR sample 6507/7-2 DST 6, the ratio is again higher ($^{51}V/(^{51}V + ^{58}Ni) = 1.000$, GOR = 2636 m^3/m^3) than in the 6507/7-2 DST 2 sample ($^{51}V/(^{51}V + ^{58}Ni) = 0.802$, GOR = 48 m^3/m^3). Furthermore, the ratio is slightly higher in, for example, the 6407/7-1 samples from Njord ($^{51}V/(^{51}V + ^{58}Ni) = 0.35$, GOR = 207 to 218 m^3/m^3) and in 6407/7-3 DST 3 ($^{51}V/(^{51}V + ^{58}Ni) = 0.30$, GOR = 389 m^3/m^3), than in the Njord samples with lower GORs, e.g. in 6407/7-4 DST 2A and 2B ($^{51}V/(^{51}V + ^{58}Ni) = 0.07–0.10$, GOR = 170 to 180 m^3/m^3) and in 6407/7-2 DST 1 ($^{51}V/(^{51}V + ^{58}Ni) = 0.12$, GOR = 180 m^3/m^3).

Possibly, this could reflect either a higher thermal stability for the V porphyrins relative to Ni porphyrins (Sundararaman & Moldowan 1993) or the affinity of Ni and V carriers, which need not only be porphyrins, to the saturated fractions, the amount of which tends to be positively correlated with GOR. Galimov *et al.* (1990), studying heavy bituminoids, found that the sum of Ni and V porphyrins made up only 16 to 36% of the metal contents in their samples. Thus, care should be exercised in interpreting vanadium/nickel ratios until the metal carriers are better characterized.

Helgeland basin core extract 6609/11-1 2561 m

A portion of a sandstone core at 2561 m from 6609/11-1 was apparently dark-stained with bitumen. A sample of this comparatively coarse sand was extracted (Fig. 45) and in a principal components plot (Fig. 46), this moderately mature core extract deviates, on the basis of the considered parameters, dramatically from the rest of the Haltenbanken petroleums. Its light hydrogen and carbon isotopic composition and waxy character could imply that it represents a terrestrially influenced oil. This interpretation is supported by the vanadium/nickel ratio, $^{51}V/(^{51}V + ^{58}Ni) = 0.074$, which suggests a basic depositional environment characteristic of terrestrially sourced low-sulphur oils. Furthermore, the lack of dibenzothiophenes and methyldibenzothiophenes in this core extract differentiates it from the other analysed Haltenbanken petroleums, as does the low aromatic content (Fig. 45).

In the light of successful ongoing studies of Palaeozoic source rocks in this basin and the symmetry between the Norwegian and the East Greenland post-Caledonian palaeogeography, we postulate that this petroleum, sampled from a bitumen-impregnated sandstone, may represent migration from a substantially more deeply buried, terrestrially influenced source rock. It is known that Mesozoic and Permian sediments in the greater Haltenbanken region have locally developed

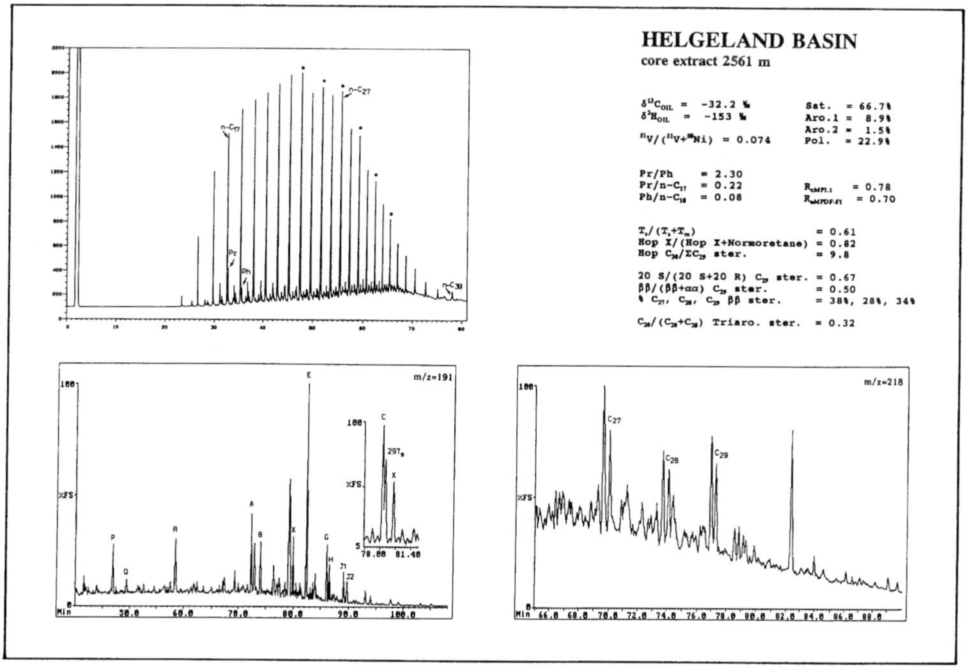

Fig. 45. The extract of core 6609/11-1 (Helgeland basin) from 2561 m. Note the light stable carbon isotope composition of the extract, its V/(Ni + V) ratio and the relatively high concentrations of *n*-alkanes relative to isoprenoids in this moderately mature bitumen. The sample is believed to be sourced from a deeper Triassic/Palaeozoic terrestrial, possibly lacustrine, source-rock system.

terrestrial or even lacustrine source-rock characteristics. However, based on knowledge about East Greenland Palaeozoic source rocks (Surlyk 1983; Surlyk *et al.* 1981, 1984*a*,*b*, 1986; Christiansen *et al.* 1990, 1992) and the occurrence of isotopically similar Devonian sourced oil in the Beatrice Field in the UK sector (Bailey *et al.* 1990; Peters *et al.* 1989), which also has a hopane/sterane ratio (Mackenzie *et al.* 1984) on a par with the Helgeland basin core extract, one could speculate that this bitumen represents oil sourced from a Triassic or Palaeozoic terrestrial source rock.

The impregnated sandstone contains a polar extract (Fig. 45) and is, in many ways, similar to a small tar mat. The occurrence itself should be studied in more detail to determine whether it represents a migrational avenue, similar to the phenomena described in cores from the Ula and Hild Fields (Karlsen *et al.* 1993; Wilhelms & Larter 1995; Skålnes *et al.* 1993), or a palaeo-OWC of a palaeo-accumulation.

Conclusions

This study has characterized the maturity and genetic relationships between reservoired petroleums in the Haltenbanken area. In general, the C_{15+} hydrocarbons represent a significant maturity suite, with low-maturity Mikkel oils and mature condensate in 6407/4-1 forming opposite extremes. Several of the condensates are otherwise characterized as evaporative moderately mature condensates, separated from a cogenerated oil during migration, which in cases may be found in the same reservoir, e.g. as an oil leg in the Midgard Field. In terms of organic facies, the bulk of C_{15+} hydrocarbons from the Haltenbanken area seem to have been generated from the Spekk Formation. However, significant vertical and lateral changes in the organic facies of the Spekk Formation are invoked, as evidenced by the systematic geographic shifts in pristane/phytane ratios, bisnorhopane contents, stable carbon isotope composition and vanadium/nickel ratios. This facies shift in the Spekk Formation is thought to reflect greater terrestrial input and also possibly more dysaerobic conditions in a proximal direction. In the distal western regions, the Spekk Formation is inferred generally to be developed as a typical sapropelic marine anoxic shale.

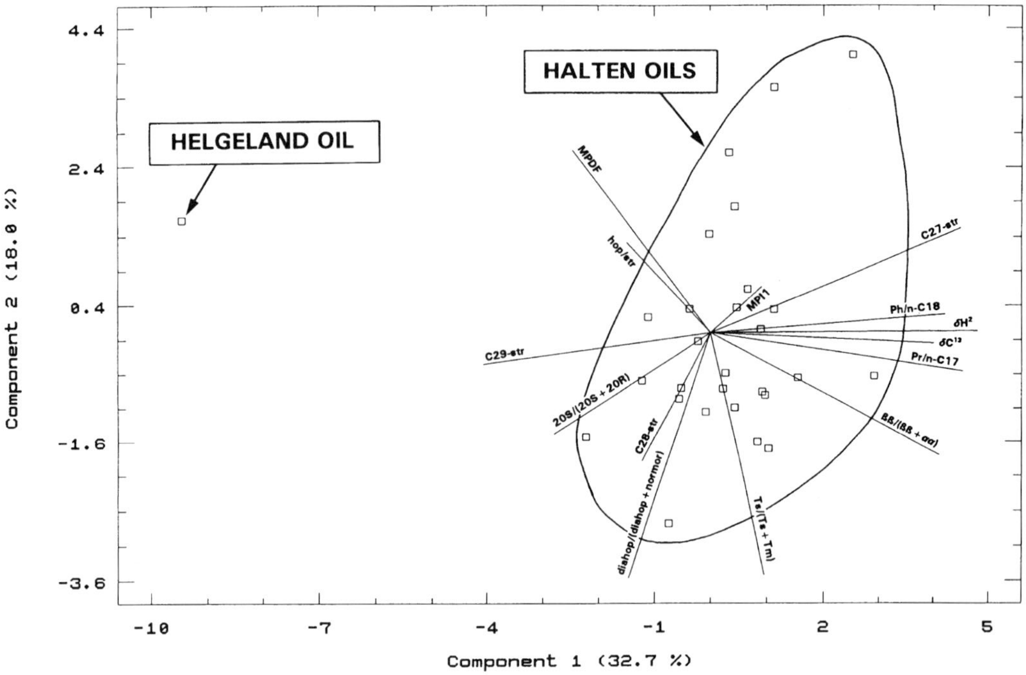

Fig. 46. Principal components plot illustrating that the Haltenbanken database is clearly separated from the Helgeland basin sample 6609/11-1 2561 m.

In most cases, except for the Draugen Field and to a lesser extent the Midgard Field, lateral migration is proposed to be of short to medium range with respect to inter-field distances. We see no evidence for remigration of C_{15+} hydrocarbons from Tyrihans Sør to Tyrihans Nord, nor to Midgard, while the Trestakk Field and the Tyrihans Sør accumulations may be part of the same migration system.

A possible terrestrially influenced or lacustrine petroleum was identified in a core extract from the Helgeland basin. The geochemistry of this core extract is different from the Haltenbanken petroleums and we suggest that there exists in this basin a mature terrestrial, possibly Triassic or Palaeozoic source rock, of sufficient thickness and quality to have expelled quantities of petroleum able to saturate a carrier system.

If the geochemistry of petroleum in diagenetically formed fluid inclusions from the structures/reservoirs could be analysed, the data could shed light on the timing of petroleum arrival at the structures and on the maturity and source facies of early generated, expelled and entrapped fluids (Karlsen *et al.* 1993). Such interpretation schemes have proved useful in other regions, e.g. in the Ula Field, Central Graben, and in the Hild Field, Northern Viking Graben (Karlsen *et al.* 1991; Nedkvitne *et al.* 1993; Skålnes *et al.* 1993). Furthermore, it is believed that such studies may be especially valuable for understanding the relationship between overpressure and seal failure of 'dry' unfilled structures in the Haltenbanken West area.

We thank the Norwegian Petroleum directorate, the operators on the licences and the licence partners for permission to publish this paper. The interpretations presented in this paper reflects those of the authors and do not necessarily coincide with the official view of the Petroleum Directorate nor of the involved institutions. Interpretations made in such an empirical study are limited by the available database, and it is therefore to be expected that interpretations made on migrational and, in particular, source rock geochemistry will be modified and refined as the available database expands. We thank two anonymous reviewers for their help in improving this manuscript. Furthermore, we acknowledge isotope measurements performed at the Institute of Energy Technics (IFE), Kjeller and Geoloab Nor, Trondheim, Norway.

References

ABBOTT, G. D. & MAXWELL, J. R. 1988. Kinetics of the aromatization of rearranged ring-C monoaromatic steroid hydrocarbons. *Organic Geochemistry*, **13**, 881–885.

ALEXANDER, R., KAGI, R. I., ROWLAND, S. J., SHEPPARD, P. N. & CHIRILA, T. V. 1985. The effects of thermal maturity on distributions of dimethylnaphthalenes and trimethylnaphthalenes in some ancient sediments and petroleum. *Geochimica et Cosmochimica Acta*, **49**, 385–395.

BAILEY, N. J. L., BURWOOD, R. & HARRIMAN, G. E. 1990. Application of pyrolysate carbon isotope and biomarker technology to organofacies definition and oil correlation problems in the North Sea basin. *Organic Geochemistry*, **16**, 1157–1172.

BANDURSKi, E. 1982. Structural similarities between oil generation kerogens and petroleum asphaltenes. *Energy Sources*, **6**, 47–67.

BARWISE, A. J. G. 1990. Role of nickel and vanadium in petroleum classification. *Energy & Fuels*, **4**, 647–652.

BEACH, F., PEAKMAN, T. M., ABBOTT, G. D., SLEEMAN, R. & MAXWELL, J. R. 1989. Laboratory thermal alteraton of triaromatic steroid hydrocarbons. *Organic Geochemistry*, **14**, 109–111.

BEHAR, F. & PELET, R. 1984. Asphaltene characterization by pyrolysis and chromatography. *Journal of Analytical and Applied Pyrolysis*, **7**, 121–135.

BELETSKAYA, S. N. 1978. Differentiation of disseminated organic matter in the process of primary migration. *In*: NAVLIKINJ, V. D. & ALIEVE, M. M. (eds) *Fundamental Aspects of the Genesis and Accumulation of Oil and Gas*. Nauka, Moscow, Russia, 52–70.

BJØRLYKKE, K., AAGAARD, P., DYPVIK, H., HASTINGS, D. S. & HARPER, A. S. 1986. Diagenesis and reservoir properties of Jurassic sandstones from the Haltenbanken area, offshore mid Norway. *In*: SPENCER, A. M. ET AL. (eds) *Habitat of hydrocarbons on the Norwegian Continental Shelf*. Norwegian Petroleum Society, Graham & Trotman, London, 275–286.

BJORØY, M., HALL, K., GILLYON, P. & JUMEAU, J. 1991. Carbon isotope variations in n-alkanes and isoprenoids of whole oils. *Chemical Geology*, **93**, 13–20.

BØEN, F. S., EGGEN, S. & VOLLSET, J. 1984. Structures and basins of the margin from 62 to 69°N and their development. *In*: SPENCER, A. M. ET AL. (eds) *Petroleum Geology of the North European Margin*. Norwegian Petroleum Society, Graham & Trotman, London, 253–270.

BUGGE, T., KNARUD, R. & MØRK, A. 1984. Bedrock geology on the Mid-Norwegian continental shelf. *In*: SPENCER, A. M. ET AL. (eds) *Petroleum Geology of the North European Margin*. Norwegian Petroleum Society, Graham & Trotman, London, 271–284.

BUKOVICS, C. & ZIEGLER, P. A. 1985. Tectonic development of the Mid-Norway continental margin. *Marine and Petroleum Geology*, **2**, 2–22.

CAMPBELL, C. J. & ORMAASEN, E. 1987. The discovery of oil and gas in Norway: an historical synopsis. *In*: SPENCER, A. M. ET AL. (eds) *Geology of the Norwegian Oil and Gas fields*. Graham & Trotman, London, 1–37.

CHRISTIANSEN, F. G., OLSEN, H., PIASECKI, S. & STEMMERIK, L. 1990. Organic geochemistry of Upper Palaeozoic lacustrine shales in the East Greenland basin. *Organic Geochemistry*, **16**, 287–294.

——, DAM, G., PIASECKI, S. & STEMMERIK, L. 1992. A Review of Upper Palaeozoic and Mesozoic Source Rocks from Onshore East Greenland. *In*: SPENCER, A. M. (ed.) *Generation, Accumulation and Production of Europe's Hydrocarbons II*. Special Publication of the European Association of Petroleum Geologists, **2**, Oxford University Press, Oxford, 151–161.

CHUNG, H. M., WINGERT, W. S. & CLAYPOOL, G. E. 1992. Geochemistry of oils in the Northern Viking Graben. *In*: HALBOUTY, M. T. (ed.) *Giant Oil and Gas Fields of the Decade 1978–1988*. American Association of Petroleum Geologists, Memoir, **54**, 277–296.

COHEN, M. J. & DUNN, M. E. 1987. The Hydrocarbon habitat of the Haltenbanken–Trænabanken area offshore Mid-Norway. *In*: BROOKS, J. & GLENNIE, K. (eds) *Petroleum Geology of North West Europe*. Graham & Trotman, London, 1091–1104.

CONNAN, J. & CASSOU, A. M. 1980. Properties of gases and petroleum liquids derived from terrestrial kerogen at various maturation levels. *Geochimica et Cosmochimica Acta*, **44**, 1–23.

CORNFORD, C., NEEDHAM, C. E. J. & DE WALQUE, L. 1986. Geochemical habitat of North Sea oils and gases. *In*: SPENCER, A. M. ET AL. (eds) *Habitat of Hydrocarbons on the Norwegian Continental Shelf*. Norwegian Petroleum Society, Graham & Trotman, London, 39–54.

DAHL, B. & AUGUSTSON, J. H. 1993. The influence of Tertiary and Quaternary sedimentation and erosion on hydrocarbon generation in Norwegian offshore basins. *In*: DORÉ, A. ET AL. (eds) *Basin Modelling: Advances and Application*. NPF Special Publication, **3**, Elsevier, Amsterdam, 419–431.

DURAND, B. 1988. Understanding of HC migration in sedimentary basins (present state of knowledge). *Organic Geochemistry*, **13**, 445–459.

——, HUC, A. Y. & OUIDIN, L. 1987. Oil saturation and primary migration: observations in shales and coals from the Kerbau wells Mahakam Delta, Indonesia. *In*: DOLOGEZ, B. ET AL. (eds) *Migration of HC in Sedimentary Basins*. Editions Technip, Paris, 173–195.

EHRENBERG, S. N. 1990. Relationship between diagenesis and reservoir quality in sandstones of the Garn Formation, Haltenbanken, Mid-Norwegian continental shelf. *American Association of Petroleum Geologists, Bulletin*, **74**, 1538–1558.

——, GJERSTAD, H. M. & HADLER-JACOBSEN, F. 1992. Smørbukk Field – a gas condensate fault trap in the Haltenbanken Province, offshore Mid-Norway. *In*: HALBOUTY, M. T. (ed.) *Giant Oil and Gas Fields of the Decade 1978–1988*. American Association of Petroleum Geologists, Memoir, **54**, 323–348.

ELLENOR, D. W. & MOZETIC, A. 1986. The Draugen oil discovery. *In*: SPENCER, A. M. *ET AL.* (eds) *Habitat of Hydrocarbons on the Norwegian Continental Shelf.* Norwegian Petroleum Society, Graham & Trotman, London, 313–316.

ELVSBORG, A., HAGEVANG, T. & THRONDSEN, T. 1985. Origin of the gas-condensate of the Midgard Field at Haltenbanken. *In*: THOMAS, B. M. *ET AL.* (eds) *Petroleum Geochemistry in Exploration of the Norwegian Shelf.* Norwegian Petroleum Society, Graham & Trotman, London, 213–219.

ENGLAND, W. A. & MACKENZIE, A. S. 1989. Some aspects of the organic geochemistry of petroleum fluids. *Geolgische Rundschau*, **78**, 291–303.

——, ——, MANN, D. M. & QUIGLEY, T. M. 1987. The movement and entrapment of petroleum fluids in the subsurface. *Journal of the Geological Society, London*, **144**, 327–347.

ESPITALIE, J., LAFARGUE, E. & EGGEN, S. 1991. Petroleum potential of terrestrial and marine organic matter in Jurassic sequences of the northern North Sea and offshore mid-Norway. *In*: SPENCER, A. M. (ed.) *Generation, Accumulation and Production of Europe's Hydrocarbons*. Special Publication of the European Association of Petroleum Geoscientists, **1**, Oxford University Press, Oxford, 49–63.

FORBES, P. L., UNGERER, P. M., KUHFUSS, A. B., RIIS, F. & EGGEN, S. 1991. Compositional modelling of petroleum generation and explusion: trial application to local mass balance in the Smørbukk Sør Field, Haltenbanken Area, Norway. *American Association of Petroleum Geologists, Bulletin*, **75**, 873–893.

GABRIELSEN, R. H., FAERSETH, R., HAMAR, G. & RØNNEVIK, H. 1984. Nomenclature of the main structural features of the Norwegian Continental Shelf north of the 62nd parallel. *In*: SPENCER, A. M. (ed.) *Petroleum Geology of the North European Margin*. Norwegian Petroleum Society, Graham & Trotman, London, 41–60.

GALIMOV, E. M. 1973. *Carbon Isotopes in Oil and Gas Geology*. Nedra, Moscow, Russia (English translation, Washington NASA TT F-682).

—— 1980. $^{13}C/^{12}C$ in kerogen. *In*: DURAND, B. *ET AL.* (eds) *Kerogen*. Editions Technip, Paris, 271–299.

—— 1985. *The Biological Fractionation of Isotopes*. Academic Press, Orlando.

GALIMOV, R. A., KRIVONOZHKINA, L. B., ABUSHAYEVA, V. V. & ROMANOV, G. V. 1990. Relationships governing the distribution of vanadium, nickel and their porphyrin complexes in petroleum components. *Petroleum Chemistry of the USSR*, **30**, 55–60.

GOWERS, M. B. & LUNDE, G. 1984. Geological history of Trænabanken. *In*: SPENCER, A. M. (ed.) *Petroleum Geology of the North European Margin*. Norwegian Petroleum Society, Graham & Trotman, London, 237–252.

GRANTHAM, P. J., POSTHUMA, J. & DE GROOT, V. 1980. Variation and significance of the C_{27} and C_{28} triterpane content of a North Sea core and various North Sea crudes. *In*: DOUGLAS, A. G. & MAXWELL, J. K. (eds) *Advances in Organic Geochemistry 1979*. Pergamon Press, Oxford, 29–38.

GRIGO, D., MARAGNA, B., ARIENTI, M. T., FIORANI, M., PARISI, A., MARRONE, M., SQUAZZERO, P. & UBERG, A. S. 1993. Issues in 3D sedimentary basin modelling and application to Haltenbanken, offshore Norway. *In*: DORÉ, A. *ET AL.* (eds) *Basin Modelling: Advances and Application*. NPF Special Publication, **3**, Elsevier, Amsterdam, 419–431.

HERMANS, L., VAN KUYK, A. D., LEHNER, F. K. & FEATHERSTONE, P. S. 1992. Modelling secondary hydrocarbon migration in Haltenbanken, Norway. *In*: LARSEN, R. M. *ET AL.* (eds) *Structural and Tectonic Modelling and its Application to Petroleum Geology*. Proceedings of Norwegian Petroleum Society Workshop, 18–20 October 1989, Stavanger, Norway, Elsevier, Amsterdam, 305–323.

HEUM, O. R., DALLAND, A. & MEISINGSET, K. K. 1986. Habitat of hydrocarbons at Haltenbanken (PVT-modelling as predictive tool in hydrocarbon exploration). *In*: SPENCER, A. M. *ET AL.* (eds) *Habitat of Hydrocarbons in the Norwegian Continental Shelf.* Norwegian Petroleum Society, Graham & Trotman, London, 259–274.

HODGSON, G. W. & BAKER, B. L. 1959. Geochemical aspects of petroleum migration in Pembina, Redwater, Joffre and Lloydminster oil fields of Alberta and Saskatchewan, Canada. *American Association of Petroleum Geologists, Bulletin*, **43**, 311–328.

HOLLANDER, N. B. 1984. Geohistory and hydrocarbon evolution of the Haltenbanken Area. *In*: SPENCER, A. M. (ed.) *Petroleum Geology of the North European Margin*. Norwegian Petroleum Society, Graham & Trotman, London,383–388.

HUC, A. Y., IRWIN, H. & SCHOELL, M. 1985. Organic matter quality changes in an Upper Jurassic shale sequence from the Viking Graben. *In*: THOMAS, B. M. *ET AL.* (eds) *Petroleum Geochemistry in Exploration of the Norwegian Shelf.* Norwegian Petroleum Society, Graham & Trotman, London, 179–183.

HVOSLEF, S., LARTER, S. R. & LEYTHAEUSER, D. 1988. Aspects of generation and migration of hydrocarbons from coal-bearing strata of the Hitra formation, Haltenbanken area, offshore Norway. *Organic Geochemistry*, **13**, 525–536.

JACKSON, J. S. & HASTINGS, D. S. 1986. The role of salt-movement in the tectonic history of Haltenbanken and Trænabanken and its relationship to structural style. *In*: SPENCER, A. M. *ET AL.* (eds) *Habitat of Hydrocarbons in the Norwegian Continental Shelf.* Norwegian Petroleum Society, Graham & Trotman, London, 241–257.

KARLSEN, D. A. & LARTER, S. R. 1989. A rapid correlation method for petroleum population mapping within individual petroleum reservoirs: Application to petroleum reservoir description. *In*: COLLINSON, J. D. (ed.) *Correlation in Hydrocarbon Exploration*. Norwegian Petroleum Society, Graham & Trotman, London, 77–85.

—— & —— 1991. Analysis of petroleum fractions by TLC–FID: Applications to petroleum reservoir description. *Organic Geochemistry*, **17**, 603–617.

——, NEDKVITNE, T., LARTER, S. R. & BJØRLYKKE, K. 1993. Hydrocarbon composition of authigenic inclusions: Application to elucidation of petroleum reservoir filling history. *Geochimica et Cosmochimica Acta*, **57**, 3641–3659.

KILLOPS, S. D. & AL-JUBOORI, M. A. H. A. 1990. Characterisation of the unresolved complex mixture (UMC) in the gas chromatograms of biodegraded petroleums. *Organic Geochemistry*, **15**, 147–160.

KROOSS, B. 1985. *Investigation of the diffusion of low molecular-weight hydrocarbons in water-saturated sedimentary rocks*. PhD thesis, RWTH Aachen, Germany.

—— 1988. Experimental investigation of the molecular migration of C_1–C_6 hydrocarbons: Kinetics of hydrocarbon release from source rocks. *Organic Geochemistry*, **13**, 513–523.

KVALHEIM, O. M., CHRISTY, A. A., TELNAES, N. & DJØRSETH, A. 1987. Maturity determination of organic matter in coals using methylphenanthrene distribution. *Geochimica et Cosmochimica Acta*, **51**, 1883–1888.

LEADHOLM, R. H., HOT, T. Y. & SAHAI, S. K. 1985. Heat flow geothermal gradients and maturation modelling in the Norwegian Continental Shelf using computer methods. *In*: THOMAS, B. M. *ET AL.* (eds) *Petroleum Geochemistry in Exploration of the Norwegian Shelf.* Norwegian Petroleum Society, Graham & Trotman, London, 131–143.

LEWAN, M. D. 1980. *Geochemistry of vanadium and nickel in organic matter of sedimentary rocks*. PhD dissertation, University of Cincinnati, USA.

—— 1984. Factors controlling the proportionality of vanadium to nickel in crude oils. *Geochimica et Cosmochimica Acta*, **48**, 2231–2238.

LEYTHAEUSER, D., SCHAEFER, R. G. & RADKE, M. 1987. On the primary migration of hydrocarbons. *In*: *Proceedings of the 12th World Petroleum Congress, Houston, 1987, Vol.* 2. John Wiley and Sons, London, 227–237.

LONGMAN, M. W. & PALMER, S. E. 1987. Organic geochemistry of mid-continent Middle and Late Ordovician oils. *American Association of Petroleum Geologists, Bulletin*,**71**, 938–950.

MACKENZIE, A. S., MAXWELL, J. R., COLEMAN, M. L. & DEEGAN, C. E. 1984. Biological marker and carbon isotope studies of North Sea crude oils and sediments. *In*: *Proceedings of the 11th World Petroleum Congress, Vol.* 2. John Wiley & Sons, Chichester, 45–56.

——, PATIENCE, R., MAXWELL, J. R., VANDENBROUCKE, M. & DURAND, B. 1980. Molecular parameters of maturation in the Toarcian shales, Paris Basin, France – I. Changes in the configuration of the acyclic isoprenoid alkanes, steranes and triterpanes. *Geochimica et Cosmochimica Acta*, **44**, 1709–1721.

——, RULLKØTTER, J., WELTE, D. H. & MANKIEWICZ, P. 1985. Reconstruction of oil formation and accumulation in North Slope, Alaska, using quantitative gas chromatography–mass spectrometry. *In*: MAGOON, L. B. & CLAYPOOL, G. E. (eds) *Alaska North Slope oil/rock correlation study.* American Association of Petroleum Geologists, Special Studies in Geology, **20**.

MELLO, M. R., TELNÆS, N., GAGLIANONE, P. C., CHICARELLI, M. I., BRASSELL, S. C. & MAXWELL, J. R. 1988. Organic geochemical characterisation of depositional paleoenvironments of source rocks and oils in the Brazilian marginal basin. *Advances in Organic Geochemistry*, **13**, 31–45.

MO, E. S., THRONDSEN, T., ANDRESEN, P., BACKSTRØM, A., FORSBERG, A., HAUG, S. & TØRUDBAKKEN, B. 1989. A dynamic deterministic model of hydrocarbon generation in the Midgard Field drainage area offshore Mid-Norway. *Geologische Rundschau*, **78**, 305–317.

MOLDOWAN, J. M., FAGO, F. J., CARLSON, M. K., YOUNG, D. C., VAN DUYNE, G., CLARDY, J., SCHOELL, M., PILLINGER, C. T. & WATT, D. S. 1991. Rearranged hopanes in sediments and petroleum. *Geochimica et Cosmochimica Acta*, **55**, 3333–3353.

NEDKVITNE, T., KARLSEN, D. A., BJØRLYKKE, K. & LARTER, S. R. 1993. Relationship between reservoir diagenetic evolution and petroleum emplacement in the Ula field, North Sea. *Marine and Petroleum Geology*, **10**, 255–270.

OLSEN, S. D., BREKKE, T., KARLSEN, D. A. & NYLAND, B. 1993. The relationship between metals determined by ICP–MS in DST oils and source rocks extracts. *In*: PARNELL, J. *ET AL.* (eds) *Geofluids 93: Contributions to an International Conference on fluid evolution, migration and interaction in rocks, Torquay, England, May 4–7, 1993*, 410–412.

PETERS, K. E., MOLDOWAN, J. M., DRISCOLE, A. R. & DEMAISON, G. J. 1989. Origin of Beatrice oil by co-sourcing from Devonian and Middle Jurassic source rocks, Inner Moray Firth, United Kingdom. *American Association of Petroleum Geologists, Bulletin*,**73**, 454–471.

PRICE, L. C. & CLAYTON, J. L. 1992. Extraction of whole versus ground source rocks: Fundamental petroleum geochemical implications including oil–source correlation. *Geochimica et Cosmochimica Acta*, **56**, 1213–1222.

PROVAN, D. M. J. 1992. Draugen Oil Field, Haltenbanken Province, Offshore Norway. *In*: HALBOUTY, M. T. (ed.) *Giant Oil and Gas Fields of the Decade 1978–1988*. American Association of Petroleum Geologists, Memoir, **54**, 371–382.

RADKE, M. 1987. organic geochemistry of aromatic hydrocarbons. *In*: BROOKS, J. & WELTE, D. (eds) *Advances in Petroleum Geochemistry*, **2**, 141–199.

—— 1988. Application of aromatic compounds as maturity indicators in source rocks and crude oils. *Marine and Petroleum Geology*, **5**, 224–236.

——, WILLSCH, H. & WELTE, D. H. 1980. Preparative hydrocarbon type determination by automated medium pressure liquid chromatography. *Analytical Chemistry*, **512**, 406–411.

REDDING, C. E., SCHOELL, M., MONIN, J. C. & DURAND, B. 1980. Hydrogen and carbon isotopic composition of coals and kerogens. *In*: DOUGLAS, A. G. & MAXWELL, J. K. (eds) *Advances in Organic Geochemistry, 1979*. Pergamon Press, Oxford, 711–725.

SAJGÓ, CS., MAXWELL, J. R. & MACKENZIE, A. S. 1983. Evaluation of fractionation effects during the early stages of primary migration. *Organic Geochemistry*, **5**, 65–73.

SCHWARZKOPF, T. & SCHOELL, M. 1985. Die Variation der C- und H-Isotoperverhältnisse in Kohlen und deren Abhängigkeit von Macerlzusammensetsung und Inkohlungsgrad. *Geologische Fortschritte-Rheinland und Westfalen*, **33**, 161–168.

SCOTCHMAN, I. C. 1991. Kerogen facies and maturity of the Kimmeridge Clay Formation in the southern and eastern England. *Marine and Petroleum Geology*, **8**, 278–295.

SEIFERT, W. K. & MOLDOWAN, J. M. 1981. Paleoreconstruction by biological markers. *Geochimica et Cosmochimica Acta*, **45**, 783–794.

SHIMOYAMA, A. & JOHNS, W. D. 1971. Catalytic conversion of fatty acids to petroleum-like paraffins and their maturation. *Nature*, **232**, 140–144.

—— & —— 1972. Formation of alkanes from fatty acids in the presence of $CaCO_3$. *Geochimica et Cosmochimica Acta*, **36**, 87–91.

SILVERMAN, S. R. 1965. Migration and segregation of oil and gas. *In*: YOUNG, A. & GALLEY, J. E. (eds) *Fluids in Subsurface Environments*. American Association of Petroleum Geologists, Memoir, **4**, 53–65.

SKÅLNES, E. 1993. *Petroleum geochemistry and filling history of the Hild Field, Norwegian Continental Shelf.* Cand. Scient. Thesis, University of Oslo.

——, PATIENCE, R., BJØRLYKKE, K. & KARLSEN, D. A. 1993. Petroleum geochemistry and filling history of the Hild Field, Norwegian Continental Shelf. *In*: ØYGARD, K. (ed.) *Advances in Organic Geochemistry, 1993*. Extended Abstracts, Falch, Oslo, 51–55.

SOLLI, H. & LEPLAT, P. 1986. Pyrolysis-gas chromatography of asphaltenes and kerogens from source rocks and coals – a comparative structural study. *In*: LEYTHAEUSER, D. & RULLKØTTER, J. (eds) *Advances in Organic Geochemistry, 1985*. Organic Geochemistry, **10**, Pergamon Press, Oxford, 313–329.

STAINFORTH, J. G. & REINDERS, J. E. A. 1990. Primary migration of hydrocarbons by diffusion through organic matter networks, and its effect on oil and gas generation. *Organic Geochemistry*, **16**, 61–74.

SUNDARARAMAN, P. & MOLDOWAN, J. M. 1993. Comparison of maturity based on steroid and vanadyl porphyrin parameters: A new vanadyl porphyrin maturity parameter for higher maturities. *Geochimica et Cosmochimica Acta*, **57**, 1379–1386.

SURLYK, F. 1983. Source rock sampling, stratigraphical and sedimentological studies in the Late Palaeozoic of the Jameson Land basin, East Greenland. *Grønlands Geologiske Undersøgelse Rapport*, **155**, 88–93.

——, CLEMMENSEN, L. B. & LARSEN, H. C. 1981. Post-Palaeozoic evolution of the East Greenland Continental margin. *Canadian Association of Petroleum Geologists, Memoir*, **7**, 611–645.

——, HURST, J. M., MARCUSSEN, C., PIASECKI, S., ROLLE, F., SCHOLLE, P. A., STEMMERIK, L. & THOMSEN, E. 1984*a*. Oil geological studies in the Jamson Land basin, East Greenland. *Grønlands geologiske Undersøgelse Rapport*, **120**, 85–90.

——, ——, PIASECKI, S., ROLLE, F., SCHOLLE, P. A., STEMMERIK, L. & THOMSEN, E. 1986. The Permian of the western margin of the Greenland Sea – a future exploration target. *In*: HALBOUTY, M. T. (ed.) *Future Petroleum Provinces of the World.* American Association of Petroleum Geologists, Memoir, **40**, 629–659.

——, PIASECKI, S., ROLLE, F., STEMMERIK, L., THOMSEN, E. & WRANG, P. 1984*b*. The Permian base of East Greenland. *In*: SPENCER, A. M. *ET AL.* (eds) *Petroleum Geology of the North European Margin.* Norwegian Petroleum Society, Graham & Trotman, London, 303–315.

THOMPSON, K. F. M. 1987. Fractionated aromatic petroleum and the generation of gas-condensates. *Organic Geochemistry*, **11**, 573–590.

—— 1988. Gas-condensate migration and oil fractionation in deltaic systems. *Marine and Petroleum Geology*, **5**, 237–246.

——, KENNICUTT, M. C. & BROOKS, J. M. 1990. Classification of offshore Gulf of Mexico oils and gas condensates. *American Association of Petroleum Geologists, Bulletin*, **74**, 187–198.

THOMPSON, S., COOPER, B. S., MORLEY, R. J. & BARNARD, P. C. 1985. Oil generating coals. *In*: SPENCER, A. M. *ET AL.* (eds) *Petroleum Geochemistry in Exploration of the Norwegian Shelf.* Norwegian Petroleum Society, Graham & Trotman, London, 59–73.

TISSOT, B. 1981. Connaissance actuelles sur les produits lourds due petrole. *Revue Institute Francaise du Petrole*, **36**, 429–446.

VIK, E. & HERMANRUD, C. 1993. Transient thermal effects of rapid subsidence in the Haltenbanken area. *In*: DORÉ, A. *ET AL.* (eds) *Basin Modelling: Advances and Application.* NPF Special Publication, **3**, Elsevier, Amsterdam, 107–117.

WEST, N., ALEXANDER, R. & KAGI, R. I. 1990. The use of silicalite for rapid isolation of branched and cyclic alkane fractions of petroleum. *Organic Geochemistry*, **15**, 499–501.

WHITLEY, P. K. 1992. The geology of the Heidrun, a giant oil and gas field on the Mid-Norwegian Shelf. *In*: HALBOUTY, M. T. (ed.) *Giant Oil and Gas Fields of the Decade 1978–1988.* American Association of Petroleum Geologists, Memoir, **54**, 383–406.

WILHELMS, A. & LARTER, S. R. 1995. Overview of the geochemistry of some tar mats from the North Sea and USA: implications for tar-mat origin. *This volume.*

YEH HSUEH-WEN & EPSTEIN, S. 1981. Hydrogen and carbon isotopes of petroleum and related organic matter. *Geochimica et Cosmochimica Acta*, **45**, 753–762.

The reservoir geochemistry of the Eldfisk Field, Norwegian North Sea

D. P. STODDART[1,4], P. B. HALL[2], S. R. LARTER[1], J. BRASHER[3,5], MAOWEN LI[1] & M. BJORØY[2]

[1] *Fossil Fuels and Environmental Geochemistry, Drummond Building, University of Newcastle, Newcastle Upon Tyne, NE1 7RU, UK*

[2] *Geolab Nor, Hornebergveien 5, PO Box 5740, Fossegrenda, 7002, Trondheim, Norway*

[3] *Phillips Petroleum Company (Norway), Norway E & P Division, PO Box 220, N-4058, Tananger, Norway*

[4] *Current address: Universitat zu Koln, Geologisches Institut, 50674 Koln, Zulpicher Strasse 49a, Germany*

[5] *Current address: Phillips Petroleum Co., Research and Services Division, 260 Geoscience Building, Bartlesville, OK 74004, USA*

Abstract: A detailed geochemical study has been performed on the North Sea Eldfisk Field chalk reservoir, which forms part of a larger regional evaluation of the Greater Ekofisk area. Utilizing rapid screening methods (Rock-Eval, Iatroscan), conventional biomarker maturity parameters, stable carbon isotope data (bulk and individual *n*-alkanes) and pyrrolic nitrogen distributions, it is shown that the Eldfisk Field petroleum is heterogeneous, both laterally on a kilometre scale and vertically on a metre scale. The chemical heterogeneities throughout the field led to a better understanding of the charging of the reservoir, the nature of communication between the Upper Jurassic source rocks and trap, and the identification of geological barriers to petroleum mixing within the field itself. The biomarker maturity and pyrrolic nitrogen compound data suggest that charging of the Eldfisk Field ocurred from the south, petroleum moving northwards up and along the Skrubbe fault zone charging the Eldfisk Alpha and Bravo structures. Although filling has largely taken place since *c.* 10 Ma BP, initial charging of the Eldfisk structure started *c.* 20 to 25 Ma BP. The data suggest that petroleum emplacement was via faults and fractures, with communication between the source rocks and trap being discontinuous. The distribution of the different petroleum populations within the field was controlled significantly by the internal field plumbing which is sedimentologically influenced. Sharp, step-like biomarker-parameter maturity variations downhole in several wells have been identified and verified statistically. These geochemial discontinuities can be correlated with specific geological features (hardgrounds and tight zones) which have prevented mixing of the petroleum column. In one case, the presence of a barrier in the reservoir has been confirmed by repeat formation test (RFT) pressure data. The identification of lithofacies variations laterally across the field and barriers vertically in the petroleum column may become important when assessing production scenarios.

Large-scale heterogeneities in petroleum composition within reservoirs have been known for some time (Sage & Lacey 1938; Schulte 1980; Hirschberg 1984). The work of England *et al.* (1987), Larter *et al.* (1991), Horstad *et al.* (1990) and Leythaeuser & Ruckheim (1989) has brought petroleum reservoir geochemistry, especially the description of petroleum reservoir heterogeneities, to the forefront of petroleum geochemistry.

England *et al.* (1987) and England & Mackenzie (1989) suggested that intra-reservoir petroleum compositional variations in parameters such as gas/oil ratios (GOR) and biomarker ratios in petroleum accumulations may be interpreted as being due to inherited source facies and maturity variations in the petroleum charge preserved during filling of the reservoir. These compositional variations are sometimes preserved on a lateral kilometre scale across communicating reservoirs over geological time but rarely on a petroleum column vertical scale. This is a result of the distance dependence of diffusive mixing times in fluids (England *et al.* 1987). These authors have calculated the order of magnitude time-scales for reservoir fluid mixing by diffusion as a function of reservoir properties. Homogenization times for inherited petroleum column heterogeneities on a kilometre scale by lateral diffusion are on the order

From CUBITT, J. M. & ENGLAND, W. A. (eds), 1995, *The Geochemistry of Reservoirs*, Geological Society Special Publication No. 86, pp. 257–279.

of 100 Ma, and for vertical diffusion on a reservoir thickness scale (100 m), 1 Ma. For reservoirs filled during late Tertiary time (i.e. Eldfisk Field), we would expect some degree of vertical homogeneity in petroleum column composition but preservation of significant lateral heterogeneity.

The compositional variations in the petroleum column may potentially be used to establish fill directions on the assumption that the most mature petroleum entering the reservoir last displaces the less mature petroleum (England *et al.* 1987). This has been used successfully by several authors: England & Mackenzie (1989) used GOR to establish potential fill directions around the Forties Field, whilst Horstad *et al.* (1990) and Larter & Horstad (1992) used conventional biomarker maturity ratios (C_{29} ααα sterane S/(S + R), C_{30} diahopane/(C_{30} diahopane + C_{29} βα hopane)) to identify tentative charging directions from maturity gradients in the Gullfaks Field. In assessing the distributions of in-reservoir maturity differences, other processes that may affect biomarker parameters, such as biodegradation and secondary migration-related phase-controlled molecular fractionations, have to be considered (Larter & Mills 1992). Because the Eldfisk Field filled as a single-phase petroleum which is not biodegraded, these considerations are thought to be of little consequence here. The persistence of vertical compositional petroleum-column variations over geological time (on a metre scale) may occur if petroleum mixing is prevented by the presence of geological barriers. Smalley *et al.* (1992) used differences in the $^{87}Sr/^{86}Sr$ ratios in the Ekofisk Field formation waters to identify potential barriers between formations. However, to our knowledge, no work has been published on the use of molecular biomarker parameters to identify potential barriers to petroleum mixing in the petroleum column. This paper describes the analyses performed on Eldfisk reservoir petroleum and describes the chemically heterogeneous nature of the fluids. The distribution of the fluids is tied to the charging of the Eldfisk Field, the nature of communication between source and trap, and geological barriers to intra-reservoir mixing. In this paper, we describe a geochemical study of the Eldfisk chalk Field which is placed in the larger context of a regional geochemical study of oils from the Greater Ekofisk area.

Experimental section

Samples used in this study include 30 drill-stem test (DST) oils from nine reservoirs in the Greater Ekofisk area (20 from the Eldfisk Field), a total of 709 reservoir core samples from 11 wells throughout the Eldfisk Field, and 50 shale samples from the Upper Jurassic Mandal, Farsund and Haugesund Formations from two wells in the Eldfisk area (Fig. 1).

The study involved a systematic reservoir geochemical study (Larter & Aplin 1995) incorporating Rock-Eval and Iatroscan screening of reservoir core extracts (Karlsen & Larter 1989), followed by more detailed molecular analysis of reservoir core extracts. Thin layer chromatography–field ionization detection (TLC–FID, Iatroscan) analysis (Karlsen & Larter 1991) and thermal desorption–gas chromatography (Bjorøy *et al.* 1985) were performed on all the reservoir core samples and from these 185 samples were selected for thermal desorption–gas chromatography mass spectrometry analysis, from which biomarker data were obtained (Bjorøy *et al.* 1991*b*). Aliphatic hydrocarbon fractions of all the Eldfisk reservoir DST oils (20) were analysed by gas chromatography–isotope ratio mass spectrometry (GC–IRMS) (Bjorøy *et al.* 1990, 1991*a*) to determine stable carbon isotopic ratios for individual *n*-alkanes.

Great care was taken in the sample handling and data acquisition procedures to ensure the reproducibility of the analyses. The standard deviations for the biomarker parameters Ts/(Ts + Tm), C_{30} diahopane/(C_{30} diahopane + C_{29} βα hopane), C_{29} sterane αββ/(αββ + ααα) and C_{29} ααα sterane S/(S + R) from replicated analyses are 1.1, 0.67, 0.76 and 0.65%, respectively. This suggests that the small gradients across the Eldfisk Field discussed below are real differences in the petroleum composition and we confirm this using statistical tests.

Nitrogen compound analysis was performed on all the Eldfisk DST oils (20), reservoir core extracts from 11 wells (35) (12 from well 2/7-B12) and Upper Jurassic source rock extracts (three) taken from well 2/7-15X. An analytical flow chart for the separation of the nitrogen compounds is given in Li *et al.* (1995) and the pyrrolic nitrogen concentrates were obtained from de-asphaltened DST oils and reservoir core and source rock extracts via a two-step open column liquid chromatographic separation scheme (Later *et al.* 1981). See Li *et al.* (1992) for detailed procedures and instrumentation.

As differences in biomarker ratios among petroleum samples are small and subtle, two statistical analysis methods were used to validate the conclusions concerning compositional variations in petroleum columns, principal components analysis (PCA) (Kvalheim *et al.* 1987; Kvalheim 1987) and Student's *t*-test (Davis 1973). For PCA, the biomarker data matrix consisted of a maximum of 44 biomarker variables selected from *m/z* 191 and 217 ion chromatograms, and a minimum of 15

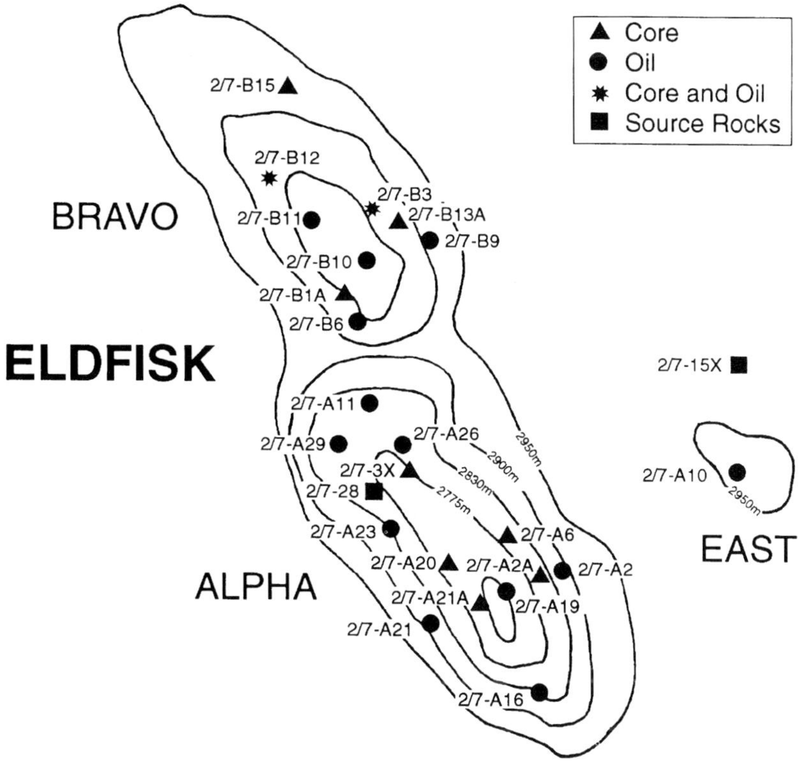

Fig. 1. Location of the wells sampled across the Eldfisk Field. The contour lines are structure contours on the Ekofisk Formation.

to 20 samples. The data were in the form of biomarker concentrations (ppm of rock) which were subsequently initially normalized to 100% and then autoscaled.

The Student *t*-test (Davis 1973) was performed in order to statistically discriminate between visually or PCA perceived different petroleum populations (based on biomarker ratios; Horstad *et al.* 1990).

Geology of the Greater Ekofisk area and the Eldfisk reservoir

The Eldfisk Field is located in the southernmost portion of the Norwegian sector of the North Sea in Block 2/7, and is one of nine petroleum-producing chalk fields in the Greater Ekofisk area (Fig. 2). The first petroleum was discovered in the Eldfisk Field structure in 1970 by the 2/7-1X well, but it was the 2/7-3X well drilled in 1972 that discovered petroleum in commercial quantities. Subsequently, an additional five exploration wells and 74 development wells have been drilled over the Eldfisk Field. Production began in 1979 from initial in-place reserves of 2300 MMSTB oil (360×10^6 m^3) and 3670 BSCF gas (100×10^9 m^3) (2900 MMBOE = 460×10^6 m^3 OE).

The Lindesnes Ridge is a NW–SE trending anticline located within the Feda Graben of the Central Trough. Movements along the Skrubbe fault, which parallels the ridge, were largely responsible for its formation. Most notable were the effects of halokinesis which contributed to the formation of the present Eldfisk Field structure (Brasher 1994). The Eldfisk Field is separated into a southern structure called Eldfisk Alpha, a northern structure called Eldfisk Bravo and a small structure called East Eldfisk.

Production from the Eldfisk Field is primarily from the Ekofisk (Danian) and Tor Formations (Maastrichtian), although the Hod Formation does contribute minor amounts in Eldfisk Alpha. The Hod Formation, of Turonian to Campanian age, is dominated by the presence of pelagic chalks, although reworked chalk units also occur. Core analysis and dipmeter/FMS (formation microscanner) logs indicate the Tor Formation to be largely comprised of highly reworked and slumped sediments. While the lower portion of the Ekofisk

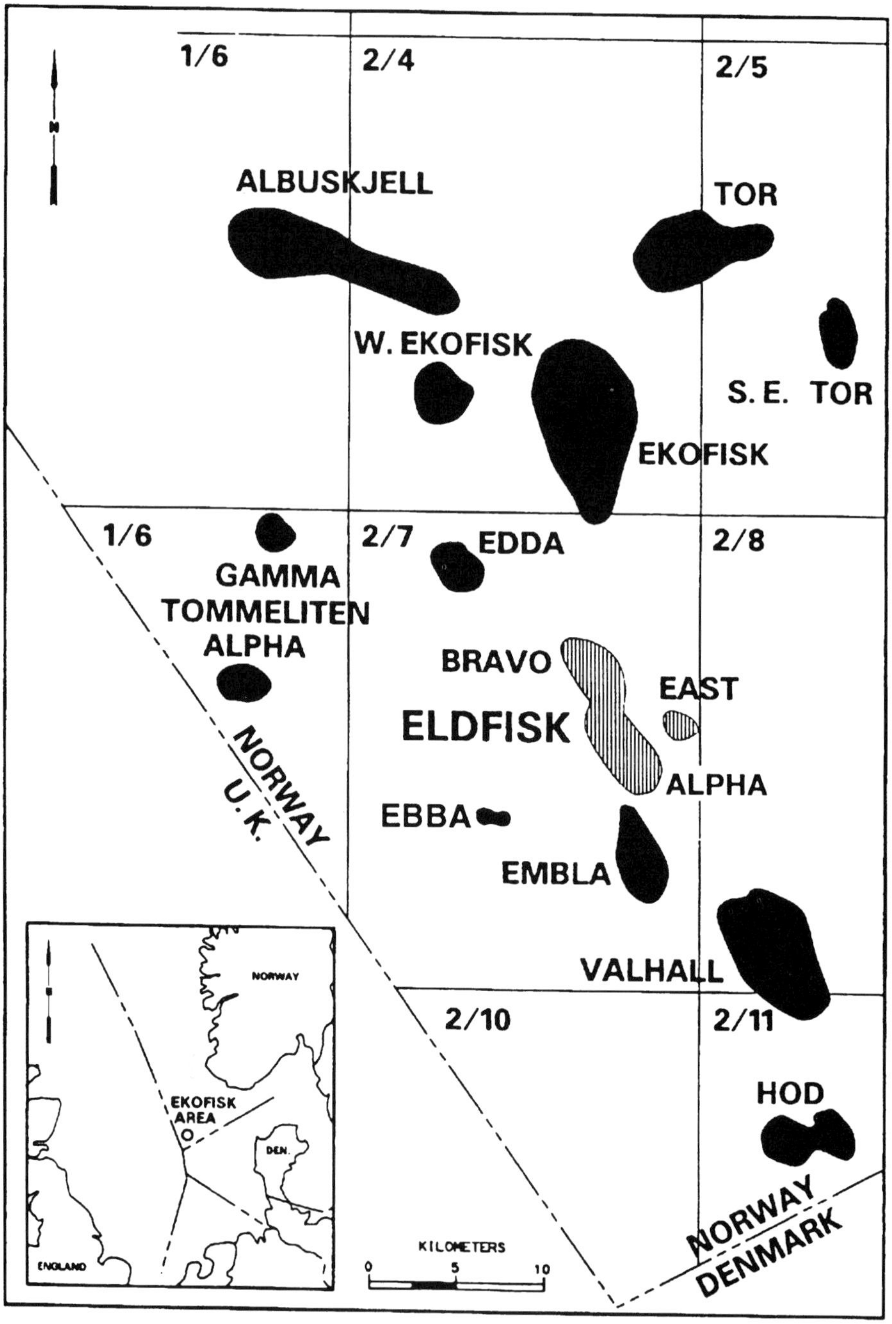

Fig. 2. Location map of the Eldfisk Field in relation to the Greater Ekofisk area and the southern Norwegian North Sea.

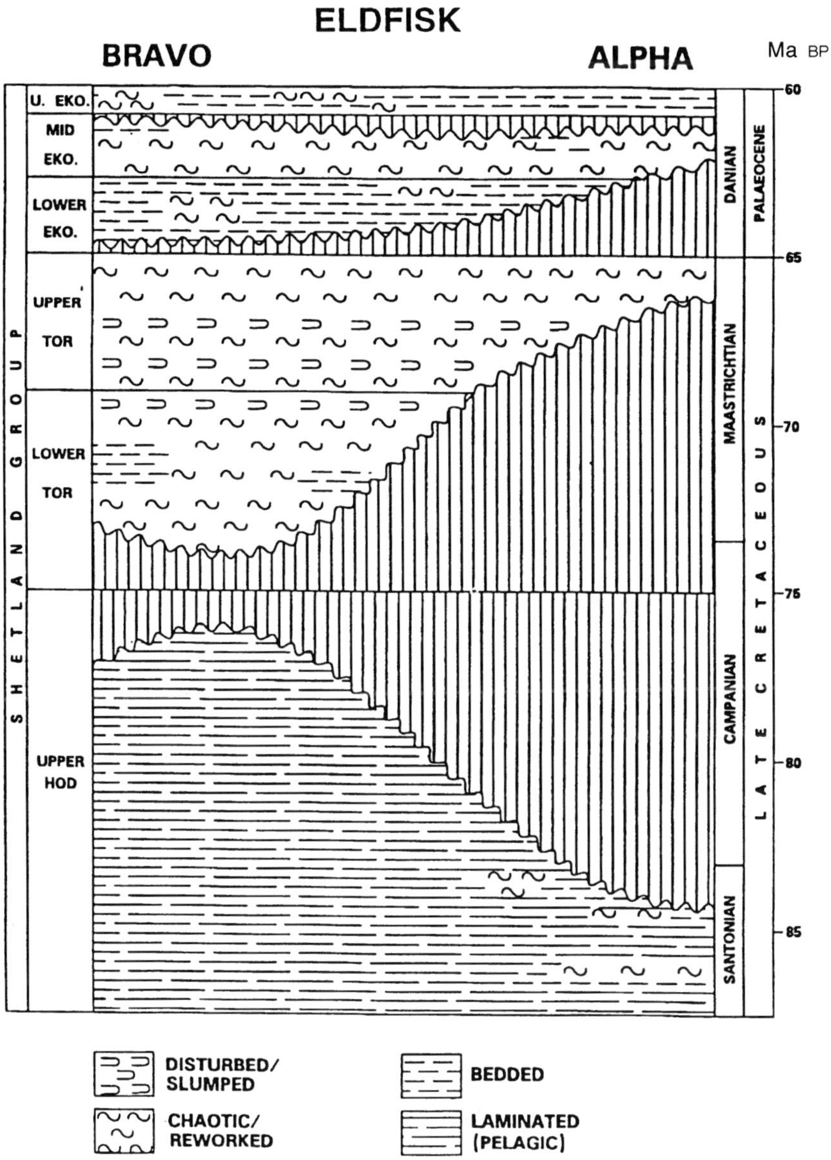

Fig. 3. Eldfisk Field area stratigraphy showing the major reservoir chalk facies.

Formation comprised bedded to laminated chalks, which reflect a more quiescent tectonic period, the mid and upper Ekofisk Formation consist of more highly reworked, thick sediment packages (debris flows and slumps) to thin distal reworked units (turbidites and mudflows) (Brasher 1994). Figure 3 shows the general Eldfisk Field stratigraphy and lithofacies character.

In the Eldfisk Field, hardgrounds have been observed associated with all the major unconformities (Fig. 3). Although often thin, these highly lithified, low porosity, low permeability units may be relevant to fluid flow in the reservoir because they could represent barriers to petroleum mixing over geological time periods or impediments to flow during production.

The Tor and Ekofisk Formations in the Eldfisk Field typically show porosities ranging from 25 to 45% and permeabilities embracing 0.01–35 mD, with fractures locally raising effective permeabilities to *c.* 140 mD. Porosity within the Hod formation seldom exceeds 30% and permeabilities are generally lower. The higher porosities are believed to result from: (a) rapid deposition of reworked chalk resulting in overpressuring, which acts to limit mechanical compaction, and/or (b) early petroleum migration, which effectively inhibits ensuing cementation and thereby reduces the effects of chemical compaction (Van den Bark & Thomas 1980; Scholle *et al.* 1983; Taylor & Lapre 1987). In areas where pelagic deposition has occurred, such as in the Hod Formation and in parts of the lower Ekofisk Formation, matrix permeability is exceedingly low (0.01–0.03 mD). In allochthonous rocks, the matrix permeability may be up to 10 mD. With the presence of open fractures, the effective permeability may be increased by an order of magnitude or more.

There is no definite oil/water contact within the Eldfisk Field and water saturation varies across the field (Michaud 1987). Among the factors which influence this are reservoir lithofacies variations and diagenesis, differential timing of migration, continued filling of the reservoir to the present day and cap-rock leakage.

Regional geochemistry of the Greater Ekofisk area

Many papers have been published on the geochemistry of oils and source rocks in this area. Except for Hughes *et al.* (1985) and Gabrielsen *et al.* (1985) most involve only general comparisons with other areas of the North Sea (Cooper & Barnard 1984; Cornford *et al.* 1983; Thomas *et al.* 1985; Spencer *et al.* 1986; Damtoft *et al.* 1986; Bailey *et al.* 1990). These authors state that the source of the petroleum in the Greater Ekofisk chalk Fields is from the Upper Jurassic Mandal and Farsund Formation shales. The organic contents of these shales have been shown to vary considerably. Van den Bark & Thomas (1980) reported TOC values ranging from 1 to 3% around the Ekofisk Field region, whereas Damtoft *et al.* (1986) reported values ranging from 0.5 to 22% in the Danish section of the Central Graben. Several organic facies have been identified in the Upper Jurassic shales around the Central Graben. Cooper & Barnard (1984) detected both algal-rich and terrestrial facies, whilst Bailey *et al.* (1990), using Rock-Eval and stable carbon isotope data from kerogen pyrolysates, recognized two main source organic facies in the Central Trough, an organic-rich, carbon isotopically heavier facies (S2 20 to 80+ kg HC/ton rock, kerogen $\delta^{13}C$ –27 to –25 per mil) and a leaner, isotopically lighter facies (S2 5 to 20 kg HC/ton rock, kerogen $\delta^{13}C$ –30 to –27 per mil).

In the margins of the Central Graben, such as the Ula Trend (Spencer *et al.* 1986), the Mandal Formation is organic-rich and oil-prone (TOC 5 to 8%, S2 25 to 30kg HC/ton rock), whereas the Farsund and Haugesund Formations are either absent or organically lean (TOC 2 to 5%, S2 5 to 10 kg HC/ton rock) and gas-prone. In the central part of the Central Trough there are similar differences between the Mandal and Farsund/Haugesund Formations, although the higher maturity level of the Upper Jurassic of the Central Trough reduces the differences between them and makes it more difficult to assess their original source potential. For example, in well 2/7-15X the TOC content is between 3 and 5% in the Mandal Formation and from 1 to 3% in the Farsund/Haugesund Formations. The lower TOC values quoted for the Central Trough, compared with the Ula Trend, are probably due to the lack of pre-oil generation Upper Jurassic well sections in the Central Trough (except in the Hod Field area). However, in parts of the Central Trough there are variations to this apparent general rule of organic-rich and oil-prone Mandal Formation versus organically leaner and more gas-prone Farsund/Haugesund Formations. In well 2/7-28 there are distinct horizons within the Farsund/Haugesund Formations which are extremely organic-rich (TOC values from 10 to 20% and Rock-Eval S2 from 50 to 100 kg HC/ton rock). Clearly, in the areas where these zones are present and highly mature, they could have been extremely prolific source rocks. It is worth noting that the Upper Jurassic shales of well 2/7-28 on the Lindesnes Ridge have always been much shallower than those in the 2/7-15X area, even during deposition of the units. This may have contributed to the source differences observed.

In the wells analysed, the Upper Jurassic source rocks are near peak oil generation, hence it is difficult to distinguish between kerogen types with ease (well 2/7-15X). However, in less mature wells, away from the Lindesnes Ridge, Rock-Eval data indicate that the Mandal Formation is predominantly Type II kerogen (i.e. oil-prone) while the Farsund/Haugesund Formations contain predominantly Type III kerogens, with some beds containing Type II or II/III kerogen (i.e. mainly gas-prone). In the Lindesnes Ridge area (e.g. well 2/7-28), the Mandal/Farsund/Haugesund Formations (including the extremely rich intervals) have mainly Type II kerogen with some Type I in the rich intervals (hydrogen indices up to 800 mg HC/g TOC). It is considered that in the Lindesnes Ridge area the whole of the Farsund and Haugesund Formations tend to be, or have been, good, rich oil-prone source rocks rather than the gas-prone sections found elsewhere in the Central Trough.

Petroleum maturity variations throughout the Greater Ekofisk area and the Eldfisk Field

Bulk and molecular geochemical data

Hughes *et al.* (1985) identified a distinct decrease in the apparent maturity of Greater Ekofisk oils over a NW–SE trend. This maturity trend can be confirmed by cross-plotting two terpane maturity parameters, Ts/(Ts + Tm) (Seifert & Moldowan 1978) and C_{30} diahopane/(C_{30} diahopane + C_{29} $\beta\alpha$ hopane) ratios (Cornford *et al.* 1986), both showing similar NW–SE decreasing maturity trends (from the West Ekofisk/Edda Fields to the least mature Hod/Valhall Fields, Fig. 4). A similar separation of the Greater Ekofisk Field petroleums is evident from a ternary plot of gross petroleum composition (Fig. 5).

In order to confirm independently the maturity relationships of the different DST oils throughout the Greater Ekofisk area and the Eldfisk reservoir itself, principal components analysis (PCA) was performed on a set of biomarker data obtained by gas chromatography–mass spectrometry (GC–MS) analysis of separated aliphatic hydrocarbon fractions.

The data matrix consisted of 30 DST oil samples with 40 biomarker variables determined for each. The first three principal components (PC1, PC2 and PC3) calculated accounted for 85.5% of the total variance (Fig. 6). From the loadings plot, there are negative correlations between the variables Ts and Tm, C_{29} $\beta\alpha$ hopane and C_{30} diahopane, tricyclic terpanes (C_{23} to C_{25}) and extended hopanes (C_{29} to C_{34}), and diasteranes and regular steranes. These relate to the conventional biomarker maturity parameters, Ts/(Ts + Tm), C_{30} diahopane/(C_{30} diahopane + C_{29} $\beta\alpha$ hopane), diasteranes/(diasteranes + steranes), and tricyclic terpanes/(tricyclic terpanes + extended hopanes), which have been used in the past as maturity indicators. Hence, it can be concluded that PC1 is predominantly maturity related, with scores correlating with conventional biomarker parameters (Ts/(Ts + Tm), Fig. 7). This confirms that the Hod/Valhall petroleums have a lower maturity than the rest of the DST oils, with the Edda and West Ekofisk petroleums being the most mature. The Eldfisk area petroleums are of intermediate maturity (Fig. 6).

The data matrix for the Eldfisk Field DST oils consisted of 38 variables and 20 Eldfisk DST oil samples. The first three principal components accounted for 56.5% of the total variance of the dataset, of which PC1 amounted to 29.2% and PC2 to 19.2%. The loadings plot for PC1 again shows negative correlations between C_{27}, C_{28}, C_{29} $\alpha\alpha\alpha$ sterane S and R isomers, C_{27}, C_{28}, C_{29} $\alpha\beta\beta$ sterane

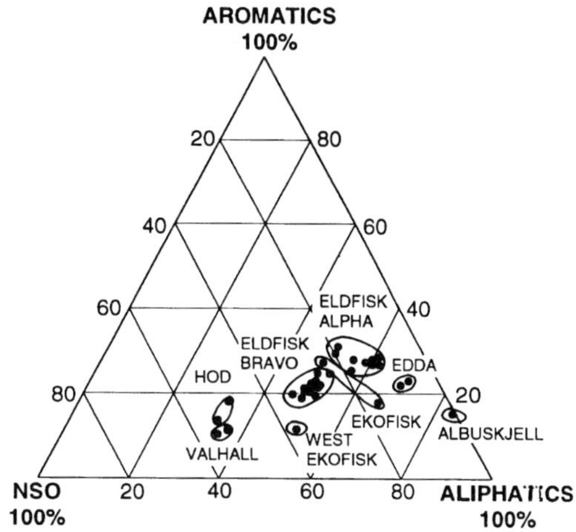

Fig. 5. Distribution of Greater Ekofisk area DST oils in relation to gross geochemical compositional data (relative concentrations of aliphatic and aromatic hydrocarbon compounds, and NSO compounds). The separation of the Greater Ekofisk Field petroleums in relation to the gross compositional data is similar to that using conventional biomarker data (Fig. 4), and PCA of the biomarker data (Fig. 6), both the latter datasets inferring a SE–NW trend of increasing petroleum maturity for the Greater Ekofisk DST oils.

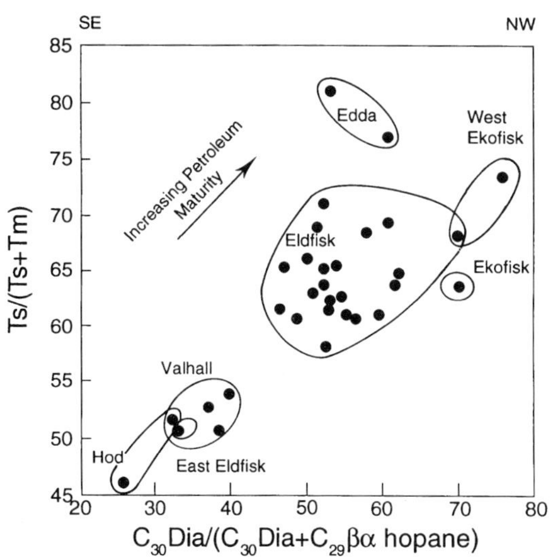

Fig. 4. A cross-plot of Greater Ekofisk area DST oils in relation to two established biomarker maturity parameters. A SE–NW petroleum maturity trend for the Greater Ekofisk area DST oils is obvious, Hod/Valhall Field DST petroleum being less mature than the West Ekofisk/Edda Field analogues.

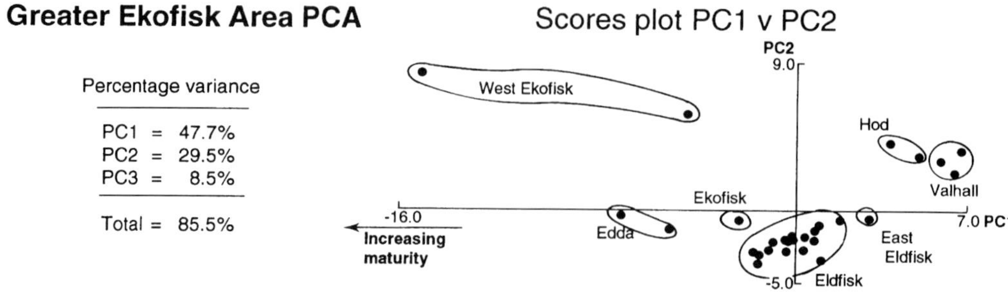

Fig. 6. Principal components analysis (PCA) scores plot (PC1 versus PC2) of the biomarker data from the Greater Ekofisk area DST petroleums. The data suggest a SE–NW increasing petroleum maturity gradient across the Greater Ekofisk area, the Valhall/Hod Field DST oils being less mature than the West Ekofisk/Edda Field analogues.

S and R isomers, Ts and Tm, and the C_{30} diahopane and C_{29} $\beta\alpha$ hopane, suggesting that PC1 is probably predominantly maturity related. From the scores plot, the data reveal that the Eldfisk Bravo structure wells probably contain less mature petroleum compared to the Eldfisk Alpha structure wells (Fig. 8). The East Eldfisk structure (2/7-A10) contains the least mature petroleum.

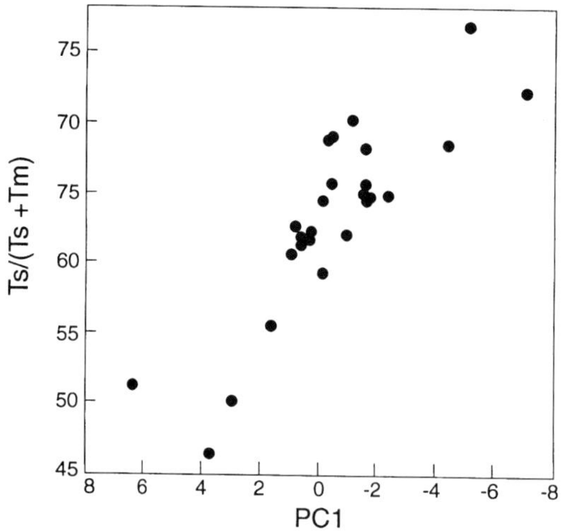

Fig. 7. Cross-plot of PC1 scores (from principal components analysis (PCA) of the Greater Ekofisk area DST oils) with the Ts/(Ts + Tm) biomarker maturity ratio. Good correlation between PC1 and Ts/(Ts + Tm) suggests that PC1 is predominantly maturity-related. Although not shown, PC1 also correlates with biomarker maturity parameters C_{30} diahopane/(C_{30} diahopane + C_{29} $\beta\alpha$ hopane) and C_{23}–C_{25} tricyclic terpanes/(C_{23}–C_{25} tricyclic terpanes + C_{29}–C_{35} $\alpha\beta$ hopanes), further supporting the conclusion that PC1 is maturity-related.

Burial and maturation history of the Eldfisk Field area

It is generally recognized that the primary source rocks for liquid petroleum generation in the Central Trough are to be found in the Upper Jurassic shale sequences. Throughout the Eldfisk Field, rift-related faulting has played an important role in controlling the current thicknesses of these Upper Jurassic shales. The thickness of the Upper Jurassic shale sequence ranges from *c.* 1 km beneath the Eldfisk Field up to 1.8 km in the northern and eastern areas of the Central Trough. Figure 9 shows a NE–SW geological cross-section across the Eldfisk Field area.

Four potential kitchens for liquid petroleum generation are distinguished (Fig. 10):

(1) to the west of the Skrubbe fault zone lies a thin sequence (*c.* 100 m) of Upper Jurassic shales;
(2) to the north and east of the Eldfisk Field lie *c.* 1.8 km of Upper Jurassic shales;
(3) directly beneath the Eldfisk Field lie *c.* 1 km of Upper Jurassic shales;
(4) to the south of the Eldfisk Field *c.* 1 km of Upper Jurassic shales are located in a *c.* 300 m downthrown fault block.

Source-rock data were available from only two of these four potential source kitchens, being from well 2/7-28, which is located in the Eldfisk Field, and well 2/7-15X to the east of the main field (Fig. 10). No well data were available from petroleum source kitchens west of the Skrubbe fault zone, within the downthrown block to the south of the field, or towards the north of the field. Consequently, three pseudo-wells (W1 and W2) and 2/7-B12 on the Eldfisk Bravo structure were modelled to cover potential source kitchens (Fig. 10). All the major fault movements took place prior to the deposition of the Tertiary shales, which

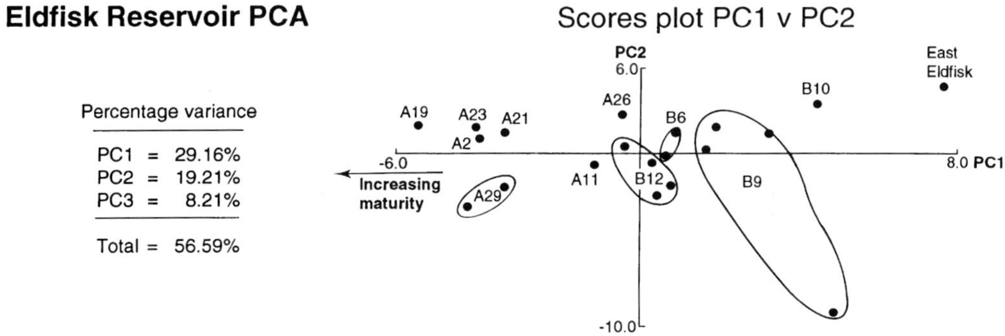

Fig. 8. Principal components analysis (PCA) scores plot (PC1 versus PC2) of the biomarker data from the Eldfisk Field DST petroleums. The data suggest that Eldfisk Alpha structure petroleums are more mature than Eldfisk Bravo analogues, with east Eldfisk petroleum being the least mature. Well prefix: A = Alpha structure, B = Bravo structure.

ultimately buried the Upper Jurassic shales into the oil window. Simple backstripping and decompaction of well-tied seismic section tops (N–S and W–E) across the Eldfisk structure (Fig. 9) enabled construction of burial history profiles for the Upper Jurassic section. No compensation for salt or fault movement was made.

The five wells were modelled using a basin modelling package (Troll) which was calibrated against vitrinite reflectance and biomarker maturity data (i.e. C_{32} $\alpha\beta$ hopane S/(S + R) and C_{29} $\alpha\alpha\alpha$ sterane S/(S + R) from wells 2/7-15X and 2/7-28). A present-day geothermal gradient for the area of *c.* 35°C km^{-1} was calculated from corrected well

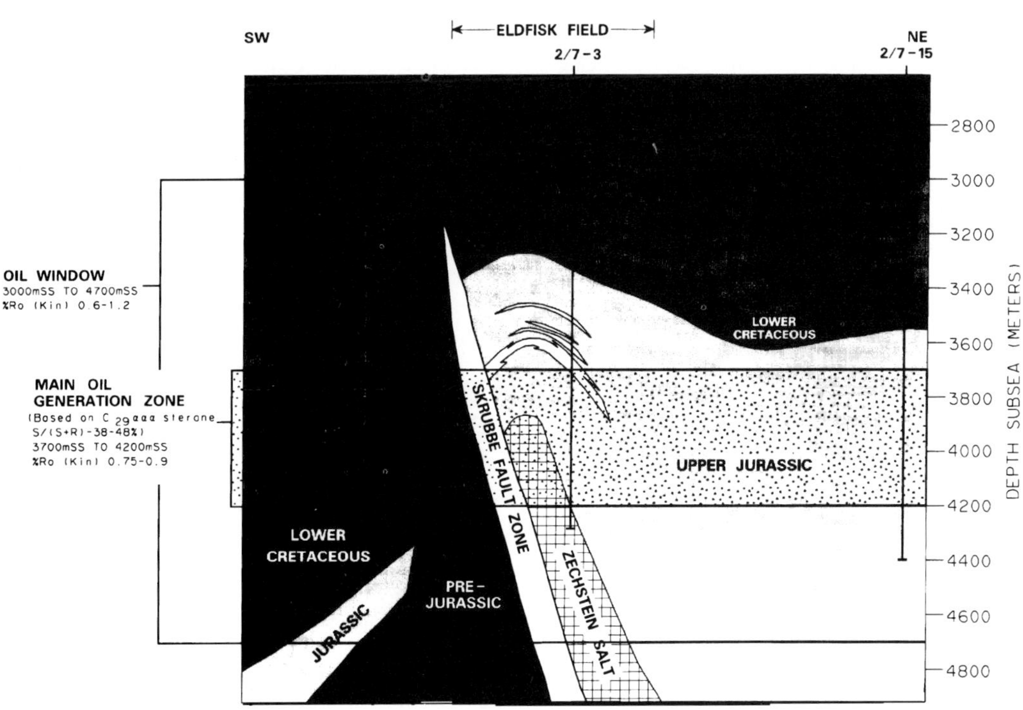

Fig. 9. Southwest–northeast geological cross-section of the Eldfisk Field area. Superimposed on to the cross-section are the depths of the oil window calculated from modelling (predicted $\%R_o$ data), and the main oil generation zone calculated by superimposing measured C_{29} $\alpha\alpha\alpha$ sterane S/(S + R) values on to predicted C_{29} $\alpha\alpha\alpha$ sterane S/(S + R) values.

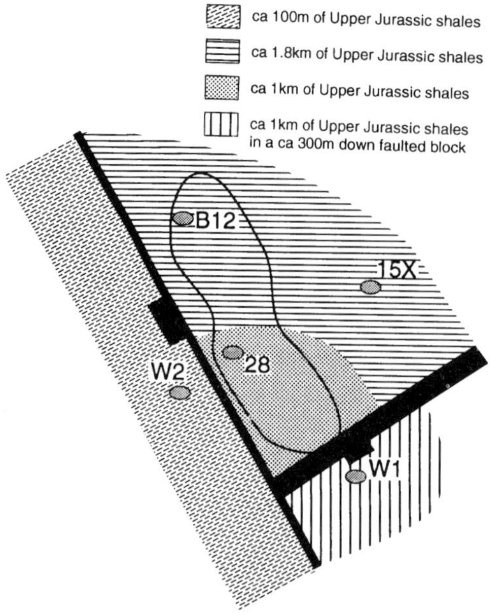

Fig. 10. Four potential petroleum-generating kitchens distinguished within the Eldfisk catchment area, and location of wells used to assess source-kitchen maturation history. Note the positions of the two faults, the SE–NW trending Skrubbe fault zone, and the SW–NE trending southern bounding fault. The faults are associated with filling of the Eldfisk Field, providing migration conduits for the transport of generated petroleum from the source kitchens to the field.

temperatures (well 2/7-15X, 150°C at 4350 mss), and a sea-bottom temperature of 6°C, representing the sediment surface temperature, was used. Figure 11 displays the subsidence of sediments through time and illustrates that the majority of the Upper Jurassic source section burial occurred with the deposition of the thick Tertiary sediment sequence. It can be calculated that the subsidence rates in the Neogene period were as high as *c.* 120 m/Ma, resulting in rapid burial of the source rocks in all source kitchens. In order to calibrate the model, a comparison of measured vitrinite reflectance data and modelled values predicted by a kinetic vitrinite reflectance model were employed (Larter 1989; Burnham & Sweeney 1989).

Qualitatively similar results were obtained from all the wells and pseudo-wells. We illustrate the modelling with well 2/7-15X, where the model produced predicted reflectance values (Fig. 12) which were higher than the actual measured data (in our experience this is a common feature of vitrinite data from North Sea Upper Jurassic shales). Measured and modelled biomarker data for the C_{32} $\alpha\beta$ hopane S/(S + R) and C_{29} $\alpha\alpha\alpha$ sterane S/(S + R) maturity parameters (Fig. 12) show a reasonable correlation with the predicted trends using the 35°C km^{-1} geothermal gradient and 2/7-15X burial curve. The measured downhole temperatures today agree broadly with the modelled reflectance values (i.e. current borehole temperature is probably the best maturity indicator). From the modelled reflectance values, it can be concluded that the oil-window depth range in the Eldfisk area today is below *c.* 3000 mss (*c.* 0.5 to 0.6% R_o, Fig. 9).

The Eldfisk DST oils have C_{29} $\alpha\alpha\alpha$ sterane S/(S + R) ratios of 41 to 50%, with reservoir core-extract data similarly ranging from 38 to 48% (probably equivalent to actual generation ranges of *c.* 0.75 to 0.9% R_o). This illustrates that the Eldfisk Field petroleum was generated from source rocks which were in the early to peak petroleum generation stage. The zones of main oil generation and expulsion from the source rocks contributing petroleum to the Eldfisk Field can be derived in broad terms by superimposing the values of the Eldfisk reservoir core extracts and DST oil data onto the modelled biomarker curve for C_{29} $\alpha\alpha\alpha$ sterane S/(S + R) maturity parameter (Fig. 12). This provides evidence that the petroleum was probably generated from Upper Jurassic source rocks around 3700 to 4200 mss. Hence, most of the source rocks surrounding the Eldfisk Field are capable of contributing petroleum to the reservoir, except for some shales (*c.* 300 m thickness) below the Eldfisk Field which lie above the main oil generation zone.

Using the burial/maturation history profile, an initial charge time of 20 to 25 Ma (early Neogene)

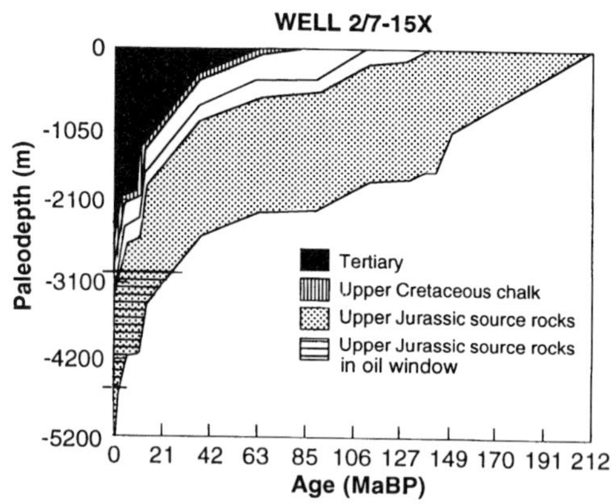

Fig. 11. Burial history profile of sediments in the Eldfisk area based on well 2/7-15X. In all the five wells modelled, the Upper Jurassic source rocks were buried into the oil window by the thick Tertiary sediment sequence (*c.* 3000 m).

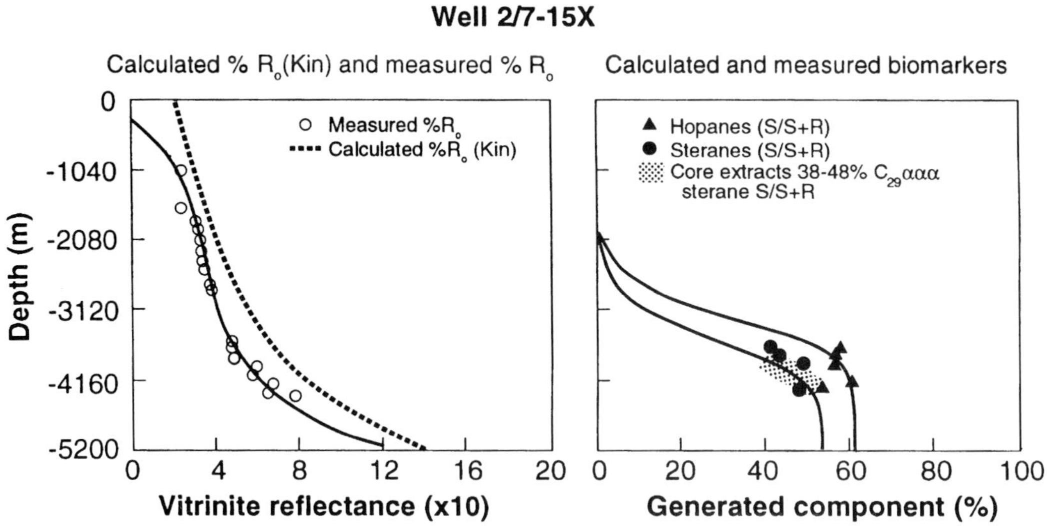

Fig. 12. Predicted and measured vitrinite reflectance and biomarker data from well 2/7-15X. Measured reservoir core extract C_{29} $\alpha\alpha\alpha$ sterane S/(S + R) values are superimposed onto the predicted values to estimate the depths of the main oil generation zone in the Eldfisk area. The main oil generation zone probably exists between 3700 and 4200 mss.

to the present day could be deduced for the Eldfisk Field (i.e. from when the Upper Jurassic entered the oil window and onwards, Fig. 11), with much of the petroleum arriving recently, most charging probably being within the last 10 Ma BP. The existence of a variable depth transitional oil–water contact across the field (Michaud 1987) may indicate that filling was recent, or that tilting of the field or cap-rock leakage has recently disturbed the petroleum column. It has been shown that mechanical stabilization of petroleum columns in high-quality reservoirs by Darcy flow is generally rapid, with the time-scales being related to the large-scale permeability of the field (England & Mackenzie 1989). The persistence of a tilted oil–water contact transition zone is not inconsistent with a field of low average permeability (0.01 to 35 mD) which is probably still filling at the present day from one side (Stoddart 1993).

Eldfisk Field petroleum composition

Lateral variations in petroleum-column composition

It has been suggested theoretically, and supported by several case studies, that lateral compositional variations in the petroleum column of even well-connected reservoirs can exist for geological time periods (England *et al.* 1987; England & Mackenzie 1989; Horstad *et al.* 1990) though there are cases where laterally homogeneous petroleum columns are found over kilometre scales (unpublished data).

Lateral petroleum-column maturity variations are apparent throughout the Eldfisk Field using conventional biomarker maturity parameters from the DST oils and the reservoir core-extract samples (Figs 13 & 14). Although trends are similar, the maturity variations seen from DST oils and reservoir core samples are not strictly comparable from well to well. This may reflect the fact that the DST oils are biased towards petroleums produced from the most permeable zones, whereas core extracts tend to show petroleum composition of the whole producing formation. Comparing the results from the bulk geochemical data (Fig. 5), PCA of the Eldfisk DST oils (Fig. 8) and biomarker data from the reservoir core samples (Fig. 13), we see that regionally the datasets show similar general results, with the Eldfisk Alpha structure petroleum being more mature than the Eldfisk Bravo analogue. Within the Eldfisk Bravo and Alpha structures, the DST oil biomarker maturity parameters appear to show a maturity increase from east to west. From the reservoir core extracts, only the Bravo structure (both Tor and Hod Formations) show an east to west increasing maturity gradient which can be verified statistically (Fig. 14). The Eldfisk Alpha structure petroleum shows no coherent maturity distribution for either the Ekofisk or Tor Formations using the reservoir core-sample data.

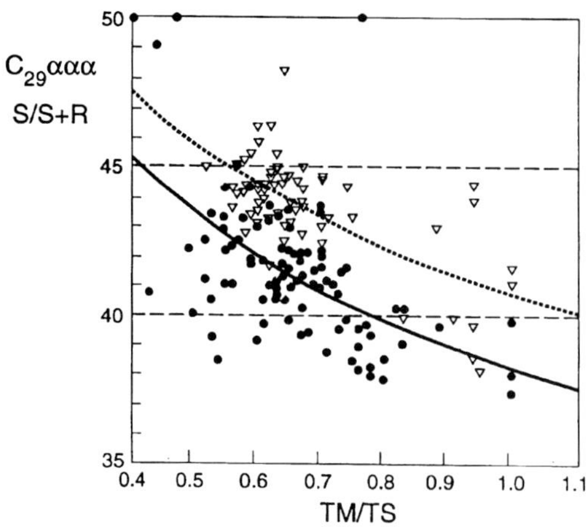

Fig. 13. A cross-plot of two established petroleum maturity parameters used to illustrate that Eldfisk Alpha structure reservoir core extracts are more mature than Eldfisk Bravo structure analogues. Polynomial regression lines are given for both the Eldfisk Alpha and Bravo reservoir core-extract datasets and are for visualization only. Combining the Eldfisk Field DST oil bulk compositional data (Fig. 5), principal components analysis of biomarker data from Eldfisk Field DST oils (Fig. 8) and Eldfisk reservoir core extract biomarker data, it can be concluded that the Eldfisk Alpha structure contains more mature petroleum than the Eldfisk Bravo analogue.

The lack of maturity gradients across the Alpha structure, as revealed by the reservoir core-extract data, may be due to the sparse coverage and biased positions of the wells, which are all centred on the crest of the structure. Taking account of both the DST oil and reservoir core-extract data, a north to south and an east to west increasing maturity gradient can be inferred across Eldfisk Alpha and Bravo structure petroleums. Relating these maturity gradients to possible fill directions throughout the Eldfisk structure infers that charging of the field probably occurred from a southern and western aspect, most likely via the southern bounding fault and the Skrubbe fault zone (Figs 9 & 10).

Petroleum source facies variations within the Eldfisk Field petroleum

Bailey *et al.* (1990) have shown that there is an organic-rich, carbon isotopically heavy organic facies (S2 20 to 80 kg HC/ton rock; kerogen $\delta^{13}C$ values from –27 to –25 per mil) in the Mandal Formation centred on the Greater Ekofisk area. This carbon isotopically heavy source zone is probably present below the Eldfisk Field in the Mandal/Farsund Formations, where there is an organic-rich zone (S2 50 to 100 kg HC/ton rock) as seen in well 2/7-28 (*c.* 3500 m). A carbon isotopically heavy petroleum (whole oil $\delta^{13}C$ value of –26.41 per mil) was produced in well 2/7-3X from a thin stringer of sandstone within Upper Jurassic shales at *c.* 3450 mss (Fig. 9). The petroleum in the Upper Jurassic sandstone was most probably generated from carbon isotopically heavy Upper Jurassic source zones in the immediate vicinity of the Eldfisk Field. Using GC–IRMS (Bjorøy *et al.* 1991*a*), the Eldfisk DST oils were tentatively separated into two petroleum populations based on variations in the $\delta^{13}C$ of their individual *n*-alkanes (Fig. 15). Though still an experimental technique, with some concern over technique reproducibility, it appears that petroleum populations defined by the technique can be broadly described with reference to other DST oils in the area. Two end-member oils can be defined; an isotopically lighter Edda Field-like end-member oil, and an isotopically heavier

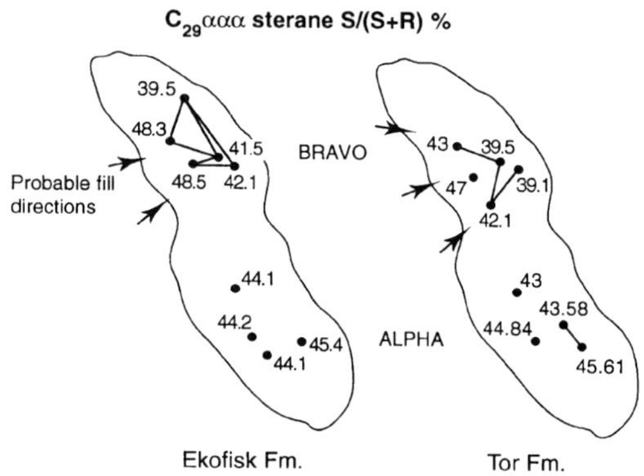

Fig. 14. Lateral petroleum maturity gradients (predominantly E–W) across the Eldfisk Field based on reservoir core-extract data. Average C_{29} $\alpha\alpha\alpha$ sterane S/(S + R) biomarker maturity values from reservoir core-extract samples of each well and formation (Ekofisk and Tor) are presented. The observed petroleum maturity gradients across the field were verified statistically using the Student *t*-test. The Student *t*-test statistically determines whether two datasets, each consisting of reservoir core-extract data C_{29} $\alpha\alpha\alpha$ sterane S/(S + R) biomarker maturity values from the same formation from adjacent wells, are statistically different at the 95% confidence level. Statistically significant differences between adjacent wells are indicated by solid black lines on the diagram. The inferred, tentative petroleum-filling directions for the Eldfisk reservoir are schematic only. Similar petroleum maturity gradients (E–W) across the Eldfisk Field (for both formations) were observed and statistically verified using the Ts/(Ts + Tm) biomarker maturity parameter. The contour line represents the Upper Ekofisk chalk Formation at 2950 m.

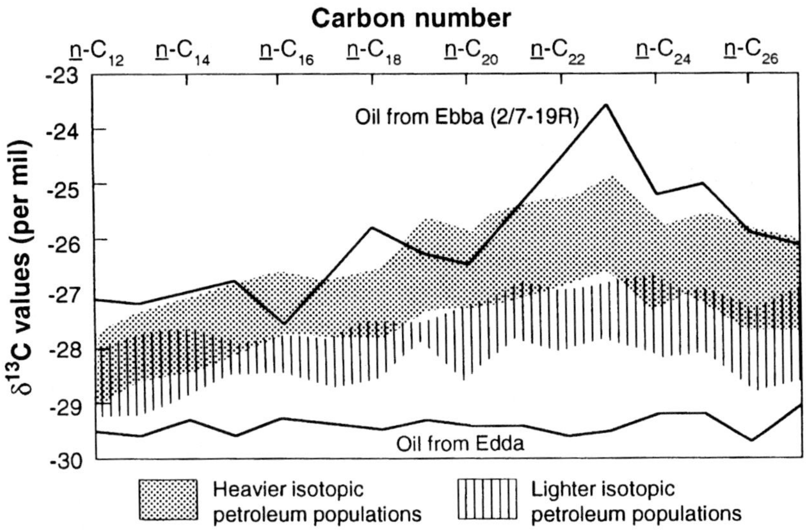

Fig. 15. Identification of two isotopically defined petroleum populations based on the $\delta^{13}C$ *n*-alkane signatures from the Eldfisk Field DST oils. Shaded bands represent two *n*-alkane isotopically defined petroleum populations, which are sandwiched between two end-member type oils, an Edda Field oil (isotopically light) and an Ebba Field oil (isotopically heavy). Isotopically heavier Ebba DST oil was probably generated from the isotopically heavier Upper Jurassic source facies. Eldfisk Field DST oils have probably resulted from mixing of the two end-member petroleum populations.

end-member from well 2/7-3 and Ebba Field (Fig. 15). The isotopically heavier petroleums are most likely sourced from the isotopically heavier source facies described previously, and have similar biomarker characteristics to this zone.

The Eldfisk Field petroleum column has probably resulted from various degrees of mixing of the two end-member petroleum populations. The distribution of the two isotopically defined petroleum populations throughout the Eldfisk Field is shown in Fig. 16. The isotopically heavier petroleum population embraces an area straddling the Alpha and Bravo structures, East Eldfisk and well 2/7-A2. The isotopically lighter petroleum population incorporates wells 2/7-A21, 2/7-A19, 2/7-A16, 2/7-A29 and 2/7-B12. The presence of an isotopically heavier petroleum population straddling south Bravo–north Alpha suggests that there may be a conduit in this part of the field controlled by lithofacies distribution within the reservoir. Such documented and confirmed distributions of petroleum in complex fields could potentially be used to monitor the efficiency of production plans for the Eldfisk Field.

The filling history of Eldfisk Field is clearly much more complex than that reported earlier for clastic siliceous reservoirs elsewhere in the North Sea (England 1990; Larter & Horstad 1992). The observation of general petroleum maturity gradients infer general south to north and west to east migration directions for the filling of the Eldfisk Field, onto which are superimposed the effects of variable-source organic facies signatures. These observations, together with the generally low

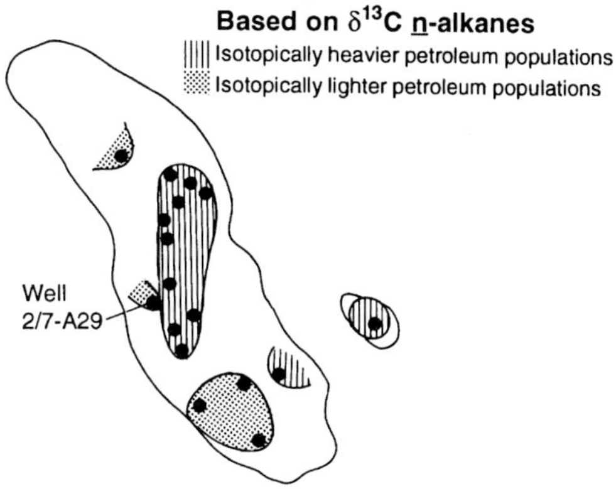

Fig. 16. Distribution of the two *n*-alkane isotopically defined petroleum populations throughout the Eldfisk Field. The *n*-alkane isotopically heavier petroleum population straddles both the Eldfisk Alpha and Bravo structures; this area of the field may represent a conduit which is controlled by lithofacies distributions within the reservoir. Well 2/7-A29 shows a different isotopic petroleum population to the surrounding Eldfisk DST oils, and may represent an isolated reservoir segment.

matrix permeability of the Eldfisk Field, imply that petroleum emplacement most probably was along faults with some fractures variably accessing different source facies.

Vertical petroleum-column compositional variations

Bulk petroleum-column geochemical variations within the Eldfisk Field

High resolution sampling of reservoir cores on a metre scale enables the production of detailed petroleum concentration logs. Iatroscan data from well 2/7-A6 show variations in petroleum concentrations which are directly related to the porosity and fluid saturation variations seen downhole with a resistivity (MSFL) log (Fig. 17). While the general petroleum content logs obtained by the resistivity logging (immovable oil-dark zone + movable oil) and Iatroscan procedures are similar, the geochemical log shows greater near-well resolution. This is related to the greater spatial integration of the logging tool. We speculate that the petroleum concentration log obtained from geochemical logging by Iatroscan/Rock-Eval may potentially aid in the positioning of production perforations in low quality, variably saturated reservoirs.

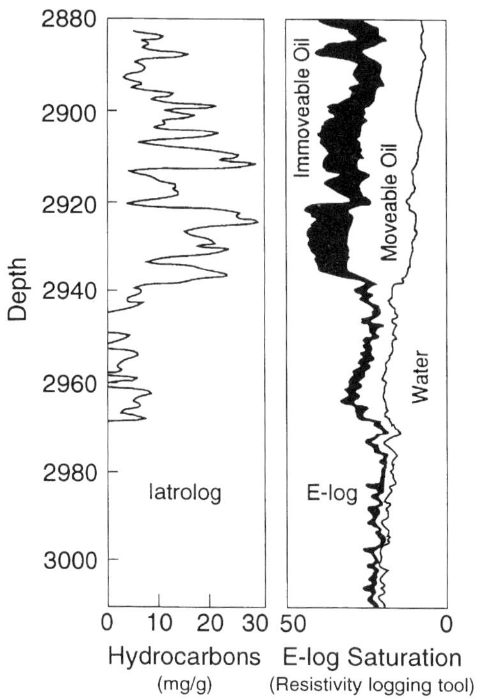

Fig. 17. Comparison of the Iatrolog and the resistivity (MSFL) log, which are used to determine petroleum concentration and fluid saturation in reservoir core, respectively. the Iatrolog shows more variation in petroleum concentration compared to the resistivity (MSFL) fluid saturation log. This results from the greater spatial integration of the resistivity logging tool. The Iatrolog may be used to aid the positioning of perforation holes in low quality reservoirs where near-well petroleum saturation is variable.

Molecular geochemical variations downhole within the Eldfisk reservoir

Visually identifiable petroleum column compositional steps or more general variations are present in four of the 11 wells sampled (2/7-A6, 2/7-B13A, 2/7-B15 and 2/7-B12), where compositional discontinuities occur either within reservoir formations or between such formations (Figs 18 & 19). Data from two wells (2/7-A6 and 2/7-B13A) are used to show how data analysis methods (visual analysis, PCA and Student *t*-tests) can be used together to aid in deciding whether different petroleum populations are present vertically in a well.

Visual representation of the data from well 2/7-A6, illustrated using a representative biomarker parameter (Ts/(Ts + Tm)), suggests that two distinct petroleum populations are present in the Tor and Hod Formations (Fig. 18). The Hod Formation could be interpreted to have a lower maturity petroleum compared to the Tor Formation, although the actual geochemical interpretation is irrelevant to the following discussion. Two petroleum population groupings are also confirmed by the PCA analysis, with the sample distribution shown on a scores plot (Fig. 18). More significantly, performing the Student *t*-test on the groupings defined visually and by PCA confirms the presence of the two petroleum populations as the *t*-statistic is greater than *t*-critical. Similarly, but more tentatively, two different petroleum populations are evident by visual examination of Fig. 19 (well 2/7-B13A), where a compositional discontinuity is observed between the Ekofisk and Tor Formation petroleums (Fig. 19). Again, the discontinuity is suspected from visual examination but confirmed with statistical methods (*t*-statistic is greater than *t*-critical, PCA broadly separating the two populations), confirming the separation of the data at the 95% confidence limit.

Barriers to petroleum mixing in the Eldfisk Field

The locations of the compositional discontinuities recognized previously can generally be related to

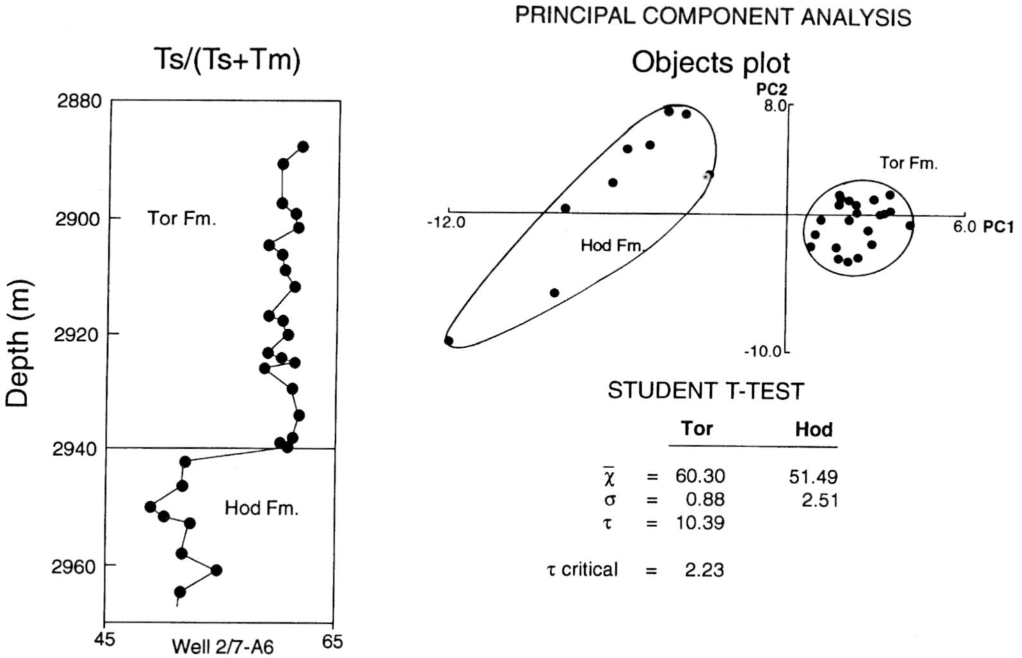

Fig. 18. Three data analysis methods (visual, PCA and Student *t*-test), illustrating that a discontinuity exists between the Tor and Hod chalk Formations in well 2/7-A6, inferring the presence of a barrier/baffle preventing diffusive mixing of petroleum. The three data analysis methods were used independently of each other to verify the existence of a barrier/baffle in the petroleum column.

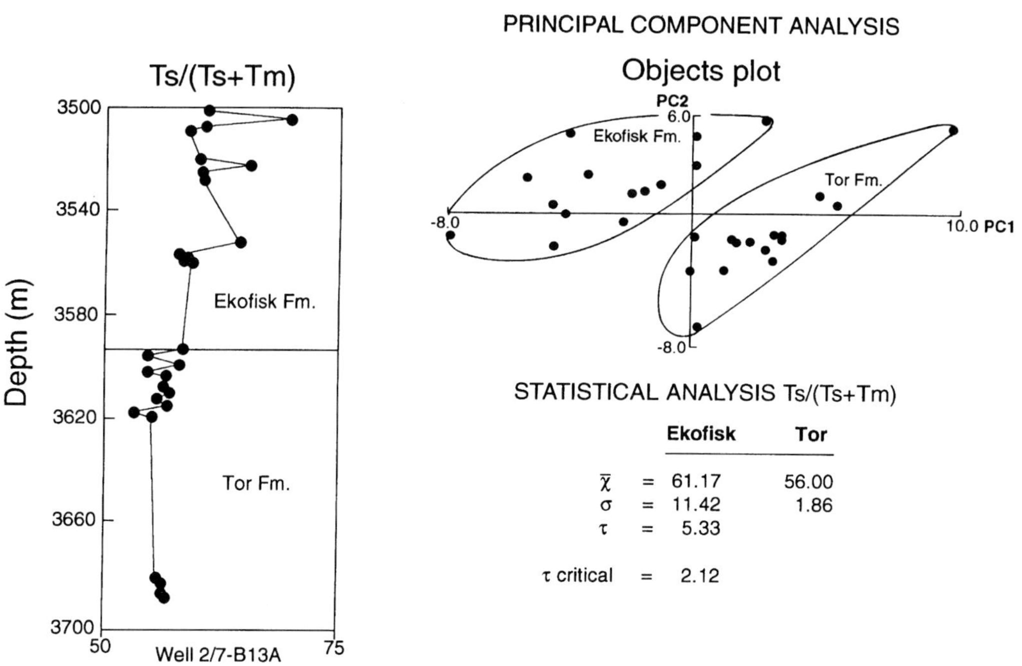

Fig. 19. Three data analysis methods (visual, PCA and Student *t*-test) used independently of each other to illustrate that a geochemical discontinuity exists between the Ekofisk and Tor chalk Formations in well 2/7-B13A. We infer the presence of a barrier/baffle in the petroleum column which has prevented diffusive mixing.

geological features, these compositional 'steps' representing barriers to petroleum mixing on a geological time-scale. England *et al.* (1987) calculated order of magnitude time-scales for vertical diffusion of petroleum species in petroleum columns in order to eliminate inherited compositional heterogeneities within reservoirs. With large-scale permeabilities (whole petroleum column average) estimated for the Eldfisk of *c.* 0.1 to 1 mD, mixing times on a reservoir column scale (100 m) would be around 1 to 2 Ma. Assuming that the majority of the filling of the Eldfisk Field occurred since 10 Ma BP, it can be concluded that the reservoir has probably had sufficient time for vertical diffusion to locally homogenize, or at least attenuate, any inherited compositional variations vertically in the petroleum column. Thus, the compositional discontinuities observed may well be due to low permeability (high tortuosity) barriers in the petroleum column. As the theoretical model is, of necessity, rather imprecise, engineering data are necessary to calibrate this approach (Larter & Aplin 1995).

In the petroleum columns of wells 2/7-A6, 2/7-B13A, 2/7-B15 and 2/7-B12, different petroleum populations occur vertically in the reservoir, indicating that barriers or baffles may exist. These barriers are related to particular geological features, such as hardgrounds associated with unconformities, or rapid lateral and vertical changes in lithofacies (thin, pelagic, distal reworked facies). A typical barrier facies is seen in the lower Ekofisk Formation (ED layer) where the relatively low oil saturation, low porosity and low vertical permeability are clearly related to more siliceous, highly bedded/laminated chalks. These ED layer features are observed in most log and core data across the Eldfisk Field.

The presence of a barrier (ED layer) to petroleum mixing in well 2/7-B13A is confirmed by repeat formation pressure data (RFT) in the well, which show a marked change in pressure (draw-down pressures) at the Ekofisk/Tor Formation boundary (Fig. 20), correlating with the molecular data-step. In several wells in the Eldfisk Field, pressure discontinuities across the ED layer are seen but they are not always accompanied by geochemical discontinuities. It is important to remember that a homogeneous petroleum column is not necessarily indicative of communicating fluids, as the same

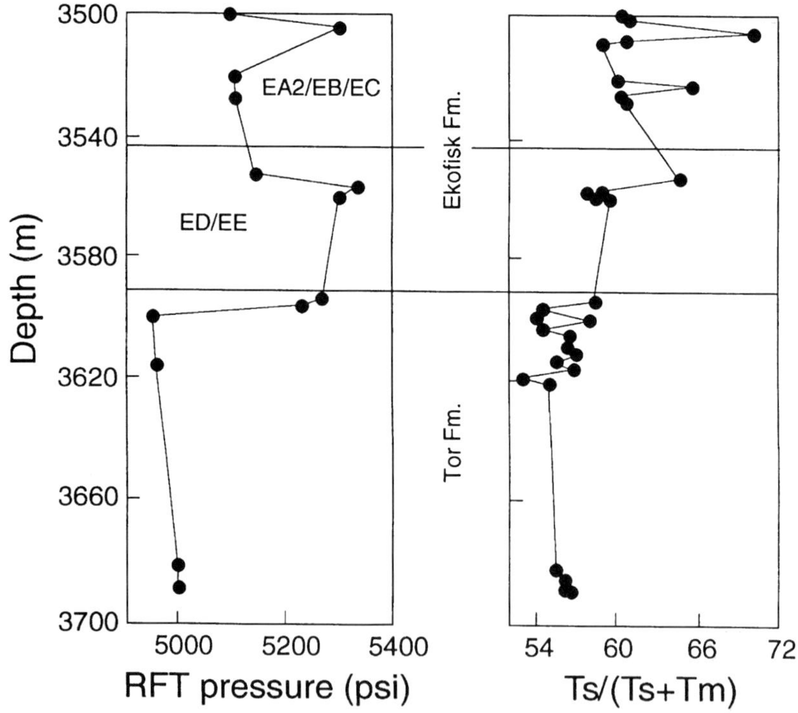

Fig. 20. Well 2/7-B13A RFT pressure data showing a sudden step at the Ekofisk/Tor boundary, indicating the presence of a barrier. Biomarker data (Ts/(Ts + Tm)) also show a discontinuity at this point. Pressure in the ED/EE layer has not been drawn down to the same extent because this represents a low permeability zone.

petroleum could have been emplaced both above and below a baffle and/or barrier (this has been observed elsewhere: Petch, S. G. pers. comm.). The geochemical anomaly in this case is substantiated by established engineering procedures, which lends credence to the use of such data to detect possible intra-reservoir barriers inexpensively at an early stage of field development, when pressure data may be sparse or where pressure steps are only visible after production starts.

Whilst a geochemical discontinuity may not necessarily indicate a production-relevant barrier in every case, correlations of this kind between engineering and geochemical data are important because RFT pressure data are not always obtained when the well or area has gone into production.

The minimum extent of the barriers or baffles in wells 2/7-A6, 2/7-B13A, 2/7-B15 and 2/7-B12 can be estimated with a knowledge of several key factors and a diffusion model for mixing. First, the reservoir was largely charged with petroleum *c.* 10 Ma BP. This should have provided sufficient time for vertical homogenization of the petroleum column. Using a reservoir tortuosity factor of $\sqrt{3}$ (which is probably too low for chalk reservoirs) and diffusion coefficient estimates for a C_{12} molecule (1.7×10^{-10} m^2 s^{-1}; England *et al.* 1987), it is calculated that over a period of 10 Ma lateral diffusive mixing would have chemically homogenized petroleum over a distance of 500 m. Diffusion would have to occur around the barrier to eliminate any chemical differences, hence the minimum length of the barrier is around 250 m. Geological considerations suggest that the barrier is much more laterally extensive.

Petroleum pyrrolic nitrogen compound distributions throughout the Eldfisk chalk reservoir

Although intra-reservoir petroleum column compositional variations in hydrocarbon fractions have previously been identified, little or no attention has been paid to variations in polar compound distributions in general, or nitrogen compounds in particular. This is, in part, due to the analytical problems associated with low nitrogen compound concentrations in crude oils (typically 0.1 to 1.0 wt%). Although the majority of the work reported on nitrogen compounds has concentrated on aspects of separation and identification (Richter *et al.* 1952; Snyder 1970; McKay *et al.* 1976; Schmitter *et al.* 1979; Later *et al.* 1981; Burchill *et al.* 1983; Frolov *et al.* 1989; Bakel & Philp 1990; Li *et al.* 1992), there is clear evidence that the distributions of organic nitrogen compounds in crude oils can be used as an indicator of filling directions or petroleum migration range tracers (Yamamoto 1992; Li *et al.* 1994; Larter & Aplin 1995).

Twenty DST oils from the Eldfisk reservoir were analysed for pyrrolic nitrogen compounds. Based on compound class (relative proportions of alkylated carbazoles and alkylated benzocarbazoles) and isomeric differences (relative proportions of pyrrolic hydrogen-shielded (Group 1) over partially shielded (Group 2) and nitrogen exposed (Group 3) isomers of alkylated carbazoles (see Li *et al.* 1995 for details of terminology)), two types of pyrrolic nitrogen distributions were identified (Figs 21 & 22).

Group A. Alkylated benzocarbazoles predominate over the alkylated carbazoles and enrichment of pyrrolic hydrogen exposed isomers, such as 3,5-dimethylcarbazole (Group 3), over pyrollic hydrogen partially shielded isomers, such as 1,4-dimethylcarbazole (Group 2), is seen. This group of DST oils contains six of the eight DST oils derived from the Alpha structure, plus DST oils from East Eldfisk, wells 2/7-B10 and 2/7-B12 (Figs 21, 22 & 23).

Group B. Alkylated carbazoles predominate over the alkylated benzocarbazoles, with an enrichment of pyrollic hydrogen partially shielded (Group 2) over exposed (Group 3) isomers of the alkylated carbazoles. This distribution accounts for nine out of the 11 DST oils derived from the Bravo structure, and both DST oils from well 2/7-A29 (Figs 21, 22 & 23).

These distributions, though not clearly understandable in detail at this time, are consistent with the major charge to the field from the south if, as suggested by Li *et al.* (1995), petroleum migration is the major control on petroleum nitrogen compound distributions. Well 2/7-A29 showed different pyrrolic nitrogen distributions and stable carbon isotope *n*-alkane signatures to the surrounding DST oils, suggesting that this may represent an isolated reservoir segment (Figs 16 & 23).

A total of 12 closely spaced reservoir core samples were taken at various depths from well 2/7-B12 to assess whether small-scale heterogeneities are present in the pyrrolic nitrogen distributions within the petroleum extractable from the core. As with the DST oils, the reservoir core extracts could be divided into two groups based on differences in the relative abundance of the compound classes present (carbazoles versus benzocarbazoles), the relative proportions of partially shielded to exposed pyrrolic hydrogen isomers, and the carbon number distributions of alkylcarbazoles. Figure 24 shows mass fragmentograms of the reservoir core extract pyrrolic nitrogen compound fractions, the large peak in each chromatogram being an internal standard (*N*-phenylcarbazole)

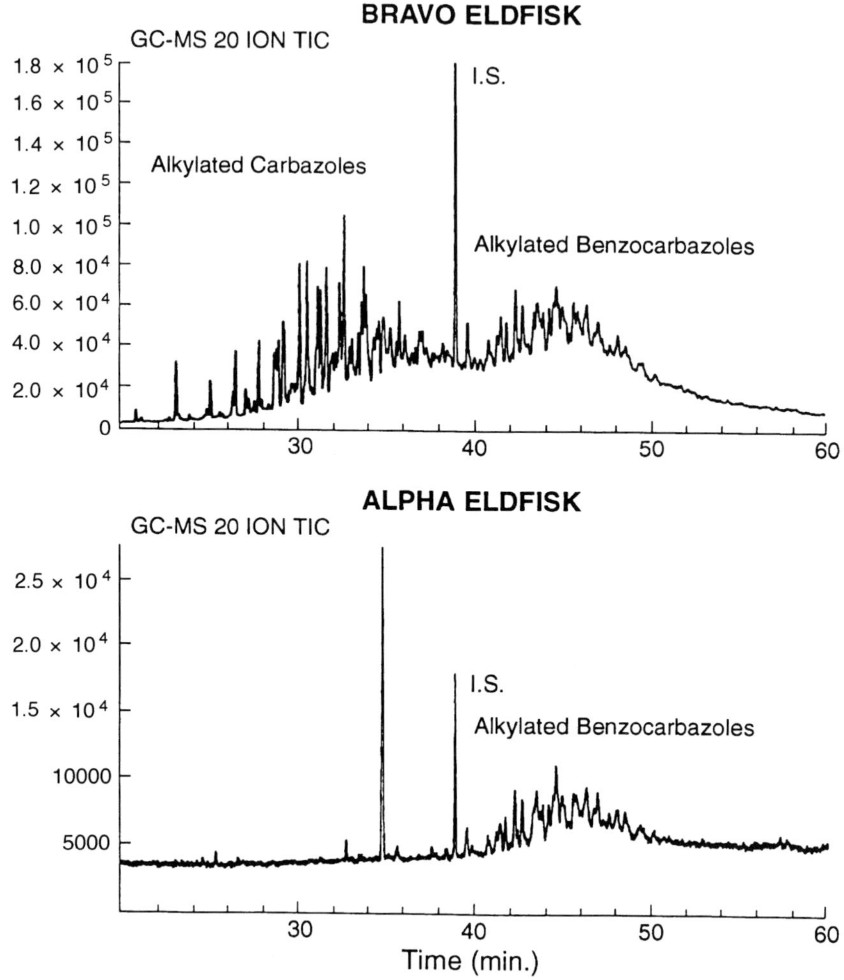

Fig. 21. Typical pyrrolic nitrogen compound distributions for the Eldfisk, Alpha and Bravo DST oils (grouped as A and B in the text). The pyrrolic nitrogen compound distributions for the reservoir core-extract data from well 2/7-B12 are similarly grouped. I.S. = internal standard (*N*-phenylcarbazole). The 20 mass ions which comprise the total ion current (TIC) chromatograms are *m/z* 167, 181, 195, 209, 223, 237, 251, 265, 279, 293 (alkylated carbazoles), 217, 231, 245, 259, 273, 287 (alkylated benzocarbazoles) and 267, 281, 295, 309 (alkylated dibenzocarbazoles).

present at constant abundance. It is clearly seen that compositional heterogeneity occurs at a metre scale throughout the petroleum column. Further, even from this single well, the nitrogen compound distributions cover the complete range found within the present Eldfisk Field DST sample suite, even though the DST oil distributions are spatially organized within the field.

Yamamoto (1992) and Li *et al.* (1995) have shown that significant differences in nitrogen species distributions occur between crude oils and source-rock extracts, and between reservoired crude oils with different migration distances. Distributions of pyrrolic nitrogen compounds in crude oils appear to be less significantly influenced by organic facies, depositional environments or maturity (Li *et al.* 1995). Primary and secondary migration effects appear, with our current, albeit small database, to be the dominant controls on the distribution of pyrrolic nitrogen compounds in petroleum. With increasing migration distance, the composition of petroleum pyrrolic nitrogen compounds changes with relative enrichment of alkylcarbazoles relative to alkylbenzocarbazoles, an increase in the proportions of pyrrolic hydrogen partially shielded to exposed isomers, and

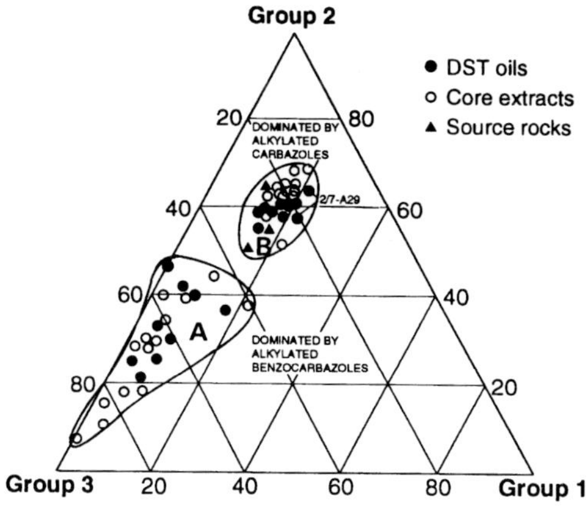

Fig. 22. Distribution of dimethylcarbazole isomers in Eldfisk DST oils, reservoir core and Upper Jurassic source rocks. The division of the classes, Group 1 (nitrogen shielded), Group 2 (nitrogen partially shielded), and Group 3 (nitrogen exposed) are based on the degree of alkyl group shielding of the nitrogen compound N–H functionality. Groups A and B are marked on the diagram and represent the two types of pyrrolic nitrogen distribution identified on the basis of pyrrolic nitrogen compound class distributions and the degree of pyrrolic hydrogen shielding.

increased abundance of more alkylated alkylcarbazole homologues compared to the less alkylated homologues (Li *et al.* 1995). These fractionation effects probably result from normal phase-adsorptive chromatographic interactions, through hydrogen and Van der Waals bonding between the pyrrolic >N–H functional group and organic (kerogen/bitumen) and mineral phases present in the petroleum transport system (Li *et al.* 1994; Larter & Aplin 1995).

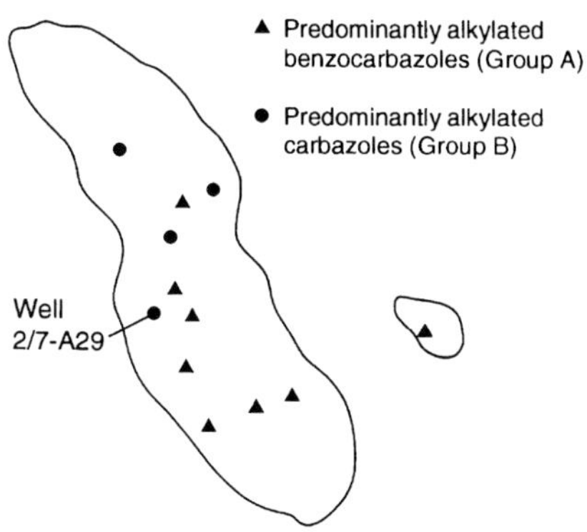

Fig. 23. Distribution of petroleums with Groups A and B pyrrolic nitrogen compound distribution. The majority of Eldfisk Bravo DST oils have pyrrolic nitrogen compound distribution similar to Group B, whilst the majority of Eldfisk Alpha DST oils have characteristics typical of Group A. Well 2/7-A29 shows different pyrrolic nitrogen distribution to the surrounding Eldfisk DST oils and may represent an isolated reservoir segment. The contour line represents the Upper Ekofisk chalk formation at 2950 m.

The majority of the DST oils from the Eldfisk Alpha structure are relatively enriched in alkylbenzocarbazoles to alkylcarbazoles, when compared to oils from the Eldfisk Bravo structure, suggesting that the latter oils have migrated further. This is consistent with the geological and geochemical evidence described previously, which indicated that the Eldfisk structure was predominantly charged from the south.

Three of the four DST oils from well 2/7-B12 show pyrrolic nitrogen distributions similar to Group B (Bravo structure oils), whilst the reservoir core extracts display the two distinct types of pyrrolic nitrogen distributions. It would appear that the core extract and DST petroleums are not in equilibrium and it might be speculated that once nitrogen compounds have been adsorbed onto rock surfaces, they do not equilibrate further by diffusive mixing. We suggest, therefore, that the variations in pyrrolic nitrogen compound distribution seen in Fig. 24 may in large part be related to first-fill phenomena, the distributions being set by the first oil to encounter a given portion of the reservoir. In this scenario, the DST (producible oils) represent pore oil rather than adsorbed oil. We are continuing to investigate the significance of nitrogen species in petroleum reservoir cores as they relate to reservoir-wetting state.

Filling history of the Eldfisk Field

The filling history of the Eldfisk Field is complex and clearly not totally defined by our study. Nevertheless, we feel we can make some statements about the processes that have occurred, accepting that in some cases the conclusions are not definitive.

The main oil-generation zone in the Upper Jurassic source rocks currently charging the field is estimated to lie between 3700 mss and 4200 mss, indicating that significant generation probably occurred from all source kitchen areas surrounding the Eldfisk structure.

The burial/maturation history of the kitchens, as typified by well 2/7-15X, suggests that the Upper Jurassic shales began generating (and thus expelling) petroleum *c.* 20 to 25 Ma BP, with most of the petroleum migrating in from 10 Ma BP onwards. The tilted transitional oil–water contact, visible throughout the Eldfisk reservoir, may be consistent with a field of low average permeability

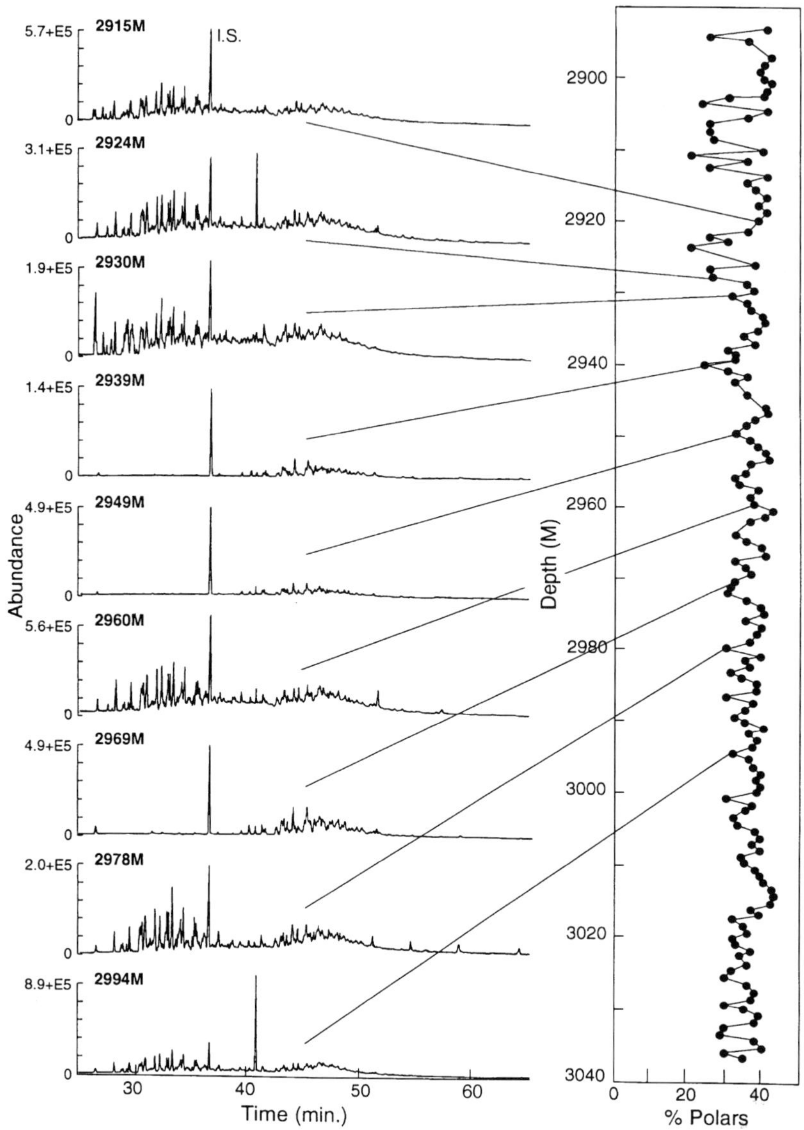

Fig. 24. Heterogeneous pyrrolic nitrogen distributions within the core extracts of well 2/7-B12. Both Group A and B pyrrolic nitrogen compound distributions are seen in the reservoir core extracts of well 2/7-B12, illustrating that core extract pyrrolic nitrogen distributions are heterogeneous on a metre scale.

and a recent, probably ongoing fill, tilting or leakage.

The majority of the Eldfisk petroleum migrated from kitchens to the south of the field which lie within the main petroleum-generating zone, probably entering the structure through the southern bounding fault and Skrubbe fault zone. This suggestion is clearly supported by the higher petroleum maturities in the Eldfisk Alpha structure compared to the Bravo analogue and by migration-related pyrrolic nitrogen distributions in the DST oils. A fill direction locally from west to east within the Eldfisk Field can also be inferred from the presence of small lateral maturity gradients, and it is probable that local sourcing of the north of the field also occurred. It appears that internal sedimentological as well as tectonic influences are important to the filling of the Eldfisk reservoir.

Conclusions

Eldfisk Field petroleum is heterogeneous, both laterally on a kilometre scale and locally on a vertical metre scale, with a complex filling history linked to faulting and fracturing which facilitated charging of the reservoir.

Initial charge time estimates for the Eldfisk Field suggest that filling could have occurred from early Neogene (20 to 25 Ma BP) onwards, with the majority of the charge entering the structure from 10 Ma BP.

Filling of the Eldfisk Field was via a system of faults/fractures which are thought to link the Upper Jurassic source rocks to the Eldfisk structure. There is probably variable communication between different areas of the source rock and the field, with certain fault systems supplying petroleum to selected regions within the structure at different times.

Pyrrolic nitrogen distributions in Eldfisk Bravo DST oils differ from those of Eldfisk Alpha in the following ways:

(1) enrichment of alkylated carbazoles relative to alkylated benzocarbazoles;
(2) increase in the proportions of pyrrolic hydrogen partially shielded isomers to nitrogen exposed isomers;
(3) increased abundances of more alkylated alkylcarbazole homologues compared to the less alkylated homologues.

It is concluded that Eldfisk Bravo DST oils have migrated further compared to the Alpha analogues. Based on the biomarker maturity data for DST oils, the majority of the Eldfisk petroleum is thought to have been generated from source rocks to the south of the field, lying within the main oil-generation zone. Subsequent migration up and along the southern bounding fault and Skrubbe fault zone then facilitated the charging of the Eldfisk Alpha and Bravo structures.

Compositional steps downhole have been identified in Eldfisk reservoir core-extract petroleums from several wells. These can, in some cases, be verified statistically. In one well (2/7-B13A), the compositional discontinuity correlates well with a significant change in RFT pressure across a stratigraphic boundary, adding credence to the use of such geochemical anomalies for early detection of possible barriers in petroleum columns.

The highly variable pyrrolic nitrogen compound distributions throughout the petroleum column of well 2/7-B12, as monitored by nitrogen compound analysis of core extracts, is probably related to first-fill phenomena, which may result from fresh oil contacting 'clean' rock surfaces during penecontemporaneous reservoir filling and faulting. The reservoir core extract nitrogen compound distributions appear to be out of equilibrium with DST oils, and the observation that Eldfisk reservoir core extracts differ in bulk composition from the Eldfisk DST oils has implications for reservoir engineering scenarios (Larter & Aplin 1995). Clearly, the England (1987) mixing model, which describes mixing of free hydrocarbon components, is not applicable when adsorptive polar species are considered.

We would like to acknowledge Phillips Petroleum Company Norway and co-venturers, including Fina Exploration Norway u.a.s., Norsk Agip A/S, Elf Petroleum Norge AS, Norsk Hydro a.s., Den norske stats oljeselskap a.s, Total Norge A.S, Elf Rex Norge A/S and Norminol A/S, for their financial and technical support and permission to publish this paper. N. Frey is thanked for useful discussion. M. Jones, P. Farrimond,P. Donohoe, Y. Hall, B. Brown, C. Jeans and I. Svindland are thanked for their assistance. The opinions expressed are those of the authors and do not necessarily represent those of Phillips Petroleum Company Norway nor co-venturers. This paper has benefited greatly from comments made by W. England (BP) and an anonymous reviewer.

References

BAILEY, N. J. L., BURWOOD, R. & HARRIMAN, G. E. 1990. Application of pyrolysate carbon isotope and biomarker technology to organofacies definition and oil correlation problems in North Sea Basins. *Organic Geochemistry*, **16**, 1157–1172.

BAKEL, A. J. & PHILP, R. P. 1990. The distribution and quantitation of organonitrogen compounds in crude oils and rock pyrolysates. *Organic Geochemistry*, **16**, 353–367.

BJORØY, M., HALL, K., GILLYON, P. & JUMEAU, J. 1991*a*. Carbon isotope variations in *n*-alkanes and isoprenoids of whole oils. *Chemical Geology*, **93**, 13–20.

——, ——, HALL, P. B., LEPLAT, P. & LØBERG, R. 1991*b*. Biomarker analysis of oils and source rocks using a thermal extraction-GC-MS. *Chemical Geology*, **93**, 1–11.

——, —— & JUMEAU, J. 1990. Stable carbon ratio analysis on single components in crude oils by direct GC-isotope analysis. *Trends In Analytical Chemistry*, **9**(10), 331–337.

——, SOLLI, H., HALL, K. & LEPLAT, P. 1985. Analysis of source rocks, reservoir rocks and cap rocks by combined thermal extraction and pyrolysis-gas chromatography. *In*: THOMAS. B. M. *et al.* (eds) *Petroleum Geochemistry in Exploration of the Norwegian Shelf.* Norwegian Petroleum Society, Graham & Trotman, London, 327–337.

BRASHER, J. E. 1994. Local tectonics and effects on sediment distribution within the Eldfisk Field. *In*: HANSLIEN, S. (ed.) *Petroleum Exploration and Exploitation in Norway*. Norwegian Petroleum Society, in press.

BURCHILL, P., HEROD, A. A. & PRITCHARD, E. 1983. Investigation of nitrogen compounds in coal tar products. 2. Basic fractions. *Fuel*, **62**, 20–28.

BURNHAM, A. K. & SWEENEY, J. J. 1989. A chemical kinetic model of vitrinite maturation and reflectance. *Geochimica et Cosmochimica Acta*, **53**, 2649–2657.

COOPER, B. S. & BARNARD, P. C. 1984. Source rocks and oils of the Central and Northern North Sea. *In*: DEMAISON, G. & MORRIS, R. J. (eds) *Petroleum Geochemistry and Basin Evaluation*. American Association of Petroleum Geologists, Memoir, **35**, 303–314.

CORNFORD, C., MORROW, J. A., TURRINGTON, A., MULES, J. A. & BROOKS, T. 1983. Some geological controls on oil composition in the UK North Sea. *In*: BROOKS, J. (ed.) *Petroleum Geochemistry and Exploration of Europe*. Geological Society, London, Special Publication, **12**, 175–194.

——, NEEDHAM, C. E. & DE WALQUE, L. 1986. Geochemical habitat of North Sea oils and gases. *In*: SPENCER, A. M. ET AL. (eds) *Habitat of Hydrocarbons on the Norwegian Continental Shelf*. Norwegian Petroleum Society, Graham & Trotman, London, 39–54.

DAMTOFT, K., ANDERSEN, C. & THOMPSON, E. 1986. Prospectivity and hydrocarbon plays of the Danish Central Trough. *In*: BROOKS, J. & GLENNIE, K. W. (eds) *Petroleum Geology of North West Europe*. Graham & Trotman, London, 403–418.

DAVIS, J. C. 1973. *Statistics and Data Analysis in Geology*. John Wiley & Sons, Inc., New York.

ENGLAND, W. A. 1990. The organic geochemistry of petroleum reservoirs. *Organic Geochemistry*, **16** (1–3), 415–425.

—— & MACKENZIE, A. S. 1989. Geochemistry of petroleum reservoirs. *Geologische Rundschau*, **78**, 214–237.

——, ——, MANN, D. M. & QUIGLEY, T. M. 1987. The movement and entrapment of petroleum fluids in the subsurface. *Journal of the Geological Society, London*, **144**, 327–347.

FROLOV, Y. B., SMIRNOV, M. B., VANYUKOVA, N. A. & SANIN, P. I. 1989. Carbazoles of crude oils. *Petroleum Chemistry, USSR*, **29**, 87–102.

GABRIELSEN, R. H., ULVØEN, S., ELVSBORG, A. & FREDRIK, O. 1985. The geological history and geochemical evaluation of Block 2/2, offshore Norway. *In*: THOMAS, B. M. *et al.* (eds) *Petroleum Geochemistry in Exploration of the Norwegian Shelf*. Graham & Trotman, London, 165–178.

HIRSCHBERG, A. 1984. Role of asphaltenes in compositional grading of a reservoir's fluid column. *Society of Petroleum Engineers*, **13171**, 89–94.

HORSTAD, I., LARTER, S. R., DYPVIK, H., AAGAARD, P., BJORNVIK, A. M., JOHANSEN, P. E. & ERIKSEN, S. 1990. Degradation and maturity controls on soil field petroleum column heterogeneity in the Gullfaks Field, Norwegian North Sea. *Organic Geochemistry*, **16** (1), 497–510.

HUGHES, W. B., HOLBA, A. G., MILLER, D. E. & RICHARDSON, J. S. 1985. Geochemistry of Greater Ekofisk crude oils. *In*: THOMAS, B. M. ET AL. (eds) *Petroleum Geochemistry in Exploration of the Norwegian Shelf*. Graham & Trotman, London, 75–92.

KARLSEN, D. A. & LARTER, S. R. 1989. A rapid correlation method for petroleum populations mapping within individual petroleum reservoirs: applications to petroleum reservoir description. *In*: COLLINSON, J. D. (ed.) *Correlation in Hydrocarbon Exploration*. Graham & Trotman, London, 77–85.

—— & —— 1991. Analysis of petroleum fractions by TLC-FID: Applications to petroleum reservoir description. *Organic Geochemistry*, **17**, 603–617.

KVALHEIM, O. M. 1987. Oil-source correlation by the combined use of principal component modelling, analysis of variance and a coefficient of congruence. *Chemometrics and Intelligent Laboratory Systems*, **2**, 127–136.

——, CHRISTY, A. A., TELNAES, N. & BJORSETH, A. 1987. Maturity determination of organic matter in coals using the methyl-phenanthrene distribution. *Geochimica et Cosmochimica Acta*, **51**, 1883–1888.

LARTER, S. R. 1989. Chemical models of vitrinite reflectance evolution. *Geologische Rundschau*, **78**, 349–359.

—— & APLIN, A. 1995. Reservoir Geochemistry: Methods, applications and opportunities. *This volume*.

—— & HORSTAD, I. 1992. Migration of petroleum into Brent Group reservoirs: some observations from the Gullfaks field, Tampen Spur area North Sea. *In*: HASZELDINE, A. C., GILES, R. S. & BROWN, S. (eds) *Geology of the Brent Group*. Geological Society, London, Special Publication, **61**, 441–452.

—— & MILLS, N. 1992. Phase-controlled molecular fractionations in migrating petroleum charges. *In*: ENGLAND, W. A. & FLEET, A. J. (eds) *Petroleum Migration*. Geological Society, London, Special Publication, **59**, 137–147.

——, BJØRLYKKE, K. O., KARLSEN, D. A. ET AL. 1991. Determination of petroleum accumulation histories: Examples from the Ula Field, Central Graben, Norwegian North Sea. *In*: BULLER, A. T., BERG, E., HJELMELAND, O., KLEPPE, J., TORSAETER, O. & AASEN, J. O. (eds) *North Sea Oil & Gas Reservoirs-II*. Graham & Trotman, London, 319–330.

LATER, D. W., LEE, M. L., BARTLE, K. D., KONG, R. C. & VASSILAROS, D. L. 1981. Chemical class separation and characterisation of organic compounds in synthetic fuels. *Analytical Chemistry*, **53**, 1612–1620.

LEYTHAEUSER, D. & RUCKHEIM, J. 1989. Heterogeneity of oil composition within a reservoir as a reflectance of accumulation history. *Geochimica et Cosmochimica Acta*, **53**, 2119–2123.

LI, M., LARTER, S. R. & FROLOV, Y. B. 1994. Adsorptive interactions between petroleum nitrogen compounds and organic/mineral phases in subsurface rocks as models for compositional fractionations of pyrrolic nitrogen compounds in petroleum during petroleum migration. *High Resolution Chromatography*, **17**, 230–236.

——, ——, STODDART, D. P. & BJORØY, M. 1992. Practical liquid chromatographic separation schemes for pyrrolic and pyridinic nitrogen aromatic heterocycle fractions from crude oils suitable for rapid characterisation of geochemical samples. *Analytical Chemistry*, **64**, 1337–1344.

——, ——, —— & —— 1995. Fractionation of pyrrolic nitrogen compounds in petroleum during migration: derivation of migration-related geochemical parameters. *This volume*.

MCKAY, J. F., WEBER, J. H. & LATHAM, D. R. 1976. Characterisation of nitrogen bases in high-boiling petroleum distillates. *Analytical Chemistry*, **48**(6), 891–898.

MICHAUD, F. 1987. Eldfisk. *In*: SPENCER, A. M. (ed.) *Geology of the Norwegian Oil and Gas Fields*. Graham & Trotman, London, 89–105.

RICHTER, P. P., CEASSER, P. D., MEISEL, S. L. & OFFENHAUSER, R. D. 1952. Distribution of nitrogen according to basicity. *Industrial Engineering Chemistry*, **44**, 2601–2605.

SAGE, B. H. & LACEY, W. N. 1938. Gravitational concentration gradients in static columns of hydrocarbon fluids. *Transactions of the American Institute of Mining and Metallurgical Engineers*, **132**, 120–131.

SCHMITTER, J. M., VAJTA, Z. & ARPINO, P. J. 1979. Investigation of nitrogen bases from petroleum. *In*: DOUGLAS, A. & MAXWELL, J. (eds) *Advances in Organic Geochemistry*. Pergamon, London, 67–76.

SCHOLLE, P. A., ARTHUR, M. A. & EKDALE, A. A. 1983. Pelagic environments. *In*: SCHOLLE, P. A., BEBOUT, D. G. & MOORE, C. H. (eds) *Carbonate Depositional Environments*. American Association of Petroleum Geologists, Memoir, **33**, 619–691.

SCHULTE, A. M. 1980. *Compositional variations within a hydrocarbon column due to gravity*. Society of Petroleum Engineers, Paper No. **9235**.

SEIFERT, W. K. & MOLDOWAN, J. M. 1978. Applications of steranes, terpanes and monoaromatics to the maturation, migration and source of crude oils. *Geochimica et Cosmochimica Acta*, **42**, 77–95.

SMALLEY, P. C., LØNØY, A. & RÅHEIM, A. 1992. Spatial $^{87}Sr/^{86}Sr$ variations in formation water and calcite from the Ekofisk chalk oil field: Implications for reservoir connectivity and fluid composition. *Applied Geochemistry*, **7**, 341–350.

SNYDER, L. R. 1970. Nitrogen and oxygen compound types in petroleum: Total analysis of an 850–1000°F distillate from a California crude oil. *Analytical Chemistry*, **41**(8), 1084–1094.

SPENCER, A. M., HOME, P. C. & WIIK, V. 1986. Habitat of hydrocarbons in the Jurassic Ula Trend, Central Graben, Norway. *In*: SPENCER, A. M. *ET AL*. (eds) *Habitat of Hydrocarbons on the Norwegian Continental Shelf*. Norwegian Petroleum Society, Graham & Trotman, London, 111–128.

STODDART, D. P. 1993. *petroleum and nitrogen compound reservoir geochemistry of the Eldfisk chalk reservoir*. Ph.D thesis, Newcastle University, UK.

TAYLOR, S. R. & LAPRE, J. F. 1987. North Sea chalk diagenesis: its effects on reservoir location and properties. *In*: BROOKS, J. & GLENNIE, K. (eds) *Petroleum Geology of North West Europe*. Graham & Trotman, London, 483–495.

THOMAS, B. M., MØLLER-PEDERSEN, P., WHITAKER, M. F. & SHAW, N. D. 1985. Organic facies and hydrocarbon distributions in the Norwegian North Sea. *In*: THOMAS. A. M. *ET AL*. (eds) *Petroleum Geochemistry in Exploration of the Norwegian Shelf*. Graham & Trotman, London, 3–26.

VAN DEN BARK, E. & THOMAS, O. D. 1980. Ekofisk: First of the giant oil fields in Western Europe. *In*: HALBOUTY, M. T. (ed.) *Giant Oil and Gas Fields of the Decade 1968–1978*. American Association of Petroleum Geologists, Memoir, **30**, 195–224.

YAMAMOTO, M. 1992. Fractionation of azarenes during oil migration. *Organic Geochemistry*, **19**, 389–402.

The reservoir geochemistry and petroleum charging histories of Palaeogene-reservoired fields in the Outer Witch Ground Graben

P. C. MASON[1], R. BURWOOD[2] & B. MYCKE[2]

[1] *Fina Exploration Ltd, Fina House, Ashley Avenue, Epsom, Surrey, UK*

[2] *Fina Exploration and Production, Fina Research SA, Zone Industrielle C, 7181 Seneffe (Feluy), Belgium*

Abstract: Post-Chalk, particularly Tertiary, reservoired oils of the North Sea show a wide range of gross composition and frequent indications of biodegradative alteration. The latter has been associated with shallower reservoir depths and complex emplacement and accumulation histories. Comprising a detailed case study, lateral and vertical variations in petroleum composition within the Eocene-reservoired Chestnut and Alba Fields in the Outer Witch Ground Graben have been evaluated using bulk properties gas chromatography (GC), GC-mass spectrometry and isotopic techniques. The oils have experienced varying degrees of in-reservoir biodegradation. They were generated at an early main phase of source-rock maturity. The migration pathway to the Eocene reservoir is thought to be controlled by selective reactivation of Jurassic faults at times of compressional movement during the Tertiary. Using this model in conjunction with subsidence-history profiles, the timing of generation has been investigated and the temperature of the Eocene reservoir at petroleum emplacement has been estimated. The dominant control on biodegradative alteration is believed to be associateded with the distance from meteoric water influx and the available oxygen budget. According to distance from the hydrostatic head, an apparent decline in biodegradation, albeit showing contradictory levels of alteration, was observed along the trend. Various in-reservoir mechanisms to account for this apparent decline, including topping-up, intermediate transformation products and differential oil-column attack, have been evaluated. The study continues with the objectives of investigating the compositional variations of petroleums reservoired within Lower Cretaceous sandstones and the charging relationships between these and the Eocene reservoirs.

The series of Middle Eocene sandstone reservoired accumulations located along a NW–SE trend across UKCS Blocks 16/26 and 22/2 are collectively termed the Alba Trend (Fig. 1). The reservoir comprises a series of sandstone bodies that are interpreted as submarine gravity-flow deposits, derived from a southeastwards prograding delta complex during Late Palaeocene through Eocene times. At the present day, the depth of burial of these sandstones is relatively shallow (6000 to 7000 ft ss) and many of the petroleums are characterized by low API gravities.

An initial study of the available crude-oil data and new information showed that the low gravities are due to in-reservoir processes, most notably biodegradation.. However, even from this limited database, the emplacement and subsequent in-reservoir alteration of these petroleums was perceived to be complex. Biodegradation had progressed to varying degrees, and a second emplacement phase of light oil and gas was suggested by certain compositional traits.

This is now disputed through subsequent application of more searching geochemical technology. While confirming the generic relationships of these petroleums, these data have also provided the means for demonstrating their respective generative maturity versus the extent of in-reservoir biodegradative alteration. Rather than a monotonously constant situation, a diversity was observed that provided an interesting insight into the sequence of generation, emplacement and alteration when placed in the context of the evolution of the Upper Jurassic source-kitchen area through time. A single charging and in-reservoir alteration episode is now thought most likely in terms of both structural and timing controls on emplacement and composition of the reservoired oils.

Regional setting of the Alba Trend sandstones

From a study of seismic data tied to biostratigraphic information, it can be established that all

From CUBITT, J. M. & ENGLAND, W. A. (eds), 1995, *The Geochemistry of Reservoirs*, Geological Society Special Publication No. 86, pp. 281–301.

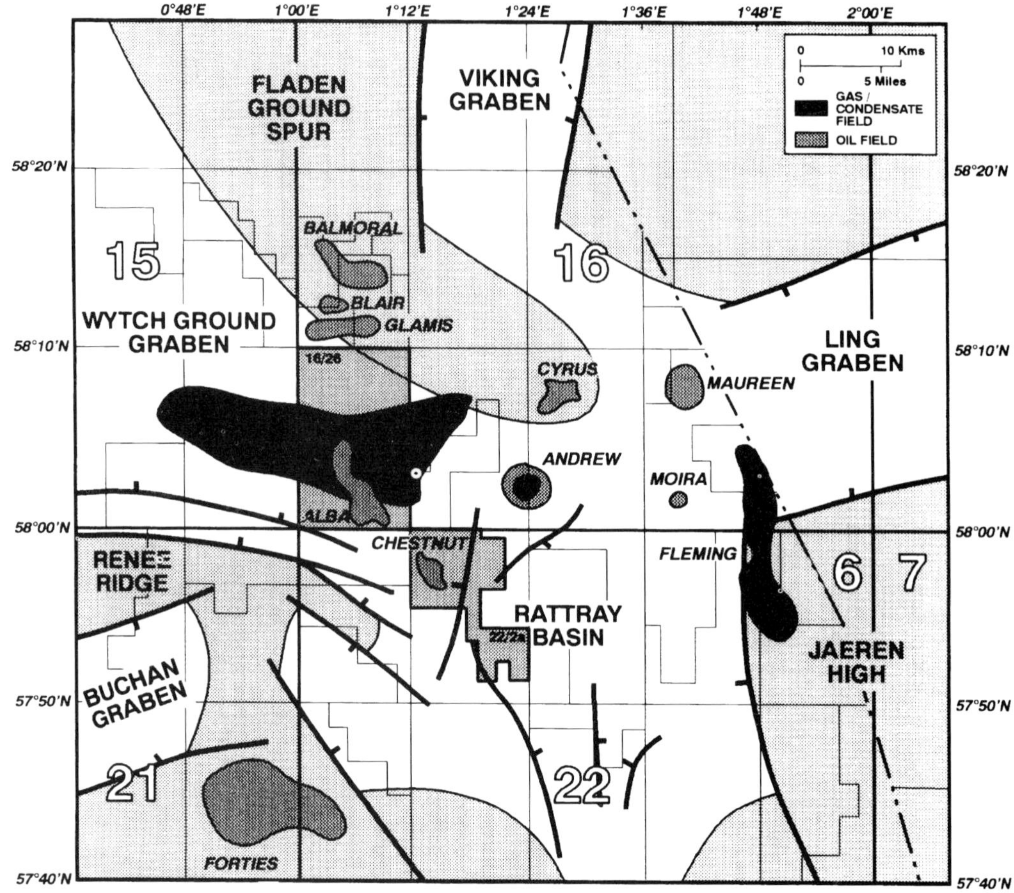

Fig. 1. Alba Trend regional setting

the Alba Trend sandstone bodies occur within the same stratigraphic sequence of the Middle Eocene. In detail, however, the sandstones show a trend of relative stratigraphic separation and younging northwestwards. The depositional model for these sands has been described by Harding *et al.* (1990) and Mattingly & Bretthauer (1992) as an aggradational submarine channel/levee complex, and by Newton & Flanagan (1993) as deposition into a previously eroded channel. The palaeogeographic setting of the Alba sequence is dominated by a significant southeastwards progradation (basinward shift) of the Middle Eocene shelf edge. Sands that were accumulating on the shelf edge were redeposited, due to shelf edge and slope failure, over the slope area and into the basin via submarine gravity flows.

The sandstones, though massive in part, are not homogeneous and change laterally and vertically in terms of mica content, fabric and facies. Shales are uncommon within the sandstone bodies. The separation of individual sandstone bodies is manifest, however, in the differing oil–water contacts encountered in the Alba and Chestnut Fields.

Crude oil provenance and characteristics

North Sea Tertiary-reservoired oils show a wide range in gross composition and properties (API gravity etc.). This is apparently related to complex migration, accumulation and in-reservoir histories (Barnard & Bastow 1991). In common with a majority of North Sea plays, an Upper Jurassic Kimmeridge Clay Formation (KCF) provenance of the Alba petroleums can be assumed (see Barnard & Bastow 1991). This is easily substantiated in terms of the biomarker and stable isotope parameters, as summarized in the composite illustration in Fig. 2. All of the petroleums

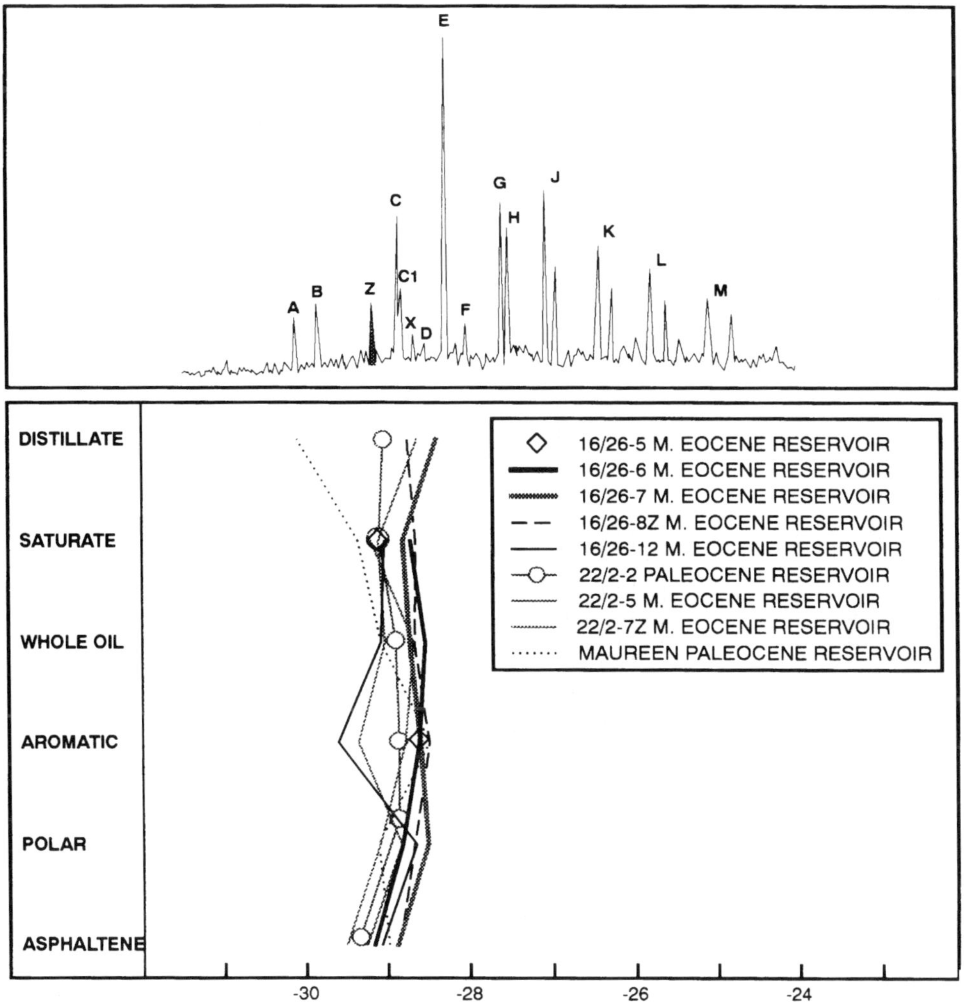

Fig. 2. UKCS: Comparison of hydrocarbon $\delta^{13}C$ isotopic profile for Block 16/26 and 22/2 Alba Trend petroleums with a Maureen benchmark, plus triterpane fragmentogram (*m/z*191) showing component Z (28,30-bisnorhopane), typical of North Sea petroleums, inclusive of the Alba Suite, derived from a Kimmeridge Clay Formation source.

examined provided indications of the diagnostic biomarker 28,30-bisnorhopane (Z) which is widely recognized as a characteristic for KCF-derived petroleums (Grantham *et al.* 1980). The present work additionally demonstrates the persistence, at least to the levels of biodegradation experienced by the study oils, of this marker component.

The co-identity of the oils is further demonstrated by the $\delta^{13}C$ isotope plots (Figs 2, 3 & 4), with the tight clustering of data leaving little doubt as to a common provenance. Comparison of these data with kerogen pyrolysate signature, typical of the KCF high versus low gamma ray response zones (Burwood *et al.* 1990) established a firm correlation with the former and most oil-prolific unit of a Viking Graben-style KCF (Fig. 3). The majority of the oils conformed to any early main phase of generation, as determined by the methylphenanthrene index procedure (Radke *et al.* 1984) and corroborated by cross-plotting with a triaromatic sterane parameter (Fig. 5).

In order to provide a basis for comparison, the Paleocene-reservoired oil from the Maureen Field (Block 16/29) was selected as the local benchmark petroleum. This oil was thought to be representative of main-phase petroleum charge from the Rattray Basin and to show little, if any, biodegradative effects.

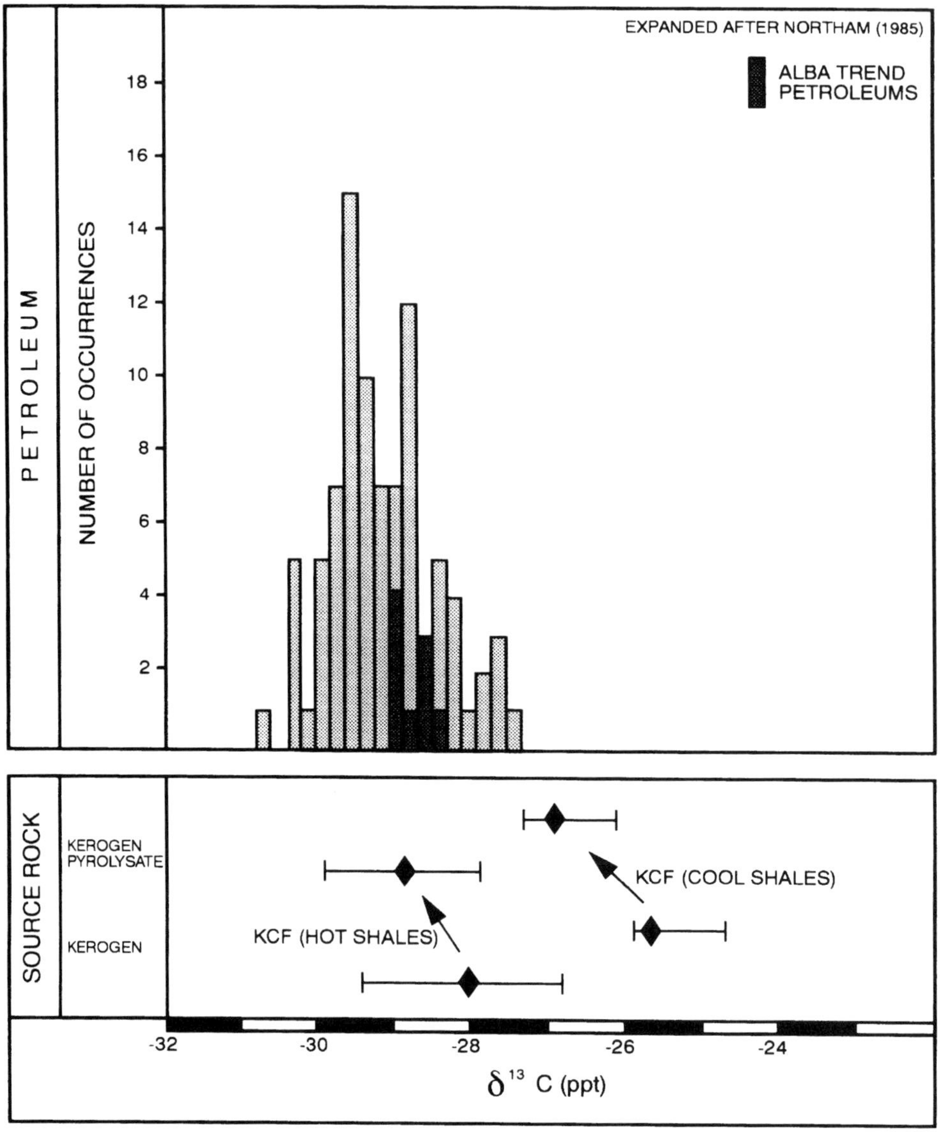

Fig 3. Relationship of Alba Trend petroleums to a selection of typical Kimmeridge Clay Formation derived petroleums.

Collated quality data for the Alba Trend oils are compiled in Table 1. Variations in API gravity, sulphur, wax and asphaltene contents are immediately apparent and are illustrated in Fig. 6. Sulphur contents are higher in the lower API oils. This is characteristic of biodegraded oils, where the enhancement in sulphur is a consequence of a concentration effect attendant on hydrocarbon consumption. Wax contents, as could be expected, show in most cases a direct variation with API gravity. Wax-range *n*-alkanes ($> n\text{-}C_{20}$), being easily biodegradable components, decline in abundance with diminishing gravity. Asphaltene contents also show a regular increase with declining gravity. This is again interpreted as largely a concentration effect mirroring that of the sulphur content.

Accepting that a lowered gravity is a measure of the extent of biodegradative alteration, this process is clearly facilitated by a shallower reservoir depth.

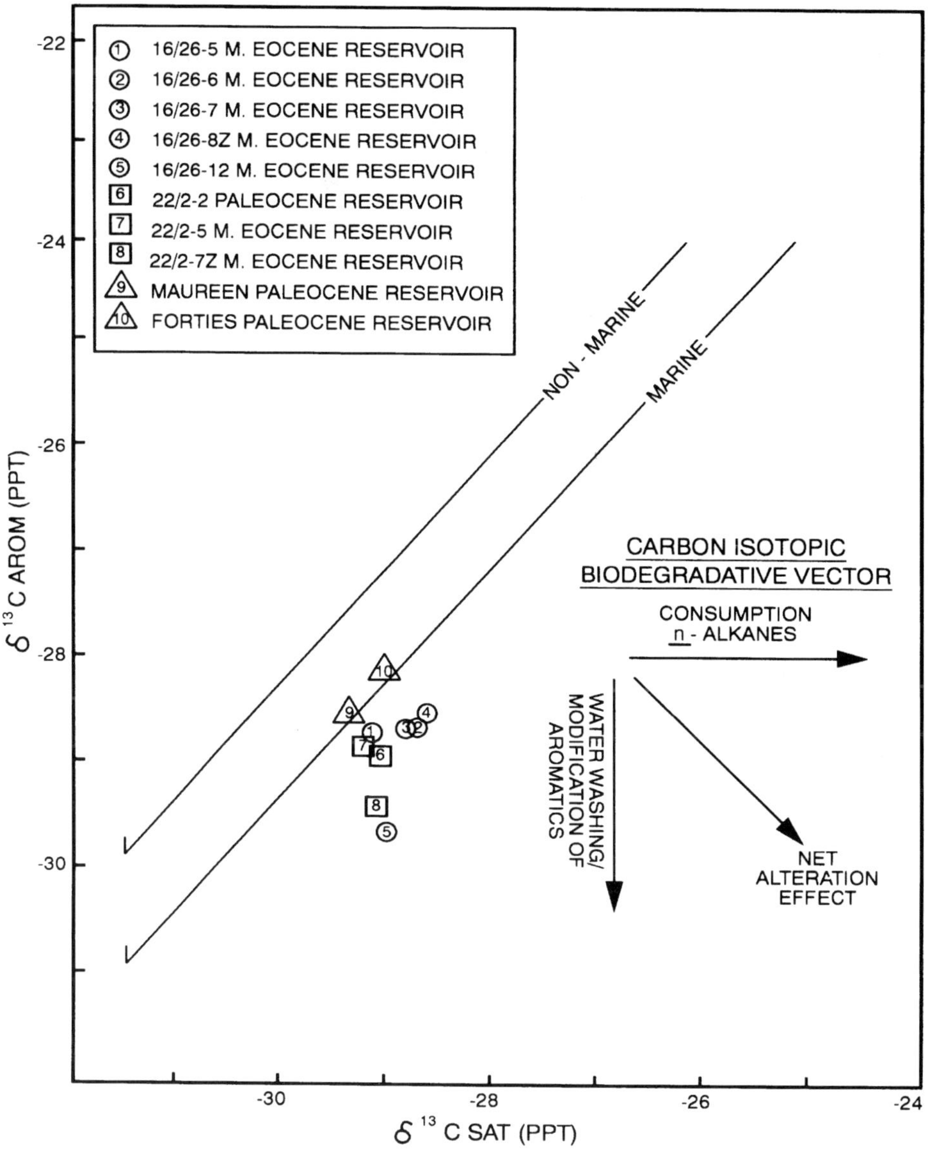

Fig. 4. UKCS: comparison of Alba Trend petroleums with a Maureen benchmark.

In turn, this is seen as a manifestation of the obvious control over the effective invasion of reservoirs by meteoric and oxygenated waters, originating and driven by a surface hydrostatic head. Reservoir temperature would also appear to exert a limiting control. It is relevant to note that present-day reservoir temperatures for the biodegraded oils range from *c.* 60°C, but do not exceed 100°C (Fig. 7). The non-biodegraded control has experienced a higher reservoir temperature (*c.* 120°C), implying that biodegradation is inoperative by that temperature. Owing to a relatively wide time-span (the Neogene) over which generation and migration was possible, and the continuing accrual of overburden, it is conceivable that these reservoirs have experienced an increasing temperature since charging. In this way, biodegradation of the deepest reservoired Alba Trend oil (22/2-2) took place at a lower than present-day temperature. In terms of the inflection in Fig. 7, it is anticipated that

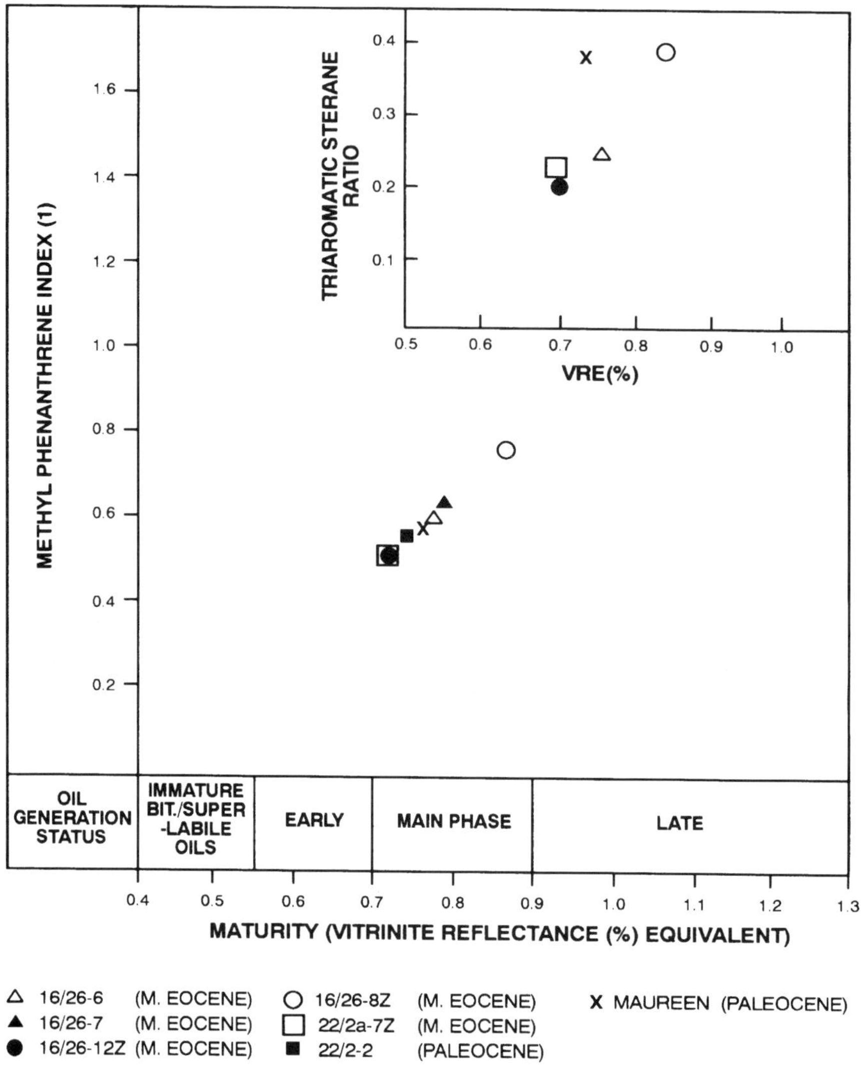

Fig. 5. Comparative maturity for Alba Trend Middle Eocene reservoired heavy oils via MPI(1) and triaromatic sterane parameters, triaromatic ratio being $C_{20}/C_{20} + C_{28}$ (*m/z* 231).

the upper limit in temperature for biodegradation to persist is less than 75°C, in practice being that of an exponential decay away from the optimum physiological value (*c.* 35°C).

Bridging gross compositional diversity, yet generic co-identity, the individual character of the Alba petroleums is best illustrated in terms of the 'whole oil' chromatograms, as displayed in Fig. 8. This demonstrates the causative reason for the observed diversity within the context of in-reservoir biodegradative alteration (cf. Connan 1984).

Starting in the southwest, a dramatic change in composition is visually apparent on progressing northwestwards along the trend. Whereas the deepest reservoired oil (22/2-2) shows an abundant *n*-alkane component, a decrease in content is already evident in the 22/2-5 and -7z fluids. In Block 16/26, further consumption of the *n*-alkanes leaves denuded residuum which, in a qualitative visual sense, shows increasing biodegradative alteration. Many authors (Burns *et al.* 1987; Horstad *et al.* 1992) note the sequential metabolism of the lighter (n-C_{10}) followed by waxy and mid-

Table 1. *Alba trend Middle Eocene-reservoired heavy oils: reservoir information and crude oil assay data*

Well (field)	Present-day reservoir temperature (°C)	Gravity (°API)	Wax content (wt%)	Sulphur content (wt%)	Asphaltene content (wt%)
Maureen Palaeocene reservoir control	117	34.0	14.2	0.31	1.56
16/26-5	61	21.0	ND	ND	ND
16/26-6	69	19.8	2.5	1.21	4.6
16/26-7	69	16.0	0.6	0.62	4.7
16/26-8Z	73	20.0	3.0	0.57	6.0
16/26-12Z	70	21.0	7.0	0.90	2.1
16/26-15	70	20.0	—	1.20	3.3
22/2A/7Z	81	28.0	8.3	0.64	1.6
22/2-5	89	28.0	17.3	1.43	3.7
22/2-2 (Palaeocene res)	97	28.0	3.0	0.64	3.9

range n-alkanes as a prelude to vigorous acyclic alkane (pristane etc.) removal.

Although the whole oil chromatograms are explicit as to n-alkane consumption, further corroborating evidence as to the effective metabolism of this component is provided in Fig. 6, where the variation in cruide oil wax content with depth is displayed. Waxes constitute the back-end n-alkane content of crude oil. Whereas a consumption gradient is evident between the Maureen benchmark, 22/2-7z, and 16/26-12z, thereafter disappearance accelerates dramatically. On this basis, the 16/26-6 and -7z (and 16/26-5 by implication) fluids, with a wax content of 1% or less,

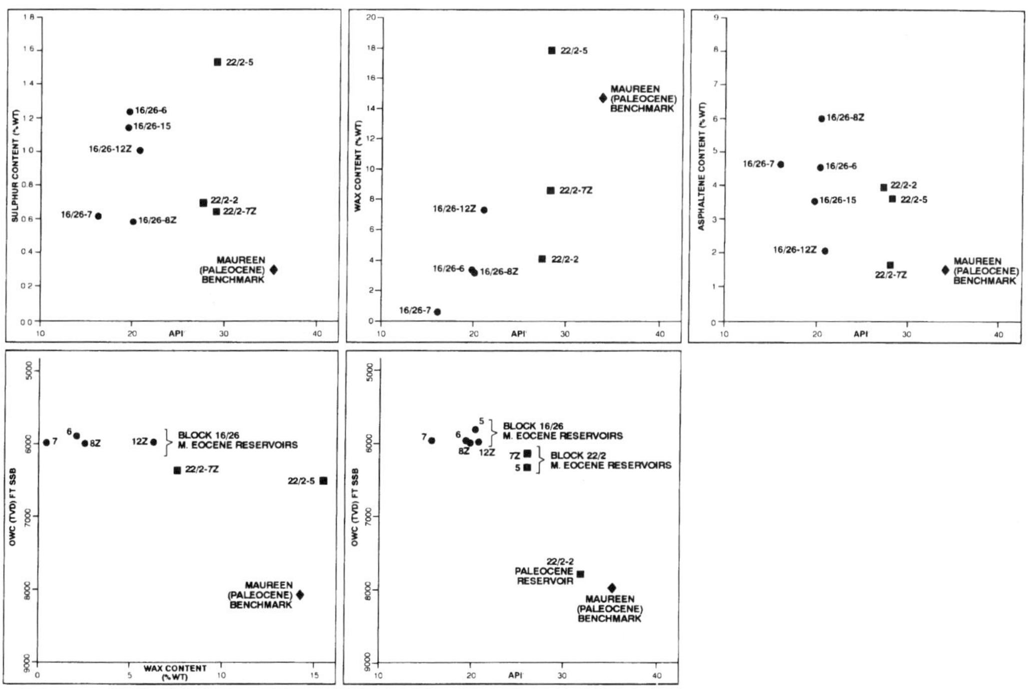

Fig. 6. Alba Trend Middle Eocene/Palaeocene reservoired oils: relationships between crude oil gravity, assay parameters (sulphur, wax, asphaltene) and reservoir depth.

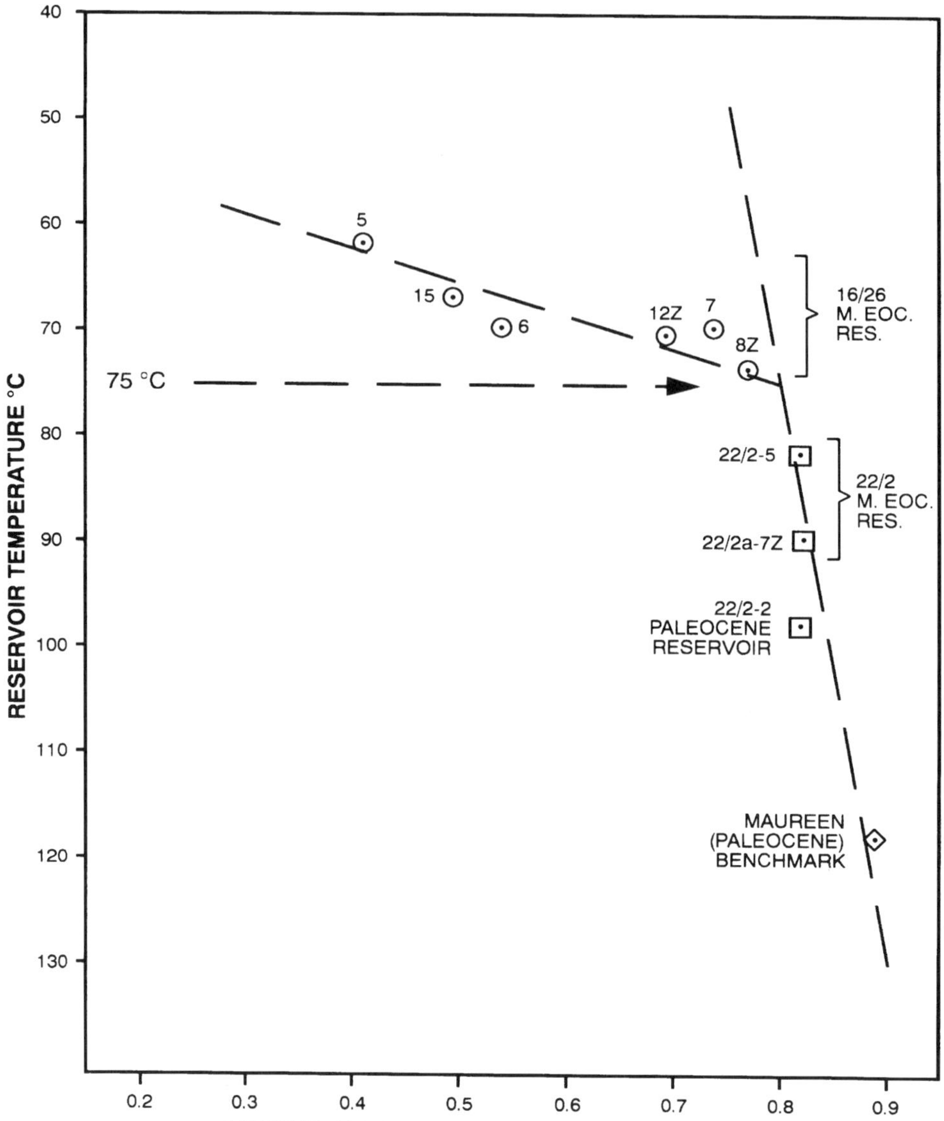

Fig. 7. Alba Trend Middle Eocene/Palaeocene reservoired oils. Variation of acyclic alkane biodegradation ratio (ABR) with reservoir temperature.

can thus be regarded as the most biodegraded petroleums of the study suite. However, where this might fall within the overall biodegradative spectrum requires further definition.

Two complementary schemes purporting to represent the extent of biodegradative alteration have been proposed by Volkman *et al.* (1983) and Philp & Lewis (1987). These are combined in Fig. 9, where an overall scheme allowing definition of the in-reservoir process to be expressed in terms of stages, ranging from onset to extreme, has been devised. Rough quantification of these stages is possible in terms of the acyclic alkane biodegradation ratio (ABR), covering the onset to moderate range, and a 25-norhopane-based desmethylhopane transformation ratio (DTR) which has been applied to the severe to extreme stage. The ABR is defined by the pristane content

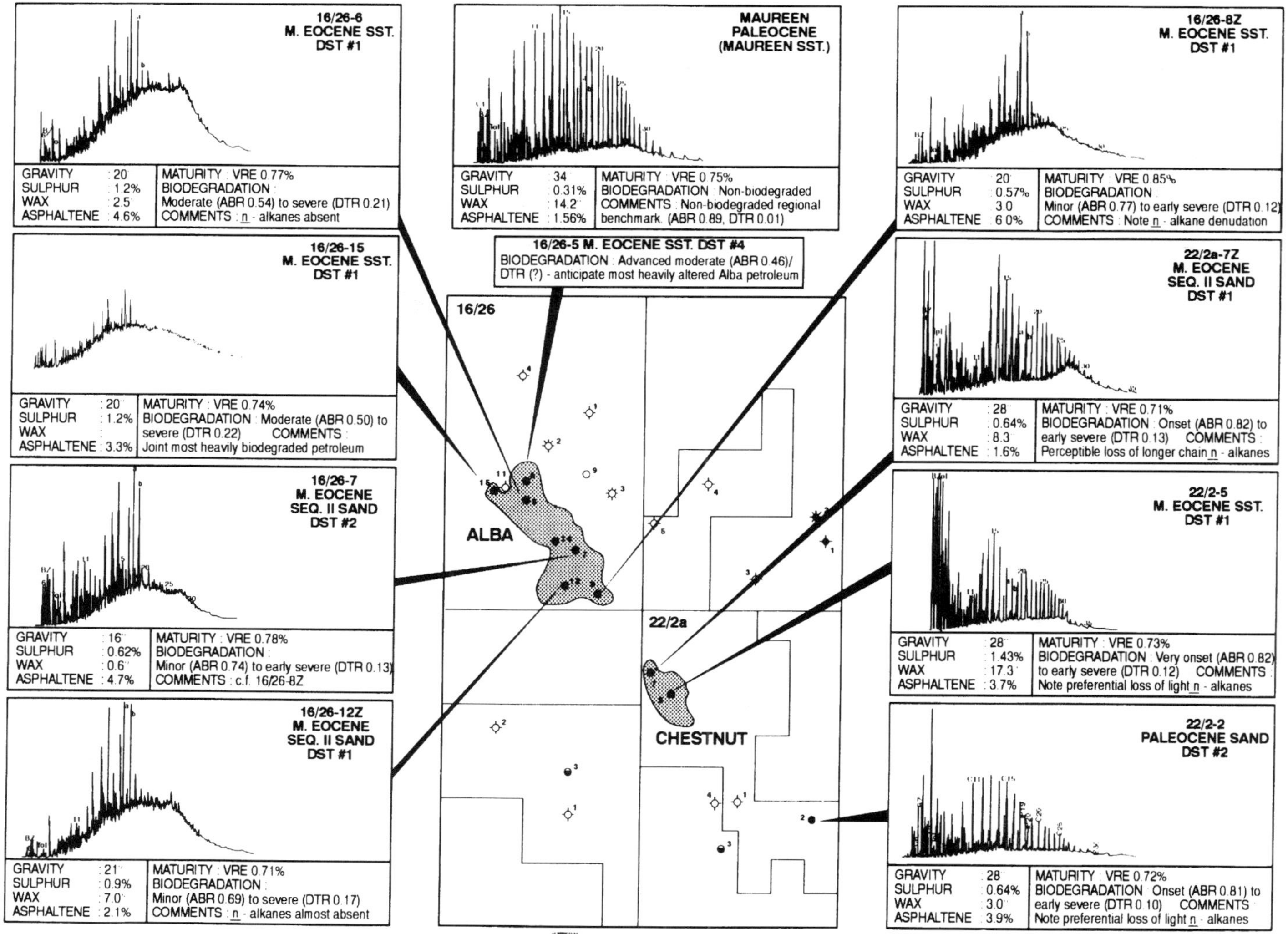

Fig. 8. Crude oil quality and progressive level of in-reservoir biodegradative alteration as illustrated by 'whole oil' chromatograms.

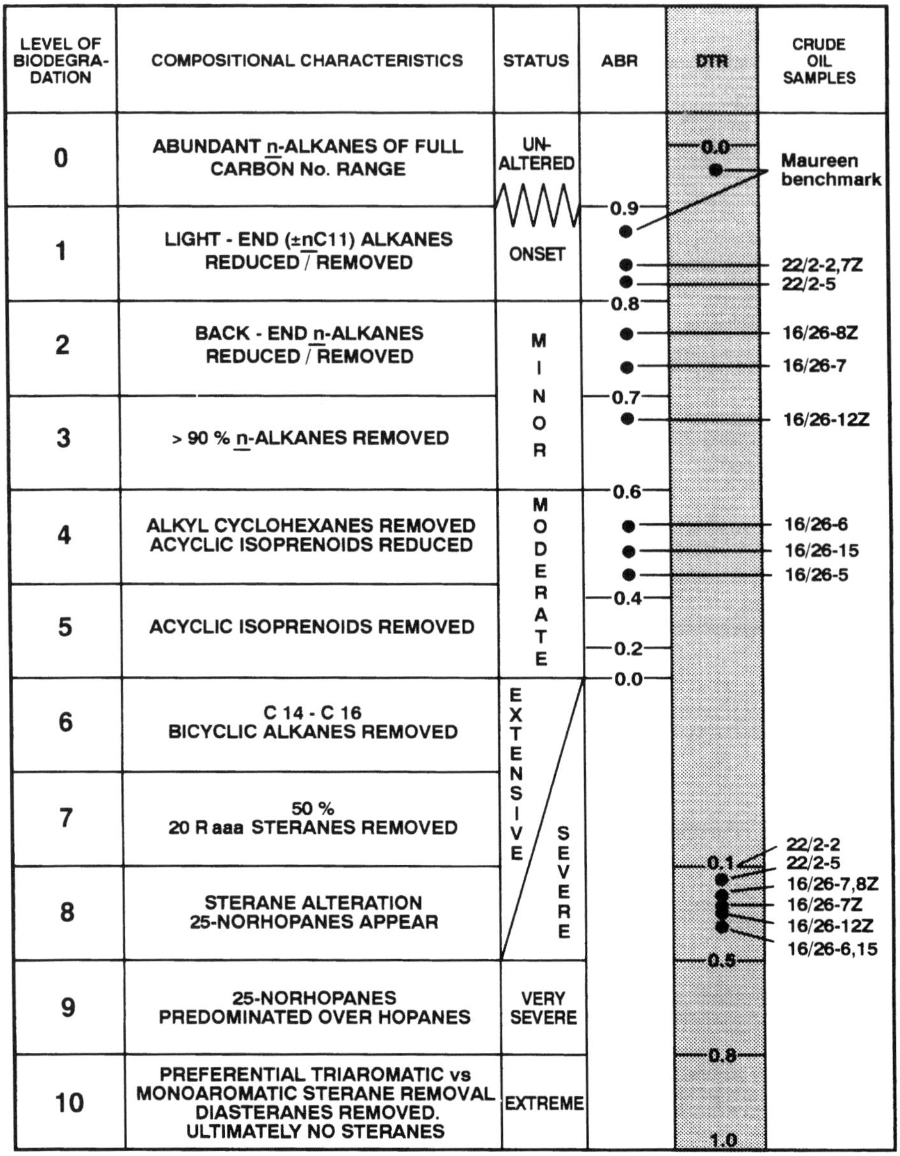

Fig. 9. Comparative stage of biodegradation for Alba Trend Middle Eocene and Chestnut Palaeocene reservoired heavy oils in terms of the acyclic alkane biodegradation (ABR) and desmethylhopane transformation (DTR) ratios.

normalized to pristane plus the subordinate 'hump', whilst the DTR used in this study comprised the C_{28} to C_{34} 25-norhopanes normalized to their precursor C_{29} to C_{35} regular 17α 21β-hopanes.

For the study suite, 25-norhopanes (cf. Blanc & Connan 1992; Peters & Moldowan 1991) were not only conspicuous in the more degraded Block 16/26 oils but were also present in those petroleums showing residual, and even abundant, *n*-alkane contents (Fig. 10). In terms of the two parameters, there is both a systematic decrease in the ABR (0.82 – 0.50), suggestive of onset to moderate alteration, and increase in DTR (0.10 – 0.22), corresponding to extensive to severe biodegradation, over a northwesterly traverse of the trend (cf. Fig. 8). These data immediately present a conundrum in terms of the conventional interpretation of biodegraded oil data, where 25-norhopanes

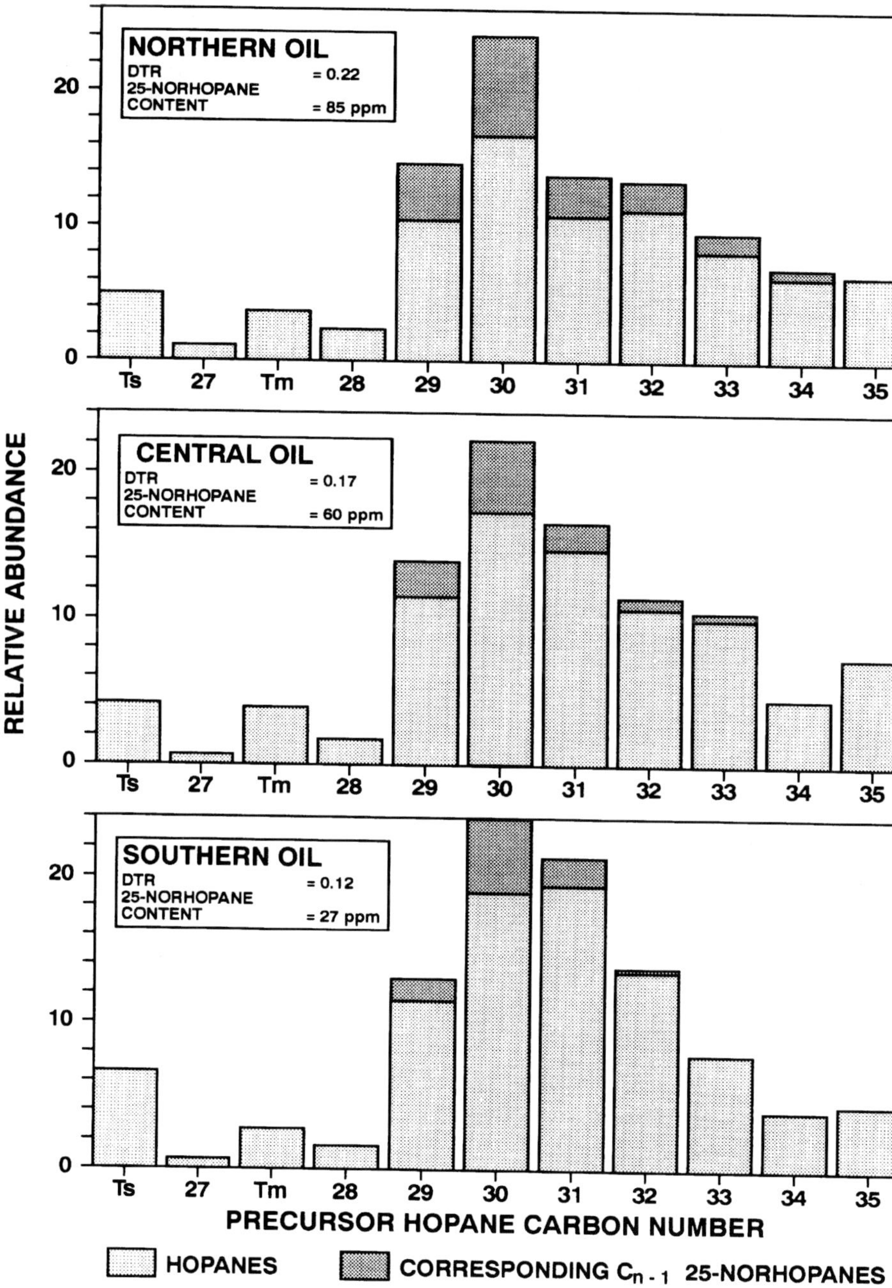

Fig. 10. Regular hopane and 25-norhopane composition of selected Alba Trend oils. Hopane and 25-norhopane contents normalized to a precursor whole oil basis using non-biodegraded diasteranes as an internal concentration maker.

apparently do not appear until after removal of the acyclic isoprenoids and >50% consumption of 20R 5α, 14α, 17α-steranes. This situation is summarized in Fig. 9, where it is evident from the incompatibility of the two parameters that the biodegradative alteration is not a simple single-phase, albeit progressive, in-reservoir procedure.

For rationalization of these observations, it is necessary to think in terms of a more complex mechanism, possibly including multiple phases of alteration. At this point, we revert to the geology and basin history to obtain an understanding of the probable timing constraints on petroleum generation, migration and alteration of the reservoir fluids n these fields.

Petroleum migration and entrapment history

Maturation History of the Upper Jurassic source rock

The Alba Trend fields lie over the contiguous basinal area of the Rattray Basin and the Outer Witch Ground Graben (Fig. 11). At the present day, the source rocks of the Upper Jurassic Kimmeridge Clay Formation (KCF) are mature across much of the study area and have reached the gas-generation stage within the deeper parts of the Rattray Basin and Witch Ground Graben.

A thermal modelling study was conducted in order to relate the maturities of the analysed oils to the timing of generation and expulsion from the source rock. Figure 11 presents a summary of the results from this study and illustrates the modelled maturity for the Upper Jurassic source rocks at end-Palaeocene, end-Oligocene and Plio-Pleistocene times. The hatching represents the interpeted area of basal Upper Jurassic sandstones and is shown to indicate the extent of the Late Jurassic basins.

At the end of the Palaeocene, immediately prior to the deposition of the reservoir sandstones, the source rock was locally mature in the basin depocentres. By end-Oligocene times, an extensive mature source-rock kitchen extended across the area, with the onset of gas generation in the deepest area. In the Plio-Pleistocene, this gas-generative area was extensively developed, as evidenced by the Britannia Field, which is a gas-condensate discovery in Lower Cretaceous sandstones underlying (in part) the Alba Trend.

However, the high levels of maturity seen in these condensates are not seen in the Alba Trend oils. Indeed the observed maturity of these oils, *c.*0.75% vitrinite reflectance equivalent, was obtained in the source rock within the likely drainage area of the Alba Trend, during mid to late Oligocene times.

Timing of petroleum entrapment

There are also a number of indicators in the reservoir geology as to the timing of petroleum emplacement and biodegradation. As noted earlier, the reservoir largely comprises unconsolidated sands, though within the water leg these sands appear to have consistently lower porosities, indicative of increased diagenesis. Indeed, a flat event can be identified on the seismic data across the Alba Field, approximately coinciding with the field oil–water contact (OWC). The seismic data also show that this diagenetic OWC event is displaced by a series of minor faults that terminate at the Early Miocene Unconformity. These observations are interpreted as evidence that the Alba Field was charged prior to the displacement of the 'diagenetic' OWC by the high-level faults at or by Early Miocene times.

Furthermore, the biodegration process requires that fresh, oxygenated (i.e. meteoric) water are available and formation temperatures are less than 65°C. The Early Miocene Unconformity provides the opportunity for the introduction of meteoric water into the reservoir, as an extensive area of the Eocene/Oligocene Shelf to the northwest of the fields would have been subaerially exposed. However, the rate of aquifer recharge may have been slow, as the Alba sandstones do not subcrop this unconformity and there is no obvious conduit to the surface with the Alba Field being stratigraphically trapped.

As part of the thermal modelling study, the reservoir temperatures in the Late Oligocene were also determined (Fig. 12). These are typically in the range 25–35°C, which is close to the optimal temperature for biodegradation in the presence of meteoric water. However, the geochemical data show that the levels of biodegradation are not extreme, particularly in the southern oils, and this is reconciled with the assertion that the rate of aquifer recharge was probably slow as there is no obvious conduit. Thus, rather than being a flood, the meteoric water was introduced more as a 'dribble' and presumably to the northern end of the system, which is closer to the exposed Eocene/Oligocene Shelf.

Migration conduit

As a final observation, there is a clear relationship between the occurrence of oil in Eocene reservoirs and inversion anticlines in the pre-Eocene section. There have been a number of apparently valid Eocene structures that have proved dry and the lack of a petroleum migration route has been inferred as the likely cause of failure. Notably, the proven accumulations (Alba and Chestnut Fields) overlie

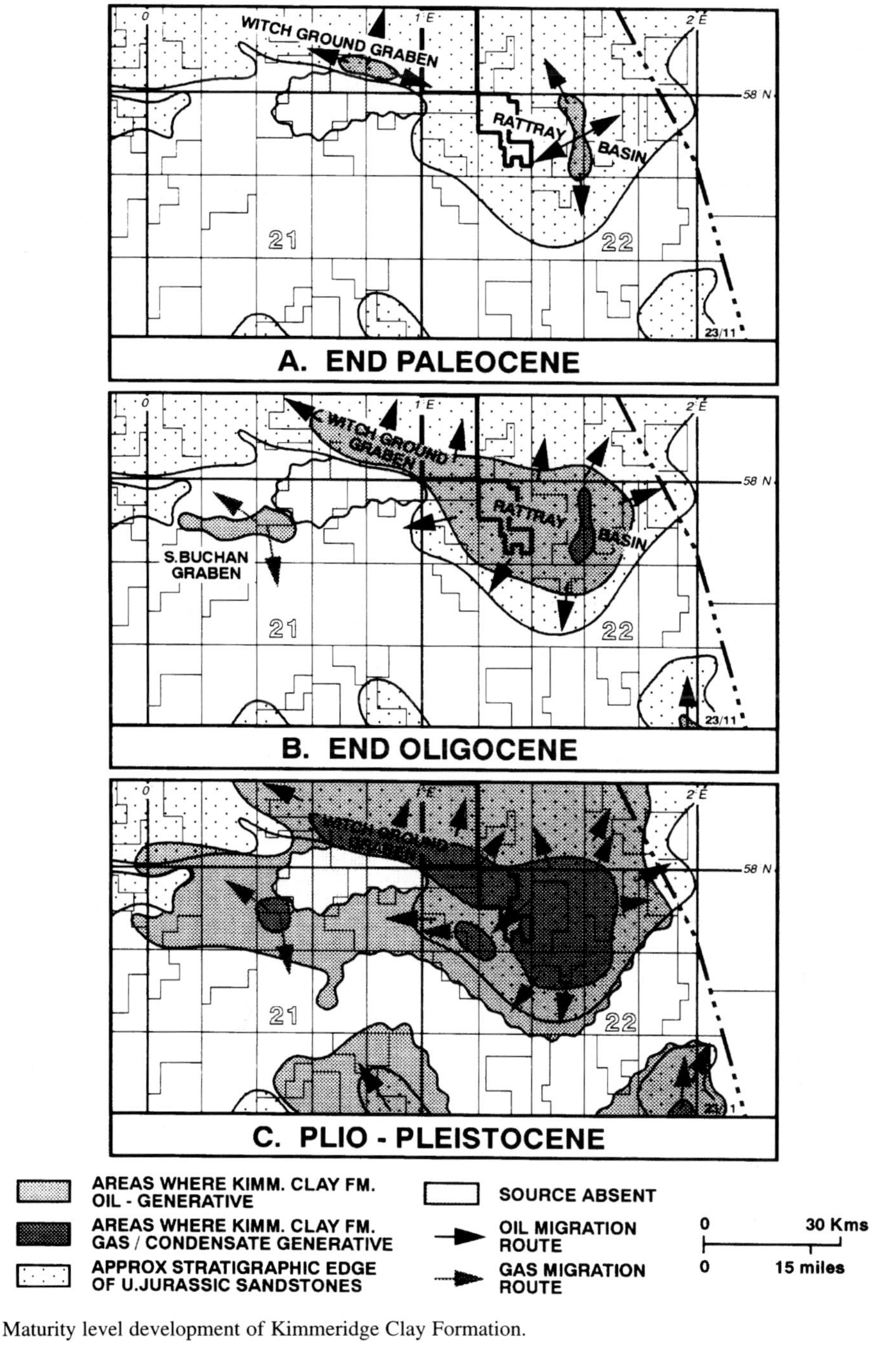

Fig. 11. Maturity level development of Kimmeridge Clay Formation.

inversion anticlines, as mapped at Top Chalk level, and generated by reactivation of deeper Cimmerian faults. These NW–SE trending faults are believed to have provided a migration pathway into the sand-rich Palaeocene section. The faults were selectively reactivated during periods of ragional strike–slip/compressional movement in Early Cretaceous (Mid-Albian), Late Cretaceous (Mid- to Late Campanian) and Tertiary (Ypresian and Early Miocene), providing open conduits to the

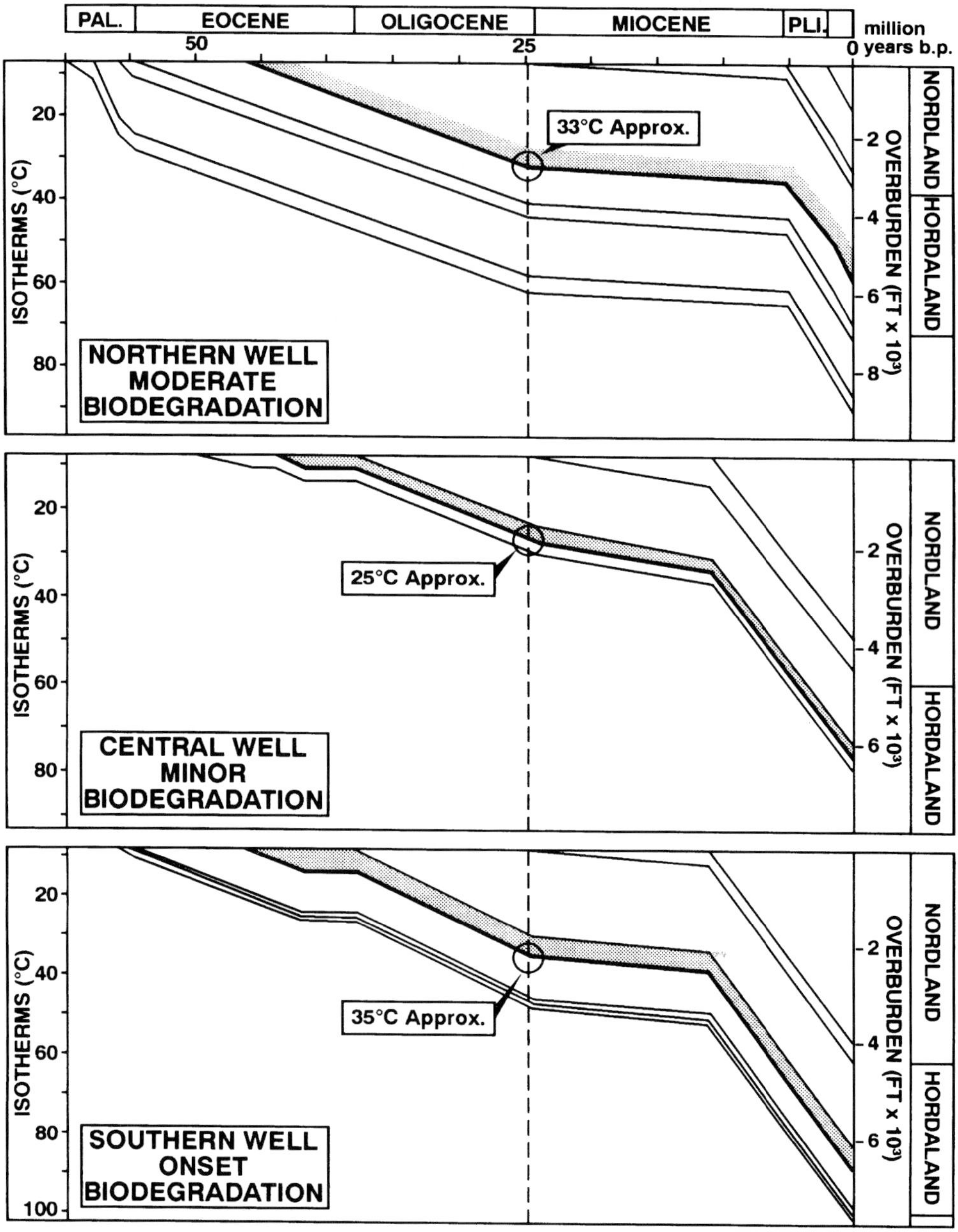

Fig. 12. Alba Trend reservoir temperatures at time of emplacement (base oil column).

Palaeocene section. The erosive downcutting of successive Palaeocene submarine fan systems results in the petroleum invariably migrating into the stratigraphically highest sandstone. Once within this carrier/reservoir horizon, lateral migration focuses the petroleum towards updip traps.

Migration from Palaeocene into Eocene sandstones is thought to occur via a suite of high-level faults associated with the Early Miocene strike–slip movements. These faults are clearly evident on the seismic data, cutting the Eocene/Oligocene interval, and appear to sole out at the top of the Palaeocene section. When mapped out, the faults

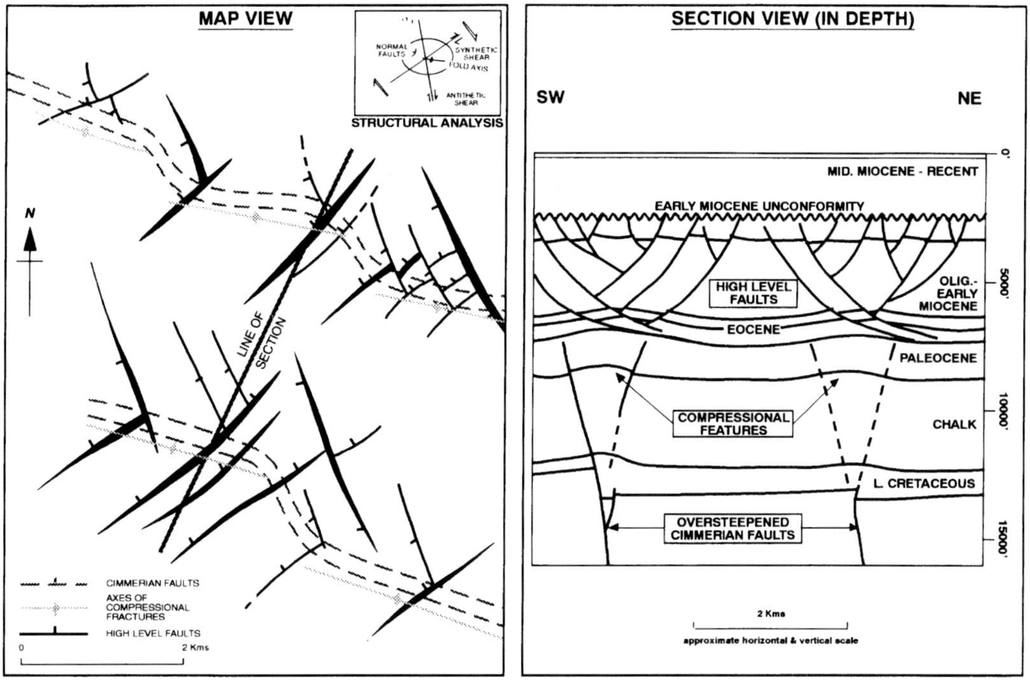

Fig. 13. Effects of Early Miocene strike–slip movements on Palaeogene section in the Alba Trend.

are shown to be broadly centred above the deeper NW–SE trending compressional features (Fig. 13). Within the study area, both the main Eocene-reservoired discoveries, the Alba Field (Block 16/26) and the Chestnut Field (Block 22/2), are located in association with such faults.

This migration model, together with the low net sandstone content of the Eocene section, predicts that only those Eocene sandstone bodies which intersect the NW–SE oriented compressional trends will have been charged during the Early Miocene strike–slip movements.

It is noteworthy that the fields of the Alba Trend overlie the laterally extensive Britannia Field. This is a gas-condensate discovery reservoired in Early Cretaceous Kopervik Formation sandstone. The implication of an Eocene migration model, dependent on sporadic fault movement, is that the Kopervik sandstone reservoir may act as an intermediate trap. Thus initial secondary migration, after explusion from the Kimmeridge Clay Formation source rocks, will have charged the Kopervik sandstones of the Britannia structure. During subsequent periods of Mesozoic fault reactivation, this structure may have leaked petroleum into the shallower section, though there is no geochemical evidence for a component of high maturity condensate in the Alba oils.

In summary, the geological and geophysical data, plus the geochemical evidence and thermal maturity modelling, all consistently point to a single period of petroleum charge to the Alba Trend Sandstones in the Late Oligocene to the Early Miocene, concomitant with biodegradation.

Discussion

In the light of the geological timing constraints on petroleum emplacement and alteration described in the previous section, three alternative models can be proposed to reconcile the geochemical conundrum of the occurrence of 25-norhopanes at relatively low levels of biodegradation.

The simplest explanation would involve the topping-up concept (Volkman *et al.* 1983; Cassani & Eglinton 1991). Here it could be invoked that an initial reservoir charge was subject to vigorous alteration, commensurate with the range of observed DTR values, being greatest to the north-west. Here, the level of alteration would not only be controlled by the influx of oxygenated waters, but also by optimum temperatures associated with shallow reservoir depths.

Subsequently, further fresh, unaltered charge entered the reservoirs under conditions where biodegradative processes were still operative. By virtue of both diminishing oxygen content of connate waters and increasing reservoir temperature, the extent of alteration for the replenished charge was limited to a moderate level in the shallower (northern), as opposed to onset conditions for the more deeply buried (southern) pods. Convective in-reservoir homogenization of the two phases of biodegradation would then give an aggregate oil with the illusion of a contradictory alteration history.

While an attractive proposition, the above interpretation would require rather special emplacement conditions. These would entail not least the counter-charging of the reservoir environment by petroleum from depth, but additionally connate waters from a surface hydrostatic head. For co-occurrence of these requirements during both phases, and particularly the second phase, of biodegradative alteration, a structurally and hydrodynamically difficult set of circumstances would be necessary. Where proposed (e.g. for Gullfaks Field, in Horstad *et al.* 1992), this mechanism was found wanting, being disputed by several considerations, not least the matter of reservoir content re-homogenization. This has led to a search for other mechanisms whereby the evident contradictory alteration history of the Alba oils could be explained. Of significance here was the apparent relationship between the two biodegradative parameters as illustrated in Fig. 14.

This is a strong argument against the proposition of independent phases of biodegradation over the Alba Trend, because the extent of 'hump' production (with coincident *n*-alkane removal) is obviously in a close relationship to the appearance of 25-norhopanes. Each of the two biodegradation processes can be described as a function of the other, indicating a direct genetic link. The data points lying off this trend represent oils (e.g. 22/2-5 and 16/26-7, -12Z) reservoired in isolated sand bodies in the Alba Trend and may behave slightly differently to the main group for that reason. Nevertheless, they still indicate the same general evolution with a higher ABR consistent with higher 25-norhopane concentrations within a deviation of 10%, which in a geochemical sense is an acceptable tolerance.

Confirmation that the ABR is a linear, evolving to an exponential, function of the 25-norhopane concentration, and vice versa, is provided by other North Sea but unrelated biodegraded oils (e.g. see Fig. 14, data from UK and Norwegian Quadrants

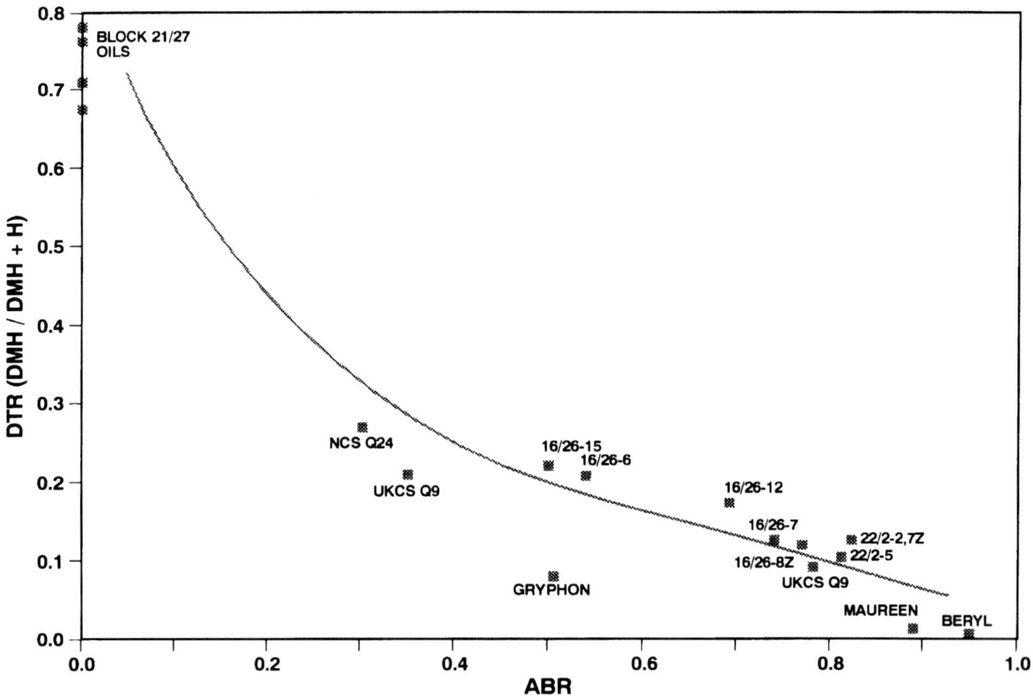

Fig. 14. Biodegradation of North Sea Tertiary-reservoired crude oils : cross comparison of acyclic biodegration (ABR) and desmethylhopane transformation (DTR) ratios.

9 and 24, respectively). These observations are supportive of a single-phase alteration process with the concerted consumption of *n*-alkanes and coproduction of 25-norhopanes.

In support of this contention, a cross-plot of the ABR and DTR parameters against a well location inter-reservoir vector (Fig. 15) again showed an inter-relationship linking the least and most altered

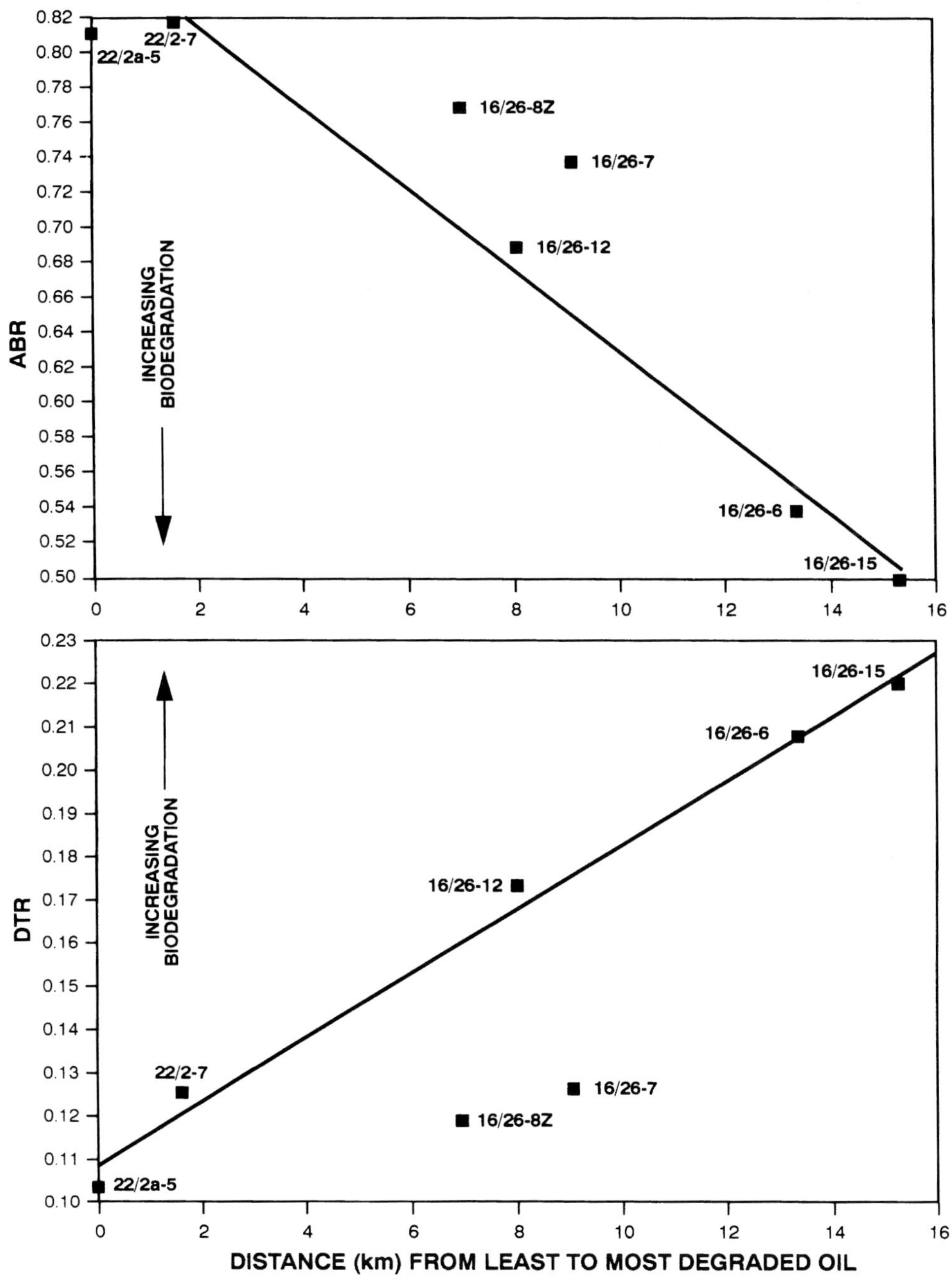

Fig. 15. Biodegradation status of Alba Trend Middle Eocene oils versus inter-reservoir distance vector.

oils, which cannot be regarded as accidental. The contemporary and present-day manifestation of this vector is a progressive increase in reservoir depth and temperature to the southeast.

Interpretation of these data suggests that biodegradation could proceed via parallel attack on (a) the *n*-alkanes and acyclic isoprenoids, and (b) the cyclic molecules such as hopanes. The hopanes, however, are initially oxidized at the C-25 position and thereby transformed into functionalized carboxylates that are not found in the hydrocarbon fraction studied for saturated biomarkers. Biological oxidation has also been proposed as a potential formation mechanism for the C-homo-26-norhopanes, recently identified in biodegraded asphalts (Trendel *et al.* 1993).

Alternatively, the methyl group at the C-10 position appears to be a preferential site for enzymatic attack on the hopane skeleton (Trendel *et al.* 1990). Only after the secondary decarboxylation at C-10 on the hopane skeleton are 25-norhopanes produced and their detection via the saturated fraction permitted. 28,30-Bisnorhopane-25-oic acid, an important intermediate in such a speculative hopane oxidation–decarboxylation process, was recently identified by De Lemos Scofield (1990) in the geological environment. Further, it is of note that the extent of biodegradation has been associated with significant increase in the acid number of substrate oils (Wolcott *et al.* 1989), with Mackenzie *et al.* (1983) reporting the presence of triterpane acids in such residuum. This defunctionalization process would undoubtedly be accelerated by the elevated reservoir temperatures accompanying continued subsidence of the structures since charging. In this way, the appearance of detectable 25-norhopanes could be temperature-mediated, with this explanation not necessitating the topping-up of a previously biodegraded residuum with fresh oil and then the resumption of biodegradation.

A third explanation of the observed dichotomy in biodegradation status derives from the fact that oil compositional studies, including the present, are most frequently conducted on fluids produced over a wide stratigraphic interval (i.e. via production or drill-stem tests). In this way, the oil analysed is an aggregate representative of a continuous and/or discontinuous oil column. For the present suite of oils, most are of a DST origin and variously cover inervals from about 30 ft to 300 ft (e.g. 16/26-7 at 304 ft, and 22/2-2 at 16 ft). Recent studies on both conventionally (e.g. Larter *et al.* 1990; Hillebrand & Leythaeuser 1992) and side-wall cored (Burwood *et al.* unpublished data) oil-productive North Sea reservoir intervals have revealed considerable oil-column heterogeneities. In the latter case, such a closely spaced study over the extent of a Palaeogene oil column in a Norwegian Quadrant 24 well has demonstrated a significant variation in oil saturation and the status of in-reservoir alteration (Fig. 16). From those data, it could be concluded that the in-reservoir process was highly dependent upon *in situ* conditions, most probably the level of interfacial oil–water contact. In the case of the Alba Trend, the filling of the reservoir is interpreted to be concomitant with biodegradation and, in such a dynamic system, the displacement of aquifer water in the reservoir by oil would lend itself to the varying degrees of alteration observed.

Thus, it could be conceived that the oil column was subject to a differential biodegradation, those horizons adjacent to the aquifer experiencing attack, whilst the body of the accumulation was protected. With convective circulation and the passage of time, a greater proportion of the reservoir content would experience alteration. In this way, a whole spectrum of altered compositions, from severe, moderate to onset and unaltered, could be anticipated according to the duration of the overall biodegradation window. Production of such an oil would result in an aggregate fluid reflecting all of these stages of alteration.

Application of this concept to the Alba/Chestnut oils suggests that the more altered, northwest culmination oils have experienced the greater biodegradation exposure, with a minimum level of moderate alteration extending to severe over the more conducive oil–water interfacial horizons of the oil column. To the southwest, the relative exposure (time and/or appropriate physiological conditions embracing oxygen plus nutrient availability and temperature) to alteration is much reduced, with the attendant level of severe attack similarly attenuated. The produced aggregate oil would reflect these constraints.

Support for this concept is provided by the DTR, which not only progressively and numerically declines to the southeast (Fig. 8) but also reflects a hopane carbon number preference on the order of attack (Fig. 10). Similarly, the absolute 25-norhopane abundances (ppm) show a trend-orientated decrease from 85 to 20 ppm when contrasted against a background level of $\leq$ 2 ppm in the Maureen benchmark oil (Fig. 10). On this basis, we would speculate that the complete northern oil column has attained at least a moderate level of alteration, and more severe attack is well underway. To the south, some of the oil column is either not biodegraded or is at onset levels only, and approximately 30 to 50% of the oil (i.e. by comparing the north–south DTR ratio and absolute 25-norhopane contents) has experienced at least moderate to possibly more severe alteration. In this way, data from the oil column aggregate petroleums analysed provide an explanation for the illusion of contra-

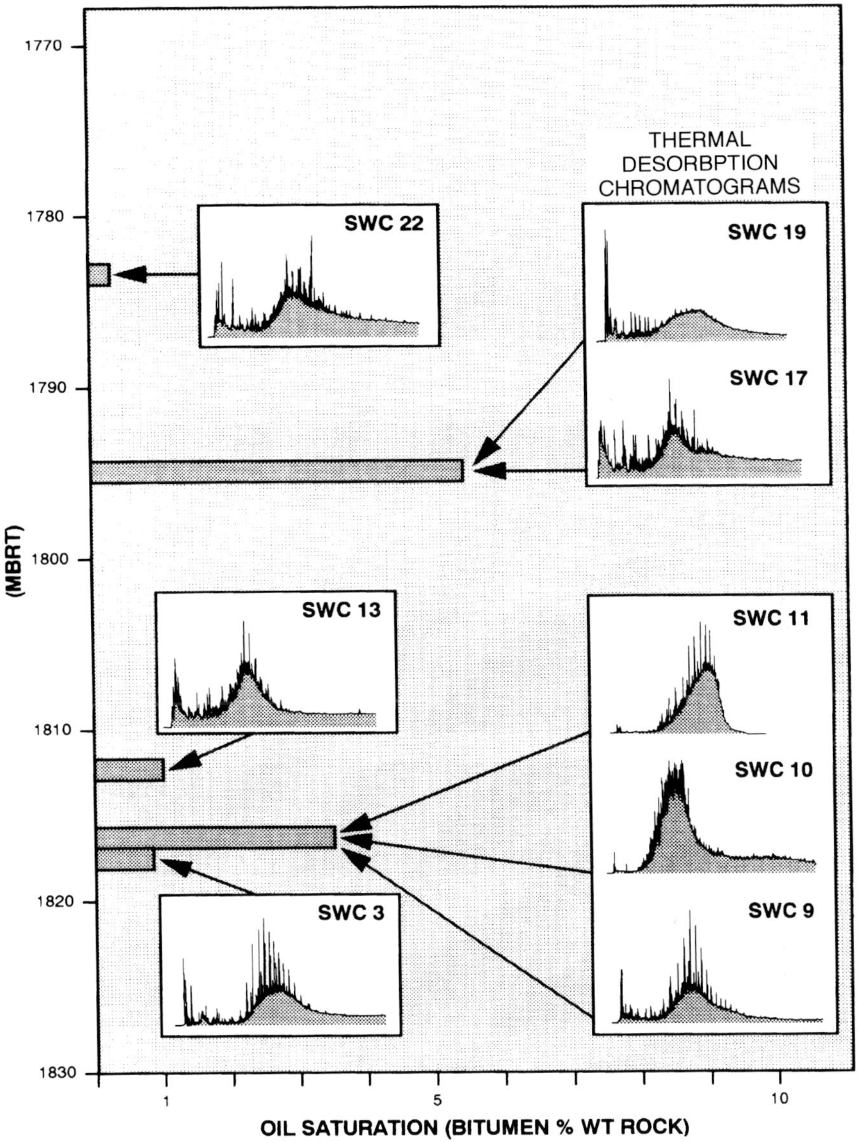

Fig. 16. Norwegian Quadrant 24 Paleogene heavy oils : compositional variation of SWC extract by thermal desorption chromatography. Note contrast in residual *n*-alkane content of SWC 9 versus 19.

dictory levels of alteration. Similarly, the ABR–DTR relationship and the reservoir depth and distance vectors (cf. Figs 14 & 15) are satisfied.

Thus, the simplest explanation for the character of the Alba/Chestnut oils involves single charging and in-reservoir alteration processes, the latter being mediated by distance from the point of meteoric water ingress. In this way, only part of the oil column for the more distal locations was subject to active biodegradation, with these processes being prematurely arrested by further subsidence and curtailment of microbial activity with increasing temperature. Conversely, at the northwestern extent of the Alba Trend, an overall shallower reservoir depth, less increase in temperature and more abundant oxygen supply resulted in an extended biodegradative window and a cumulatively more altered residuum.

Conclusions

The low API gravity, Palaeogene-reservoired oils of the Alba Trend are an early main phase generative product of the Kimmeridge Clay Formation and show variable degrees of in-reservoir biodegradative alteration.

Thermal maturity modelling showed that the most likely timing of oil emplacement to the Palaeogene reservoirs occurred in the Late Oligocene to Early Miocene. This supports the geochemical evidence that the oils represent a single period of hydrocarbon fill. The Early Miocene also represents a period of probable aquifer recharge, as large areas of the Palaeogene Shelf were subaerially exposed at this time. The resulting introduction of meteoric waters and the low reservoir temperatures at this time provided the opportunity for the biodegradation of the oils to proceed.

The oils are more altered towards the northwest along the trend, where reservoirs are shallower and cooler. As an overprint, all the oils showed apparently incompatible indications of early versus more severe alteration, as witnessed by the co-occurrence of abundant to residual *n*-alkanes in concert with significant C-10 25-norhopane contents. A genetic relationship between the removal of *n*-alkanes in concert with significant C-10 25-norhopane contents. A genetic relationship between the removal of *n*-alkanes and the appearance of the norhopanes was deduced, however, to exist.

From a number of alternative explanations considered to account for the illusion of contradictory levels of alteration, the model of differential biodegradation during filling of the reservoir was favoured. By virtue of the aggregate nature of the drill-stem test samples analysed, these oils would represent the spectrum of alteration levels achieved over an extended oil column. The overall control on biodegradative alteration was deduced to be oxygen availability, as mediated by distance from the point of meteoric water input to the reservoir, and in-reservoir oil–water interfacial contact.

We thank the management of Fina Exploration and Production Ltd and the partners of the licence groups involved in the Alba Field and the Chestnut discovery for permission to publish this paper. Formative input and helpful discussions with M. J. Cope are gratefully acknowledged.

References

BARNARD, P. C. & BASTOW, M. A. 1991. Hydrocarbon generation, migration, alteration, entrapment and mixing in the Central and Northern North Sea. *In*: ENGLAND, W. A. & FLEET, A. J. (eds) *Petroleum Migration*. Geological Society, London, Special Publication, **59**, 167–190.

BLANC, Ph. & CONNAN, J. 1992. Origin and occurrence of 25-norhopanes: a statistical study. *Organic Geochemistry*, **18**, 813–828.

BURNS, B. J., BOSTWICK, T. R. & EMMETT, J. K. 1987. Gippsland terrestrial oils – Recognition of compositional variations due to maturity and biodegradation effects. *APEA Journal*, **27**, 73–84.

BURWOOD, R., JACOBS, L. & PAULET, J. 1990. Kerogen pyrolysis–carbon isotope technology: Application to source–oil correlation problems. *Review of Paleobotany and Palynology*, **65**, 367–377.

CASSANI, F. & EGLINTON, G. 1991. Organic geochemistry of Venezuelan extra-heavy crude oils, 2. Molecular assessment of biodegradation. *Chemical Geology*, **91**, 315–333.

CONNAN, J. 1984. Biodegradation of crude oils in reservoirs. *In*: BROOKS, J. & WELTE, D. (eds) *Advances in Petroleum Geochemistry*. Academic Press, London, Volume 1, 299–335.

DE LEMOS SCOFIELD, A. 1990. *Nouveaux marquers biologiques de sédiments et pétroles riches en soufre: identification et mode de formation*. Thèse de l'Université Louis Pasteur, Strasbourg.

GRANTHAM, P. J., POSTHUMA, J. & DE GROOT, K. 1980. Variation and significane of the C_{27} and C_{28} triterpane content of a North Sea core and various North Sea crudes. *In*: DOUGLAS, A.G. & MAXWELL, J.R. (eds) *Advances in Organic Geochemistry 1979*. Pergamon Press, Oxford, 29–38.

HARDING, A. W., HUMPHREY, T. J., LATHAM, A., LUNSFORD, M. K. & STRIDER, M. H. 1990. Controls on Eocene submarine fan deposition in the Witch Ground Graben. *In*: HARDMAN, R. F. P. & BROOKS, J. (eds) *Tectonic Events Responsible for Britain's Oil and Gas Reserves*. Geological Society, London, Special Publication, **55**, 353–367.

HILLEBRAND, T. & LEYTHAEUSER, D. 1992. Reservoir geochemistry of Stockstadt oilfield: compositional heterogeneities reflecting accumulation history and multiple source input. *Organic Geogemistry*, **19**, 119–131.

HORSTAD, I., LARTER, S. R. & MILLS, N. 1992. A quantitative model of biological petroleum degradation within the Brent Group reservoir in the Gullfaks Field, Norwegian North Sea. *Organic Geocehmistry*, **19**, 107–117.

LARTER, S. R., BJØRLYKKE, K. O., KARLSEN, D. A. *et al.* 1990. Determination of petroleum accumulation histories: Examples from the Ula Field, Central Graben, Norwegian North Sea. *In*: BULLER, A. T. *et al.* (eds) *North Sea Oil and Gas Reservoirs-II*. Graham and Trotman, London, 319–330.

MACKENZIE, A. S., WOLFF, G, A. & Maxwell, J. R. 1983. Fatty acids in some biodegraded petroleums. Possible origins and significance. *In*: BJORØY, M. *et al.* (eds) *Advances in Organic Geochemistry 1981*. Wiley, Chichester, 637–649.

MATTINGLY, G. A. & BRETTHAUER, H. B. 1992. The Alba

Field – A Middle Eocene deepwater channel system in the U.K. North Sea. *In*: HALBOUTY, M. T. (ed.) *Giant Oil and Gas Fields of the Decade 1978–1988*. American Association of Petroleum Geologists, Special Publication, **54**.

NEWTON, S. K. & FLANAGAN, K. P. 1993. The Alba Field: evolution of the depositional model. *In*: PARKER, J. R. (ed.) *Petroleum Geology of NW Europe: Proceedings of the 4th Conference*. Geological Society, London, 161–174.

NORTHAM, M. A. 1985. Correlation of Northern North Sea Oils: the different facies of their Jurassic source. *In*: THOMAS, B. M. (ed.) *Petroleum Geochemistry in Exploration of the Norwegian Shelf*. Graham and Trotman, London, 94–99.

PETERS, K. E. & MOLDOWAN, J. M. 1991. Effects of source, thermal maturity and biodegradation on the distribution and isomerisation of homohopanes in petroleum. *Organic Geochemistry*, **17**, 47–61.

PHILP, R. P. & LEWIS, C. A. 1987. Organic geochemistry of biomarkers. *Annual Review of Earth and Planetary Sciences*, **15**, 363–395.

RADKE, M., LEYTHAEUSER, D. & TEICHMULLER, M. 1984. Relationship between rank and composition of aromatic hydrocarbons for coals of different origins. *Organic Geochemistry*, **6**, 423–430.

TRENDEL, J. -M., GRAFF, R., WEHRUNG, P., ALBRECHT, P., DESSORT, D. & CONNAN, J. 1993. C(14α)-Homo-26-nor-17α-hopanes, a novel and unexpected series of molecular fossils in biodegraded petroleums. *Journal of the Chemical Society, Chemical Communications*, **5**, 461–463.

——, GUILHEM, J., CRISP, P., REPETA, D., CONNAN, J. & ALBRECHT, P. 1990. Identification of two C-10 demethylated C_{28} hopanes in biodegraded petroleum. *Journal of the Chemical Society, Chemical Communications*, **5**, 424–425.

VOLKMAN, J. K., ALEXANDER, R., KAGI, R.I. & WOODHOUSE, G. W. 1983. Demethylated hopanes in crude oils and their applications in petroleum geochemistry. *Geochimica et Cosmochimica Acta*, **47**, 785–794.

WOLCOTT, J. M., SASSEN, R. & CONSTANT, W. D. 1989. Correlation of the physical and chemical properties of three South Louisiana crudes with their degree of biodegradation. *Society of Petroleum Engineers*, Paper No. **SPE 18483**, 227–236.

Sour gas and water chemistry of the Bridport Sands reservoir, Wytch Farm, UK

ANDREW C. APLIN[1] & MAX L. COLEMAN[2,3]

[1] *Fossil Fuels and Environmental Geochemistry Postgraduate Institute: NRG, Drummond Building, The University of Newcastle, Newcastle upon Tyne, NE1 7RU, UK*

[2] *BP Exploration, Chertsey Road, Sunbury-on-Thames, Middlesex, TW16 7LN, UK*

[3] *Postgraduate Research Institute for Sedimentology, University of Reading, Whiteknights, Reading, RG6 2AB, UK*

Abstract: Co-produced waters and gaseous H_2S were sampled at four wells at Wytch Farm oil field, UK, over a one year period. The chemical and stable isotopic (H, O) compositions of the waters were determined, plus the sulphur isotopic composition of dissolved sulphate and gaseous sulphide. Formation waters contain about 70 000 mg l^{-1} total dissolved solids (TDS) and are evolved meteoric waters; there is a 10% fieldwide variation in TDS. H_2S was formed by (bacterial?) reduction of formation-water sulphate. A high but variable fraction of the sulphide has been subsequently lost, perhaps by reaction with abundant Fe-bearing minerals. At least some, and possibly a significant fraction, of the sulphide in the reservoir is dissolved in formation water. Simple mass-balance models show that when reservoir sulphide is predominantly within formation water, the concentration of H_2S in produced gas will increase dramatically with increasing water cuts. In this case, increasing concentrations of H_2S in produced gas do not necessarily imply that souring has been caused by production practices. Seawater breakthrough is recognized and quantified in one well using chemical and stable isotopic data. K, Mg and SO_4 are all lost from injected seawater during its passage through the reservoir. Geochemical modelling and isotopic data fail to identify the fate of the missing sulphate, although the isotopic data tentatively suggest that some may have been converted to sulphide. However, the levels of sulphide produced from the breakthrough well are no higher than in other wells, suggesting that reduction of seawater sulphate contributes a small percentage of fieldwide H_2S production.

H_2S (sour gas) is toxic, corrosive and generally downgrades the commercial value of a discovery. Particularly disconcerting is the common phenomenon whereby H_2S concentrations in produced gas increase during field production; this is known as reservoir souring and, because it has not been accounted for during production planning, can significantly influence field profitability. Potential problems include the replacement of well casing and wellhead facilities, the possibility of pipeline shutdowns and increased refining costs.

Origin of H_2S in reservoir fluids

Effective management of any souring problem requires the definition of both its origin and extent. This information forms the basis for the assessment of future trends in H_2S production and also the most cost-effective remedial treatments. There are several ways in which H_2S can be generated on both geological and/or human time-scales, the relevance of which will depend on reservoir temperature and production practice (Orr 1977; Trudinger *et al.* 1985; Marsland *et al.* 1989):

(1) bacterial reduction of dissolved sulphate, below 105°C;
(2) thermochemical reduction of sulphate by hydrocarbon gases, above *c.* 140°C;
(3) dissolution or reductive dissolution of mineral phases such as iron monosulphide (FeS) or pyrite (FeS_2);
(4) electrochemical reduction of bisulphite (oxygen scavenger) by steel well casings;
(5) thermal decomposition of organic sulphur compounds, primarily above 175°C.

Effect of oil–water partition

Another important situation in which reservoir souring may occur is where sulphide is present naturally in the reservoir but is partitioned strongly into formation water (Whittingham & Jones 1987, Fig. 1). The reservoir will appear to be sweet during the early phase of water-free production, but

From CUBITT, J. M. & ENGLAND, W. A. (eds), 1995, *The Geochemistry of Reservoirs*, Geological Society Special Publication No. 86, pp. 303–314.

increasing levels of H_2S will accompany increasing water cuts. Because higher water cuts are often associated with the breakthrough of injected (sea)water, souring is often (but perhaps wrongly) ascribed to the bacterial reduction of seawater sulphate.

When H_2S is present naturally in the reservoir, future souring trends can be predicted using a simple mass-balance approach, if the partition of H_2S in the reservoir fluids is known. For a reservoir containing undersaturated oil (i.e. no gas cap) and assuming that all reservoir H_2S is flashed to the gas phase under surface conditions:

$$(\text{Mass } H_2S)_{\text{produced}} = V_g[H_2S]_g = (\text{Mass } H_2S)_{\text{oil}} + (\text{Mass } H_2S)_{\text{water}} \quad (1)$$

where V_g = volume produced gas, and $[H_2S]_g$ = concentration of H_2S in produced gas.

Equation 1 can be modified to account for specific partition of sulphide in oil and/or water. If all the reservoir H_2S is associated with oil:

$$V_g[H_2S]_g = V_o[H_2S]_o \quad (2)$$

If V_o is in barrels, V_g in scf, $[H_2S]_o$ in mg l^{-1} and $[H_2S]_g$ in ppm (v/v), then:

$$V_g(H_2S]_g = 3711V_o[H_2S]_o \quad (3)$$

$$[H_2S]_g = (3711/\text{GOR})[H_2S]_o \quad (4)$$

where GOR = gas/oil ratio. If all the reservoired H_2S is associated with water:

$$V_g[H_2S]_g = V_w[H_2S]_w \quad (5)$$

For the same measurement units as previously:

$$[H_2S]_g = \text{WOR}(3711/\text{GOR})[H_2S]_w \quad (6)$$

where WOR = water/oil ratio.

If H_2S is produced from oil and water, we can add the contributions from both:

$$[H_2S]_g = \text{WOR}(3711/\text{GOR})[H_2S]_w + (3711/\text{GOR})[H_2S]_o \quad (7)$$

If H_2S is equilibrated between oil and water in the subsurface, Equation 7 becomes:

$$[H_2S]_g = \text{WOR}(3711/\text{GOR})1/k[H_2S]_o + (3711/\text{GOR})[H_2S]_o \quad (8)$$

where k = partition coefficient between oil and water at subsurface conditions.

The effect on H_2S production of the distribution of sulphide between oil and water within the reservoir is shown in Fig. 1. If sulphide is contained only in the oil phase, H_2S concentrations in produced gas are constant and independent of water cut (Fig. 1a). The situation where sulphide is partitioned between oil and water is more difficult

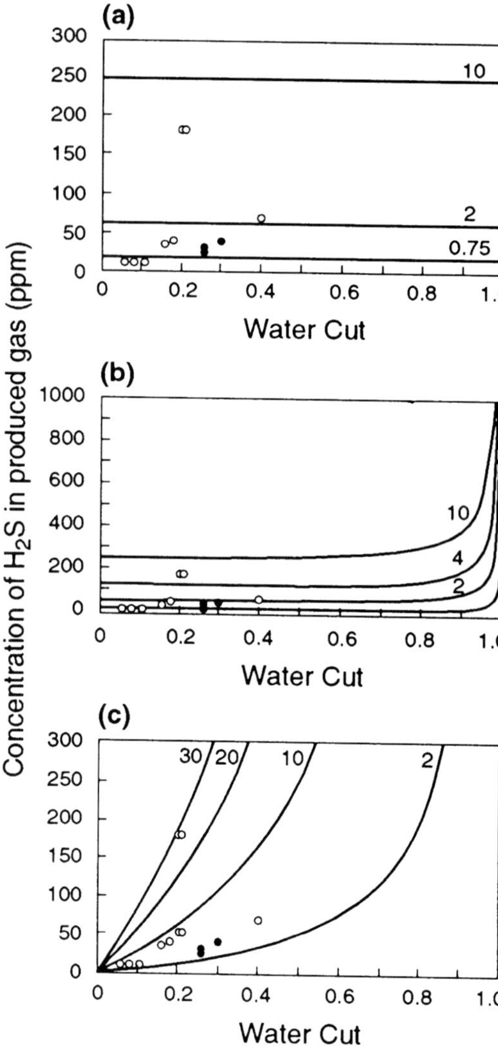

Fig. 1. Modelled concentrations of H_2S in produced gas as a function of water cut in produced liquids. (**a**) All the *in situ* reservoir sulphide is associated with oil. (**b**) Sulphide is partitioned between oil and formation water in a ratio of 2.5 : 1. (**c**) All the *in situ* reservoir sulphide is associated with formation water. Contours in (**a**) and (**b**) represent concentrations of dissolved sulphide in mg/l oil, and those in (**c**), concentrations of dissolved sulphide in water. Where *in situ* sulphide is water-associated, H_2S concentrations rise rapidly with increasing water cut. Open circles indicate data for Wytch Farm wells which have not suffered seawater breakthrough; filled circles indicate data from a well in which seawater breakthrough has occurred.

to quantify because the equilibrium partition coefficient of H_2S between oil and water is unknown and will vary according to pressure, temperature and the chemical composition of both phases. For illustrative purposes, we assume an oil–water partition coefficient (k) of 2.5, similar to that for decane and distilled water (Selleck *et al.* 1952). In this case, the H_2S content of produced gas only increases significantly at high water cuts (Fig. 1b). Where sulphide is contained uniquely in formation water, H_2S concentrations in produced gas increase exponentially with water cut (Fig. 1c). This is due to the injection of an increased mass of produced sulphide into a decreasing volume of produced gas. Higher GORs will therefore damp the rate of increase of H_2S concentrations with increasing water cuts (Fig. 2).

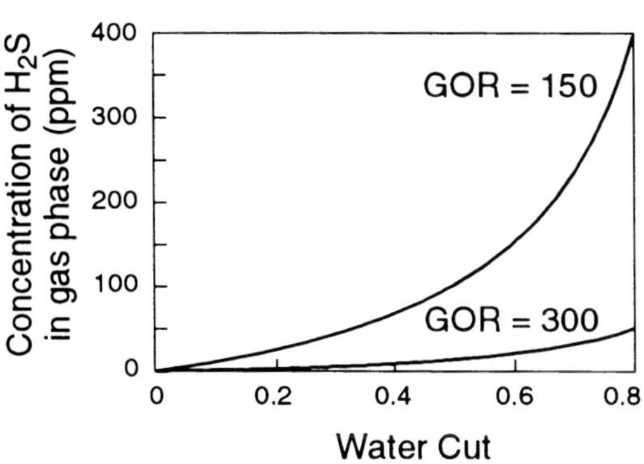

Fig. 2. Modelled concentrations of H_2S in produced gas as a function of water cut, demonstrating the effect of GOR on H_2S abundance. In this case, *in situ* reservoir sulphide is wholly associated with formation water, at a concentration of 4 mg l^{-1}. At high GORs, H_2S flashed from water to gas is diluted by the larger volumes of sweet gas.

Equation 1 and Equation 8 can be modified to account for sulphide associated with seawater breakthrough:

$$\begin{aligned}(\text{Mass } H_2S)_{\text{produced}} &= V_g[H_2S]_g = \\ &= (\text{Mass } H_2S)_{\text{oil}} + (\text{Mass } H_2S)_{\text{water}} \\ &\quad + (\text{Mass } H_2S)_{\text{seawater}} \qquad (9)\end{aligned}$$

The effect of seawater breakthrough on H_2S production will depend on the relative amounts of sulphide in oil, formation water and seawater. For example if the seawater is less sour than the formation water, then the H_2S concentration in produced gas will decrease as seawater production increases.

Stable sulphur isotopes

It is extremely difficult to pinpoint the mechanism by which reservoir H_2S formed and thus define the cause of reservoir souring. The main geochemical label that H_2S carries is the isotopic composition of its sulphur atoms: the $^{34}S/^{32}S$ ratio, expressed as $\delta^{34}S$. Unfortunately, it is rare that the sulphur isotopic composition of H_2S uniquely identifies its source (Fig. 3). Ancillary geochemical data plus relevant reservoir geological and production data are usually required to constrain the origin of H_2S. Isotopic analyses of other sulphur-bearing phases are important because, combined with a knowledge of isotopic fractionations attendent upon their conversion to H_2S, this may rule out some formation mechanisms. The most obvious and important analysis is that of dissolved sulphate in formation

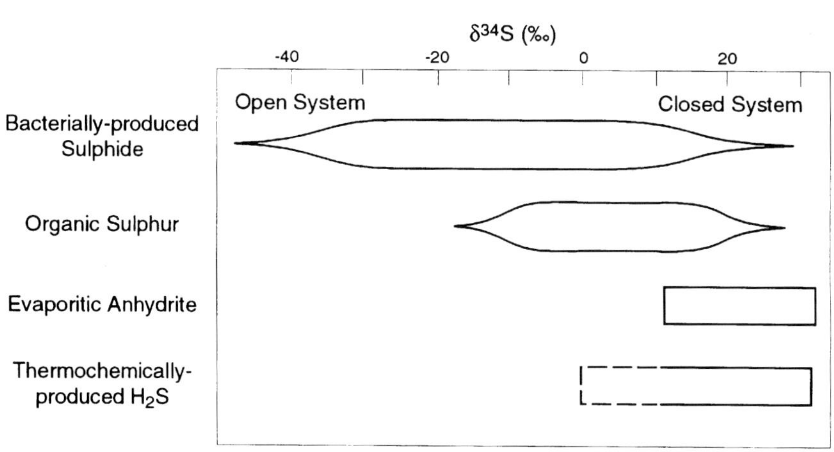

Fig. 3. Range of sulphur isotopic composition of H_2S according to formation mechanism. In many cases it is impossible to determine the origin of H_2S from isotopic composition alone.

waters. Reduction of sulphate, especially by bacteria, involves a kinetic isotopic fractionation which results in the production of H_2S which is isotopically lighter (i.e. more negative $\delta^{34}S$) than precursor sulphate. Correspondingly, the isotopic composition of remaining sulphate becomes increasingly positive. The way in which the isotopic compositions of sulphate and sulphide change with continued sulphate reduction in a closed system can be quantified using the Rayleigh distillation equations. More extensive, closed-system reduction leads to isotopically heavier sulphide and sulphate so that analysis of both species in co-produced fluids is advantageous.

Wytch Farm

Wytch Farm is located on the southern English coast (Fig. 4) and is the largest onshore oilfield in western Europe containing 300 million barrels of oil in place. Most of the oil occurs in the Triassic Sherwood Sandstone. The subject of this paper, the Lower Jurassic Bridport Sands, contain 30 million barrels of undersaturated oil with a GOR of 150 scf/bbl. The reservoir occurs at a depth of about 900 m, at a temperature of 40°C and a pressure of 1500 psi.

The Bridport Sands Formation is thought to have been deposited on the mid-continental shelf. It is a very fine sandstone with about 10% clay minerals, including the Fe-bearing minerals chlorite and berthierine. Authigenic siderite ($FeCO_3$) is common and comprises up to 10% of the rock, and pyrite (FeS_2) typically comprises 1 to 2% of the rock. Laterally extensive (kilometre scale) calcite-cemented horizons, up to 1 m thick, occur at intervals of 1 to 2 m throughout the reservoir and severely reduce large-scale vertical permeability.

Production from the Bridport Sands started in 1979 and was at a level of 6000 bbl/day in 1984. Seawater injection started in November 1980 at Well D10 and had broken through to Well D9 by 1984 (Fig. 4).

Sample acquisition and analysis

Well-head samples of co-produced waters and H_2S were taken from Wells B3, B8, D4 and D9 on three occasions over a one year period. One additional sample was taken from B3 one year later. H_2S was precipitated as CdS by bubbling well-head gases through a 0.05 N solution of cadmium acetate and acetic acid, and was then converted to Ag_2S by the addition of $AgNO_3$. In the laboratory, Ag_2S was purified with concentrated ammonia solution and deionized water and analysed for its isotopic composition using Robinson & Kusakabe's (1975) method. Analytical precision was *c.* 0.2‰.

Waters were separated from petroleum gravitationally, sometimes with the use of small amounts of commercial de-emulsifiers. Any remaining petroleum was removed by repeatedly extracting the water with Arklone, a commercial dry-cleaning fluid. Tests using standard waters showed that this

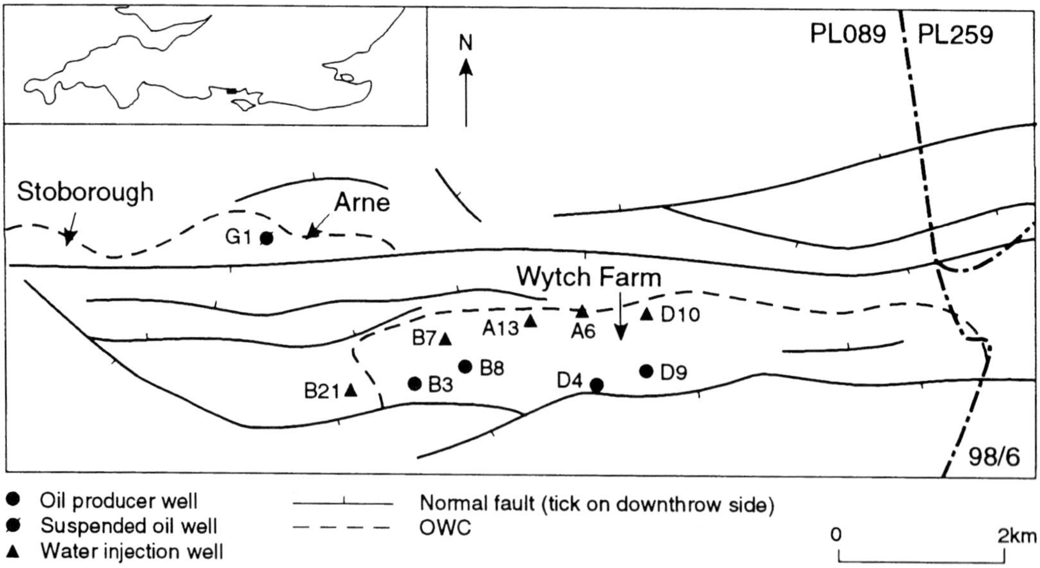

Fig. 4. Map of the Bridport Sands reservoir, showing the location of the key injector and producer wells.

Table 1. *Chemical and isotopic composition of formation waters and injection waters, Bridport Sands reservoir, Wytch Farm*

Sample + sample date	Na	Cl	K	Mg	Ca	Sr	Ba	HCO_3	pH	δD (‰ SMOW)	$\delta^{18}O$ (‰ SMOW)
B3 18/7/84	27 910	48 240	125	650	1730	229	11	145	7.6	−34.4	−4.7
B3 7/9/84	27 330	47 300	125	637	1790	221	14	285	7.6	−35.6	−5.4
B3 5/9/85	27 020	46 691	115	610	1730	210	7	226	7.2	−35.4	−5.3
D4 18/7/84	25 710	44 330	119	597	1610	189	10	155	7.8	−33.4	−5.3
D4 7/9/84	25 550	43 900	115	577	1570	181	14	180	7.5	−34.1	−5.0
D4 5/9/85	25 120	43 261	120	570	1590	176	6	241	7.2	−38.0	−5.2
B8 18/7/84	25 550	44 290	116	605	1710	207	11	135	7.6	−33.6	−4.8
B8 5/9/85	25 798	44 911	107	637	1726	202	18	357	7.1	−33.6	−5.2
D9 18/7/84	20 160	34 430	96	590	1140	129	10	165	7.8	−23.7	−3.5
D9 7/9/84	19 630	33 600	105	596	1150	127	11	245	7.6	−24.9	−3.2
D9 5/9/85	17 140	29 155	100	600	965	99	5	388	7.5	−21.3	−2.8
Injected seawater	10 540	18 740	366	1160	399	7	6	130	8.2	−1.6	+0.2
Injected seawater	10 170	18 550	365	1210	416	7	6	145	8.1	−2.3	0.0

Ion concentrations in mg l^{-1}

treatment does not affect the chemical or stable isotopic composition of waters. Having quantitatively removed dissolved salts by high temperature vacuum distillation, the hydrogen and oxygen isotopic compositions of the waters were determined using the methods developed by Coleman *et al.* (1982) and Epstein & Mayeda (1953). Analytical precision was *c.* 1‰ for hydrogen and 0.2‰ for oxygen. Dissolved sulphate was precipitated as $BaSO_4$ and its isotopic composition analysed using Coleman & Moore's (1978) method. Analytical precision was *c.* 0.2‰.

Cation analyses were performed using ICP-AES and anion analyses by ion chromatography to an accuracy of better than 5%.

Water chemistry

Formation waters in the Bridport Sands are broadly twice as saline as seawater and, like formation waters from many sedimentary basins (e.g. Sheppard 1986), have hydrogen and oxygen isotopic compositions which form a cluster to the right of the meteoric water line (Table 1, Fig. 5). Without a more detailed chemical characterization, it is impossible to define uniquely the origin of these waters. On the basis of their isotopic compositions, they can, however, be distinguished both from seawater ($\delta D = \delta^{18}O = 0‰$) and from present-day meteoric water ($\delta D = -40‰$, $\delta^{18}O = -6.3‰$). Extrapolating the trend of the isotopic data back to the meteoric water line suggests that one component of the water is a meteoric water which must have entered the basin when the area was located at a slightly lower latitude than at present (Dansgaard 1964). High salinities indicate the interaction of the water with either evaporites (which occur deeper in the basin) and/or the residues of evaporated seawater ('bitterns'). Enrichment of waters in ^{18}O is generally ascribed either to reaction between water and rock minerals, such as calcite or clays, and/or by the involvement of bitterns (Clayton *et al.* 1966; Suchecki & Land 1983; Knauth & Beeunas 1986).

Waters produced from the Bridport Sands display a limited degree of compositional heterogeneity. In particular, waters produced from Well B3 are slightly more saline than those produced from Wells B8 and D4, which are located respectively 1 km and 4 km from B3 and which produce water with similar chemical compositions. These data add to the increasing body of evidence that both oil columns and water columns in reservoirs can be compositionally heterogeneous (England 1990; Larter & Horstad 1992; Smalley & England 1992; Coleman 1992). Because diffusive mixing over the vertical scale of a typical reservoir is geologically rapid, vertical variations in the composition of aquifer waters can be used to infer laterally extensive barriers to vertical fluid flow. In the Bridport Sands, compositional variations are seen laterally, on the scale of kilometres. However, because diffusive mixing times of solutes over kilometres are on the order of 10 to 50 Ma (Smalley & England 1992), one cannot infer vertical barriers from lateral variations in fluid composition.

The breakthrough of injected seawater at Well D9 is observed on a plot of Na against Cl, on which the breakthrough waters fall on a mixing line between the seawater and the cluster of formation waters (Fig. 6). Because injected Na and Cl behave conservatively in the reservoir, the proportion of injected water in produced water could be accurately quantified if the composition of formation water is constant. In the Bridport Sands, however, the composition of waters produced from different wells on the same day (Table 1). Analysis of the stable isotopic composition of the waters shows that these variations are not due to seawater

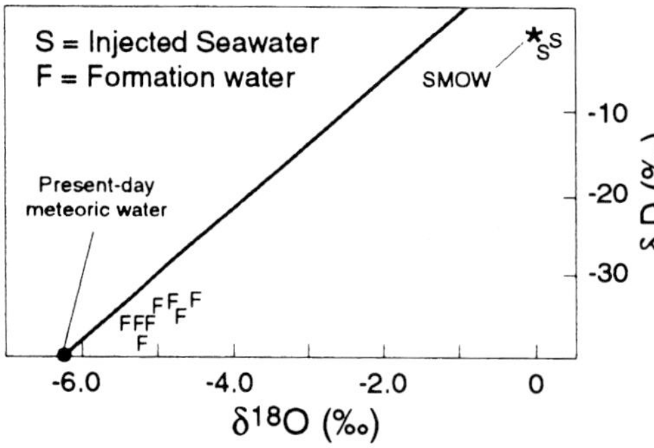

Fig. 5. Hydrogen and oxygen isotopic composition of formation waters from the Bridport Sands reservoir and of injected seawater.

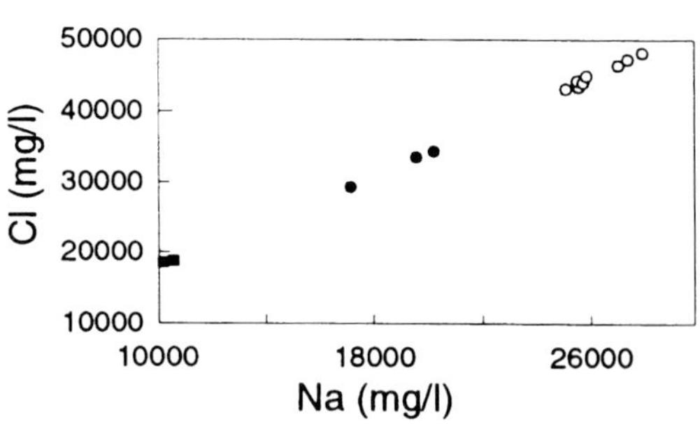

Fig. 6. Detection of seawater breakthrough using a cross-plot of Na versus Cl concentrations in produced waters. Open circles, formation waters; filled squares, injected seawater; filled circle, breakthrough waters.

breakthrough but reflect real heterogeneity in the indigenous formation waters. Plots of δD and $\delta^{18}O$ versus Cl (Fig. 7) show that the waters produced from Well D9 may not be simple mixtures of injected seawater and a single formation water, but may contain small amounts of another water. The origin of this water cannot be identified but might be local groundwater (δD = –40‰, $\delta^{18}O$ = –6.3‰) or an isotopically lighter and/or less saline formation water. It is worth recording that these subtle but geochemically important variations in water composition were only discerned because stable isotopic analyses were made in addition to the usual suite of ten ions.

Quantification of seawater breakthrough allows one to check which ions have behaved conservatively during the mixing of injection and formation water. In this case, variations in the composition of formation water preclude precise quantification, but plots of K, Mg and SO_4 versus Cl show that all these ions have been removed in the reservoir. The removal mechanisms are unclear; the geochemical modelling code PHREEQE (Parkhurst *et al.* 1980) does not predict the precipitation of any K- or Mg-bearing minerals in amounts great enough to account for the observed changes. It is possible that K is removed by cation exchange on the abundant clay minerals in the reservoir, but the fate of Mg is not clear.

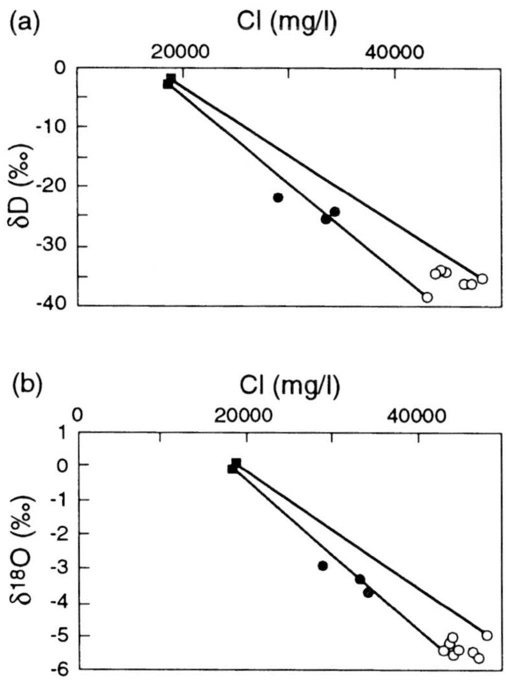

Fig. 7. Mixing lines on plots of δD and $\delta^{18}O$ versus Cl in produced waters suggest that waters in the seawater breakthrough well may not be a simple, two-component mixture of formation water and injected seawater. A third water may be present, either recent groundwater (δD = –40‰, $\delta^{18}O$ = –6.5‰) or a compositionally distinct formation water. Symbols as in Fig. 6.

Mixing calculations using Cl as a conservative tracer suggests that 10 to 20% of injected sulphate is lost in the reservoir. The fate of the sulphate is important as precipitation will lead to well-bore or reservoir scaling whereas reduction to H_2S may lead to reservoir souring. PHREEQE predicts that the only sulphate mineral to precipitate by mixing formation water and injection water is barite, but the concentrations of Ba in the formation fluid are probably insufficient to account for the observed loss of sulphate. Reduction of 15% sulphate to H_2S and the quantitative production of the H_2S in produced gas would give a concentration of H_2S of around 1000 ppm, 20 to 40 times greater than concentrations measured at Well D9. Only 0.5% of injected sulphate needs to be reduced to give the levels of H_2S observed in produced gas. These data imply that either the modelling codes do not accurately describe the geochemical system in the Bridport reservoir and/or that significant amounts of sulphate have been reduced to H_2S, which has subsequently been lost by precipitation with iron minerals within the reservoir. In the next section, we address these possibilities through the use of stable sulphur isotopes.

Sulphur geochemistry

The isotopic composition of sulphate dissolved in the Bridport Sands' formation waters varies between +29.3‰ and +51.5‰ (Table 2). All these values are much more positive than any plausible source of sulphate (e.g. present-day seawater = +21‰ (Rees *et al.* 1978); Jurassic seawater = +16‰ (Claypool *et al.* 1980)) and indicate that some of the sulphate has been reduced to sulphide within a system which was essentially closed to continuous addition of fresh sulphate. Reduction could have been either thermochemical or bacterial and the isotopic data cannot absolutely differentiate between them. However, thermochemical sulphate reduction is extremely sluggish at temperatures below 140°C and is unlikely to be relevant in the Bridport Sands. Furthermore, thermochemically produced H_2S is most commonly only a few per mil lighter than precursor sulphate (Krouse *et al.* 1988), whereas the H_2S in the Bridport Sands is 20 to 30‰ lighter than coexisting sulphate. Isotopic differences in excess of 20‰ are typical of bacterial systems (Goldhaber & Kaplan 1974) and suggest that H_2S naturally

Table 2. *Abundance and isotopic composition of sulphur species in fluids injected into and produced from the Bridport Sands reservoir, Wytch Farm*

Well number	Sample data	Water cut (%)	H_2S (ppm)	$\delta_{34}S_{H_2S}$ (‰)	SO_4^{2-} (mg l^{-1})	$\delta^{34}S_{SO_4}$ (‰)
B3	18/7/84	20	180	+13.5	55	+45.9
B3	7/9/84	21	180	+14.4	60	+51.5
B3	5/9/85	18	40	+0.2	45	+29.3
B3	10/8/86	–	–	+34.0	–	+58.2
D4	18/7/84	8.5	11	–	180	+32.7
D4	7/9/84	11	12	+4.5	195	+33.8
D4	5/9/85	6	12	+6.8	165	+32.7
B8	18/7/84	16	35	+3.3	185	–
B8	5/9/85	40	70	+16.0	80	+38.2
D9	18/7/84	26	25	–	725	+24.9
D9	7/9/84	26	31	–4.6	740	+25.3
D9	5/9/85	30	40	–?	985	+26.6
Injected seawater + scavenger	–	–	–	–	2580	+23.3
Injected seawater	–	–	–	–	2250	+21.3

present in the Bridport Sands is the result of bacterial sulphate reduction.

In a closed, homogeneous system, reduction of sulphide results in precise and quantifiable trends between the concentration and isotopic composition of reactant and product. These relationships are expressed mathematically by the Rayleigh distillation or fractionation equations, shown graphically on Fig. 8. If the sulphur geochemistry of the Bridport formation waters were a closed, homogeneous system, the Rayleigh curves show that we would expect to see the following general trends in the sulphur data:

(1) a negative trend between the concentrations of sulphate and H_2S;
(2) a negative trend between the concentration and isotopic composition of sulphate;
(3) a positive trend between the concentration and isotopic composition of H_2S;
(4) a positive trend between the isotopic compositions of H_2S and sulphate;
(5) a positive trend between the isotopic composition of sulphate and the extent of conversion of sulphate to H_2S (expressed as $H_2S/(H_2S + SO_4)$ in produced fluids.).

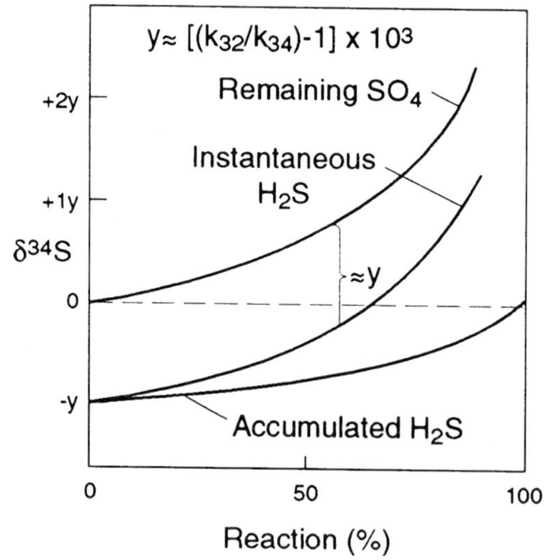

Fig. 8. Isotope fractionation resulting from the closed-system, unidirectional reduction of sulphate to sulphide (Rayleigh distillation). k_{32} and k_{34} are the respectiv e rate constants for the reduction of $^{32}SO_4$ and $^{34}SO_4$.

Only trend 4 is observed (Table 2, Fig. 9). The main ways in which the Bridport fluids depart from a closed system are the extent to which sulphate has been reduced and, particularly, the extent to which H_2S has been removed from the fluids. The former can be inferred from the isotopic composition of dissolved sulphate and the latter can be inferred by calculating the isotopic composition of the total sulphur in produced fluids as:

$$\delta^{34}S_{\Sigma S} = ((\delta^{34}S_{SO_4}M_{SO_4}) + (\delta^{34}S_{H_2S}M_{H_2S}))/(M_{H_2S} + M_{SO_4}) \quad (10)$$

where M = moles produced per unit time (e.g. one day).

If no H_2S has been lost from the fluid, the isotopic composition of the total sulphur will be the same as that of the total sulphate initially present in the fluid. We cannot precisely define the isotopic

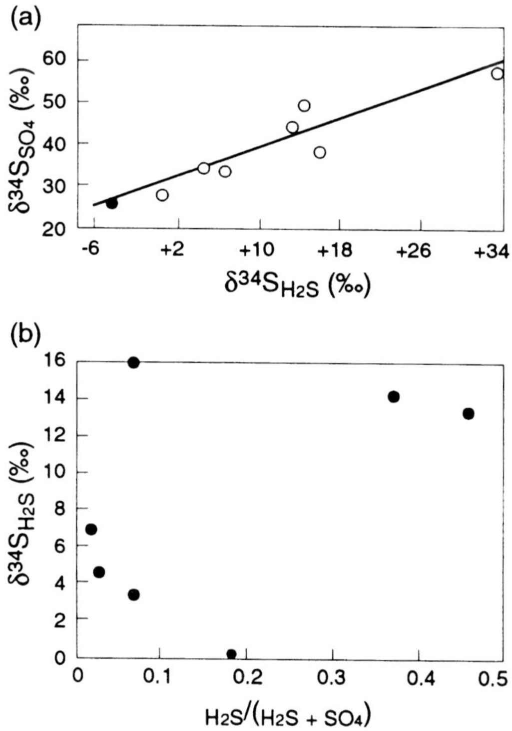

Fig. 9. (**a**) The positive correlation between $\delta^{34}S_{SO_4}$ in formation water and $\delta^{34}S_{H_2S}$ in co-produced gas suggests that the H_2S was formed by reduction of sulphate in formation water. The filled circle is a sample from Well D9, which has suffered seawater break-through. In a truly closed sulphate reduction system, in which no sulphate is added and no sulphide removed, a plot of $\delta^{34}S_{H_2S}$ versus the extent of reduction (i.e. $H_2S/(H_2S + SO_4)$) would show an exponential increase. The Bridport Sands reservoir cannot be described in this simple way (**b**).

composition of the initial sulphate but, because it is likely to be derived from seawater (or palaeo-seawater) or by sulphide oxidation, an absolute maximum of 21‰ can be specified. The isotopic composition of total produced sulphur is between +23.9‰ and +37.9‰ (Table 3, Fig. 10), indicating that H_2S has been lost from all the samples, but in different amounts.

The mechanism by which H_2S has been lost from the fluids cannot be determined, but it is relevant that there are abundant detrital and authigenic Fe-bearing minerals in the reservoir with which H_2S could react to form FeS or FeS_2. The stability of these minerals with respect to the Fe-bearing minerals siderite and berthierine could be assessed using geochemical modelling programs such as EQ3/6, but only if a more detailed chemical analysis of the fluid were available. In particular, we need to know the redox state of the fluid which, for example, influences the relative stability of FeS and FeS_2.

There is only one measurement of the isotopic composition of H_2S produced from the seawater breakthrough well and the value (–4.6‰) is neither sufficiently distinct nor similar to the other values to prove or disprove that sulphide is being produced by reduction of injected seawater sulphate. The isotopic composition of this H_2S is lighter than other measured values, and it is possible that this represents a mixture of reservoir H_2S, with $\delta^{34}S = 0$ to +16‰, and H_2S derived from the reduction of seawater sulphate, with $\delta^{34}S = -10$ to -30‰ (Goldhaber & Kaplan 1974). It is also noticeable that the isotopic composition of dissolved sulphate increases as the extent of seawater breathrough increases. This contradicts one's expectations based on simple mixing between seawater sulphate at 21 to 23‰ and formation water sulphate with a more positive isotopic composition. This might indicate that reduction of seawater sulphate has occurred. However, the concentration of sulphide in the produced water, calculated assuming that all the gaseous H_2S was originally in the formation water, is similar to that in natural formation waters in the Bridport Sands (Fig. 1). If H_2S has been produced from seawater sulphate, either the amounts are small or much of it has been lost, perhaps by reaction with Fe-bearing mineral phases, or also by partitioning into immovable oil.

Part of the initial rationale of this project was to establish the source of the H_2S in the Bridport Sands in order to predict how H_2S production might alter through time. Much of the H_2S is present naturally and has been formed by the (bacterial?) reduction of sulphate within *in situ* formation water. It is crucial to predictive models of H_2S production to determine the partition of sulphide between oil and water at reservoir conditions, but this is not straightforward. H_2S is more soluble in hydrocarbon than water, although the partition coefficient will depend on the composition of both phases, as well as pressure–temperature conditions (Selleck *et al.* 1952). Figure 1 shows that as water cut increases, levels of H_2S will increase much more rapidly if the H_2S is wholly contained in formation water compared to a situation where dissolved sulphide is partitioned between oil and water according to an equilibrium partition coefficient. In the present work, we do not know how much sulphide was present in the initially produced oil and cannot therefore make sensible, predictive calculations about future H_2S production. This illustrates the need to obtain baseline geochemical data early in field appraisal and production. However, the concentration of H_2S in produced gas shows a general

Table 3. *Gas, water and sulphur production data, Bridport Sands reservoir, Wytch Farm*

Sample + sample date	Water prodn (m^3/day)	Gas prodn (m^3/day)	Mass gas (kg/day)	SO_4 ($M\ l^{-1}$)	H_2S ($mg\ kg^{-1}$)	Total moles SO_4	Total moles H_2S	H_2S/SO_4	$\delta^{34}S_{SO_4}$ (‰)	$\delta^{34}S_{H_2S}$ (‰)	$\delta^{34}S\Sigma_S$ (‰)
B3 18/7/84	3.63×10^2	4.58×10^4	3.27×10^4	5.7×10^{-4}	180	208	173	0.83	+45.9	+13.5	+31.2
B3 7/9/84	4.35×10^2	4.16×10^4	2.97×10^4	6.25×10^{-4}	180	272	157	0.66	+51.5	+14.4	+37.9
B3 5/9/85	3.69×10^2	4.68×10^4	3.34×10^4	4.7×10^{-4}	25	345	39	0.14	+29.3	+0.2	+23.9
B8 18/7/84	1.79×10^2	2.80×10^4	2.00×10^4	1.9×10^{-3}	35	345	26	0.06	–	+3.3	–
B8 5/9/85	7.46×10^2	2.99×10^4	2.13×10^4	8.3×10^{-4}	70	622	44	0.11	+38.2	+16.0	+36.7
D4 18/7/84	2.03×10^2	6.38×10^4	4.56×10^4	1.9×10^{-3}	11	381	15	0.04	+32.7	–	–
D4 7/9/84	2.54×10^2	6.11×10^4	4.36×10^4	2.0×10^{-3}	12	516	15	0.02	+33.8	+4.5	+33.0
D4 5/9/85	4.19×10^2	5.73×10^4	4.09×10^4	1.7×10^{-3}	12	720	14	0.02	+32.7	+6.8	+32.2

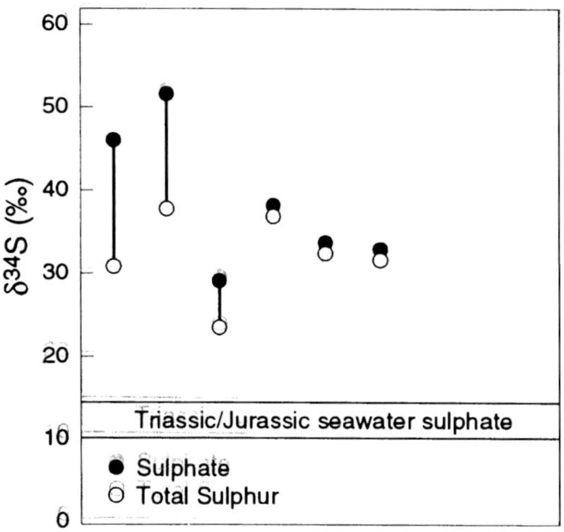

Fig. 10. The isotopic composition of total sulphur in produced fluids is much more positive than the present-day or palaeoseawater sulphate from which we infer it was derived. This suggests that much of the earlier formed, isotopically light sulphide has been lost from the fluid subsequent to its formation. A greater degree of loss is indicated by the approach of the isotopic composition of total sulphur to that of remaining sulphate.

increase with increasing water cut (Table 2). We interpret this to mean that a significant fraction of the sulphide within the reservoir is associated with formation water and this information could be included in predictive souring models. The variability in the concentrations of H_2S produced from well B3 suggests, however, that the concentration of sulphide in the formation water is also variable and this will compromise the accuracy of any souring predictions. We do not know exactly why the concentrations of sulphide appear to be higher in the vicinity of Well B3 but the positive isotopic composition of dissolved sulphate plus the relatively high H_2S/SO_4 ratios in produced fluids suggests that a smaller fraction of bacterially produced sulphide has been subsequently lost from this section of the reservoir. An obvious next step towards solving this problem would be to examine the Fe-bearing minerals in the reservoir for signs of sulphidation.

Conclusions

Most or all the H_2S in the Bridport Sands reservoir was produced by (bacterial?) reduction of sulphate within natural formation water. A high but variable fraction of this sulphide has been subsequently lost, perhaps by reaction with Fe-bearing minerals in the reservoir. Much of the variation in the H_2S levels of produced gases can be ascribed to differences in the extent to which sulphide has been lost from the reservoir fluids.

Some, perhaps most, of the *in situ* sulphide is associated with formation water. Simple mass-balance calculations show that in reservoirs where the sulphide is associated mainly or wholly with formation waters, the H_2S content of produced gas will increase dramatically with increasing water cuts, independently of production practice. Knowing the partition of sulphide between oil and water in the initial reservoir is therefore critical to accurate souring predictions.

Seawater breakthrough has occurred in one well and can be identified and quantified using the chemical and isotopic composition of the produced water. Quantification of breakthrough suggests that 10 to 20% of injected sulphate has been lost within the reservoir. Its fate is unclear but geochemical models and isotopic data suggest that some may have precipitated as barite and some may have been reduced to sulphide. However, the sulphide concentrations in the breakthrough well are no higher than in other wells, implying that reduction of injected seawater sulphate is a relatively unimportant process.

This work was performed when A.C.A. was with BP Research. Help from the Wytch Farm operators, Marg Savell and Jonathan Evans is gratefully acknowledged, as is the geochemical modelling expertise of Chris Osborne and Gordon Macleod. BP are thanked for permission to publish this study, which was expertly reviewed by Chris Clayton and William England.

References

Claypool, G. E., Holser, W. T., Kaplan, I. R., Sakai, H. & Zak, I. 1980. The age curves of sulfur and oxygen isotopes in marine sulfate and their mutual interpretation. *Chemical Geology*, **28**, 199–260.

Clayton, R. N., Friedman, I., Graf, D. L., Mayeda, T. K., Meents, W. F. & Shimp, N. F. 1966. The origin of saline formation waters: I. Isotopic composition. *Journal of Geophysical Research*, **71**, 3869–3882.

Coleman, M. L. 1992. Water compositioone formation. *In:* Kharaka, Y. K. & Maest, A. S. *(eds) Water Rock Interaction*. A. A. Balkema, Rotterdam, 1109–1112.

—— & Moore, M. P. 1978. Direct reduction of sulphates to sulphur dioxide for isotopic analysis. *Analytical Chemistry,* **50**, 1594–1595.

——, Shepherd, T. J., Durham, J. J., Rouse, J. E. & Moore, G. R. 1982. Reduction of water with zinc for hydrogen isotope analysis. *Analytical Chemistry,* **54**, 993–995.

DANSGAARD, W. 1964. Stable isotopes in precipitation. *Tellus*, **16**, 436–468.

ENGLAND, W. A. 1990. The organic geochemistry of petroleum reservoirs. *Organic Geochemistry*, **16**, 415–425.

EPSTEIN, S. & MAYEDA, T. K. 1953. Variation of ^{18}O content of waters from natural sources. *Geochimica et Cosmochimica Acta*, **4**, 89–103.

GOLDHABER, M. B. & KAPLAN, I. R. 1974. The sulfur cycle. *In*: GOLDBERG, E. D. (ed.) *The Sea, Volume 5*. J. Wiley & Sons, Chichester, 569–665.

KNAUTH, L. P. & BEEUNAS, M. A. 1986. Isotope geochemistry of fluid inclusions in Permian halite with implications for the isotopic composition of ocean water and the origin of saline formation brines. *Geochimica et Cosmochimica Acta*, **50**, 419–433.

KROUSE, H. R., VIAN, C. A., ELIUK, L. S., UEDA, A. & HALAS, S. 1988. Chemical and isotopic evidence for thermochemical sulphate reduction by light hydrocarbon gases in deep carbonate reservoirs. *Nature*, **50**, 419–433.

LARTER, S. R. & HORSTAD, I. 1992. Migration of petroleum into Brent Group reservoirs: some observations from the Gullfaks field, Tampen Spur area, North Sea. *In*: MORTON, A. C., HASZELDINE, R. S., GILES, M. R. & BROWN, S. (eds) *Geology of the Brent Group*. Geological Society, London, Special Publication, **61**, 441–452.

MARSLAND, S. D., DAWE, R. A. & KELSALL, G. H. 1989. *Inorganic chemical souring of oil reservoirs*. Society of Petroleum Engineers, Paper No. **18480**.

ORR, W. L. 1977. Geologic and geochemical controls on the distribution of hydrogen sulphide in natural gas. *In*: CAMPOS, R. & GONI, J. (eds) *Advances in Organic Geochemistry 1975*. Enadimsa, Madrid, 572–597.

PARKHURST, D. L., THORSTENSON, D. C. & PLUMMER, L. N. 1980. *PHREEQE – a computer program for geochemical calculations*. United States Geological Survey, Water Resources Investigation, **80–96**.

REES, C. E., JENKINS, W. J. & MONSTER, J. 1978. The sulfur isotopic composition of ocean water-sulphate. *Geochimica et Cosmochimica Acta*, **42**, 377–382.

ROBINSON, B. W. & KUSAKABE, M. 1975. Quantitative preparation of sulfur dioxide for $^{34}S/^{32}S$ analyses from sulphides by combustion with cuprous oxide. *Analytical Chemistry*, **47**, 1179–1181.

SELLECK, F. T., CARMICHAEL, L. T. & SAGE, B. J. 1952. Some volumetric and phase behavior measurements in the hydrogen sulfide–water system. *Industrial and Engineering Chemistry*, **44**, 2219–2226.

SHEPPARD, S. M. F. 1986. Characterization and isotopic variations in natural waters. *In*: VALLEY, J. W., TAYLOR, H. P. Jr & O'NEILL, J. R. (eds) *Stable Isotopes in High Temperature Geological Processes* Mineralogical Society of America, Reviews in Mineralogy, 165–183.

SMALLEY, P. C. & ENGLAND, W. A. 1992. *Assessing reservoir compartmentalisation during field appraisal: how geochemistry can help*. Society of Petroleum Engineers, Paper No. **25005**.

SUCHECKI, R. K. & LAND, L. S. 1983. Isotopic geochemistry of burial-metamorphosed volcanogenic sediments, Great Valley sequence, northern California. *Geochimica et Cosmochimica Acta*, **47**, 1487–149.

TRUDINGER, P. A., CHAMBERS, L. A. & SMITH, J. W. 1985. Low-temperature sulphate reduction: biological versus abiological. *Canadian Journal of Earth Sciences*, **22**, 1910–1918.

WHITTINGHAM, K. P. & JONES, T. J. 1987. Reservoir souring. *Royal Society of Chemistry*, **67**, 228–243.

Index